BEHAVIOR-GENETIC ANALYSIS

McGRAW-HILL SERIES IN PSYCHOLOGY

CONSULTING EDITORS
NORMAN GARMEZY
HARRY F. HARLOW
LYLE V. JONES
HAROLD W. STEVENSON

BEHAVIOR-GENETIC ANALYSIS

Edited by
JERRY HIRSCH
Professor of Psychology and Zoology
University of Illinois

McGraw-Hill Book Company
New York   St. Louis
San Francisco
Toronto
London
Sydney

TO ROBERT CHOATE TRYON,
Trailblazer and Teacher

FOREWORD

It is now fashionable, among some biologists as well as among popular writers, to declare that we are witnessing a revolution in biology. This is perhaps a little more melodramatic than the facts warrant, unless you are prepared to stretch the argument to mean that biology has been in a state of permanent revolution for more than a century, since Darwin. It is, of course, indisputable that the advances, particularly in molecular biology, have been spectacular, especially in the postwar period. We may go so far as to surmise that the mid-twentieth century will stand in the history of science as the time when biology forged ahead to become an equal of the physical sciences, which dominated the scientific scene for centuries, at least from Galileo to Einstein. The philosophical and methodological foundations of biology have not, however, been changed by recent discoveries. These foundations continue to be the Descartian mechanistic reductionism and the Darwinian evolutionistic compositionism. Both are equally important in the molecular and in the organismic level biology.

What of the future? It is generally safe to expect that most exciting scientific discoveries will be made in most unexpected places. It does not, however, follow that scientists should study things at random, in the hope that something interesting may turn up. On the contrary, scientists endeavor to gain knowledge and understanding in order to guide them to problems most likely to yield new knowledge and new understanding. In the biological sciences, investigations on the molecular level are at present attracting most attention, most money, and most students. And yet in recent years there has also been some vigorous growth in the studies of animal behavior, particularly of genetics and behavior. The interest and importance of such studies are evident. After all we, men, are animals, though a very special kind of animal. Human behavior is an outgrowth, or a uniquely specialized form, of animal behavior. Understanding the behavior of simpler creatures may help us better to understand human behavior, even if the difference will prove to be quantitatively so large as to amount to a qualitative difference.

The genetics and evolution of behavior are at the same time very old and very new subjects of study. Man has been associated with his domesticated animals so closely and for so long that he could not fail to be impressed by the individual and breed differences in behavior. The individual and group differences in the behavior of humans are also too strikingly obvious to escape notice. Yet it is only quite recently that reasonably precise quantitative methods for analytic, rather than anecdotic, description of these differences have been found. This book edited by Professor Hirsch is concerned with these methods and with the results of their application.

Although it is a collective work of nineteen authors, it is unlike so many symposium volumes in which the different chapters are disparate both in the level of scientific preparation they assume in the reader, and too often also in the style and quality of the writing. This collective effort has evidently been very carefully planned and coordinated. Were the names of the authors not indicated at the beginning of each chapter, one might perhaps take the book as a whole for a systematic presentation by a single extraordinarily versatile writer.

Theodosius Dobzhansky
The Rockefeller University

PREFACE

In the summer of 1959 I proposed holding a conference on heredity and behavior to R. C. Tryon and G. E. McClearn. The correspondence between himself and B. E. Ginsburg that McClearn immediately showed us made it clear that several people had been thinking along the same lines.

After consultation with the National Science Foundation and the Center for Advanced Study in the Behavioral Sciences, a committee consisting of Benson E. Ginsburg, Jerry Hirsch (chairman), and Gerald E. McClearn, later joined by Howard F. Hunt, was formed to make arrangements for one or more meetings. A list of prospective participants and a tentative agenda were proposed, and in due time financial support for two meetings was obtained from the National Science Foundation. Invitations were extended, a detailed agenda was prepared, and two 3-week meetings were held in August, 1961, and in August, 1962, at the Center for Advanced Study in the Behavioral Sciences, Stanford, California.

At the conference a rich fund of information, both technical and general, was exchanged. Occasionally almost an entire day was devoted to detailed discussion of research from a single laboratory. Other days the work of several laboratories was discussed. At still other sessions, panels of specialists reviewed formal knowledge, theoretical issues, and methodological problems in genetics, in behavior study, and in behavior genetics. That the meetings afforded us all an oportunity to learn surprisingly more than we had anticipated was the consensus of the participants. To quote one of the conferees, "Diversity without acrimony was the strength of the conference."

Participants in one or both of the meetings are listed below:

Gordon Allen	David Yi-Yung Hsia
Peter Broadhurst	Howard F. Hunt
Jan H. Bruell	John A. King
Ernst W. Caspari	Daniel S. Lehrman
L. Erlenmeyer-Kimling	Gardner Lindzey
Benson E. Ginsburg	Aubrey Manning
David A. Hamburg	Gerald E. McClearn
Eckhard H. Hess	R. C. Roberts
Jerry Hirsch	Walter C. Rothenbuhler
	William R. Thompson

In addition, about a score more people then at the Center for Advanced Study, Stanford University, the University of California, San Jose State College, or temporarily visiting the area were invited to participate at individual sessions. They contributed in an important way to the success of the discussions.

As a direct result of those meetings, two possibilities initially put forward in the conference proposal as tentative suggestions have now been realized. First, a summer institute in behavior genetics, in which several of us participated as teachers, was organized in 1964 for students of biology and the social sciences. It received financial support from the Training Branch of the National Institute of Mental Health and was administered at the University of California, Berkeley, by G. E. McClearn, under the sponsorship of the Committee on Genetics and Behavior (now called Biological Bases of Social Behavior) of the Social Science Research Council (the two previous summer meetings had played an important role in launching that SSRC committee, as well). Second, the group represented in this volume undertook to make more generally available the material it now contains.

Lastly, I wish to acknowledge my debt to Mrs. Gayleen Andrews for her great care in seeing this volume through all its phases from typescript to index and to Professor Harry F. Harlow for his critical reading of the manuscript and for his valuable suggestions about its organization.

Jerry Hirsch

CONTENTS

CONTRIBUTORS

Peter L. Broadhurst
Department of Psychology, University of Birmingham, Birmingham, England.

Jan H. Bruell
Department of Psychology, Western Reserve University, Cleveland, Ohio.

Ernst W. Caspari
Department of Biology, University of Rochester, Rochester, New York.

John C. DeFries
Department of Dairy Science, University of Illinois, Urbana, Illinois.

Benson E. Ginsburg
Behavior Genetics Laboratory, University of Chicago, Chicago, Illinois.

David A. Hamburg
Department of Psychiatry, Stanford University Medical School, Palo Alto, California.

Jerry Hirsch
Departments of Psychology and Zoology, University of Illinois, Urbana, Illinois.

David Yi-Yung Hsia
Department of Pediatrics, Northwestern University School of Medicine, Chicago, Illinois.

John A. King
Department of Zoology, Michigan State University, East Lansing, Michigan.

Gardner Lindzey
Department of Psychology, University of Texas, Austin, Texas.

Aubrey Manning
Department of Zoology, University of Edinburgh, Edinburgh, Scotland.

Gerald E. McClearn
Department of Psychology, University of Colorado, Boulder, Colorado.

Eugene Roberts
Department of Biochemistry, City of Hope Medical Center, Duarte, California.

R. C. Roberts
Institute of Animal Genetics, University of Edinburgh, Edinburgh, Scotland.

Walter C. Rothenbuhler
Department of Zoology and Entomology, Ohio State University, Columbus, Ohio.

Judith Shirek
Department of Anthropology, University of California, Berkeley, California.

James N. Spuhler
Department of Anthropology, University of Michigan, Ann Arbor, Michigan.

William R. Thompson
Department of Psychology, Queens University, Kingston, Ontario, Canada.

Sherwood L. Washburn
Department of Anthropology, University of California, Berkeley, California.

☐ The nature of the effect of genetics upon behavior is one of the basic problems of psychology, in that no . . . explanation of behavior is complete without heredity. It is also one of the basic problems of biology. . . . No theory of evolution of animals is complete without a consideration of behavior . . . (Scott, 1964).

INTRODUCTION

Behavior-genetic analysis is the approach to the study of organisms and their behavior that combines the concepts and methods of genetic analysis, based on knowledge or control of ancestry, with the concepts and methods of behavioral analysis from psychology and ethology, based on knowledge or control of experience (see pp. 425–426).

Prior to the Roots of Behavior symposium (*Science*, 130:1344, 1959) and the volume resulting from it (Bliss, 1962), the pattern of thinking that motivated most behavioral-science research was pre-Mendelian, in fact pre-Darwinian—what Mayr (1959) so aptly calls typological thinking. There was neither an appreciation of genotypic diversity nor of its relationship to those ubiquitous "individual differences" that were buried in the error term of too often inappropriate statistical analyses. Furthermore, many who learned textbook genetics, even those who actively studied genetics and behavior, couched their interpretations of relations between heredity and behavior in the causality-laden reductionistic terminology that characterizes physiology and physiological psychology. Where the physiologist uncovered the hormonal, neuronal, etc. (=causal) bases of behavior, others now searched for the genetics *of* behavior (Hirsch, 1964, 1965, 1967).

It will be apparent from much that appears in this volume, however, that a typological-reductionist-causal terminology still has a high association value in our collective verbal-habit hierarchies. Nevertheless, the cumulative effect of the thinking and research that have become behavior genetics is leading us wherever possible (with allowances for the fallibility of human memory and fluency of expression) to eschew naïve reductionism.

Inspired by the triumphs of Newtonian mechanics physiologists, psychologists, and more recently ethologists embarked upon a causal analysis of behavior. Their conceptual model was that of *the* machine whose operations are to be explained in terms of the functioning of *the* component parts. While the operations of *a* machine certainly are to be explained in terms of *its* component parts, today we realize both the uniqueness and the diversity of the "machines" that are studied in the biosocial sciences. The characteristics of their components vary throughout populations and across species. Every individual is unique—a fact which today is perfectly well understood within the framework of mechanistic science (see Hirsch, 1962, 1963). Individual uniqueness is the ontogenetic result of the particular balance of components which arises first of all because of a unique genetic endowment at conception (monozygotes excepted) and then because of an idiosyncratic developmental history (monozygotes included).

Our failure to appreciate sooner the limitations of the typological-reductionist approach can be attributed, among other things, to (1) the long fight required to establish the validity of mechanistic analysis and to eliminate vitalism and (2) the heredity-environment controversy and the fight to win recognition for the new science of genetics, which, when first established, fell directly in line with the prevailing typological-reductionist pattern of thinking. Before we understood the concepts of population genetics, it might be that we were incapable of appreciating the limitations of the traditional thought pattern. Later it was also necessary to exonerate genetics of responsibility for the more unsavory aspects of the eugenics movement and the claims made on its behalf by elitists, racists, and fascists, all the while guarding against the dangers of a rampant environmentalism that led to Lysenkoism.

This volume considers several of the kinds of knowledge, both substantive and formal, on which behavior-genetic analyses must be based. It presents neither an introduction to a so-called field, nor a comprehensive summary of its literature, nor a review of human research and its methodology. Those tasks have been performed by Fuller and Thompson's pioneering text (1960; see reviews by Caspari, 1961a, and by Hirsch, 1961a) and by the volume under Vandenberg's editorship (1965) that resulted from the 3-day Louisville conference organized at the suggestion of R. B. Cattell.

Part I places behavior study in an evolutionary perspective. We consider the behavioral changes that reflect evolution and the role of behavior as a factor in evolution. Caspari distinguishes two problems in the study of evolution: analysis of the mechanisms responsible for the changes that comprise evolution and reconstruction of the sequences of forms that have appeared. Washburn and Shirek present a discussion of human evolution that stresses the importance of tool use, both in the evolution of bipedalism among the primates and in the development of man's large brain—a valuable example of the feedback relation between behavior and our gene pool. King considers the relations of ecological and developmental genetics to mammalian behavior and, most importantly, the kind of feedback from behavioral variation to the gene pool that is now amenable to investigation. Manning examines the literature on behavioral evolution in insects, emphasizing how allelic differences could effect small changes in nervous thresholds. And Rothenbuhler's discussion focuses on the Hymenoptera. It provides a detailed survey of an extensive literature dealing with bee genetics and bee behavior as well as a description of his own elegant analysis of two genes involved in hygienic behavior.

Part II considers the mechanisms that intervene between genes and behavioral phenotypes. Caspari reviews fundamental aspects of gene activity as well as developments relating the genome to neuronal activity and to memory. Ginsburg describes research strategies for behavior-genetic analyses, using as illustrations work from his laboratory on the relations between audiogenic seizures, heredity, and the morphology and histochemistry of the hippocampus. Next, Hamburg examines the relations of hormone metabolism (adrenocortical) to heredity and to psychological stress. And Hsia discusses inborn errors of metabolism as the natural units involved in several human behavioral phenotypes. Then,

Eugene Roberts develops a model of the synapse which he relates to significant studies of the chemicals involved in nervous transmission.

Because of the complexity of behavior and because polygenic systems are implicated in many complex phenotypes, Part III is an exposition of quantitative genetic analysis. R. C. Roberts begins with an extensive review of the fundamentals of quantitative genetics. Next, for geotaxis, Hirsch describes behavior-genetic analysis at the chromosome level of organization. Then, Bruell and Broadhurst discuss the analysis of quantitative traits in populations and describe the diallel method, with Bruell showing the evolutionary implications in behavior-genetic data. In doing so, Bruell examines the relations between inbreeding and heterosis and between laboratory populations and natural populations.

Part IV deals with conceptual and methodological issues of general import. In Genes, Generality, and Behavioral Research, McClearn discusses some limitations on the scope of the inferences that we can make. He relates them to our ability to specify and control the biological makeup of the organisms whose behavior we study. Then, from the point of view of modern biometry, DeFries makes a retrospective evaluation of the extant literature on experimental behavior-genetic analyses for quantitative traits. R. C. Roberts shows the interesting opportunities that behavior study now offers to genetics. Thompson considers some important differences between behavior genetics and other kinds of genetics and how they may be related to our understanding of the complexities of personality and intelligence. Then, Spuhler and Lindzey examine the race concept and present their evaluation of its place in the study of behavior.

Finally, the last chapter provides a fundamental and completely general treatment of race (p. 431), an overview and, it is hoped, synthesizing commentary on the place of behavior-genetic analysis in the biosocial sciences.

PART I BEHAVIOR AND EVOLUTION

CHAPTER ONE
INTRODUCTION TO PART I AND REMARKS ON EVOLUTIONARY ASPECTS OF BEHAVIOR
Ernst W. Caspari

INTRODUCTION

All biological phenomena can be considered from two points of view: mechanism and evolution. These two points of view do not exclude each other, and both should be considered in the complete analysis of any biological phenomenon. It is therefore necessary that one part of this book should be devoted to the evolutionary aspects of behavior.

The theory of evolution has been one of the main pillars on which the structure of biological theory has been erected. It deals fundamentally with two interrelated problems. First, the mechanisms by which evolution proceeds must be described, analyzed, and experimentally confirmed. Secondly, the evolutionary history of the species that exist now and have existed in the past has to be reconstructed. While these two goals of the theory of evolution are conceptually closely interrelated, the methods used for the study of the two aspects are so different that it is convenient to distinguish them from each other by the use of different terms. I shall therefore restrict the word *evolution* to the analysis of evolutionary mechanisms and use for the study of the history of species and other taxonomic categories the word *phylogeny*.

The mechanisms by which evolution proceeds are now reasonably well established. The theory, which was originally developed by mathematical models, is supported by a large amount of experimental and observational evidence. The theory and the supporting evidence have been collected and integrated in numerous books, of which those by Dobzhansky (1962) and by Mayr (1963) should be mentioned here, since both contain a large amount of material on the topics discussed in this book.

GENERAL THEORY OF EVOLUTION

The modern theory of evolution is based on our knowledge of the behavior of genes in populations. The basic conceptual advance in this field was the recognition that a sexually reproducing population, a "Mendelian" population, can be regarded as a collection of genes, a

3

gene pool, in which the genes are reshuffled every generation. It is therefore possible to abstract from the individuals that are, so to speak, attached to the genes and to describe the population in terms of gene frequencies in the gene pool.

The basic facts of population genetics are so well known that it appears unnecessary to repeat them here. Suffice it to state that evolutionary processes, under the theory of population genetics, may be defined as changes in the composition of the gene pool in time. From an evolutionary point of view, it is therefore most important to recognize the mechanisms by which the frequency of genes in a gene pool can be altered. Several mechanisms have been proposed and demonstrated, but among these natural selection is of such paramount importance that it is sufficient to consider only this aspect in these introductory remarks.

Natural selection, in its most general form, can be expressed as the fact that usually two alleles of the same gene are not transmitted to successive generations with the same frequency. The probability of one allele's being transmitted to the next generation, as compared with its partner, is designated as its "adaptive value." The adaptive value of any gene is dependent on the phenotypic characteristics that it imparts on its carriers in homozygous and heterozygous conditions. If the adaptive value of one of the homozygotes is highest, the corresponding allele will be favored by natural selection, and the other allele will be gradually eliminated. If the heterozygote is superior in adaptive value to either homozygote ("heterosis") both alleles will be kept in the gene pool, and a stable genetic polymorphism will result.

The adaptive value of a gene is not constant. It depends on environmental factors and on the other genes present in the same population, its genotypic milieu. Therefore, changes in environmental conditions will generally introduce changes in the adaptive value of some of the genes present in the population. They will, therefore, if continued, lead to changes in the composition of the gene pool, i.e., according to our earlier definition, to evolutionary changes. These evolutionary changes are generally in an adaptive direction; i.e., the resulting new gene pool will have a higher fitness in the new environment, will be better adapted, than the old gene pool had been. It is important to realize that genetic adaptation of this type can proceed with considerable speed, provided the selective pressure is sufficiently strong and the generation time of the organism sufficiently short. Examples are the rapid adaptation of many insect species to insecticides, as observed in nature, and the cases of evolutionary changes of populations induced in the laboratory (e.g., Dobzhansky, 1947; Ayala, 1965).

While we generally assume a one-to-one relationship between gene and phenotypic character, this holds strictly only for those characters which are, in a developmental sense, "close" to the gene, e.g., protein structure and the constitution of cell surface antigens. (See chapter on gene action.) All other characters are influenced by several genes, frequently by a large number. Such polygenic systems are thoroughly discussed in Part III. At this point, it should be mentioned only that the genetic control of a character in a population depends on the genetic structure of the population rather than on the genetic nature of the character itself. If, for instance, a particular character is dependent on a certain number of genes, only one of which is represented in a particular population by a pair of alleles, the others being homozygous, then it would appear that the genetic variation of the character in this population was depending on this one gene pair. In other populations of the same species, however, a different gene or more than one gene might be present in several allelic forms. It should be obvious that these populations would react in genetically different ways if the same selective pressure were exerted on them. Different genes or gene complexes become fixed in different populations. Therefore, the result of selective pressure on a gene pool depends as much on the initial potentialities of the gene pool as it does on the quality of the selective stimulus.

Since many characters, as described by morphological or behavioral methods, are dependent on polygenic systems of genes, the relation of these genes to each other in the production of the phenotypic character is interesting. This matter is discussed by others in Part III. In the present context it is of importance to note that the effects of different genes on a complex character such as adaptive value are by no means always additive, but the interaction is frequently much more complex. What is meant by the above statement is that the effect of a particular gene does not depend only on the environment but also on the other genes present in the same population, the genetic milieu. A certain gene may have a favorable effect in the presence of some specific other genes in the population, but not of all others. The case of hygienic behavior of the honeybee described by Rothenbuhler (Chapter 5) offers a particularly simple example of this fact. Neither one of the two genes involved would, by itself, alter the fitness of the population in the presence of a damaging environmental factor, infection with *Bacillus cereus*. But both genes in combination greatly increase the fitness of the population. These two genes offer, therefore, a perfect example of what Dobzhansky has called "coadaptive gene complexes." Selection for coadaptive gene complexes, coadaptation, appears to be the rule in natural selection, and selection for individual genes may be regarded as a borderline case, interesting because of its simplicity but frequently confusing because it omits the complicating systemic nature of the gene pool.

The main principle of evolutionary theory is therefore simple as far as individual pairs of alleles are concerned, but it becomes highly complex when whole gene pools are taken into account. The reaction of a particular population to a specified selective stimulus is not predictable unless the total constitution of the gene pool is known, which cannot be accomplished in practice. Even if it were known, there is some doubt whether the reaction of the gene pool would be completely predictable (Dobzhansky and Pavlovsky, 1953).

The discussion up to this point has dealt with evolutionary changes which lead to adaptation of the population to changes in the environment. Another aspect of evolutionary changes, that of speciation, must be mentioned. Speciation consists in the breakup of a gene pool into two discrete gene pools, species. The mechanisms involved and the consequences of the process have been exhaustively and brilliantly discussed by Mayr (1963). It is sufficient to indicate here that speciation, by definition, is dependent on the occurrence of a mating barrier inside a population which inhibits the exchange of genes between groups of members of the original population. This mating barrier may be geographical, ecological, physiological, behavioral, or of several other types. The mechanism does not matter, as long as gene exchange is inhibited or at least greatly diminished. The two resulting gene pools will differentiate along different lines, giving rise to different species.

The two aspects of living organisms mentioned, their adaptedness to the environments in which they live and their division into reproductively isolated species, are very striking to every observer of nature. Both of these phenomena find their answer only in the context of the theory of evolution. They constitute indeed the most serious problems with which workers in this field are concerned at the present time.

THE ROLE OF BEHAVIOR IN
SELECTION AND SPECIATION

Adaptive value, as defined in the preceding section, is a highly complex character which embraces almost all the characters and activities of an organism. In order to analyze it further, Wright (1949) has divided it into three components: viability, defined as the probability of a genotype's surviving to reproductive age; fertility, the probability of a genotype's producing offspring; and fecundity, the average number of offspring contributed by a genotype to the next generation. To these three components, Spiess and Langer (1961, 1964) have added rate of development and maturation for mating. While the influence of these components depends to a large degree on the ecological conditions in which the organism lives, there is no doubt that they may be in some cases important factors of fitness, as shown by Spiess and Langer for some chromosome rearrangements of *Drosophila persimilis*.

The behavior of an animal may be related to several of the above mentioned components of fitness. It would be expected to be especially effective in the determination of the first two components. The effectiveness of mating is obviously very important for the fertility of crossbreeding organisms, and mating behavior constitutes its main determinant in free-living animals. Considerable attention has therefore been given by population geneticists and evolutionists to the genetic determination of mating behavior, particularly in insects, and the chapter by Manning quotes numerous examples. It may be stated here only that in the investigation of the adaptive value of mutants it has frequently turned out that their pleiotropic effects on mating behavior rather than on viability are the predominant cause of their lowered fitness.

It may also be expected that viability is to a certain degree dependent on behavioral characters. Behavior is an active way by which an animal adapts to its environment. Behavior leads to the active selection of favorable habitats and food sources and to escape from dangers such as enemies. Examples are given in all chapters in Part I, but it appears that the genetic determination of this aspect, possibly because it appears so obvious, has not been investigated as thoroughly and systematically as mating behavior in insects. On the other hand, it appears that behavioral characters will have less influence on fecundity and rate of development, which do not involve strong interaction with the environment.

The influence of a gene on behavioral characters may, then, constitute a major factor in its adaptive value and may determine whether a particular allele will be kept in the population or eliminated. On the other hand, the genetically determined characteristic behavior of a species may have a strong influence on its ecology and population structure. Behavioral characters, therefore, in turn, influence the structure of a gene pool and its potentialities for change. This point has been elaborated in the chapter by King and does not need any further discussion here.

Mention should be made of the modifiability of behavior. It is well known that behavior of an individual can be modified by environmental conditions and particularly by previous stimuli and behavioral activities of the same animal. Complex learning in its various forms is often assumed to be restricted to the higher vertebrates, but the chapter by Rothenbuhler gives impressive examples of learning by bees. Nevertheless, modifiability of behavior by learning is particularly pronounced in mammals and birds. There is good evidence for a genetic determination of learning ability. But the evolutionary implications do not seem to have been extensively studied.

Seiger and Kemperman (unpublished) have therefore recently investigated, by means of theoretical models, the possible influence of imprinting in birds on evolutionary processes. They make the simplifying assumptions that birds mate only with partners who phenotypically resemble their parents on which they have been imprinted in early life and that similarity or dissimilarity is determined by one pair of alleles for which the population is polymorphic, e.g., a color gene. It turns out that this situation leads to the breakup of the population into two separate populations, each one homozygous for one of the two alleles, i.e., to speciation. It is not clear whether such an oversimplified model has any direct bearing on situations found in nature. But the case of the snow goose and the blue goose (Cooch and Beardmore, 1959) may well be an example since assortative mating has been observed (Manning, quoted in Mayr, 1963, p. 469). However, other possible explanations of this case have been proposed and cannot be excluded at the present time.

The latter case indicates that behavioral characters may have an influence not only on the evolutionary processes that lead to adaptation but also on those leading to speciation. The evidence that genetically controlled mating and habitat preferences may be important factors in the formation of species is supported by a large amount of experimental material, particularly from Drosophila. Even in cases where a behavioral character is associated with cross-infertility in the establishment of a mating barrier, it can sometimes be demonstrated that assortative mating is the primary source for the breakup of the original population. In a case investigated by Ehrman (1964) it has been shown that as yet incomplete reproductive isolation of two populations of *Drosophila paulistorum* was initiated by genetically determined mating preferences and secondarily reinforced by a cytoplasmically determined sterility of the hybrid males.

THE PHYLOGENY OF BEHAVIOR

After Darwin established the theory of evolution by natural selection, problems of phylogeny occupied the center of interest of biologists for about 50 years. In the past four decades, interest in phylogeny has abated in favor of an interest in mechanisms, even in evolutionary studies. Phylogeny is primarily a historical discipline. It attempts to understand the present as a result of processes which have gone on in the past. The reconstruction of the past offers serious methodological difficulties which are the same in the phylogeny of organisms, in human history, and in cosmology. They are due to the fact that all conclusions are based on indirect and often fragmentary evidence and are not subject to confirmation by experiment. All statements concerning history do not, therefore, have the same degree of certitude as statements concerning processes and mechanisms directly observable at the present time.

Nevertheless, methods have been worked out in all historical sciences which lead to valid conclusions if applied with proper caution. The methods used in phylogeny at the present time have been discussed and exemplified at a Symposium on Principles and Methods in Phylogeny (Caspari, 1963a) sponsored by the American Society of Naturalists. There are fundamentally two methods used: the comparative study of similarities and dissimilarities in extant organisms and the more direct but fragmentary evidence obtained from fossils. It should be mentioned that the more modern methods by which it is attempted to elucidate phylogenetic relationships, such as the study of chromosomes, of proteins, and particularly of the DNAs of related organisms, are extensions of the comparative method at more fundamental levels. They are, however, expected to give, by reason of their closeness to the basis of evolutionary processes, i.e., changes in the genetic material, a more nearly correct picture of the actual phylogenetic relationships than the comparison of morphological and physiological characters.

From a taxonomic point of view, different species differ in their behavioral activities and potentialities just as much as they do in their morphological characters. Mayr (1958) has pointed out that taxonomy could be based just as well on behavioral as on morphological characters and that in many cases behavioral analysis would give more refined and reliable results. Since the taxonomy of an animal group reflects its phylogenetic relationships, it should be possible to use for the study of the evolution of behavior criteria that are in principle similar to those used for the investigation of morphological characters. The similarity of behavior patterns inside a group of related organisms enables us to use the comparative method for the study of these characters.

It may be assumed that, from the point of view of the phylogeny of behavior, only the comparative method is applicable; it has indeed been extensively used and given rise to many fascinating problems. But the use of fossil evidence for behavioral characters is by no means impossible, as is impressively shown in the chapter by Washburn and Shirek for the evolution of human behavior. Here, preserved artifacts play a very large role, but it should be noted that, from the structure of the bones and from the circumstances of preservation, far-reaching conclusions can be drawn concerning the behavior of extinct organisms.

In the elucidation of the phylogeny of behavior of groups that rarely give rise to good fossils, such as birds and insects, the comparative method plays the predominant role. Rothenbuhler's discussion of the social behavior of different species of bees offers a good example of the potentialities of the comparative method in behavior studies. The topic is discussed on a more theoretical level in the chapter by Manning, who points out that the comparative method permits isolation of "units" of behavior which, in the course of evolution, can be reshuffled and modified and put to different functional uses. The genetic nature of these units of behavior identified by comparative observation constitutes one of the most important problems in understanding the phylogeny of behavior.

CONCLUDING REMARKS

Part I, on the evolution of behavior, contains four chapters, two of which deal primarily with evolutionary mechanisms, whereas the other two are focused on problems of phylogeny. It will be seen that these two sets of problems cannot be neatly separated from each other, but the main emphasis of the discussions follows this division.

Furthermore, the two chapters on evolutionary mechanisms deal especially with these groups of organisms that have been most intensively investigated from the point of view of behavior genetics: insects and higher vertebrates. In the two chapters dealing primarily with phylogeny, the same dichotomy into insects and vertebrates has been followed. In this case, the highest and most complex type of social behavior found in each one of the two groups has been chosen for intensive treatment, that of the honeybee and of mankind.

CHAPTER TWO
HUMAN EVOLUTION[1]
Sherwood L. Washburn and Judith Shirek

INTRODUCTION

Evolution is the result of changes in the gene frequencies of populations. Behaviors leading to reproductive success are favored by natural selection, and the genetic bases of these successful behaviors are incorporated into the gene pool of the population. There is a feedback between behavior and its biological base, so that behavior is both a cause of changing gene frequencies and a consequence of changing biology. This relationship is appraised in much more detail in other chapters of this volume, and it has recently been reviewed, with particular reference to man, by Caspari (1961b, 1963b) and Dobzhansky (1962). A more general account of the relations of genetics, evolution, and systematics is given by Mayr (1963).

Emphasis on variable populations and behavioral-structural complexes is leading to a reformulation of many of the problems of human evolution. For example, it used to be argued that the hand of an ape could not evolve into a human hand because there was a trend in ape evolution leading to longer and longer fingers and a shorter thumb. This argument overlooked the great variation in the hands of contemporary apes (Marzke, 1964), depended on the idea of orthogenesis, and was based on the notion that the human hand had evolved to its present form long ago. In contrast to this earlier typological, orthogenetic approach, we recognize today that the evolution of hands was affected by the changing selection pressures that came with bipedalism and tool use. The earliest hominid hand shows very apelike features in the basal phalanges of the fingers and a thumb intermediate between that of contemporary apes and man (Napier, 1962). Remains of this hand, discovered by Leakey in Olduvai Gorge, Tanganyika, clearly show that many of the features that distinguish the human hand evolved long after bipedalism and tool use. The characteristic features of the human hand, and of the areas of the brain that control it, evolved in response to new selection pressures,

[1] This is part of a program on primate behavior supported by Public Health Service Grant MH 08623. We wish to thank Phyllis C. Jay and Jane B. Lancaster for their comments and criticism.

and the form of the hand is not a frozen relic to be used in evolutionary argument without regard to the changing way of life of our ancestors.

As can be seen from the example of the hand, the contribution of genetics to the study of long-term human evolution is to suggest a model, that is, to give the rules that guide the interpretation of the fossils. Problems are posed by fossils, and without fossils the course and rate of evolution could not be determined. But, even with a considerable number of well-dated fossils, there will be no agreement among scientists unless interpretation is attempted within a comparable frame of reference. If what evolved was successful behavior, then understanding of the evolution of our species can be achieved only by the reconstruction of the behaviors of past populations. The more well-dated fossils that are available and the more that is known of the behavior of the living primates, the more reliable the reconstruction of the past will be. But it must be stressed that understanding comes from the reconstruction of past populations and not directly from the fossils themselves. This is because the ancestral forms were alive when they took part in the evolutionary process and it was their reproductive success, not some feature of a bone as such, that determined the course of evolution.

This chapter will discuss human evolution in terms of the relationships of apes and man, of brain and behavior, and of coming to the ground.

APES AND MEN

In the past the overemphasis on anatomical features led, on the one hand, to the creation of numerous genera of fossil apes and men and, on the other, to the exclusion of almost every known fossil from human ancestry.[2] Straus (1949) reviewed many of these theories and discussed the anatomical difficulties inherent in the arguments for deriving man from an apelike ancestor (deriving the Hominidae from the Pongidae). However, recent biochemical and cytological investigations[3] have shown that living men are most closely related to the apes (Pongidae) and especially to the African apes (genus *Pan*), including

[2] Harrison et al. (1964) give an excellent, brief discussion of this problem. They point out that the specimens from almost every individual site have been placed in a separate species or even genus, which results in the assignment of specimens that can usefully be included in *Australopithecus africanus* and *A. robustus* to five separate genera. The specimens which Simpson (1963) or Le Gros Clark (1964) assigns to a single genus, *Australopithecus,* have been placed in as many as seven genera by different scientists. In each case the justification has been that the new specimens are not anatomically the same as previously known specimens assigned to *Australopithecus.* Aside from the overemphasis on minor anatomical points and the lack of attention to variability (Schultz, 1963), the interpretation of the anatomical differences depends on the reconstruction of the way of life of the fossils. If the members of the genus *Australopithecus* were bipedal, savanna-living, tool-using, hunting creatures, it is likely that they occupied extensive ranges with only racial variation. This view is supported by the similarity of the two jaw fragments from Java, called *Meganthropus,* which are very like the comparable parts of *A. robustus* from South Africa (Robinson, 1962).

[3] The latest evidence is summarized and evaluated in the three symposia edited by Buettner-Janusch (1963–1964), by Napier and Barnicot (1963), and by Washburn (1963).

both the chimpanzee and the gorilla (Simpson, 1963). Conclusions from the latest information agree with those of Huxley in 1863; among living primates men and apes are most closely related. The support given this theory by the latest techniques greatly strengthens the position taken by Gregory, Keith, Hooton, Schultz, and many others, and, unless some radically new discoveries are made, it is no longer necessary to consider theories that postulate that man is descended from a monkey, tarsier, or even more primitive form.

Given the high probability that the Hominidae arose from the Pongidae, it is evident that the differences between the families are primarily in the brain and in the manner of locomotion; it is evident in the fossil record that bipedalism came first and that it was not until much later that the brain increased three or four times in volume to typically human size. The general sequence of events has been discussed elsewhere (Washburn and Avis, 1958; Spuhler, 1959; Washburn and Howell, 1960), but the important point in this sequence is that stone tools and evidences of hunting are found with the earliest bones of *Australopithecus*.

Robinson (1962) gives 450 to 550 cc as the capacity of the skulls of both the large and small species of *Australopithecus* from South Africa, and the skull of the large species from Olduvai has a capacity of 530 cc (Tobias, 1964). The smaller species from Olduvai has a cranial capacity estimated as between 640 and 720 cc (Tobias, 1964). As Le Gros Clark (1964) has pointed out, taking all the evidence into account, the forms from Olduvai should be included in the genus *Australopithecus*, and, particularly because some populations of *Australopithecus* of the Lower Pleistocene probably evolved into *Homo* of the Middle Pleistocene, the discovery of intermediate forms should be expected.[4]

The representatives of the genus *Homo* from the Middle Pleistocene all have much larger cranial capacities. According to Weidenreich (1943), the capacities of the five best-preserved skulls from Peking range from 915 to 1,225 cc, and estimates from more fragmentary specimens suggest that the range of the population may have been 850 to 1,300 cc, with a mean in the neighborhood of 1,050 cc. The capacities of the three skulls of Java man are 775, 900, and 935 cc, according to Weidenreich (1943).

These figures are significant to genetics because they show that most of the differences between the brains of apes and of men evolved

---

[4] Some fragmentary remains of these intermediate forms may already have been discovered. The fossils first called *Telanthropus* (Robinson, 1954, 1963) and the later specimens called *Homo habilis* (Leakey et al., 1964) may be regarded as either the end of *Australopithecus* or the beginning of *Homo*, and the distinction may mean very little if one evolved into the other. However, the remains are exceedingly fragmentary, and it must be remembered that the rates of evolution of the various parts of the body are not the same and that it is not yet certain that any fossils earlier than Java or Peking man had a fully human bipedal-locomotor adaptation. There may well have been a period of from two to four hundred thousand years in which the main events in human evolution were the transitions from *Australopithecus* to *Homo*, and the continued emphasis on the separation of the two groups is merely obscuring the evidence for human evolution.

in response to the new selection pressures that came with the human way of life: bipedalism, tool use, and hunting. The brain did not evolve first for some unknown reason and then make possible the discovery of the human way; the human way and the structural basis for that way evolved at the same time and in a feedback relation to each other. This conclusion is greatly strengthened by examination of the cortex of the brain of man. As shown by Penfield and Rasmussen (1950), the sizes of the areas of the motor or sensory cortex primarily concerned with a particular function are proportional to use. For example, in man's brain the areas concerned with the motor control of the thumb and hand are greatly enlarged over the comparable areas in the brain of a chimpanzee. Selection for hand skill has altered the proportional representation of this part of the body in the cortex. This interpretation is supported by the fact that the cerebellum has also increased in size and that there the areas concerned with the hand are proportionally large also. Selection for increasing hand skill came with tool use and affected not only the proportions of the fingers and thumb and the muscles that move them but also the parts of the brain controlling their action. In summary, the brain not only increased greatly in size; it increased also in such a way as to make specifically human behavior possible. The structure of the brain that makes memory, planning, and language possible is also the result of the feedback relation between social evolution and its structural base.

BRAIN AND CULTURE

The interrelationships of the size and complexity of the brain with the behavior it makes possible may be illustrated by language. Apes cannot be taught to talk, because they lack the necessary neural mechanisms, although they can make a wide variety of noises and there is nothing in the structure of the ape mouth or larynx to prevent speech. In marked contrast, human beings learn a language easily. What has evolved in human beings is the structural basis for the ability to learn, a linking of auditory and motor speech areas, a family of unlearned responses making possible human language, and increased control of the vocal apparatus (Bastian, 1965; Penfield and Roberts, 1959). For the individual, the brain makes language possible. For the species the success of sound symbols introduced new selection pressures that changed the evolution of the brain, and, in many populations over many thousands of years, selection built in the mechanism necessary for language as it exists today.[5] The evolution of human society is so

[5] Language may well follow the same model as tool using, and there may have been many thousands, or even millions, of years between the first use of verbal symbols and the evolution of languages of the complexity of those in use today. The success of the first use of sounds to convey more restricted, symbolic meanings changed selection pressures on the nervous system, muscles, and other structures that make language, in the modern sense, possible. Today we see the results of this evolutionary interaction between speech and the structures that make it possible. Bryan (1963) attributed man's ability to speak to preadaptation. (Erect posture, changes in the larynx, mouth, tongue, and nose-mouth relations prepared the way for speech.) The

closely linked with that of language that one cannot be considered without the other. The efficient transfer of information makes complex social life possible, and the more that is known about the societies of monkeys and apes, the greater appears to be the role of language in the matrix of human behavior. It is doubtful if the greatly increased human capacities for memory, planning, and cooperation would have been of selective advantage without at least the beginnings of language.

Human society has depended upon the brain and has evolved in a feedback relation with it. The functions of the brain that are linked with society may be thought of as the "social brain," that is, the complex of parts that make human social life possible and distinguish it from the social life of monkeys and apes. Although the human social way is learned, this learning is possible only because the human brain evolved in response to selection for successful society. Any particular social system—like any particular language—is learned, but *the ability to participate in complex culture is biological.*

Viewed in this way, specialization of roles in society is one of the most important factors in human evolution and changes the whole pattern of relationship between biology and culture. As long as each human individual had to learn the whole behavior pattern of his group, the evolution of the brain set a limit to the evolution of society. But once roles were specialized so that the individual needed to learn only one set of skills, to understand only a part of his culture, then social complexity could evolve without further evolution of the brain. It is probable that this change began primarily with agriculture and has proceeded at an accelerating pace ever since. The primitive hunter had to know the way of his tribe, including religion, folklore, economic and social skills, and the skills of war. Except for the difference in the roles of males and females, he had to be a master of all his tribal culture, and these are the conditions under which the human brain evolved. With specialization, society may be complex but the individual need learn only a small part of the culture. In this way the evolution of complex culture is freed from the limitations of the individual organism.

From the standpoint of what the single human individual must know, modern civilization need not be more complex than tribal society. The final burst in social evolution, beginning with agriculture and vastly accelerating with the industrial-scientific revolution, is dependent in part on the freeing of culture from the limitation of being comprehensible to the individual actor and so allowing rapid change to great complexity without the necessity for further evolution of the nervous system of the human participants. This situation in which culture is vast and complex but the individual need learn only a small fraction of it in order to participate may be illustrated by language. A dictionary

---

major importance of changes in the central nervous system is minimized. This is another example of the traditional thinking which supposes that the structure evolves first, making the behavior possible. We believe that language and the structures making it possible evolved together in a feedback relation and that there must have been a long evolution between noise-making apes and the speech of *Homo sapiens.* The demonstration, that contemporary apes cannot make human sounds, merely shows that they have not shared in this part of human evolution.

gives over 550,000 English words. This is an edited list and might be supplemented by many archaic terms and special technical vocabularies. For example, a medical dictionary gives over 80,000 technical medical terms alone. Certainly our culture is using well over a million words, but, for an individual participating in a part of this culture, a few thousand words suffice.

In view of the great importance of the brain in human evolution, it would be interesting to be able to relate the detailed biological evolution to the archeological succession. But unfortunately the only direct evidence is cranial capacity, which is a poor guide to function (Mettler, 1956). A partial answer to the problem may come from relating the evolution of the brain to the rate of cultural change, rather than only to cultural complexity. Although the evidence is scanty and the absolute dates are still under discussion, the following correlation is at least suggestive. Very simple stone tools existed in the Lower Pleistocene (J. D. Clark, 1964). It seems safe to say that members of the genus *Australopithecus* made at least some of these tools, hunted, and lived bipedally in the open savanna away from trees. This stage of human activity probably lasted for a minimum of 2 million years with very little evolution. With the Middle Pleistocene, some 500,000 years ago, members of the genus *Homo* made complex tools, killed large animals, and made fire, and there is substantial and steady evolution in the archeological record. In the past 40,000 years the rate of cultural evolution increased vastly; this increase seems to correlate with the presence of *Homo sapiens*. Over the broad expanse of the past 2 or 3 million years, rate of cultural evolution and cultural complexity seems to correlate with the kind of hominid.

In summary, the notion that the evolution of the brain parallels the evolution of culture seems to fit the fossil record reasonably well. Because brain and society evolved together, it is no accident that the human brain makes possible the basic human social skills. What evolved were the structure and physiology that make complex human language and social learning possible. The biological base, therefore, does not determine the form of any particular culture; it is a necessary prerequisite to all cultures.

HUMAN ORIGINS

Speculative as these conclusions may be, there is far more evidence for them than for the reconstruction of the preceding stage in human evolution. There is very little evidence to connect *Australopithecus* of the Lower Pleistocene with any particular group of Pliocene apes (that is, to connect the Hominidae with the Pongidae). As indicated earlier, the latest evidence suggests that man is particularly closely related to the African apes, but the fossil fragments that are the most similar are those of *Ramapithecus* of India (Simons, 1963a). Simons notes that *Ramapithecus* shares many features with *Kenyapithecus* of Africa, which, unfortunately, is also represented only by very fragmentary remains. These forms are from the early Pliocene or the end of Miocene, some 10 million years before the earliest specimen of *Australopithecus*.

The question of whether man originated in Africa or Asia has been debated for many years, but even this phrasing of the question is misleading. Both the fossil record and genetics suggest that the area of origin and the meaning of the term origin be reexamined (Simons, 1963b).

Fossil apes have been found in Europe, Africa, and Asia, and the separation of these areas is geologically recent. Simons (1963a, 1963b) has pointed out the similarity of such forms as *Pliopithecus* (Europe) and *Limnopithecus* (Africa) in addition to the similarity of *Ramapithecus* and *Kenyapithecus*. During the Miocene and early Pliocene, forests extended from Europe to eastern Asia and to Africa; there was no barrier for apes, and a single genus might be represented by species in India and Africa (Simons, 1964). The habit of automatically placing African and Asiatic forms in separate genera is unwarranted. Indeed, a single species of monkeys, *Cercopithecus aethiops*, occupies an area as wide as the distance from eastern India to east Africa. The origin of the Hominidae from the Pongidae need not have taken place in one local area, as is traditionally assumed. Instead, this transition may have occurred in populations of widely distributed successful apes. The entire area which linked India to Africa was occupied by apes, and the transition may well have taken place in this huge area over a period of some millions of years.

To seek a local area of origin is as much a relic of typological thinking as to classify almost every new fossil in a new genus. The desiccation and climatic change that subsequently separated the forests of India and those of Africa removed something on the order of 5 million square miles from the habitat that could be occupied by arboreal apes. This reduction took place over a period of some 10 million years. If a genus of apes (perhaps of the *Ramapithecus-Kenyapithecus* group) evolved into *Australopithecus*, the rate of that evolution would be a new genus in something of the order of 8 million years, a span of time that is about average for mammals (Simpson, 1953b). If we assume that the species of ape that evolved into *Australopithecus* occupied only 2 million square miles (which is less than many species of monkeys) and that there was a density of 10 per square mile (which is low), there would have been 20 million animals in the transitional populations at any one time. If we assume a generation time of 10 years, then 800,000 generations separate populations of *Australopithecus* from populations of early Pliocene apes. Naturally, calculations of this sort may be very far from the actual facts, but they are introduced to correct the impression that the origin of man occurred necessarily in one restricted small place or in a short period of time.

BIPEDALISM AND TOOLS

Whatever the actual time required to separate the Hominidae from the Pongidae, the main behavioral differences between populations of apes and of *Australopithecus* are bipedalism of the human kind and the use of tools. All known members of the Hominidae are adapted for life on the ground away from trees, and coming to the ground has been regarded as a crucial stage in the transition from ape to man. Various reasons have been suggested for this descent from the trees, but, if we can judge by the behavior of contemporary monkeys and apes, feeding on the ground is the primary one. Different groups of Old World monkeys and apes have left the trees and taken to feeding on the ground. The macaques and baboons offer an example of very successful ground living. Geladas, patas monkeys, and one langur (*Presbytis entellus*) show comparable feeding adaptations. *Cercopithecus aethiops* is primarily a ground feeder, and many other species of *Cercopithecus* come to the ground to feed occasionally. Most of the gorilla's food is obtained on the ground (Schaller, 1963).

The very successful species *Cercopithecus aethiops*, one of the most abundant and widely distributed monkeys in Africa, shows the advantages and problems of coming to the ground in perhaps the clearest form. These vervets are distributed from West Africa to South Africa along the edge of the forests, in savanna where there are trees, and along rivers. They feed in the trees and on the ground, and as one watches them it is obvious that a troop is brought to the ground far more often for feeding than for moving from tree to tree, although they do that too. As we have indicated, observations of the contemporary primates strongly suggest that the primary motivation for arboreal monkeys to come to the ground is food. Fruits, buds, grasses, insects, and the like offer immediate rewards to monkeys or apes that come to the ground, and the great importance of ground feeding should not be overlooked in accounting for descent from the trees.

It has been suggested that, as the climate became drier and forests decreased, apes came to the ground simply to walk to the nearest trees (Hockett and Ascher, 1964). This view stresses the conservative side of evolution, but evolution is also opportunistic (Simpson, 1949), and observations of contemporary monkeys and apes indicate that ground feeding may be an important motivation. Of the living primates, the chimpanzee's behavior fits the "coming to the ground to stay in the trees" model best. Chimpanzees are chiefly fruit eaters, and they walk along the forest floor when moving from one feeding area to another (Goodall, 1963; Reynolds and Reynolds, 1965). If an animal moves from one group of trees to another in order to feed in the trees, it must keep its arboreal-climbing adaptation. The chimpanzee adapts to ground life by quadrupedal knuckle walking (Tuttle, 1965) and maintains its arboreally adapted pelvis and foot, keeping climbing efficiency. Maintaining efficient climbing adaptation precludes the possibility of evolving a nongrasping, weight-supporting foot. An animal that is ground-living only in order to move from one tree to another must keep its fundamental arboreal adaptations. Adaptation to ground feeding, although compatible with both bipedal-locomotor efficiency and the loss of climbing adaptation, is not enough to explain the evolution of bipedalism. Ground-living gorillas and baboons are quadrupedal, demonstrating that the evolution of bipedalism cannot be explained solely in terms of adaptation for coming to the ground.

Desiccation and restriction of forests are an inadequate explanation not only of bipedalism but of coming to the ground as well, since the edge of forest, open forest, and savanna with trees were most extensive during wet times. Consider the length of the forest borders, when they extended from Europe to eastern Asia and to South Africa. As the areas of rain and temperate forest increase, so does the periphery, and with it the likelihood of the evolution of partially or fully ground-dwelling species such as *Cercopithecus aethiops*. There is every reason to believe, both from the fossil record and from the climatic conditions, that there were many more kinds of apes in the end of the Miocene and the beginning of the Pliocene than there are today. Both because of numbers and because of the length of forest borders and the extent of open forest, there were far more opportunities for ground-living forms to evolve in wetter periods than in drier ones. However, a species can adapt only to the actual conditions under which it is living, not to long-term trends. For example, if the climate became drier at the rate of 1 inch per century, rainfall would be reduced from 60 to 10 inches in 5,000 years. What would, in sum, amount to a catastrophic climatic change would not be detectable during the lifetime of even the longest-living primates. Annual and local variations in climate in the short run are far greater than long-term trends, and it is to these actual local conditions that the species must adapt. The long-term trends increase or decrease the area available to a species, but the adaptation of the species must be to local, short-term conditions.

In summary, many Old World monkeys and apes have come to the ground. This is most likely to occur when the climate is wet and the forests are of maximum extent. Ground feeding is the most likely explanation for this move. Bipedalism might have happened only once and is not to be explained merely on the basis of providing a means for ground living.

The specialized bipedal locomotion of the Hominidae seems to be the result of the success of tool using. Numerous authors have stressed the importance of bipedal locomotion in freeing the hands for carrying (see recent reviews of this subject by Hewes, 1961, and by Hockett and Ascher, 1964). Carrying (whether tools, weapons, food, or infants too immature to cling) has been of the greatest adaptive importance to the successful evolution of man, and it is probable, on the basis of the field studies as well as the evidence of fossils, that tool-using behavior came first and the evolution of bipedalism was in response to the success of this new pattern of behavior. The traditional statement of the problem has been that bipedalism freed the hands and that the locomotor pattern evolved first—and for unexplained reasons. Recent evidence suggests that the beginnings of tool use came first, that the success of this pattern of behavior changed selection pressures, and that bipedalism then evolved in response to these new pressures.

We think of "carrying" in the broadest possible way, including carrying weapons, which changed the relation to predators and other human groups; carrying tools, which changed food supplies and habits; carrying food, water, and protective garments. When the hands are free, carrying infants, the wounded, and the sick is no problem. Biped-alism freed the hands for a wide variety of skills, making possible the human way of life. This point of view may be stated in the following way. In many populations of apes over some millions of years minimal tool use was present. In some of these populations the carrying of tools, and the products of tool use, became sufficiently important so that selection favored those groups of apes in which bipedal locomotion was more efficient. Bipedalism permitted the evolution of skillful, prac-ticed tool use, and, as the locomotor pattern evolved in response to the new pressures, more effective tool use also evolved. Locomotion and tool use affected each other (were in a feedback relationship), and each is at once cause and effect of the other. The use of tools was not made possible by a preceding bipedal adaptation, nor was tool use a simple discovery. Probably it was the result of repetitive events in thousands of populations of apes, of several species, over millions of years. The evidence for this point of view will now be examined.

Tool using by monkeys and apes has been reviewed by K. R. L. Hall (1963a and 1963b), and it is remarkable how little evidence there is for the use of tools among nonhuman primates. Although many monkeys and apes easily use sticks and stones in captivity, evidence from field studies suggests that under natural conditions such behavior is rare or absent. Certainly objects are used no more than by many nonprimates, and, if it were not for its possible relation to human behavior, object manipulation among the primates would be of no more interest than the utilization of objects by some species of birds or mammals (K. R. L. Hall, 1963a). In fact, it is surprising to see mon-keys, like baboons, skillfully groom, pick small objects, dig and clean food, without making any effort to use easily available sticks and stones to help in any of these tasks.

The chimpanzee provides the one exception to this statement con-cerning tool use by primates (Goodall, 1964). It is almost certainly no accident that it is this ape, which in many other ways is the closest of the living apes to man, that makes substantial use of objects. Goodall reports that chimpanzees throw stones and use sticks, branches, and leaves. She observed chimpanzees throwing stones toward baboons as a part of agonistic displays. They break off sticks and vines, pre-pare them, and use them to get termites out of their holes (during one season of the year only). The chimpanzees break these sticks or pieces of vine used for catching termites or fire ants at a suitable length and remove the side branches; several sticks may be prepared and carried to the nest where they are to be used. Chimpanzees use leaves to clean the body and as sponges. By chewing leaves until a pulpy mass is formed, the chimpanzee makes an efficient sponge that is used to get drinking water from pools. Young animals were observed to make tools too short or in other ways unsuitable, and it appears that learning plays a substantial part in the development of this skill.

These observations suggest that selection of material and its prepa-ration are as old as any concept of tool use. The frequently postulated evolutionary stage in which natural objects are used prior to being modified by the user is not supported by these observations. Objects are used for utilitarian purposes (getting termites, ants, water, cleaning, breaking nuts) and in agonistic displays. Display, such as branch shaking, is a common supplement to threat gestures by primates. Orangs may even break off a branch and drop it (Schaller, 1961).

Hall's review of these behaviors makes it clear that there are two possible ways for a weapon to evolve, either as a consequence of the use of larger sticks for utilitarian purposes or from the repeated discovery that a stick waved in display may do damage if it actually hits the animal against which the display is directed. Similarly, rocks thrown in the direction of another animal during display may occasionally hit it and the result of actual hitting be discovered.

The many uses of objects by the chimpanzee offer a broad base of differential use, and from such an array of behavior more efficient tool using might evolve. Goodall's observations show the artificiality of traditional speculation on whether weapons may have evolved before other tools or the use of clubs before stones; these are human categories, and they do not relate to the behaviors of the chimpanzee. Equally artificial is the discussion of whether carrying meat came before other kinds of carrying, for chimpanzees carry fruit, meat, and objects. The field studies clearly show the extent to which evolutionary hypotheses have been distorted by the tacit expectation that human categories will be useful in the description of ape behavior. Tool using by human beings differs from that of the apes in the degree of skill and in the extent of the result.

Probably the greatest importance of bipedal locomotion is that it permits the animal to move while holding an object, so that skilled use may be learned at leisure. It is skill, rather than mere carrying, that is made possible by the freeing of the hands from locomotor functions. During the lifetime of the individual the preparation for skill is play, and young children enjoy playing with objects; apparently this play is essential for the development of adult motor skills. Among the nonhuman primates juvenile play soon becomes almost exclusively interpersonal; its importance lies in the development of social skills (fighting, sex, dominance relations, grooming, and affective behaviors).

SUMMARY

The view of human evolution presented here suggests that the characteristic features of the human body evolved at the same time as the human way of life. The evolution of behavior and of the structures related to behavior are two facets of the same process. Further, it is suggested that this evolution took place in many populations, in large areas, over millions of years, for many reasons. It is most unlikely that the transition from the Pongidae to the Hominidae occurred in one place or for one reason, such as desiccation. Behavior genetics shows that in evolution behavior precedes structure, that new structures result from new selection pressures, and that such changes are most likely in large species that are partially divided in a wide variety of adaptive niches. The diversity and distribution of apes in the early Pliocene suggest that this may have been the time of the origin of the Hominidae, and the fragmentary fossils called *Ramapithecus-Kenya-pithecus* may represent the ancestral species. It is suggested that bipedalism evolved with the success of tool using and that the brain evolved with the success of the complex human way of life based on skills, complex social life, and language.

CHAPTER THREE
BEHAVIORAL MODIFICATION OF THE GENE POOL[1]
John A. King

INTRODUCTION

Behavior is modified and influenced by genes. Behavior is also active in modifying the frequency and expression of genes in a population. Perhaps no other general character of a species, including its morphology and physiology, has such influence in altering the genetic character of its populations. Often only large deviations from the normal morphological or physiological characteristics of the species significantly reduce viability or fertility of the aberrant individual, whereas even slight behavioral deviations from the norm can affect the union of the gametes in the population, the number of young produced and brought to sexual maturity, the flow of genes within and between populations, as well as the survival and continuation of the gametes of each individual. Breeding patterns, assortative mating, courtship, parental care, social tolerance, migration, ingestion, shelter seeking, and agonistic behavior are the behavioral patterns associated with changes in the gene frequencies of populations. The study of the relationships between behavior and changes in the gene frequency of populations may be designated the "ecological genetics of behavior," which emphasizes the effect behavior has first upon the dynamics of populations and ultimately upon the genetic constitution of populations.

Behavior can possibly alter the expressivity of the genes or their effect upon the phenotype in a manner that affects the composition of the gene pool. Because the genotype of an animal is not expressed in a vacuum, such environmental factors as nutrition, temperature, and substrate conditions affect the expression of the genotype in the resulting phenotype. Behavior can determine the nature of these environmental factors during ontogeny. Furthermore, the phenotypic expres-

[1] The work reported in this chapter was supported by PHS Research Grant MH-05643 from the National Institute of Mental Health, Public Health Service.

sion of genes affecting behavior may be determined by the behavior itself. This area, which emphasizes how a particular array of genes is expressed in the phenotype, may be designated as "developmental genetics of behavior." This includes those genes primarily affecting morphology and physiology as well as those genes most directly related to behavior.

Ecological genetics (Ford, 1964) and developmental genetics (Markert, 1965) are established scientific disciplines which apply to behavior as well as other organismic characters. The experimental material for ecological and developmental genetics of behavior, however, is indirect and widely scattered throughout the diverse disciplines of genetics, embryology, ecology, and psychology. The purpose of this chapter is to illustrate the organization of ecological and developmental genetics of behavior with related experiments and observations. Ecological genetics of behavior will be treated first, followed by a discussion of developmental genetics of behavior.

ECOLOGICAL GENETICS OF BEHAVIOR

Population dynamics and population genetics are well-established areas which require no review here (see Slobodkin, 1961; Li, 1955). Although their relationship to behavior is not well defined, Blair (1953), Klopfer (1962), and Wynne-Edwards (1962) have made substantial contributions to our knowledge of this relationship. In this section, the effect behavior has upon the dynamics of populations will be emphasized, since behavior has only an indirect effect upon the genetic structure of populations. The procedure followed in this section will illustrate behavioral contributions to the temporal, spatial, and sex-age distributions of the population, with brief reference to their effect upon the gene pool.

Temporal Distribution

The temporal distribution of a population depends upon the balance between natality and mortality. If natality exceeds mortality, the population will grow in size and usually become more dense. Decline results from an excess of mortality over natality. Over the course of time any population will show excesses in both natality and mortality, and the population will oscillate. The genetic composition of the population is altered by the genetic constitution of those individuals which contribute most to population growth and of those which are preferentially affected by its decline.

Natality   The primary source of recruits to the population is the birth of individuals within the population. Before birth of the offspring, the gonads in the parents must ripen, and the gametes must unite; both processes are determined to some extent by behavioral characters (Tinbergen, 1951).

Darling (1938) was one of the first investigators to propose that fertility in colonial birds may depend upon external stimulation arising from the social interaction of colony members. The subject of exteroceptive stimuli affecting reproductive processes in birds and mammals has been reviewed by Lehrman (1961). That a basic physiological process, such as ovulation, depends upon behavior may be illustrated by house mice (Whitten, 1956). When mice were first paired for mating, fewer than the expected one-third mated on the first and second nights, while many more than the expected number mated on the third night. Apparently the females were not exhibiting their regular 3-day estrous cycle prior to pairing. After the females were united with males, the regular cycle occurred and estrus appeared on the third night when most matings took place. Males confined to a small wire basket within the female's cage induced estrus. The placing of females in a cage recently contaminated by males similarly induced estrus. Some characteristic of the male, most likely his odor, was essential for the regular occurrence of estrus. Not only did the male induce estrus in female mice, but the occurrence of estrous cycles can be inhibited by the presence of other females (Whitten, 1959). When females were kept together in groups of 30, most mice failed to exhibit the normal estrous cycle. The perceptual stimuli provided by other individuals of the same species are capable of inducing or inhibiting the physiological processes involved in fertility, one of the first requirements for the maintenance and increase of the population (Parkes and Bruce, 1961).

Another critical stage determining the growth and size of populations is the union of the gametes. Among sexually reproducing species, this involves the orientation of potential mates to each other, the synchronization of their sexual receptivity, the inhibition of activities which conflict with mating, and some form of mate selection (Tinbergen, 1951). These functions are fulfilled by courtship behavior. Failure of any one function prevents fertilization, with a subsequent loss of the gametes to the population or at least differential fertility. The actual loss or possible gain in the amount of natality due to variation in courtship behavior is difficult to assess. In situations that disrupt courtship patterns, such as overcrowding, copulatory behavior is frequently absent or reduced, or the copulations do not result in fertilization (Calhoun, 1949).

More important in inducing changes in the composition of the gene pool than the relative gain or loss of fertility brought about by variation in courtship is the differential fertility resulting from mate selection, provided that the selection is correlated with or is determined by genetic characters. In homozygous matings, recessive mutants will appear in the phenotype, upon which selection can act. The sexual selection described by Darwin was largely discredited when many of the displays attributed to courtship were found to be threat displays. The pendulum has swung again in the other direction, and now both threat and courtship are recognized as contributing to differential fertility (J. M. Smith, 1958). Courtship is a complex type of behavior which includes aggressive and escape components as well as copulatory responses (Andrew, 1961).

Sexual isolating mechanisms among sympatric species preserve the genetic integrity of each species but have little or no effect in altering the gene pool within a species (Spieth, 1952) unless they fail, in which case hybridization may occur (Dobzhansky, 1941). Within a species, differences in courtship behavior may alter gene frequencies. Several mutants of *Drosophila melanogaster* exhibit different courtship patterns which contribute to differential fertility in population cages with wild-type flies (Bastock, 1956; Merrell, 1949). Merrell (1953) examined the change in the frequency of four sex-linked recessive mutant alleles by combining the mutants with wild-type flies in population bottles. By starting with a gene frequency of 0.5 in the mutants, the departure from random mating would be indicated by an excess of the mutant gene in the males as compared with the females. Populations sampled each month (estimated generation length = 24 days) over a period of 6 to 30 months revealed three of the four mutant genes (yellow, cut, raspberry) decreasing in frequency and no significant change in one (forked). Some of the replicate populations (nine for each mutant) totally eliminated the mutant gene after a few generations. These changes in gene frequencies could also result from factors other than courtship, since the behavior of the flies was not observed. However, the previously established fertility and viability of the mutants indicated little difference from the wild type, whereas mating success was different. This effective elimination of mutants, rather than the maintenance of equilibrium, strongly implicates courtship as the most important causal factor for differential productivity and the consequent changes in gene frequencies.

In addition to the effect of courtship and aggression upon natality, these behavior patterns are also responsible for sexual isolation and the establishment of breeding systems. The excellent studies on sexual isolation between species involving many different classes of animals precludes the necessity of describing examples (see Perdeck, 1958; Blair, 1958; Marler, 1960; also see chapter by Manning). Sexual isolation is certainly a most conspicuous way in which behavior can initiate and maintain populations of different genetic structure. The effect of breeding systems will be discussed later in relation to sex distribution.

After the gametes have matured and united, the behavior associated with care of the offspring has its effect upon the growth and size of the population. Parental care, like courtship, cannot be isolated from other types of behavior affecting the survival of the offspring. Southwick (1955) observed that aggression, gregariousness, intermingling of sexes, nest destruction, and communal nesting were interrelated in their effect upon litter survival among house mice. Some of these factors are density-dependent. As the density of mice in a particular population increased, communal nesting and gregariousness increased to the detriment of the numbers of young surviving. The amount of aggression exhibited by each of the six mouse populations varied, as did the population density; however, there was no consistent relation between the size of the population and the number of fights per hour. One of the smallest populations exhibited the most aggression. When approximately one aggressive encounter per hour occurred for each mouse, litter survival was lowered to the extent that population growth was retarded or ceased altogether. Aggression affected litter survival by creating social unrest and stress and by direct killing of the young during fights between adults. The actual number of deaths attributable to such aggression was difficult to establish, but in communal nests the percentage of survival was as low as 13 percent. Parental care in many other species is particularly vulnerable to changes in social milieu, often disrupting those patterns which normally lead to the survival of the young (Sawin and Crary, 1953).

Intrinsic control mechanisms for maintaining population size in natural populations have been reviewed by Wynne-Edwards (1962) and Christian and Davis (1964). Wynne-Edwards postulates a feedback mechanism from population density to behavior, whereby an increase in population density produces certain behavior patterns that limit further increase (King, 1965). Christian and Davis more specifically suggest that an endocrine feedback mechanism is responsible for the limitation of population growth among mammals.

Natality and population growth are more than simple correlates of behavior; they are the products of behavior, from the maturation, release, and union of the gametes to the age when the young enter the population as independent fertile organisms. Population size, or N, which plays such an important role in the calculation of genetic changes in populations, is a figure that depends to a large degree upon the behavior of the organism. On the other hand, the number of individuals in the population is also important in modifying some patterns of behavior.

Mortality   Mortality is almost the mirror image of natality. Behavior that does not lead to birth and survival leads to death. This may occur at any time in the life cycle, with the gametes often more vulnerable to mortality than the individuals. We have already discussed some aspects of behavior related to natality, from the production of gametes to that stage in the life cycle when individuals become independent of their parents. We can proceed from this stage by regarding those behavior patterns not leading to natality as contributing to mortality.

Individuals of different species become independent at various stages. Many species which never encounter their parents depend upon the behavior of the mother in her choice of a location to deposit the eggs. Except for this provision, the survival of the young depends primarily upon their own behavior. They must locate food sources, escape from enemies, and adjust to the physical environment in order to survive. In many species, mortality is greater at this early period than at any other period in their life cycle. The usually smaller size of the young increases the number of potential predators, their structural immaturity prevents the elicitation of adult responses, and their lack of experience and learning often causes them to make nonadaptive responses. The same mortality-causing factors in the young are also operative in the adult, although sometimes in different form. The higher mortality of the immature and their greater potential for modifying behavior than the adult make immature animals particularly interesting to examine behaviorally.

Behavior is variable and often modifiable through processes of fatigue, adaptation, habituation, and learning. In the course of development, certain responses become fixed by habit formation. Other responses maintain a certain level of plasticity, which enables the animals to apply past learning to new situations—a learning set. The actual effect of learning upon survival in natural populations is difficult to measure, although few observers would question its importance. One widespread phenomenon throughout much of the animal kingdom is the attachment an animal exhibits to its home or home range. In contrast to this attachment is the apparent stress, timidity, inactivity, or confusion animals exhibit in a strange environment. Such effects of a strange environment have been reported for several invertebrates, such as planaria (Best and Rubinstein, 1962) and cockroaches (Woolpy and Schaefer, 1962) as well as many vertebrates (Barnett, 1963). Attachment to the home range is climaxed by homing behavior. The essential difference between the home range and a strange area is the knowledge an animal acquires of the home. Learning of some type occurs in this process of familiarization. To what extent does it affect mortality?

Our best evidence is the comparison between mortality of animals in their home range and those unfamiliar with the same region. Unfortunately mortality figures are difficult to establish for wild populations. The absence of identifiable individuals from the study area is the most common indicator, for dead animals are rarely found and predation is seldom observed. Emigration is the most probable alternative to mortality to account for missing individuals, and both mortality and emigration provide similar results as far as the genetics of the population is concerned. Blair (1940) saturated an area inhabited by a pair of deer mice with 45 mice from another area. After a week he was able to recover the original pair familiar with the home range, and only 6 of the 45 introduced mice remained. Traps placed around the area of release added only 10 more mice during the following 3 weeks of trapping. Approximately 64 percent of the mice introduced into strange territory suffered mortality within a month, while the two resident mice remained and produced a litter. In a somewhat similar experiment, 60 laboratory-reared deer mice were introduced into an uninhabited, isolated field provided with food and nest boxes on three different occasions. At the end of 9 days, a mean of less than 25 percent of the mice survived; most of them disappeared during the first 4 days (King and Eleftheriou, 1957). These observations are not adequate tests of the hypothesis presented here, but they indicate that mice in a strange environment are extremely vulnerable to mortality factors.

Young animals independent of their parents are in many respects similar to introduced animals, since they also enter a strange environment. Often they are more naïve than displaced adults. Among many small rodents mortality is exceedingly high after weaning at the period when they leave the nest. Hoffmann (1958) found mortality as high as 82 percent among postweaned voles (*Microtus*) in a population he studied. Although Snyder (1956) found mortality to be age-independent among white-footed mice over 5 weeks of age, his regression lines show a steep decline of surviving mice with increasing age. Many life tables made for a number of vertebrates illustrate that mortality is independent of age after sexual maturity, whereas mortality prior to mating is dependent on age. The reasons for high mortality among the young are certainly multifold, but among higher vertebrates, at least, the lack of knowledge of the home range is an important cause. I have stressed learning, especially in the young, as a primary factor in mortality; however, all behavior from the simplest taxis to true insight is fundamental to the survival of animals in all age classes.

Oscillation   Natality and mortality affecting different genotypes differentially are the basis for all genetic changes in a population. However, certain combinations of these two factors demand special consideration. Growth and decline are of special interest, because after a population crash, the reconstituted population may represent only a small percentage of the genotypes originally in the parent population. In his study of vole populations, Frank (1957) found years in which only 1 percent of the original population survived to reconstitute the subsequent population. A pilot study of deer mice surviving the winter under seminatural conditions revealed a similar resurgence of the population from one or two females. The genetic consequences of these extreme fluctuations in size of population are clear (Elton, 1946; Wright, 1931), but how does behavior contribute to these oscillations?

The role of behavior in population oscillations is not firmly established, but behavior is strongly implicated in the fluctuations of several species of small rodents (Christian and Davis, 1964). Growth of the population, although dependent upon behavior, is usually stimulated by favorable conditions of food and shelter. Initial increases occur in the spring and continue late into the fall or early winter. Winter mortality tends to be relatively low during periods of population growth, and so by the following spring a large number of sexually mature mice contribute their progeny to the population. A population asymptote is reached usually within 3 or 4 years, at which time the population density is high. Then a rapid, precipitous decline occurs until only a small fraction of the population from the preceding year remains. These oscillations have been considered cyclic, occurring at regular intervals of 3 to 4 years among voles. The empirical evidence substantiates the cyclic nature of these populations, but so far no mechanism for its cause has been demonstrated to differ significantly from what might occur in a random series of numbers (Cole, 1957). Cyclic phenomena, however, are less important to possible genetic changes than individual oscillations, regardless of their periodicity.

During a population increase, the chances for individuals to come into contact with each other also increase. Each contact results in some kind of social interaction, which tends to become more and more competitive as the resources of the environment are depleted. Social stress contributes to other organic stresses experienced by the animals, and the general adaptation syndrome is initiated (see chapter by Hamburg). Continued activation of the adrenal cortex upsets the regulatory mechanisms of the pituitary and causes hypertrophy of the adrenal cortex. Adrenal size increases, and gonads decrease in size. Reproduction is reduced, and, under continued social stress, animals succumb in the exhaustion stage of the syndrome (Christian, 1959; Frank, 1957). Social stress may continue to affect the reproductive performance of subsequent generations (Christian and LeMunyan, 1958; Chitty, 1960). Field observations and laboratory experiments tend to confirm that this interaction of population oscillation and behavior, principally social behavior, is the determining factor in the oscillation.

Spatial Distribution

The genetic alteration of a population distributed throughout time as it grows with natality and declines with mortality and infertility is matched in importance by its spatial distribution. Most population genetic models began with the assumption of panmixia in which the chances of animals breeding are independent of their spatial distribution. Random distribution is merely a statistical convenience and not a biological fact, as model builders are aware. An animal at one extreme of the range of the population has little chance of breeding with another at the opposite edge of the range. Even animals in geographic proximity may have little opportunity to mate. These spatial barriers to mating depend largely upon such behavioral factors as habitat selection, territoriality, and group cohesion. Furthermore, the flow of genes from one population to another results from migratory behavior, homing, and dispersal.

Habitat Selection   Those individuals selecting the same habitat in a region of heterogeneous habitats not only increase their chances of interbreeding, but they also tend to encounter similar forces of natural selection. Our knowledge of habitat selection is almost exclusively limited to the recognition that it occurs. The perceptual cues involved and the effect of learning in making the selection are only partially understood. Selection of a habitat is so tightly intermingled with the adaptation of a species to it that we cannot readily decipher the causal relationships (Pittendrigh, 1958). The structural capacities of animals may lead them to select the habitat where these adaptations are used, or the reverse may be true. An animal's ability to learn the perceptual cues and the motor responses suitable for a particular habitat further complicates the processes of habitat selection (Wecker, 1963). Species competition for the most effective habitat utilization also contributes to variations in spatial distribution (Sheppe, 1961; Klopfer, 1962).

Species differences in habitat selection have little effect upon the genetic characteristics within a species unless this is the only reproductive barrier between them. Within a species, different habitat preferences have not been found except among subspecies which interbreed only at the periphery of their range. Neverthless, an inter-breeding population does occupy diverse habitats. Until the differences in habitat occupancy can be shown to result from natural selection or some other factor, the mechanisms bringing it about will remain obscure. The attachment and return of animals to their natal homes suggest that learning may also be a factor in their selection of a habitat (Terman, 1963). Animals tend to select those habitats which most closely resemble their natal homes (Wecker, 1963). Insofar as the preference is affected by the genotype, a behavioral feedback to the genotype will occur simply by keeping a given population restricted to its habitat. Evidence for ecological isolation bringing about speciation is scarce and weak (Mayr, 1963), although it must be expected to contribute to differences in gene frequencies of local populations (Ford, 1964). Ecological barriers are disrupted too frequently by the temporal factors of growth and decline, which cause genetic swamping and extinction, to become permanent. (See Hinde, 1959, for a review.)

Territoriality   An evenly spaced distribution of animals may result from resident individuals preventing prospective invaders from entering the home range. Defense of the home or territory restricts the breeding aggregate to the resident individuals, often just one pair. It also limits the effective size of the breeding population because animals without a territory rarely breed (Nice, 1941). Animals within a territory may breed repeatedly, producing many offspring which possess genotypes derived from the limited parent population. The distribution of ter-ritorial species is one of the most important factors limiting panmixia (Wynne-Edwards, 1962).

The behavior involved in territoriality is well known (Carpenter, 1958). It usually involves the selection of a particular site by the male, which he defends against other males. Later females are attracted and courtship proceeds, while interlopers are excluded, often by the female as well as the male. Young are raised in the territory and usually disperse when independent. Afterwards the territory may be abandoned or retained for subsequent broods. Many kinds of territories have been recognized, varying from the immediate vicinity of the nest to large foraging ranges. Some include only a pair of individuals whereas other territories are maintained by a group of individuals, in which some members may be more active in its defense than others. Territories may be semipermanent or transitory, and even moving ter-ritories have been considered. Each variation has its special contribution to genetics, depending largely upon the breeding pattern.

Group Cohesion   Of the three possibilities for spatial distribution, random, even, or grouped, grouping occurs most often, and cohesion among individuals is one of the principal factors responsible for it. Each group of animals of a particular species is usually separated from other groups. Groups or aggregations often arise as a result of attraction to a suitable habitat in a heterogeneous landscape, but many species exhibit grouping in a homogeneous environment. Among some or-ganisms, especially plants, clumps result from lack of mobility or asexual reproduction, and the clumping of progeny about the parent may have important genetic consequences. Lack of adequate dispersal can also cause the clumping of some animals, but more often animals are attracted to each other and remain together through some type of cohesion. This has often been called gregariousness. P. J. Clark (1956) has been able to separate mathematically the causes for a grouped distribution into regional differences in habitat and interactions between the organisms. The genetic effect of grouping is brought about by segmenting a population into small, inbreeding fragments, a condition extremely suitable for rapid genetic change (Wright, 1931).

The mechanisms (vision, audition, olfaction), the causes (sex, defense, infant care), and the results (reproduction, protection, genetic) of group cohesion vary considerably among species and remain an enigma for many familiar species. Schooling by fish is one of the most conspicuous examples of group cohesion. Recently Shaw (1960, 1961) investigated the ontogeny and visual mechanisms of schooling among several species of fish. She found that the eye of schooling fish is best adapted to the perception of small movements and sharpened contrast, rather than to acuity in perceiving a sharply defined image (Baylor and Shaw, 1962). Fry of schooling fish at first tend to avoid each other when they approach head on, but as they mature, they develop a head-to-tail approach and join into formation. Apparently the difference in visual stimuli between head on and head to tail affects their readiness to form a school. Fry raised in isolation, if placed together when they reach an appropriate size, will school shortly afterwards. If the fry are initially kept together during the head-on approach period and then isolated for a while, they take longer to form a school than those raised in isolation. These investigations suggest that some visually reinforcing image initially brings the fish together, although other sensory cues may keep them in a school once it is formed. Scott (1945) proposed that the cohesion of sheep is largely the result of the lambs' following their mothers during their nursing period. Attractiveness related to sexual behavior is conspicuous among seasonal breeders, which group during rut. Among primates, which are sexually active throughout the year, sexual behavior has been considered essential for cohesion (Carpenter, 1934), although recently sexuality has been questioned as the primary cohesive factor (DeVore, 1965).

Certain selective advantages of group cohesion arise in locating food, in detecting and defending against predators, in modifying the environment, and in caring for the young. These advantages, combined with the genetic alterations resulting from grouping in a population, may accelerate further evolution in the direction of adaptive grouping. This principle in the ecological genetics of behavior can be expressed thus: *Genes contributing to those behavioral characteristics of the population which affect the genetic equilibrium will most rapidly experience alterations in their frequency.*

Migration and Dispersal   Areal movements of populations or of individuals among populations are important for permanent genetic change. Geographical ranges expand, new habitats are invaded, and interbreeding between populations occurs through various types of movement. The ecological and evolutionary complexity of movements is clearly observed in cases of the introduction of exotic species by man (Elton, 1958). Zoogeography is particularly concerned with geographical movements of species and their subsequent extinction or proliferation. Movements of a species occurring through dispersal place representatives of the species under the selective pressures of the invaded environment. The new environment, with its special selective pressures, can alter gene frequencies or induce changes in the expressivity and penetrance of genes carried over from the old environment. If the invaded region is already occupied, the newcomers introduce their genotypes and interbreed with the residents. Wright (1931) has calculated the genetic consequences of some migration between isolated populations of moderate size.

The behavior involved in animal movements is a subject unto itself. Seasonal migration, characteristic of birds, has captured the interest of behaviorists and physiologists, although such migrations are of less genetic importance than those involved in permanent displacement. In dispersal or emigration, social factors are often responsible for the movement. Young are driven from their natal range; males compete for mates and cause the defeated to emigrate into new regions. Some individuals initiate their own dispersal, which enables them to encounter new sources of food, shelter, or mates. An aimless restlessness characterizes some individuals at certain times of the year, and they wander into strange environments where they may become lost. The motives for many movements are difficult to establish, and their genetic effects have not been precisely measured.

Sex and Age Distribution

Age   The composition of a population with respect to the distribution of ages among its constituents is one determinant for the size of the effective breeding population. Sexually immature or senile animals do not breed, and the actual breeding aggregate is usually much smaller than the total number of individuals. Age distribution is an important factor in natality which contributes to genetic alterations, as previously mentioned. Behavior that prevents or retards breeding severely limits the number of young entering the population. Aggressive behavior and sometimes hypersexuality may interfere with optimal breeding. Similarly, a lack of adequate parental care reduces the number of young. On the other hand, if populations are composed predominantly of sexually immature individuals, their growth will be arrested. Age-dependent behavior, such as play, dispersal, fighting, and mating, often characterizes large segments of the population and thus affects its function as a genetic unit.

Sex   Sexual composition is a further factor limiting the size of the effective breeding population. An abundance of either sex can restrict the fertility of the population just as does age. Another feature of the sex distribution is the type of breeding pattern or mating system established by the animals (Wright, 1921). The genetic consequences of monogamy are quite different from those of polygamy or promiscuity (Selander, 1965). Various breeding patterns are used by animal husbandmen to bring about desired genotypes. Most of the breeding patterns used for selecting and fixing favorable genotypes in our domestic animals are not commonly found in nature. Full-sib matings, for example, bring genetic fixation more rapidly than half-sib matings or repeated backcrosses to a parental line. Homozygosity increases to a greater extent in some mating systems than in others (Naylor, 1962).

The behavior exhibited by animals in establishing various types of mating systems is well known. Monogamous matings are the rule among birds that establish a territory. Many mammals and fish exhibit similar behavior. Among deer mice, monogamy is less rigid but usually results from the sedentary habits of the pair occupying the same nest and home range. Polygamy is established by a single aggressive male driving others away from his harem. Seals and ruminants provide the clearest examples of this type of mating system. Promiscuity has been described for groups of howling monkeys, in which a receptive female moves from one male to the next, often satiating each. Father-daughter types of mating frequently prevail among rapidly maturing species that occupy a home range. The sons of voles are driven off by the father or they wander away at weaning, while the daughters remain in the burrows and homes, often to be mated by the father (Frank, 1957). A similar system has been suggested for the guinea pig (King, 1956). Many other types of mating systems prevail in nature, and each species exhibits a great deal of variability. Sometimes even the most rigidly maintained harem is opened to peripheral males while the dominant sire is engaged or spent in combat.

Conclusion

As an area of scientific inquiry, behavior genetics should include the contribution of behavior to alterations of the gene pool as well as the action of genes upon behavior. Throughout this section, we encountered many concepts lacking experimental support. We understand many behavioral mechanisms involved in population dynamics, and we know how population dynamics can affect gene frequencies, but we do not have enough precise and definitive measurements on actual genetic changes resulting from behavior. Our best material comes from courtship and sexual selection, as exemplified by Merrell's work. Definitive measures of genetic changes resulting from migration, aggression, natural mating systems, differential fertility, and mortality are lacking. When the ecological genetics of behavior can be presented without the generalities which have appeared in this section, the subject will have become established.

DEVELOPMENTAL GENETICS OF BEHAVIOR

Developmental genetics of behavior is comparable to physiological genetics of behavior insofar as both approaches are concerned with the mechanisms by which genes act to bring about a particular phenotype. In physiological genetics, gene action occurs through enzymes which affect the metabolism of tissue nutrients and ultimately produce the phenotype (see Chapter 6). One of the best behavioral examples is provided by phenylketonuria as presented in Chapter 9 by Hsia. In developmental genetics of behavior, the action of the genes depends upon the environment's providing, not nutrients as in physiology, but stimuli or reinforcers necessary for the gene to be exhibited in the phenotype. Although the environment does not alter the genes themselves, it can affect their phenotypic expressivity and penetrance. This effect may be particularly pronounced in epistasis, in which several genes contribute to a particular behavioral trait. Once a particular combination of genes is expressed in the phenotype by reinforcement of the responses, selection can operate on the most adaptive patterns. These patterns of behavior tend to become fixed in the genotype through canalizing selection.

Evidence for the genetic determination of developmental rates in morphology and behavior will be examined first in this section, and an attempt will be made to relate these differential growth-rate patterns to concurrent changes in the environment. Next, the concept of reinforcement will be introduced in order to illustrate how particular patterns of behavior may become expressed in the phenotype. The last discussion in this section will be concerned with selection processes necessary to fix the phenotypic character in the genotype.

Developmental Rates

As an organism develops, it increases in size and its tissues and organs differentiate and mature. The changes in one part of the organism are intricately related to other parts; examples are the formation of a lens from the ectoderm above the outgrowing eye cup and the specificity of the mesoderm of the limb bud. The temporal and spatial relationships of cells, tissues, and organs in the course of development have been investigated by experimental embryologists, and genetic control of some of these processes has been established (Cock, 1966). Similar developmental processes occur in the central nervous system (Detwiler, 1936); less is known regarding their interactions and genetic control (Fuller, 1951). Since behavior depends upon these morphological and physiological changes occurring during the development of the organism, we can expect behavior to reflect them. When a particular behavioral response is first exhibited by a young animal, the response immediately interacts with the environment, which shapes it into mature patterns.

Genetic Control of General Developmental Rates   Some animals develop faster or slower than others under similar environmental conditions. Species differences in rates of development have been described for three species of *Peromyscus* (McCabe and Blanchard, 1950). One species is further advanced in development at birth than the other two species, but it develops more slowly during the first postnatal month. The other two species are similar at birth, but one species matures more rapidly than the other. Within a species, similar differences in developmental rates between subspecies may occur. The eyelids of *Peromyscus maniculatus bairdii* separate almost 5 days earlier than in a related subspecies, *P. m. gracilis* (King, 1958). Genotypic differences between species and subspecies are apparently responsible for these differences in developmental rates because cross-fostering has no appreciable effect upon them. Although most differences in developmental rate are controlled by many factors, single genes may also be responsible, such as the dwarf gene affecting the pituitary gland and retarding the development of the house mouse (Snell, 1929).

Developmental changes in the morphology and biochemistry of the central nervous system may also proceed at different rates. The size of the brain, the number and density of cortical neurons (Eayrs and Goodhead, 1959), the quantity of enzymes and oxidative agents (Hamburgh and Flexner, 1957), the amount of lipides (Sperry, 1955) and other chemical constituents (Folch-Pi, 1955), and the activity of γ-aminobutyric acid (Roberts et al., 1951) change during the course of development (see Driscoll and Hsia, 1958, for review of enzyme systems). Comparative studies of these developmental changes among related genotypes have not been carried out to the extent that they can be attributed to genetic factors, although adult differences between inbred strains of mice suggest their occurrence (Gluecksohn-Waelsch, 1955; Maas, 1962). One comparative study of the development of cholinesterase activity in two races of deer mice (*P. m. bairdii* and *P. m. gracilis*) suggested genetic differences in the rate of development of this enzyme system (Eleftheriou, 1959). Further investigations will certainly lead to the discovery of developmental differences in the morphology and biochemistry of the central nervous system between animals of different genotypes. Perhaps the genetic factors will be isolated, such as the single recessive gene in phenylketonuria, which blocks the metabolism of phenylalanine (Knox, 1955).

Genetic Control of Differential Developmental Rates   The overall differences in rates of development among animals of various genotypes are probably less significant to adult form and behavior than differential rates. Slowly developing animals eventually reach maturity, and their mature characteristics need not radically depart from those of their faster developing relatives. Differential rates of development, however, may permanently alter the form and behavior of an animal because some adult characteristics mature earlier or later than other characteristics. The effects of allometric growth (change of proportions with increase in size) on adult form have long been recognized and quantified by mathematical formulas (Thompson, 1917; J. S. Huxley, 1932). Differential-growth concepts have also been applied to chemical changes in animals (Needham, 1934) and to behavior (Rensch, 1959).

Multiple factors appear to be responsible for differential-growth rates affecting morphological characters of strains, subspecies, or species of animals. Among several subspecies of *Peromyscus maniculatus*, differential-growth rates of the tail and body length produce proportional differences in the adults (Dice and Bradley, 1942). The skulls and skeletons of two subspecies of this species also indicate different allometric-growth patterns for each (McIntosh, 1955). When an allometric analysis was applied to the facial and cranial bones of the same two subspecies, racial differences became apparent and suggested similar differences in the underlying neural mass (King and Eleftheriou, 1960). Further evidence for genetic determinants in the k factor of the allometry formula is provided by differences in femur growth of *Notonecta* (Misra and Reeve, 1964). Artificial selection for relative growth rates among domestic animals has produced allometry, e.g., the white breast muscles of turkeys. Similarly, selection for high and low ratios of wing to thorax length in *Drosophila* has resulted in an immediate and sustained response (F. W. Robertson, 1962). Single factors responsible for aberrations of allometry are recognized in such mutants as the grey lethal gene in mice, which has many pleiotropic effects on differential growth of different parts of the skeleton (Grüneberg, 1952).

The differential rate of development in the nervous system is also influenced by heredity. A lack of differentiation in the retina of the rodless-retina mutant of the house mouse illustrates genetic control of a developmental process affecting part of the nervous system (Keeler, 1927). The shortening of the basicranium in the congenital-hydrocephalus mutant of the mouse forces the brain to bulge dorsally and rearranges the size and proportions of many neural structures in the process (Grüneberg, 1952). Unidentified genetic factors appear responsible for different allometric patterns in the neocortex and paleocortex of large and small strains of house mice (Harde, 1949). Although genetic differences in the biochemical development of neural tissues have been indicated (Eleftheriou, 1959), strain differences in adult mice, such as those demonstrated by Caspari (1960) and Maas (1962), are possibly the result of differential rates of accumulation of neurochemical substrates during development.

These cases of genetic differences in general and differential developmental rates have been cited briefly to serve as (1) morphological analogies to the development of behavior and (2) possible explanations for behavioral development, which should ultimately depend upon maturational stages of the nervous system. The importance of relative growth rates in the central nervous system as determinants of behavior has been recognized by Rensch (1958). He has proposed that the positive allometry of the neocortex is a result of an evolutionary increase in body size. Although this relationship of allometric growth in the nervous system to behavior is compatible with the present formulation, the stress here is upon the allometric maturation of behavior, which is significant in evolution.

Rate of Behavioral Development   Since behavioral development depends upon morphological development, particularly in the nervous system, parallel differences would be expected to occur in the overall and differential rates of behavioral development among animals of varying genotypes. Our studies of two subspecies of *Peromyscus maniculatus* revealed the *P. m. bairdii* matures more rapidly in locomotor responses than *P. m. gracilis*. On an elevated maze young *bairdii* moved about and consequently fell off at a younger age than *gracilis* (King and Shea, 1959). This difference in locomotor ability was further tested by placing the young mice in water in order to examine the development of swimming (King, 1961). Again *bairdii* developed more rapidly than *gracilis* when measured by their ability to escape from the water. In contrast to locomotion, clinging responses appeared to develop at about the same rate for both subspecies. *P. m. gracilis* is a semiarboreal species, and the neonatal ability of the animals to cling to objects is probably related to their adult climbing performance (Horner, 1954). The differential rate of development of the locomotor and clinging responses in the two subspecies suggested the following experiment.

Young mice of each subspecies were tested at 2-day intervals from 6 to 20 days of age. Separate groups of mice were used at each age. An 8-inch-diameter activity wheel enclosed in a sound-deadened box was used as the test apparatus. Each mouse was given three consecutive 5-minute tests, starting with 5 minutes of freewheeling, then 5 minutes with the wheel braked, and finally with the wheel rotated by a 4-rpm motor. The first period of freewheeling measured locomotion, and the last period of forcibly rotating the wheel measured clinging. *P. m. bairdii*, as expected, exhibited more locomotion at an earlier age than *gracilis*. The developmental rate of clinging, however, was similar in both subspecies until 12 days of age, after which *gracilis* clung more than *bairdii*. Although the allometric formula was applied to the data, scalar difficulties demanded a more conservative analysis. An analysis of variance revealed that *gracilis* clung more than *bairdii*, the change over days being significant, and there was a significant subspecies-days interaction. The component of trends indicated significant differences in the linear, quadratic, and cubic curves for the development of both subspecies. The locomotor and clinging responses of each subspecies develop at different rates, and the relative rates of development are unique for each subspecies, which suggests that genetic factors are involved.

Reinforcement

Phenotypic Expression of the Genes   Responses[2] developing at differential rates do not occur in a vacuum. They require an environment for their expression, and the environment is not passive. The way responses are shaped and molded into adult patterns of behavior depends upon the reinforcing properties of the environment. Reinforcement[3] may also be derived from other responses of the animal or even from the response itself. The probability of the environment or other responses reinforcing a particular response changes simultaneously with the

---

[2] *Response* as used here is noncommitted in respect to its stimulus origin; i.e., it may be either emitted or elicited. Emitted responses presumably have an endogenous stimulus.

[3] *Reinforcer* can best be defined operationally as any stimulus which increases the probability that a particular response will be exhibited. A *negative* reinforcer decreases the probability of a response.

developmental probability that a response will be elicited. The development of behavior thus depends upon the time when a response can first be expressed, or the time when it is most likely to be reinforced by the environment, and on its temporal relationship to other responses that may impede or enhance its further development. Although genetic factors may be responsible for the initial exhibition of an incipient response, its complete phenotypic expression depends upon the extent to which it is reinforced.

Differential Probabilities of Response and Reinforcement   A behavior pattern has a changing probability of being exhibited at different stages in development. As a result of differential-growth rates, the probability of each behavior pattern or the operant level of each response changes in relationship to every other response. Once a particular response is reinforced, the probability of its repetition increases, which again increases the probability of its being reinforced. This positive feedback between response and reinforcement has an effect upon other responses which are first being elicited in development and changing in operant levels. Later-developing responses compete with those already developed and increasing in rate. The developmental potentialities of some responses are thus preempted and tend to diminish in frequency unless they affect other responses. The constant strengthening of some responses and the elimination of others constitute the process of canalization.

Simultaneously with the developmental sequence in the exhibition of behavioral responses, the environment also provides changing probabilities of reinforcing each response. Although two responses may be exhibited at approximately the same developmental age, the probability of the environment equally reinforcing both is low. Before a response can be fully developed, the probabilities of its being elicited and of the environment reinforcing it must coincide. Regardless of the appropriateness of the environment for reinforcement of a particular response, no chance for reinforcement will occur if the response is not emitted in the first place. Likewise the exhibition of a response in a nonreinforcing environment will tend to bring about extinction of the response. The effects of early experience upon adult behavior are logically a consequence of a particular response being emitted at the age when the environment is most likely to reinforce it. Critical periods may then be considered as those periods in the development of a behavioral character when the probabilities of a response being emitted and being reinforced are the greatest. One way of discovering these periods has been to prevent reinforcement of a response when it is emitted with some frequency or to reinforce responses not usually reinforced in the life of the animal (Scott, 1962). Although the probability of emitting a particular response changes throughout life, most responses are repeated frequently enough during the life cycle to be later reinforced in a rich environment. This may account for the lack of perseverance in many early experience studies. Continuously reinforced responses are most likely to perpetuate the experiences gained early in life.

Types of Reinforcement   The differential rates of development of specific behavioral patterns and the changing probability of the environment's reinforcing them are further complicated by the nature of the reinforcement process itself. One response may reinforce another response, depending upon the rate of each (Premack, 1959). Responses elicited at a higher rate can reinforce those elicited at a lower rate. In a situation where drinking was contingent upon wheel running and vice versa, Premack (1962, 1963) was able to demonstrate that a rat's drinking could be reinforced by running and that running could be reinforced by drinking, depending upon the rates of both controlled through deprivation schedules. In the course of development, a response emitted early with a relatively high frequency can reinforce a later response; or an early response which received no environmental reinforcement can be later reinforced by a late-developing response, which was environmentally reinforced. Adult behavior then arises from a combination of intrinsic and extrinsic reinforcers which are interdependent and vary in the developmental sequence with the differential rates of the maturing responses.

The possibility that a response may reinforce itself adds further complication to the developmental sequence. Demonstration of self-reinforcing responses cannot be separated experimentally from the kind in which one response reinforces another. Noningestive reinforcers, such as light (Kish, 1955), vision (R. A. Butler, 1953), exploration (Montgomery and Segall, 1955) and activity (Kagan and Berkun, 1954), suggest that these stimuli or responses may be intrinsically reinforcing (Brant and Kavanau, 1965). However, most of these demonstrations require the animal to make some type of neutral or low-operant-level response (such as bar pressing) prior to the other response. This situation can be interpreted as well by Premack's paradigm (1959). Most intrinsic reinforcers have a quality of activity or novelty, from which other responses cannot readily be separated. It may be possible to distinguish genetic differences from developmental differences in the reactions to intrinsic reinforcers.

Genetic differences are apparent in the reaction of various species of deer mice (*Peromyscus*) to sand. When provided a choice, most species select a sand-bottom cage over one with sawdust. The amount of sand removed from a tunnel, however, distinguished at least one subspecies from the other members of the genus that have been tested (King and Weisman, 1964). Over a 24-hour period, the median amount of sand removed by *P. m. gracilis* was 0.1 pound, while *P. m. bairdii* removed 5.9 pounds, *P. floridanus* 8.8 pounds, and *P. leucopus* 17.0 pounds. Some mice removed as much as 90 pounds of sand from the tunnel in a single night. Individual variability was great, and only the *gracilis* mean was significantly lower than the means of the other species. The differences are compatible with our knowledge of the life history of the mice, since *gracilis* is semiarboreal whereas the other species tend to be terrestrial. *P. floridanus* inhabits subterranean burrows in the sandy soils of Florida. The *P. leucopus* sample was from the first laboratory-raised generation of captured wild mice, which suggested that other factors may also have contributed to the differences. The sand dug from the tunnel dropped through a screen floor and provided no obvious reinforcement, other than possibly novelty or a means of escape. A similar response has been observed in domestic mice, which failed to exhibit extinction of sand digging when food was removed from the other side of a sand barrier (Earl, 1957).

The reinforcing properties of the sand-digging response were tested by making the availability of sand contingent upon a mouse pressing one of two bars, one providing a limited quantity of sand and the other, nothing. When the mouse pressed the reinforcing bar, an L-shaped tunnel was filled with sand, which had to be dug out before more sand could enter the tunnel at the next pressing of the bar. Over a period of 12 nights, the bar providing sand was pressed significantly more often than the nonreinforcing bar. Although the mice failed to show a position reversal during the next 12 nights, their level of bar pressing significantly exceeded that of a no-sand control group (King and Weisman, 1964). Since sand digging can serve as a sufficient reinforcement to learn a position discrimination, it is likely that many other species-specific responses are of the same nature, i.e., "self-reinforcing" for want of a better term. That species differ in the amount of reinforcement they receive from responses elicited at different operant levels remains to be demonstrated.

The extent to which genes are expressed in the behavioral phenotype depends upon the temporal sequences in the exhibition of responses, the probabilities of responses being reinforced by the changing environment, reinforcement derived from other responses, and the self-reinforcing value of the response. Once the expressivity of the genes is altered through this particular combination of genetic and environmental factors, selection can operate on them. Behavior that enhances or depresses the phenotypic expression of the genotype, as well as behavior that increases or decreases the allowable amount of variability of gene action, becomes exposed to selection.

Selection

The variability of the genotype exposed to selection pressure through the development of behavior can result in evolution. Since evolution occurs at the level of populations, the developmental sequence previously described must be repeated frequently in each generation over many generations. Canalizing selection (Waddington, 1961) provides a mechanism for ensuring the frequent phenotypic exposure of adaptable genotypes, and genetic assimilation may occur. Other processes, such as the Baldwin effect (Simpson, 1953a), pleiotropy, and correlated characters, may also contribute toward the evolution of behavioral patterns (Schmalhausen, 1949).

Canalizing Selection   The development of phenotypic characters results from the balancing of tendencies to become modified and to resist modification by the environment and the genotype (Waddington, 1957, 1961). Some characters are well buffered from environmental or genetic changes during the course of development, whereas other characters are extremely vulnerable to such changes. The extent to which a character is affected by environmental or genetic alterations is a result of selection canalizing the development into narrow or broad channels. In behavior, canalizing selection has brought about some responses which are reinforced by a variety of stimuli and other responses which are reinforced only by key stimuli, or releasers (Tinbergen, 1951). Through canalizing selection, response variability can be channeled into stereotyped patterns, which are species-specific or even "instinctive." Not only can selection make a variable response stereotyped, but it can also change the direction of the variability caused by environmental or genetic modifications. For example, a species that varies its feeding habits among herbs and grain, depending on modifications in its environment or genotype, can be selected for a wider range of food preferences in either a herbivorous or granivorous direction.

Genetic Assimilation   Canalizing selection operating upon a response modified by the environment tends to accumulate the genetic factors that enable its modification. The effect of this selection can be measured by returning the species to its original environment after several generations have elapsed in an environment producing new responses. The classical examples of genetic assimilation are the change in wing venation of Drosophila after pupal heat shock (Waddington, 1953), enlargement of anal papilla in Drosophila larvae raised in a salt media (Waddington, 1959), and the assimilation of the dumpy phenocopy (Bateman, 1959).

A behavioral example in *Peromyscus maniculatus* lends itself to this interpretation, although the genetic factors are obscure (Wecker, 1963). *P. m. bairdii* is a grassland race of this widely distributed species. V. T. Harris (1952) found that laboratory-raised individuals preferred a simulated grassland habitat to a simulated forest habitat. Descendants of Harris' stocks were maintained in the laboratory for 10 years, or about 12 to 20 generations removed from the original stocks. Wecker (1963) studied habitat selection of these descendants in a seminatural situation and found that they made no selection when offered a choice of field or forest. However, if these semidomesticated mice were raised in the field, they selected the field habitat significantly more frequently than the forest. Recently captured mice raised in the laboratory, on the other hand, performed like those studied by Harris; i.e., they selected the field. Apparently a genetic change had occurred among the mice raised in the laboratory for several generations. The inherent tendency to select the field habitat had been lost, but it could be regained by an early exposure to the environment that had previously imposed selection for this trait.

Wecker's experiment revealed that genetic assimilation occurred under natural conditions, and the relaxing of selection in the laboratory for several generations restored the mice to a more variable and environmentally vulnerable genotype. The variety of habitats occupied by this species of *Peromyscus* suggests that the generalized genotype of these mice is readily changed to a specific habitat selection by assimilation. The ability of the generalized genotype to reacquire the specialized habitat selection during development is particularly appropriate to the reinforcement of developing responses.

Baldwin Effect   If an animal acquires an adaptive trait, any genetic change compatible with the acquired character tends to become fixed or at least selection will not operate against it. This and related concepts have been called the "Baldwin effect" (Simpson, 1953a) after Baldwin (1896) and Morgan (1900), both of whom attempted to deal with the problem in terms of inheritance of acquired characters. The Baldwin effect is probably a special case of genetic assimilation. Waddington (1961) has reviewed it along with other allied concepts. Animals that learn an adequate pattern of behavior may be said to exhibit the Baldwin effect in particular and genetic assimilation in general if the behavior permits either the manifestation of alternative phenotypes for already existing genetic combinations or a particular assemblage of genetic factors in the absence of selection against them.

Pleiotropy and Correlated Characters   If a single gene influences several characters (pleiotropy) or if several genes are closely linked or otherwise bound together by their effect on the phenotype (correlated characters), selection for one character may influence other characters. Artificial selection for body size in laboratory mice tends to increase litter size and reduce wildness (Falconer, 1953). Although a reduction in wildness may not be adaptive in wild mice, thus limiting the advantages of selection for body size and larger litters, it is possible to conceive of situations in which reduced wildness would have no adverse effects or even advantageous effects.

The same principle applies for the selection of other behavior traits. An animal with the genetic potentiality for acquiring a new behavior pattern, which confers a selective advantage, may exhibit correlated physical or behavioral traits which had little or no selective value before the animal acquired the behavior. Under new selective forces imposed by the behavior, the correlated characters may exhibit a selective advantage. For example, if climbing ability and a dark pelage are correlated in mice, any mouse acquiring climbing skills may also exhibit a pelage color which may be more cryptic in a forest habitat. Selection then operating on the protective pelage coloration would tend to fix the behavior genes involved in climbing. The fixation of the pigment genes could also occur if climbing was more advantageous than coloration under the selection pressures imposed by the new environment.

Conclusion

Developmental genetics of behavior can be approached physiologically, morphologically, and behaviorally. The behavioral approach is closely related to the ecological study of the genetics of behavior because it emphasizes the selective forces operating during behavioral development. Before selection can operate, the genotype must be expressed in the phenotype. Phenotypic expression of the behavioral genotype is enhanced through schedules of reinforcement provided by the environment, by other responses, and by the self-reinforcing properties of the responses themselves. Sequential development of behavior is particularly important in determining the quality and quantity of reinforcement a response receives. With the operation of canalizing selection during development, genetic assimilation of initially acquired responses may occur. This is a feedback mechanism in which the behavior of a species alters its genotype and the genotype alters its behavior through the mediation of selection pressure. Animals also learn to adapt to new habitats with different types of selective influences which may in turn alter the genotype. Morphological and physiological traits correlated with behavior through pleiotropy or linkage can also produce genetic alterations when the acquisition of a behavior pattern exposes animals to new selection pressures.

CHAPTER FOUR
GENES AND THE EVOLUTION OF INSECT BEHAVIOR
Aubrey Manning

AN INTRODUCTION TO THE PROBLEM

This chapter will draw together what is known about the effects of genes upon insect behavior and will examine how far this knowledge helps in understanding its evolution. There is no attempt to provide a complete review of the literature, and much that is relevant has been omitted.

At the outset it is necessary to consider what we mean by saying that behavior is inherited. The phrase can imply only that a potentiality is transmitted. An animal possesses the potentiality to perform such and such behavior, given a particular stimulus situation and having been exposed to a particular range of environments in the period prior to testing.

For some types of behavior this environmental range may be quite broad. It matters rather little what the animal's previous experience has been; the same behavior appears at the same sort of stimulus and looks much the same when performed. Such behavior is characterized by rigid patterns of movement or posture and is often evoked by special key stimuli in the environment. It has often been called "instinctive" (e.g., Tinbergen, 1951). In spite of the numerous objections leveled at this term because of past misuse, it has no adequate substitute and is still useful, especially in behavior genetics. Instinctive behavior must be a property of the inherited structure of the nervous system, hormones, etc. This is not to imply that it is not capable of modification, but the basic properties of the sensorimotor patterns and often the responsiveness to particular stimuli are laid down with the developing nervous system. They must depend in some fashion on the physical and functional connectivity of neurons.

At the other end of the scale, behavior may be almost completely dependent in its form and elicitability upon the animal's individual history and exhibit all those characteristics we may subsume under the term "learning." This too must be related to the inherited structure of the nervous system, but in a less direct manner than instinctive behavior. We know that, just as a phylogenetic series can be traced through the vertebrates, showing increasing size and complexity of brain structure, so there is a corresponding increase in the ability for complex learning. All that can be stated is in terms of potentialities. Neither rats nor monkeys normally manifest the behavior associated with the solution of triple ambiguity problems, but the nervous system of the monkey possesses the inherited ability to organize such behavior; that of the rat does not (Harlow, 1958).

Between these two extremes of purely instinctive and purely learned behavior there are all possible intermediates. Most overt behavior depends on elements of both. The interactions between the two are sometimes complex for, as Hinde and Tinbergen (1958) point out, it may be a predisposition to learn particular things that is inherited. Various workers, e.g., Hebb (1953), Lehrman (1953) and Verplanck (1955), have suggested that the instinct-learning dichotomy is no longer a useful one for the study of the development and organization of behavior. For instance, Lehrman has shown that the knowledge that a piece of behavior is inherited does not enable one to predict the nature of its ontogeny or to separate it rigidly from acquired behavior. Nevertheless, the identification of instinctive behavior is still useful for genetic analysis, because it is so much simpler to study the inheritance of those patterns that are well buffered against environmental fluctuations during development and in adult life.

The insects are a particularly appropriate group for this kind of work. Their short life-span with attendant lack of parental care and their compact nervous system have favored the evolution of rigid patterns that require no learning. Insects can learn, but usually such learning serves only to modify the orientation of instinctive responses which are themselves unchanged. The Hymenoptera (ants, bees, and wasps) show extraordinary facility for this type of learning, and it plays an important part in their normal life. By contrast, the Diptera (two-winged flies) exhibit little trace of learning under natural conditions. Indeed, it is extremely difficult to demonstrate experimentally even simple conditioning of a fly; Frings (1941) might be the only worker to report success. The life of Diptera is tied, completely successfully, to a series of rigid, inherited responses to food, mate, and shelter. The ability to modify these responses rapidly apparently confers no noticeable selective advantage.

In general, then, the insect-behavior geneticist is able to study the effects of selection or gene substitution against a clear background of inherited behavior, with the experience of individual insects playing a relatively minor role. He has to choose "units" of behavior for genetic analysis, and behavior is often so complex and diffuse that this is likely to be a difficult task. Various aspects of this problem are discussed by Ginsburg and by Thompson in this volume. With much of insect behavior one has a natural grouping into units—the "fixed action patterns" of ethologists—which are distinct and relatively invariable in form (see Hess, 1962). Each of these often consists of a series of muscle contractions which it may be possible to break down further into units. The work of Rothenbuhler (1958 and in this volume) shows that interesting and meaningful results can be achieved by selecting fixed action patterns which are functionally distinct and which turn out to be genetically distinct also. Certainly the empirical approach has much to commend it. At a certain stage in the history of systematics Darwin's definition of a species as "what any competent taxonomist chooses to call one" served its purpose. At this stage in behavior genetics we might well use a comparable definition of a behavior unit.

THE EFFECTS OF GENES
ON NERVOUS STRUCTURE

Sperry (1958) has suggested that inherited changes in the nervous system may be due to changes in (1) size, (2) number, (3) connectivity, and (4) excitatory properties of nerve cells. It can be argued, justifiably, that this is but one way of considering genetic effects on the nervous system and that it may be more realistic to choose a completely different system. Nevertheless, Sperry's categories serve to emphasize types of genetic effect whose behavioral results must be profound but which are often overlooked. Although it is a problem of the greatest importance for behavioral evolution, we have little knowledge of how genes operate to affect behavior via the first three categories. These are effects on nervous structure, and we know very little of how far the nervous system shows the structural variability characteristic of all morphological features.

We are unlikely to be able to identify and relate altered structure to altered behavior except in extreme cases. No one is surprised that congenital microcephalics have altered behavior or that the various "waltzer"-type genes of the mouse (Grüneberg, 1952) have postural and locomotor effects commensurate with the middle-ear abnormalities they produce.

An equally profound behavioral result may be expected when the number of chromosome sets is altered. Tetraploid animals, which have twice the normal number, are usually much the same size as normals, but their cells are approximately twice normal size (Fankhauser, 1945, 1955). This means the animals have only one-half the normal number of cells, brain cells included. It is fairly easy to obtain salamanders with three, four, five, or even eight sets of chromosomes and, although they are often rather sickly animals, they are potentially of great interest for behavior studies. Fankhauser et al. (1955) and Vernon and Butsch (1957) have shown that triploid and tetraploid salamanders with $33\frac{1}{3}$ and 50 percent reduction in cell number, respectively, have markedly impaired learning ability. It would be interesting to know what happens to their instinctive behavior.

Haploid animals, with only a single chromosome set, have a doubled cell number. Among vertebrates they show very reduced viability, perhaps because of the total unmasking of deleterious recessive genes. This makes the Hymenoptera of particular interest because, with their method of sex determination, males normally develop from unfertilized eggs and are haploid in origin. Presumably the ancestral Hymenoptera had males with doubled cell number, but selection has favored genes that increase cell size and reduce cell number to normal. Haploid-male cells are about the same size as those of diploid females among *Habrobracon* (Speicher, 1935) and the honeybee (Oehninger, 1913). The situation among honeybees is complicated by the normal occurrence of endopolyploidy, where the chromosome number of various tissues or organs is multiplied (Merriam and Ris, 1954).

As far as genetic effects on the structure of the insect nervous system are concerned, there is only the work of Power (1943). He showed that the gene *bar* which reduces the number of ommatidia in Drosophila also affects the structure of the brain. Fibers develop centripetally from the retinal cells, and there are consequently many fewer in bar-eyed flies. This reduction extends beyond the first synapses in the visual system into the internal glomerulus and even beyond, though with diminishing effect. Thus, bar-eyed flies are deprived not only of receptors but also of the corresponding parts of the visual nervous system. Many genes which affect sense organs must have similar repercussions on the nervous system. The bristle genes of Drosophila, which affect tactile receptors primarily, spring to mind, and also the gene *antennaless*. The behavioral effects of this gene (Begg and Packman, 1951) go far beyond those produced by amputating the antennae of normal flies (Manning, 1959a). Part of this difference may be due to effects on the antennal nerves and sensory centers as well as general debilitation.

THE EFFECTS OF GENES ON BEHAVIOR

Because there is so little information on the structural effects of genes, we must rely mostly on behavioral descriptions of gene action. The examples in insects differ widely in the degree to which they have been analyzed, but they point toward one obvious conclusion. The commonest effect that mutations have upon behavior is to alter, not the nature of the patterns involved, but their threshold and the frequency with which they are performed.

In Drosophila, for example, no gene has been described which affects the qualitative form of a behavior pattern. There is no behavioral equivalent of a morphological mutant like *aristapedia* where an antenna is replaced by an abortive leg [at least, if we except the anomalous *transformer* gene which changes a genotypic female into a male with male behavior (Sturtevant, 1945)]. Rather, there is the equivalent of many bristle genes that alter the length and number of bristles but retain their basic form.

Various genes of *Drosophila melanogaster* have been shown to have a deleterious effect on the mating success of male flies which carry them, e.g., *ebony* and *vestigial* (Rendel, 1951); *white* (Reed and Reed, 1950; Petit, 1958); *yellow* (Bastock, 1956). It is worth examining one of the best-analyzed examples in more detail. Bastock (1956) has shown that the reduced success of *yellow* males is due to the poorer stimulating effect of their courtship behavior. *Melanogaster's* courtship can be divided into four basic behavior patterns: (1) "orientation," in which the male stands close to or follows the female, (2) "vibration," in which he rapidly vibrates that wing closest to the female's head, (3) "licking," in which he extends his proboscis and licks the female's ovipositor, and (4) "attempted copulation," in which he tries to mount her (Bastock and Manning, 1955). Vibration and licking are certainly the most important elements in stimulating the female and causing her to become receptive. *Yellow* males perform these and other elements in a precisely normal fashion. They also court as persistently as normal males, but their courtship has a smaller proportion of vibration and licking and is therefore less stimulating.

The *yellow* gene's effect is thus a subtle one, and we have no idea where the gene operates in the chain between the male's perceiving the female and performing various sequences of muscle contractions. Similarly, William and Reed (1944) have shown that various genes produce small, but significant, changes in the normal wing-beat frequency of Drosophila. The stimulus-response chain is simpler than in the case of courtship, but the genes could still operate via any of a large number of combinations of sense organs, nerves, or muscles.

Apart from single-gene effects, changes in the genotype as a result of selection or domestication also have behavioral repercussions. For example, Bösiger (1960) finds that the reduced mating success of various mutant stocks of *D. melanogaster* is not due to the marker genes but results from the accumulation of a large number of genes during many generations of domestication. Ewing (1961) shows that the genotypic changes produced by selection for large and small body size, again in *D. melanogaster*, also affect the courtship behavior of males. Small flies have a higher proportion of vibration and licking in their display than large or control flies. Ewing demonstrates convincingly that this is not a direct result of the genetic changes that have altered size. It is produced by secondary selection within the culture bottles of small flies which "compensates" for their reduced wing area and the reduced stimulating ability that follows.

Clearly, some very rapid evolution can take place in a Drosophila bottle. Various artificial-selection experiments show how much variability for genes that affect behavior is present in ordinary populations.

Hirsch and his coworkers (see Hirsch, 1962) have shown how readily accessible to selection are the numerous genes which affect the levels of simple photo- and geotactic responses. Their experiments are described in more detail elsewhere in this volume. A similarly marked response to selection was found by Manning (1961), using the mating speed of Drosophila. Fifty pairs of virgin flies were introduced into a bottle and pairs removed as they began to copulate, the fastest and slowest being used for breeding. With unselected control flies, 25 pairs had mated in the bottle after about 6 minutes. After seven generations of selection, this was reduced to some 3 minutes by the fast-mating lines, while the slow-mating lines took 30 minutes or more. Natural selection clearly keeps mating speed in normal populations quite close to the maximum.

In this case, selection changed the behavior of both sexes, but primarily the males. They showed a changed pattern of courtship comparable to that produced by the *yellow* gene described above. The genes which had accumulated in the fast-mating lines increased the frequency with which the vibration and licking elements were performed. In the slow lines these elements occurred less frequently than normal. Mating speed is a complex and somewhat arbitrary character, and it was only to be expected that sexual behavior was not the only thing affected by selection. Changes in the general locomotor activity of the flies had just as great an effect upon mating speed. These changes were not in the same direction as those in sexual behavior. The slow-mating lines were slow partly because they were so intensely active when put into the mating bottle that it was many minutes before any began courting. Conversely, the fast-mating flies were very sluggish in all but sexual responses. Unselected controls were intermediate in both respects and probably had an optimum "balance" between sexual and general activity. The genetic basis of these two aspects of behavior is certainly fairly distinct, and they can be changed independently as a result of strong selection. Their essential independence is further demonstrated by the fact that, with a different selection technique, slow-mating lines of flies were produced which also showed lowered general activity (Manning, 1963).

This short account of gene effects on Drosophila behavior serves to emphasize the lability of behavioral thresholds. It is tempting to speculate that genes may operate directly on the nervous system in a manner related to Sperry's fourth category listed above, that of the excitatory properties of neurons. This might involve changes in membrane permeability, enzyme secretion, or anything that can affect a neuron's threshold. In no case do we have any of the relevant details of gene action so that little is gained by attempting to be more precise.

The evidence that genes can affect neural thresholds directly is purely circumstantial, but the slow-mating flies of Manning (1961) might be considered as an example. Their poor mating performance does not appear to be the result of gene action on the flies' general metabolism or muscular efficiency. The same muscles that work at low intensity in a sexual situation are involved in high-intensity activity in other situations. Again, their sense organs may be impaired, but none operate exclusively in a sexual context and there is no other sign of impairment. In many respects the nervous system itself is the most plausible site for gene action.

It is not surprising that this should be so. The nervous system is in constant activity; the second-to-second changes in behavior are not produced by the initiation and termination of impulse trains in particular tracts. Rather, as a response to changing sensory inflow, there are changes in the frequencies of both excitatory and inhibitory impulses in tracts that are continuously active at some level. The infinite subtlety of threshold changes in such a system provides plenty of scope for gene action. In addition, there are important parts of the nervous system whose main function seems to be modifying the level of discharge in others. The supra- and subesophageal ganglia of insects appear to exercise this kind of control over the lower segmental centers of the thorax and abdomen. It is in the latter that the motor coordination of discrete behavior patterns is organized, but their discharge is dependent on the interplay of descending facilitatory and inhibitory impulses from the brain (see Vowles, 1961a, 1961b). For example, if the head of a male praying mantis is removed, it begins incessant walking movements. If it encounters a female, incessant copulatory movements begin, which may end with normal and successful copulation (Roeder, 1935). Roeder et al. (1960) have shown that this behavior is the result of greatly increased spontaneous activity in the thoracic and abdominal ganglia of the mantis following the severance of connections with the brain.

THE BEHAVIOR OF GYNANDROMORPHS

At this point, when considering the neurological basis of insect behavior, it is relevant to consider briefly one special type of genetic aberration which yields some interesting behavioral information. Gynandromorphs are mosaic individuals, some of whose cells are genetically male and others female. Unlike vertebrates, in insects the gonads do not produce a hormone that coordinates the development of secondary sexual characters. Thus transplanting ovaries into a castrated-male-insect larva does not affect the development of normal male characters and behavior. Sex is determined entirely by the chromosomes, and it is perfectly possible to have a male head and thorax joined to a female abdomen containing fertile ovaries. The various genetic situations that give rise to gynandromorphs are described by Sinnott et al. (1958).

The behavior of gynandromorphs has been described only in two types of Diptera, *Drosophila* (Morgan and Bridges, 1919; Hollingsworth, 1955) and the housefly, *Musca domestica* (Milani and Rivosecchi, 1954), and in two types of Hymenoptera, the parasitic wasp, *Habrobracon juglandis* (Whiting, 1932), and the honeybee, *Apis mellifera* (Sakagami and Takahashi, 1956). In addition, the behavior of "intersex" Drosophila (which show a varying mixture of male and female characters but are not true gynandromorphs) has been described by Sturtevant (1920) and Hollingsworth (1959).

Gynandromorphs are interesting because they provide a unique means of assessing the relative roles of the brain and the more peripheral centers. How does an insect behave, say, when its brain is male, but its thoracic and abdominal ganglia are female?

It is impossible as yet to give a clear-cut answer to this question, for the evidence is incomplete and conflicting. The evidence from *Drosophila* and *Habrobracon* agrees in finding that the sex of the brain determines behavior. Whiting's (1932) evidence is particularly interesting. In *Habrobracon* both sexes have distinctive behavior and, in particular, female behavior is not revealed merely by an absence of male patterns, which is often all that can be observed in *Drosophila*. Only female *Habrobracon* mount, palpate, and sting caterpillars which are their normal hosts; males ignore them. Whiting observed 62 gynandromorphs and, in those which had a head of one sex and abdomen of the other, found a clear dependence on the head. Thus wasps with female heads and male abdomens would mount and palpate caterpillars, try to sting them, and even make egg-laying movements with their completely male abdomens.

Whiting had some wasps which were male on one side and female on the other; their brains thus contained both types of cell. Such wasps showed signs of both male *and* female behavior, though rarely in a complete fashion. Left/right gynandromorphs also had a tendency to show inappropriate responses to external stimuli. Presented with a caterpillar they sometimes made male courtship movements or would try to sting a female wasp instead of courting her.

Such a full correspondence between the sex of the brain and behavior might not be expected, in view of the neurophysiological evidence given above. The insect brain operates primarily as a general modifier on behavior patterns which are themselves organized in the thoracic and abdominal centers. We might predict that these lower centers would sometimes "break through" in a gynandromorph to reveal their own potentialities.

Something approaching this has been found in gynandromorph houseflies and honeybees. Milani and Rivosecchi (1954) have a few observations on *Musca* and report that sex behavior follows the sex of the abdomen. Flies with female heads on male bodies successfully attempted copulation with normal females. Sakagami and Takahashi (1956) have records of the behavior of some 40 gynandromorph honeybees with a varying mixture of male and female parts. They could not observe specifically sexual behavior because the female workers are sterile and there is very little characteristic drone behavior. However, drones normally move about the hive in a more sluggish manner than workers. On the whole they found that this character in gynandromorphs agreed with the degree of "maleness" of the head. This was not true of the complex series of social patterns such as cell cleaning, mutual feeding, and hive ventilation, which are normally performed only by workers. Sakagami and Takahashi found that even gynandromorphs with completely male head and female body showed many of these patterns in completely normal form. This is the more striking in that the structure of the brain is very different in the two sexes. Drones have enormous eyes and correspondingly enlarged visual centers, and their corpora pedunculata or mushroom bodies are much reduced compared with those of a worker's brain.

Sakagami and Takahashi suggest several possible explanations for their results. Considering them alongside the other gynandromorph evidence, it seems certain that both male and female insects inherit a nervous system capable of all the behavior shown by the species. This is also the case among vertebrates where the hormonal balance determines which type of behavior is evoked. Some other kind of switch mechanism must operate in insects. In *Drosophila* and *Habrobracon* the switch appears to operate primarily upon the brain which in turn evokes behavior of the corresponding sex from the lower centers. However, in *Musca*, a relative of *Drosophila*, and the honeybee, a relative of *Habrobracon*, the sex of the brain has apparently less influence. These inconsistencies between relatives are unexpected, and we must simply wait for more detailed information.

HOW ARE GENE-CONTROLLED CHANGES RELATED TO THE EVOLUTION OF INSECT BEHAVIOR?

We have seen that the usual effect of genes is to alter behavior in a quantitative, rather than a qualitative, fashion. This dichotomy is not absolute. If, during evolution, a series of quantitative changes gradually reduces the performance frequency of a behavior pattern to zero, a qualitative change in the animal's behavior repertoire has resulted. Further, it can be argued that, if the units chosen for analysis are small enough, any change is a quantitative one. Nevertheless, the dichotomy is useful when considering the manner in which behavior evolves.

We must now examine how the divergence of behavior between species and races may be related to gene action. Can we understand behavioral evolution in terms of natural selection operating over a long period on small inherited changes?

We need data from a range of closely related species, and fortunately there have recently been a number of excellent comparative studies of various insect groups. Among the investigators, we may list Spieth (1952) who contributed a study of the courtship displays of some 100 species of Drosophila, Blest (1957) who analyzed the defensive displays of a number of Saturniid and Sphingid moths, and Crane who studied the analogous displays of some Mantids (1952) and the courtship of Heliconiid butterflies (1957a). Still within the bounds of the Arthropods, Crane also made comparative studies of courtship by Salticid spiders (1949) and fiddler crabs (1957b). All these workers have studied a range of related species, and all have been particularly interested in the evolution of the behavior they describe.

In all these groups it is found, not unexpectedly, that the instinctive-behavior repertoire is quite conservative. There is a limited set of behavior units, just as morphological features are also limited. It is clear that, while distant relatives may show behavior differences which appear to be "qualitative," closer relatives differ in a more simple "quantitative" way. This point has already been stressed by Tinbergen (1959a) and relates very well to what is known of gene action.

We can examine the nature of the behavior changes within a group in more detail. Nearly all the behavior under consideration has the function of communicating something to conspecifics or to potential predators. Historically, responses that serve such a function have been derived from various sources in the insects' repertoire. And they have been modified to make them distinctive and conspicuous. This type of evolutionary change has been called "ritualization" (Tinbergen, 1952). Blest (1961) provides a full discussion of the ritualization concept and, following him, the basic features of the process may be considered under two headings.

Changes in the Releasing Mechanism

In many cases we find that related species have different thresholds for the production of homologous responses to the same stimulus. Some moths of the genus *Automeris* perform a defense display in response to tactile stimulation, but their thresholds vary. Different parts of the display vary in their thresholds independently of each other (Blest, 1957). Crane (1952) describes similar threshold differences for the defensive displays of Mantids. Again, *Drosophila* species appear to differ in their sexual-behavior thresholds as measured by the latency from the introduction of the sexes to the beginning of courtship (Spieth, 1952; Manning, 1959b). Differences of this type are very easily related to genetic changes. They exactly resemble the changes produced in *Drosophila* by artificial selection (Manning, 1961).

Often relatives differ in the dominant sensory modality concerned in evoking homologous responses. *Drosophila* species vary greatly in their dependence on visual stimulation during courtship. *D. subobscura* never mates in the absence of light, but its close relatives *pseudo-obscura* and *persimilis* are little affected by darkness (Wallace and Dobzhansky, 1946). Spieth and Hsu (1950) describe a parallel case in the *melanogaster*-species group. At one end of the scale is *auraria*, like *subobscura* quite inactive sexually if kept in the dark. At the other end is *melanogaster* itself, which is scarcely affected. *Melanogaster's* closest relative, *simulans*, is strongly affected by the absence of light. These two species form a sibling pair which must have diverged very recently, yet already they show considerable differences in behavior. The courtship of both types of male is quite similar, but while *melanogaster* females are most responsive to the chemical and tactile stimuli the male provides, *simulans* responds mainly to the visual stimuli. *Simulans* females probably have lowered thresholds somewhere in the visual system, and they also have more visual receptors. Their eyes have more ommatidia than those of *melanogaster*, and this means more neurons in the visual nervous system (Manning, 1959b).

Crane (1949) has found a comparable situation among Salticid spiders, where a series based upon increasing light dependence can be traced. Here, however, the dependence on vision is not restricted to the sexual situation (we do not know that it is in *D. simulans*) but extends through all the spiders' behavior.

Changes in Coordination

This is essentially a blanket category. We have too little knowledge of the mechanisms underlying the performance of an instinctive-behavior pattern to be able to classify coordination changes meaningfully. Morris (1957) and Blest (1961) consider more fully a number of changes which might come under this heading. The "aim" of ritualization has been to produce a distinctive, unambiguous signal. One conspicuous way in which homologous patterns differ between relatives is in the degree to which various elements of the pattern are emphasized.

Male fiddler crabs of the genus *Uca* all show a rhythmic claw waving in their courtship display. The genus falls into two groups behaviorally. One accentuates the lateral movement of the claw during the wave; the other accentuates the vertical component. Species also differ in the degree to which the body is raised on the ambulatory legs during the wave (Crane, 1957b). Brown (1965) describes how the courtship wing vibration of various members of the *Drosophila obscura* species group differs in the degree to which the trailing edge of the wing is lowered. Female Heliconiid butterflies show similar variations in their wing movements during courtship (Crane, 1957a). These are Arthropod examples, but numerous cases of this type are described for birds, where some displays are very fully analyzed (e.g., Tinbergen, 1959b).

A second type of coordination change which may accompany ritualization is an increase or decrease in the speed of a movement. Blest (1957) describes variation in display speed among moths. Lindauer (1957a) compares the speed and rhythm of the waggle dance by the four species of *Apis* (honeybees). To indicate a distance of 100 meters to the food source, the domestic honeybee, *A. mellifera*, shows 10 runs per 15 seconds. *A. dorsata* shows 9 runs to indicate the same distance, *A. indica* shows 7.5, and *A. florea* only 6. The same kinds of variations are found between different strains of the domestic honeybee (Boch, 1957).

These are but a few examples of some of the more important changes which have occurred repeatedly during the microevolution of behavior; a more complete review is provided by Manning (1965). It is perfectly reasonable to consider them as a result of the accumulation of small quantitative effects produced by gene mutation. Changes in the threshold of responses are, as we have seen, typical results of mutation. The exaggerations of particular parts of a movement or changes in its speed may well result from threshold changes also. A group of muscles is active earlier or later in a sequence and maintains its activity for a longer or shorter time. The time occupied by sequences may be varied and so on. If correct, these interpretations require that genes are able to exert their effects at particular sites in the nervous system or upon particular sense organs or muscles. This speculation can be checked with suitable material, and there is some circumstantial evidence in favor of it; e.g., Rothenbuhler (1958) has found that different parts of a behavior sequence are affected individually by different genes. Certainly the nervous system is diverse enough in histology and biochemistry to make limited gene action quite feasible.

THE INHERITANCE OF
BEHAVIOR PATTERNS THEMSELVES

So far we have considered only how existing behavior patterns are modified by genes. We must now consider the inheritance of "whole" units of behavior. Instinctive behavior is usually efficiently adapted to an animal's normal environment. It is probable that some of its adaptiveness has been attained by means of "genetic assimilation" (Waddington, 1961). This is the process of accumulating genes that enable the organism to make an adaptive response to an environmental stimulus. The threshold for this response becomes lowered over successive generations, provided the environmental stimulus is consistently present. Eventually individuals are produced which develop the response even in the absence of the original stimulus.

Waddington has demonstrated genetic assimilation experimentally for some morphological and physiological responses to external stimuli by *Drosophila*. In behavioral terms it could lead to acquired behavior patterns becoming inherited ones. For example, Thorpe and others have studied a form of "larval conditioning" to host caterpillars in the parasitic Ichneumonoid, *Nemeritis* (Thorpe and Jones, 1937; Thorpe, 1938), and also the conditioning of *Drosophila* to contaminants in food (Thorpe, 1939). Here the insects have an inherited preference for the normal situation, but their aversion to abnormal stimuli can be significantly reduced by exposure to them during the larval period. Consistent selection for the insects that responded best might well lead to the development of strains that showed inherited preferences for the new situation. A start has been made on experiments of this type, using Drosophila and peppermint oil as a contaminant in their food (Moray and Connolly, 1963; Moray and Arnold, 1964). Although there are some signs of assimilation, the situation is complicated by the relative toxicity of peppermint oil.

Whether or not it arose in this way, the inheritance of any behavior unit is likely to be controlled by many genes. Caspari (1958, 1963c) and others have suggested this previously, and all the argument of the preceding sections supports this view. We can only envisage the construction of the necessary neural mechanism by many small steps. Large, sudden changes are almost certain to be disadvantageous both for their physiological effects and because they are likely to be maladaptive behaviorally.

Unfortunately, it is very difficult to obtain direct evidence of the inheritance of behavior patterns. In hybrids the behavior patterns rarely show any sign of breaking up into smaller units, although these might be impossible to recognize anyway. As Caspari (1963c) points out, F_2 hybrids and backcrosses are needed if units of behavior that correspond to genetic units are to be isolated. Only rarely can these be obtained, save from species which are so closely related that their behavioral repertoires are very similar. Such species usually differ by genes that affect only the frequency of performance of patterns common to both. Even if a pattern is apparently absent in one parent species, the necessary neural mechanism may be present, though with a very high threshold. "Scissoring," a courtship movement typical of *Drosophila simulans* males, is not normally seen in *D. melanogaster*. Nevertheless, it can be evoked under abnormal conditions (Manning, 1959b). Sometimes a behavior pattern present in both parent species is absent in the hybrid. Ehrman (1960) describes how the female hybrids between two subspecies of *D. paulistorum* are completely unreceptive to the courtship of males and never mate. This too may be a result of a greatly elevated threshold.

If the performance frequency of a behavior pattern in the hybrids is intermediate between those of the parent species, and if the F_2 generation shows a full range of frequencies, we can argue little about the pattern's actual inheritance. We know only that multiple loci affect its performance thresold. Sometimes a single locus seems to determine whether or not a pattern occurs. For example, Hörmann-Heck (1957) studied the inheritance of a number of the courtship patterns in two crickets, *Gryllus campestris* and *G. maculatus*. Some of them appeared to be controlled by a single locus, but this is only a shorthand way of saying that this locus controlled the performance threshold of the pattern in a rather switchlike manner. The underlying mechanisms must depend on numerous loci for their development, although selection may have caused these loci to become linked, perhaps within an inversion, so that they are inherited as a block. Systems of this type have been shown to control the inheritance of color and mimicry patterns of Lepidoptera (Sheppard, 1961).

An important advance in the study of behavioral inheritance may come from the analysis of animal sounds. The sound spectrograph enables an exact record to be made of a most complex, ritualized series of muscular movements. The breakdown of parental patterns in hybrids may be detected with far more certainty than is normally possible. There is already some work with hybrids among grasshoppers (Perdeck, 1958) and doves (Lade and Thorpe, 1964), although F_2 hybrids and backcrosses are difficult to obtain because of the near sterility of the F_1 hybrids.

GENES AND SEXUAL ISOLATION

Hitherto we have considered how insect behavior has been changed by evolution. However, the influence is not in a single direction. In conclusion, we must give some account of how behavior influences the course that evolution takes, and the study of sexual isolation has been foremost in this connection. Sexual isolation may be defined as the reduction of hybridization by behavioral barriers to mating between species or strains.

Sexual-isolation studies gained great impetus from the publication of two books, Dobzhansky's *Genetics and the Origin of Species* (1937) and Mayr's *Systematics and the Origin of Species* (1942). One central problem was how divergent species emerge from a common ancestral population. Must populations be geographically isolated before divergence can occur? How stable are the genetic differences which have arisen between populations in isolation if they subsequently meet?

Some barrier to prevent hybridization must be present if previously isolated populations are not to mingle once more. Müller (1942) suggested that sexual isolation could arise as one consequence of genetic divergence while two populations are still separated. When they meet again a barrier already exists and they do not interbreed. Dobzhansky considered this unlikely but envisaged sexual isolation arising rather quickly after the populations meet. If divergence has proceeded far enough to render the hybrids at a disadvantage, selection will favor the rapid evolution of a sexual-isolation mechanism.

Evidence favoring each point of view has been put forward, but nearly everybody now agrees that they are not mutually exclusive and must reinforce one another in many cases. It was important to discover how small a genetic divergence could produce sexual isolation. Many people have looked for isolation between different strains of the same species, different inbred lines, and between stocks differing by only a single gene. Nearly all the work is on Drosophila, and Patterson and Stone (1952) give a very full account of the literature to that date. Much of it suffers from a total disinterest in behavior as such. Often isolation has been detected simply by scoring the genotype of the progeny, with no direct observation. In some cases where the insects have been watched, the courtship of a male Drosophila has been considered simply as a vehicle by which he expresses the degree of his sexual isolation from the female he is courting. Thus males that copulate at random with females of their own and a foreign strain have nevertheless been described as showing "courtship discrimination" between them.

Sexual isolation among insects is based on a variety of sensory discriminations. In many *Drosophila* species it is primarily chemical, either from contact chemoreceptors on the tarsi or via the antennae (Miller, 1950; Spieth, 1952; Manning, 1959a). Females usually discriminate more strongly than males. It is not disadvantageous for males to be aroused by a wide range of stimuli for they can mate many times, but females mate less often (sometimes only once), and their choice of mate is critical.

Chemical differences between populations are almost certain to arise at a very early stage in divergence and thus are well adapted to form the basis for isolation. Kessler (1962) shows that isolation based upon contact-chemical differences has already developed between geographical races of *D. paulistorum*, which must be of very recent origin.

Visual discrimination against foreign mates is probably rare in insects; they do not possess fine enough form vision. Among the insects that use visual stimuli in courtship, such as butterflies, the males approach female models of almost any color (Tinbergen et al., 1942; Stride, 1957; Magnus, 1958). But when they come closer, they usually court only those models of roughly the correct color. Chemical stimuli are concerned in the final stages of butterfly courtship, and these are presumably more critical.

Perdeck (1958) describes a remarkable example of sexual isolation between two sibling grasshopper species (*Chorthippus brunneus* and *C. biguttulus*), which is based entirely on sound stimuli. The females are attracted to their males by their distinctive songs. If they are artificially lured into the vicinity of foreign males, they show no subsequent discrimination at all.

Genes appear to affect sexual isolation in a typically quantitative fashion. Laboratory stocks of *Drosophila melanogaster* and *D. simulans* vary in the degree of isolation between the species (Barker, 1962). Koopman (1950) showed by artificial selection that *D. pseudoobscura* and *D. persimilis* normally show considerable variability for genes affecting the sexual isolation between them. By eliminating hybrids from mixed populations, he was able to increase isolation markedly within a few generations, mostly by increasing the discrimination of *persimilis* males.

We have seen that single-gene mutations often reduce the mating success of males that carry them. However, it is less common for them to affect the receptivity of females; thus mutant males may be at a disadvantage both with mutant and normal females. This "one-sided mating preference" in *D. melanogaster* has been described for the mutants *yellow* (Merrell, 1949; Bastock, 1956), *raspberry* (Merrell, 1949), and *white* (Reed and Reed, 1950). The extent to which this situation could represent an initial stage in the evolution of true sexual isolation has been disputed. Merrell (1953), for example, argues that it has little relevance because the mutant males are always at a disadvantage in mixed populations. He has indeed shown that various genes are eliminated at the rate predicted from the mating deficiency of the carrier males.

In an isolated population, however, a change in the courtship behavior of one sex immediately imposes a new selection pressure on the other. Bastock (1956) finds that *yellow* females taken from stocks that have carried the gene for many generations are more receptive than normal females. This difference helps to "compensate" for the courtship deficiency of *yellow* males, which now inseminate *yellow* females at nearly the same rate as normal males. This increased receptivity is not an effect of the *yellow* gene itself. It has been produced by selection in *yellow* stocks where the courtship of males is poor, but the usual selection pressure to mate and lay eggs quickly remains.

If in an isolated population there appears a gene that is generally advantageous but has an effect comparable to *yellow* on the courtship of males, selection will not favor an increase of female receptivity beyond a certain point. The reasons for this appear to be twofold in *Drosophila*. Firstly, a degree of unwillingness enables females to "sample" the courtship of males and to discriminate against those which are deficient. Smith (1958) has discussed this type of sexual selection and shows how it enables *D. subobscura* females to avoid mating with inbred males of low fertility. Bösiger (1960) puts forward similar ideas. Secondly, if female receptivity increases too far, they may accept foreign males and be rendered effectively sterile since they mate only once.

Would, then, a gene such as *yellow*, if it confers other advantages, spread only rather slowly through a population, having to overcome its behavioral disadvantage? Bastock argues that selection favors females that respond preferentially to some aspect of their males' courtship which is less affected by the gene. If vibration and licking are reduced, as by *yellow*, the females might respond more to visual aspects of the orientation part of the display. This, in turn, will result in selection for males which accentuate this part of courtship. Eventually this mutual selection process will lead to sexual isolation from other populations which still rely most on vibration and licking. The divergence of *D. melanogaster* and *D. simulans* from a common ancestor is most reasonably explained in this way. *Simulans* males are more sluggish and court less actively than *melanogaster*, but their females respond to visual aspects of the display and, if anything, accept them more easily than *melanogaster* females accept their males (Manning, 1959b). Such a system agrees well with what is known of the behavior differences between other closely related *Drosophila* species. It means that quite small genetic changes might lay the foundations for sexual isolation in geographically isolated populations. These will be quickly strengthened by selection if they subsequently meet and have disadvantageous hybrids.

The effects which genes have on habitat selection also have repercussions on sexual isolation. Quite often the isolation found between species in the laboratory seems insufficient to account for the rarity of naturally occurring hybrids. *D. pseudoobscura* and *D. persimilis* are a case in point. Clearly their sexual isolation is reinforced by the fact that where they are sympatric they choose different microhabitats within an area (Pittendrigh, 1958). Waddington et al. (1954) have demonstrated that genetic divergence such as occurs in domestic Drosophila stocks has a direct effect on habitat selection and presumably Drosophila choose those environments where they survive best. Kalmus (1941) shows that, under particular circumstances, genes that are normally disadvantageous may show improved survival. *Yellow*, for example, survives starvation in a moist atmosphere better than wild types.

Differences in habitat selection and survival ability are of greatest importance in those areas where two closely related species overlap. In the absence of competition they may have a wide range of microhabitats but, when they compete, each species is forced to occupy only those where it can survive better than its relative. Among birds there are good examples of habitat selection expanding and contracting according to competition. The willow warbler (*Phylloscopus trochilus*) and its close relative the chiffchaff (*P. collybita*) both breed over a wide area of Europe. The willow warbler normally lives in low bushes and scrub whereas the chiffchaff occupies trees and tall bushes. In the Canary Islands, the willow warbler is absent; there the chiffchaff frequents both types of habitat (Lack and Southern, 1949). Similarly, in the Sierra Nevada of California, *Drosophila pseudoobscura* is forced, by competition with *D. persimilis*, into woodlands that are drier than those it prefers in the more eastern and southern portions of its range where the latter species is absent (Pittendrigh, 1958).

Even within a Drosophila population cage in the laboratory there may be a sufficient range of habitats for microgeographical isolation to exist between strains. Knight et al. (1956) and Crossley (1963) have attempted to select for sexual isolation between different mutants of *D. melanogaster*. They mixed stocks which were genotypically similar, except that one was homozygous for the gene *vestigial*, which greatly reduces wing development, and the other for *ebony*, which affects body color and also impairs vision. They imposed selection for isolation by removing the hybrids, which are phenotypically wild, each generation. Both found that the stocks did not always mix and mate at random in population cages and jars. One result of the presence of the *ebony* and *vestigial* genes was to cause the two stocks to react differently to light and other features of the containers when they were first introduced as virgins. This microhabitat selection was certainly increased by their selection and reinforced any true sexual isolation.

Another example is that given by Hovanitz (1953) who describes how a gene affecting wing color in the butterfly *Colias eurythme* also changes its optimum temperature and light intensity for flight. White females are most active in the early morning and just before sunset, when temperature and light are low. Yellow and orange females show a peak of activity at noon. The white gene does not affect the color of males, but if it affects their activity in a similar way to that of females, it may well influence the frequency of mating between the different genotypes. Genes that affect habitat selection in this way are to some extent "auto-isolating," and this is bound to influence, and perhaps accelerate, the evolution of sexual isolation.

CONCLUSION

The speculation-to-fact ratio in this discussion has inevitably been rather high. Yet we can understand in principle if not in detail how insect behavior has evolved by the accumulation of small inherited changes. One urgent need is for more physiological data on how genes affect the nervous system and its operation. Behavior genetics is a rapidly expanding field, and it is certain that insects will prove as valuable here as they have in other branches of biology in elucidating mechanisms that are of universal importance.

CHAPTER FIVE
GENETIC AND EVOLUTIONARY CONSIDERATIONS OF SOCIAL BEHAVIOR OF HONEYBEES AND SOME RELATED INSECTS[1]
Walter C. Rothenbuhler[2]

INTRODUCTION

Insects are so successful on this planet that fully 675,000 species have evolved—many more than all other animal species combined. They live essentially everywhere except in the sea. They are adapted to the utilization of foods ranging from wood to blood. A few species store supplies of food, but others die or enter some quiescent state when the natural supply is exhausted. Some are parasitic, others are free-living. Some are sessile, while others crawl, walk, and fly. Most live alone during almost their entire existence, whereas a few cannot survive apart from a colony. In reproduction, in hatching, and during growth and development, insects engage in the behavior necessary to the maintenance of the species. Their behavior reaches spectacular com-

[1] Supported in part by Research Grant HD 00368-01 from the National Institute of Child Health and Human Development, Public Health Service.

[2] I am grateful to L. Lance Sholdt for his dedicated assistance in the preparation of this chapter and to Dr. Rolf Boch and Dr. C. D. Michener who graciously and helpfully took time to read and criticize it.

plexities, particularly in the cases of the social species. This chapter is concerned primarily with the behavior of one insect, the Western honey-bee, *Apis mellifera* L., which, to be better understood, is considered in the light of some facts regarding insects in general, reproduction in the order Hymenoptera, and some of the behavior of the other honey-bees, the bumblebees, and the stingless bees.

GENERAL INSECT BIOLOGY

Morphology, Physiology, and Classification

The adult insect body has a chitinous exoskeleton and is divided into three segments: head, thorax, and abdomen. The head bears a pair of antennae, the eyes, and the mouthparts. The thorax, composed typically of three segments, usually bears three pairs of legs and two pairs of wings. The abdomen consists of many segments with no legs or wings, but it usually carries the reproductive organs. Internally the typical insect has a tubular digestive system; a long tubular heart extending from head to abdomen; a tubular and saclike system of tracheae for respiratory organs which extends from the spiracles in the body wall to the smallest internal organs; reproductive organs opening at the posterior end of the body; a complex, extensive muscular system; and a nervous system. For further information on insects, see Stefferud (1952), Ross (1956), Imms (1957), Borror and DeLong (1964), or other general textbooks.

The central nervous system of an insect consists typically of a brain, subesophageal ganglion, and a series of ganglia and their connectives, which together constitute the ventral nerve cord, lying on the floor of the thorax and abdomen. There are gross variations in this system from one group to another, involving size of the parts and concentration of ganglia into a smaller number of centers. The volume of the brain relative to the body varies, being 1/4,200th in *Dytiscus* (predacious diving beetle), 1/3,290th in *Melolontha* (beetle), 1/280th in *Formica* (ant), and 1/174th in *Apis* (the honeybee) (Imms, 1957). Brain size and shape differ in the three castes of the honeybee itself (Snodgrass, 1956). For detailed treatment of the anatomy of the nervous system, the following should be consulted: Snodgrass (1935, 1956), Vowles (1961a), Schmitt (1962), and Bullock and Horridge (1965).

The physiology of the insect nervous and sensory systems is presented by Wigglesworth (1950), Roeder (1953, 1958, 1963), Vowles (1961b), Patton (1963), Dethier (1963), Markl and Lindauer (1965), and Bullock and Horridge (1965).

Some recent reviews of behavior with emphasis on insects are found in Alexander (1964), Baerends (1959), Carthy (1965), Jander (1963), Lindauer (1962, 1965), and Haskell (1966). The social biology of ants is reviewed by Wilson (1963) and of termites by Weesner (1960).

The systematic classification of the insects to be discussed is given in Figure 5.1. Reference to various insects will be made in such a way as to permit their identification in this chart.

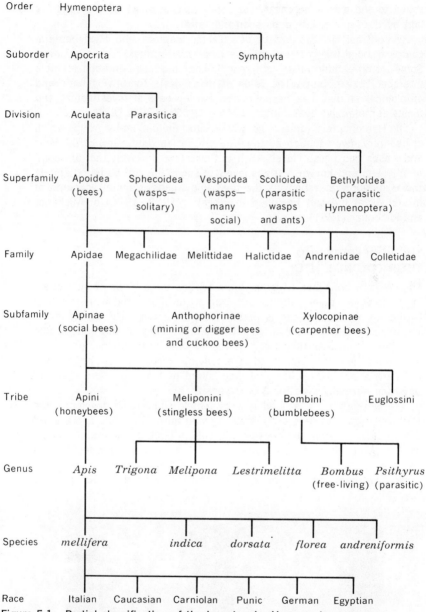

Figure 5.1 Partial classification of the insect order Hymenoptera.

Solitary and Social Life

Most insects live alone and consequently display a limited number of complex behavior patterns. Individuals of such species find food and ingest it; escape predators; and respond to light, temperature, and other elements of the physical environment. In reproduction they become social to the extent necessary for copulation, after which the eggs are laid by the female. Life ends soon thereafter.

By contrast, in two orders of insects, Isoptera and Hymenoptera, complex social life is found. All the Isoptera (termites) live in colonies. Some termites build nests with specific characteristics and considerable interspecific variability. The value of these nests for phylogenetic and ethological studies has been pointed out by several investigators and nicely summarized by Schmidt (1955a, 1955b, 1964).

The Hymenoptera (ants, bees, wasps, etc.) include many groups which are solitary. Even though all the ants are social, only a few of the bees and wasps can be so classified. It is these few, however, that are commonly seen and known. Various levels of social life can be found among the Hymenoptera, especially the bees and wasps. General accounts of insect social life can be found in Wheeler (1923, 1928) or Michener and Michener (1951).

THE HONEYBEES AND
THEIR COLONIAL LIFE

The Western honeybee (*Apis mellifera* L.) has three kinds of individuals in its colonies: queen, drone, and worker, as illustrated in Figure 5.2. Ordinarily only one queen is present in each colony. The queen is the reproducing female—the mother of all other bees in the colony. She originates from a fertilized egg, is reared in a large cell hanging vertically from the comb, and is fed royal jelly in massive quantities throughout her developmental life. She may lay up to 1,600 or more eggs per day. She normally lives for 2 or 3 years.

The workers are a second type of female which also develops from fertilized eggs. Workers are reared in small horizontal cells and are fed a food different from the queen's and only in limited amounts. Although workers start life with the same genetic endowment as queens, they become differentiated from queens in morphology, physiology, and behavior. Workers are present in numbers ranging from 7,000 or 8,000

Figure 5.2   Drone, queen, and worker honeybee, in order from left to right.

Figure 5.3 A comb showing brood in bottom center, cells of pollen above, and sealed cells of honey in upper corners.

to 60,000 or 70,000 in a normal colony. Their life-span in the summer is about 6 weeks, but during the winter it may be extended to 5 or 6 months (Maurizio, 1959).

Drones are male bees and develop from unfertilized eggs. They are reared in horizontal cells somewhat larger than those in which workers are reared. Their only known function is the obvious one of mating with the queen. They are normally present in a colony from early spring until late summer or fall in numbers ranging from a few up to several thousand. Their average length of life is unknown, but Howell and Usinger (1933), after study of a few drones, gave 54 days as an indication of the length of drone life.

The natural nest of the honeybee is composed of several vertical combs built inside a cavity such as a hollow tree, the walls of a house, or a wooden box. Each comb is composed of a vertical midrib and cells built horizontally on each side (Figure 5.3). In these cells both honey and pollen are stored and young bees are reared. Ordinarily the brood is reared in the lower parts of the combs, pollen is stored around this "brood nest," and honey is stored above it.

In modern beekeeping, combs are contained inside wooden frames suspended in a hive and consequently can be inspected and moved at the will of the beekeeper (Figure 5.4). The honeybee colony is open to a very wide variety of experimentation by those who understand the biology of bees. The culture of bees for the honey and wax produced, for the pollination services they render, or for recreation is known as apiculture or beekeeping. There are many books devoted to this subject, some of which may be cited: Grout (1963), Root (1962), Eckert and Shaw (1960), and Barth (1956). Some skill in handling bees is necessary for their successful use experimentally, but the necessary *minimum* of such skill can be acquired in a matter of a few weeks or months, preferably under the guidance of one who is already conversant apiculturally.

Figure 5.4 An open hive showing the arrangement of frames, with one frame and its comb removed from the hive.

Apis indica Fabr., the Indian or Eastern honeybee, is similar to *Apis mellifera* and indeed is classed as one of its subspecies by some authorities. However, the colonies of *A. indica* are somewhat smaller, and the bee itself is only about two-thirds or three-fourths the size of *A. mellifera*. Like *A. mellifera*, *A. indica* builds a multiple-combed nest inside a cavity.

Use of a nest cavity by these species is in contrast to the habits of *A. dorsata* Fabr., the giant honeybee, which builds a single-combed nest under a cliff, tree limb, or eave of a building. *A. dorsata* colonies contain about 4,000 to 5,000 individuals. Workers and drones are reared in cells of about the same size, but queens are reared in larger cells at the bottom of the comb.

Apis florea Fabr., the little honeybee, also builds its single comb in the open, usually fastened to a tree limb. Population size of the colony is usually 4,000 to 5,000 individuals, about the same as that of *A. dorsata*. The three kinds of individuals are reared in three different-sized cells.

Apis andreniformis Smith, also a small species, is distinctly different from *A. florea* (Michener, personal communication). Very little is known of its biology.

These five species, which probably originated in the region of India, are the only presently acceptable members of the genus *Apis*; of the group, only *A. mellifera* is found in most temperate and tropical parts of the world. *A. indica* has spread somewhat eastward to China and Japan, but *A. dorsata*, *A. florea*, and *A. andreniformis* are restricted to their ancestral homeland.

REPRODUCTION

Mating Behavior

Reproduction of a species that lives only in colonies is necessarily on two levels: the production of new individuals and the production of new colonies. Mating behavior is antecedent to both, since new individuals are prerequisite to new colonies.

It has been known since the late 1700s that the queen and drone mate outside the nest in flight (Park, 1949), but only a few observations have been reported of matings with the participants in free flight. For many years it was supposed that the queen mated with only one drone. Since a successful mating was thought always to result in detachment of the drone's genital structures and their firm implantation in the queen (called a mating sign by beekeepers), it was reasoned that a queen could not mate more than once on a given flight. Although it was known that a virgin queen would leave the hive for two or three flights, it was thought that she engaged in only one actual mating flight.

Mating behavior of honeybees is presently under intensive investigation, and spectacular discoveries have been made in the past decade or so. The number of times a queen mates was one of the first problems investigated. By confining virgin queens to the hives except when the investigator was observing the entrance, it was established by W. C. Roberts (1944) that a majority of queens go on more than one flight and return with a mating sign. Triasko (1951), by measuring the volume of semen in the genital tract of queens returning from mating flights and comparing this with the volume produced by a single drone, estimated that each queen mates with four or five drones per flight. Confirmatory evidence as well as additional information has been presented by Taber (1954); Woyke (1955); Alber, Jordan, Ruttner, and Ruttner (1955); Peer (1956); Ruttner (1956); Taber and Wendel (1958); Gary (1963); and others. Taber and Wendel concluded, after a reanalysis of several sets of published data and the analysis of one set of new data, that queens usually mate with 7 to 10 drones. The genetic implications of multiple mating will be discussed on page 71.

Since honeybee matings have generally been prevented when the participants were subjected to any sort of confinement or restraint, successful use of restraint has come as a welcome surprise. Gary (1962, 1963) has tethered queens by attaching a small thread to the dorsal side of the thorax with quick-drying acetate cement. When suspended from a line stretched between two tall poles, such tethered queens flew readily and some mated, apparently normally. This technique brings several questions on mating behavior within range of experimentation.

It has been learned that the queen bee has a sex-attractant complex the primary source of which is the mandibular glands. If this complex is extracted from a queen and presented on filter paper at a height of at least 15 feet, drones in the vicinity are attracted. By tethering virgin queen bees as described above, Zmarlicki and Morse (1963) have been able to show that drones congregate in some areas but not in others. In a detailed study of mating behavior, Gary (1963) obtained some evidence that congregations of drones fly at rather specific heights which are variable from day to day. Ruttner (personal communication), using caged queens suspended from balloons and by trapping and marking drones, has obtained evidence of drone congregation areas and some knowledge of drone travels. The time of day that drones fly is also variable (Taber, 1964), but in general it is confined to early- and mid-afternoon hours.

Since queens and drones are not attracted to each other in the nest, finding each other outside in flight presents some problems. That they can come together from great distances has been demonstrated. When colonies containing the different sexes were separated by 8 miles, 42 percent successful matings occurred; 25 percent successful matings occurred when the separation was 10.1 miles (Peer, 1957). However, such queens were considerably older than normal before mating.

The Reproduction of Individuals

Queen honeybees lay two kinds of eggs: unfertilized and fertilized. How she controls whether or not an egg is fertilized has been a point of concern. It seems probable that the mechanism involves a simple apposition of the egg to the orifice of the spermathecal duct by the valve fold, a structure in the floor of the vagina (Adam, 1913). An even more basic question remains as to why one egg is so held and another one not. Inasmuch as a queen can lay fertilized eggs in queen cells and drone cells as well as in worker cells, cell size does not provide the exclusive stimulus for a simple reflex activity. More than one simple factor is involved in this activity, and once this is admitted, the mystery disappears and a search for the factors can be initiated.

Fertilized and unfertilized eggs produce females and males, respectively, among all Hymenoptera except for a few species, including *Apis mellifera capensis* (Anderson, 1963), in which some or all unfertilized eggs give rise to females (White, 1954; Suomalainen, 1950). Such a haploid origin of males and diploid origin of females present both cytological and genetic problems.

Maintenance of chromosome number in the face of meiosis is one of these problems. Since males have haploid reproductive cells, further reduction of chromosome number cannot be made. For a long time it was thought that in the honeybees, as well as in most other Hymenoptera, there was in spermatogenesis an abortive first meiotic division. This abortive division was seen as having no division of chromosomes and an unequal division of cytoplasm, resulting only in the pinching off of a small cytoplasmic bud. Meiosis II involved both chromosomal and cytoplasmic division, which is equal in most Hymenoptera and results in the formation of two sperms. In the Apidae, however, the cytoplasmic division is unequal and produces only one sperm. Recently, White (1954, p. 329, following Walker, 1949) prefers to drop the abortive-first-division concept and say that there is a single mitotic division only, between the last gonial division and the formation of sperms.

Soon after the genic-balance theory of sex determination was developed, it was realized that genic balance could not explain sex determination in the Hymenoptera. Whereas in most species (*Drosophila* being a good example) a different proportion of sex chromosomes to autosomes could be found in the male and in the female, in the Hymenoptera no such difference in balance between sex chromosomes and autosomes could be brought into being simply by the observable difference in ploidy: diploidy or haploidy, i.e., fertilization or the lack of it. The first break came when the Whiting school of *Habrobracon* genetics (reviewed in Whiting, 1943) announced the multiple sex-allele hypothesis for *Habrobracon juglandis* Ashmead (now correctly called *Bracon hebetor* Say). This mechanism has been shown to be the probable explanation of sex determination in a related species, *Habrobracon brevicornis* (Speicher and Speicher, 1940), but it cannot without modification be extended throughout the Hymenoptera. There are certain species in which mating of closely related individuals is the rule, and the expected 50 percent inviability of progeny or diploid males is not found (Schmieder and Whiting, 1947).

Increasing evidence has been accumulated for the multiple sex-allele hypothesis for the honeybee. The expected inviability has been found (Mackensen, 1951, 1955), and even though no genetically marked, fully diploid, mature males have appeared, mosaic males carrying some diploid male tissue have been identified (Rothenbuhler, 1957; Drescher and Rothenbuhler, 1964).

It has been generally assumed that individuals homozygous for sex alleles have simply died in an embryonic or early larval stage. Consequently, it has come as an incredible surprise that behavior of adult bees is involved. Woyke has reported from studies on progeny of mated queens showing 50 percent brood inviability as follows: (1) that the eggs hatch and living larvae are present (1962); (2) that 50 percent of these larvae are male and 50 percent female (1963a); and (3) that the male larvae from fertilized eggs are eaten alive by worker bees and consequently never live to maturity in the colony (1963b). Woyke has reared some of the larvae produced from a "low-survival-rate brood" in an incubator. Both females and males were produced (1963c). He indicates that a description of these diploid drones is forthcoming in another paper. What stimulus-response mechanisms are involved in this ingestion of diploid male larvae by adult bees, but not of diploid female or haploid male larvae, is a compelling question.

One of the characteristic features of a highly developed social unit is division of labor. Division of a species into two sexes accomplishes this in a basic way. Social insects have additionally divided the labor according to morphologically differentiated castes (and temporally, according to the age of the individual, as will be seen later in detail). Although the termites have subdivided both sexes into castes, the Hymenoptera have restricted caste development to females alone. These female castes are queen and worker, but there is a wide variety of subcastes among worker ants and various degrees of separation into castes among the more primitively social bees and wasps (Michener, 1961).

A time-honored question concerns the basis of caste determination in social insects, and two possible answers have generally been advanced: trophogenic and blastogenic. Restating these in modern terms and changing them slightly, one can say environmental and genetic. Among the honeybees, caste determination is clearly environmental (and indeed trophogenic) since a difference in nutrition brings about the large and striking difference between queens and workers. Young larvae transferred from worker cells to queen cells, where they receive different food, develop into queens (Weaver, 1957b; Townsend and Shuel, 1962).

Caste determination in social insects, when it has been studied, has with one group of exceptions been found to be due to environmental agents. These are active at various times, ranging from egg to adult. Michener (1961), Brian (1957), and Weaver (1966) review the problem of caste determination in various insects.

The exceptions to environmental determination are in the genus *Melipona* in which queens and workers are produced intermixed in identical cells. Kerr (1950a and 1950b) invoked a genetic determination of castes. In one species, *Melipona marginata* Lep., under favorable conditions one-fourth of the brood cells produced queens. In several other species one-eighth produced queens. This fraction of queens can be explained by assuming in the first case that queens are heterozygous at two caste-determining loci as, for instance, *AaBb*, whereas workers are homozygous at one or both of these loci. Drones are hemizygous at these and all other loci, and the mating of any one kind of drone (e.g., *AB*) would lead to progeny from fertilized eggs in the proportion of one double heterozygote (queen) to three single or double homozygotes (workers). The same reasoning applies to the species that produce one-eighth queens but in that case three loci are involved in caste determination and queens are heterozygous at all three.

Such a system would be expected to lead to excessive queen production. In the favorable part of the year when so many queens are produced, they are simply killed within a couple of days after emergence. In the unfavorable part of the year, some other system comes into operation, and instead of one-fourth or one-eighth of the brood cells producing queens, none or only a few do so. Such limited production of queens was not understood. Kerr stated that the hypothesis was not as fully tested as was desired (Kerr and Laidlaw, 1956). One is certainly led to wonder whether or not a lack of food in the unfavorable part of the year might play a part. Some such environmental factor might be associated with the residue of the original, wholly genetic mechanism of caste determination.

Genetic Questions on Social Hymenoptera

Maintenance of Genetic Variability   It is well known that most mutant genes are to some extent deleterious. When they become homozygous, their possessor is often less well adapted than its wild-type siblings. Since male Hymenoptera are haploid, it seems that every mutant gene would be subject to an immediate selection screen. Therefore, very little genetic variability would be found in this order of insects. Work with honeybees and Habrobracon has not revealed such to be the case.

Although somewhat greater variation has been found in some morphological characters of drones than of workers (reviewed by Phillips, 1929), it does not seem as great as might be expected. The maintenance of genetic variability by sex-limited expression of genes and by overdominance phenomena has been considered in several papers (Kerr, 1951; Kerr and Kerr, 1952; White, 1954; and Drescher, 1964).

Genetic Composition of Colonies   Multiple mating of queen bees, discovered recently, was described in a previous section. Such multiple mating is of utmost importance in the analysis of colony behavior (Rothenbuhler, 1960; Hamilton, 1964). A colony is not a genetically recognizable unit. It is a "superfamily" made up of a number of variously sized "subfamilies." Although the queen is the mother of every bee in the superfamily, each subfamily has a different father. The contribution of a father to each of his offspring is genetically identical but there need be no familial relationship whatsoever between the various drones with which the queen mated, unless, of course, individual drone swarms are composed largely of drones from a single colony. In view of the amount of drifting of drones from one colony to another, this seems unlikely. Sperms from the different drones with which a queen has mated are not thoroughly mixed in the queen's spermatheca. Two samples of progeny taken at different times may contain widely different percentages of offspring from a single drone (Taber, 1955). Certain data indicate, however, that it is rare for the sperm of a single drone to be nonapparent in any sizable sample of offspring (Tucker and Laidlaw, 1965). Since normal segregation occurs in oogenesis, the genetic similarity of the queen's contribution to each of her offspring would be expected to depend upon her own state of heterozygosity or homozygosity.

Prerequisite to any genetic study of a naturally occurring colony of insects is a knowledge of mating habits. Did the reproducing female mate with more than one male? It must also be ascertained whether or not there is only one reproducing female in the colony. The presence of a mother queen and her daughter (supersedure queen) together in a colony is not a rare occurrence among honeybees. It is a point worth keeping in mind in the study of other insect colonies.

Reproduction of the Colony

For the more highly social bees, reproduction of individuals does not suffice for reproduction of the species. Reproduction of colonies must also occur, and various mechanisms for this have been evolved. In cases in which the colony itself survives the winter, a division of the colony is involved in its reproduction. This is known as swarming and occurs among all honeybees and stingless bees.

Swarming by the honeybee usually occurs in the early summer after the colony population has been greatly increased, from the low point in late winter, by intensive brood rearing. Swarming follows the construction of queen cells and the rearing of young queens to the point of the late larval or early pupal stage. At this time, the old queen plus half (or more) of the worker bees leave the hive and cluster on a nearby bush, tree, or some similar object.

In the meantime, perhaps for several days, scouts have been searching for a new nesting place (Lindauer, 1957b). After some possible places are found, the swarm engages in a decision-making process (Lindauer, 1961) and moves off to one of them. More details on swarming can be found in any major book on bees, among which Butler (1954) or Ribbands (1953) might be cited in particular.

Stingless bees also swarm, with the difference that the old queen stays with the parent colony and a new queen leaves with the swarm. Before the swarm leaves, the new nest location is found, and some provisions are moved from the old nest to the site of the new one.

Bumblebees and most other social bees and wasps found new colonies in a different way. In fact, no old colony endures throughout the entire year, and the colonies existing in a summer were all newly founded in the spring. Only the young queens survive the winter, and each one founds a new colony when she emerges from hibernation.

SOCIAL LIFE

Acquisition of the Factors of Sociality and the Development of Social Life

All insect species having two sexes active in reproduction have divided the labor of life and must communicate sufficiently to ensure mating. This would seem to be a primitive phase of social life. Most insects, however, have evolved no further in sociality and consequently are classed as nonsocial species. A few insects display one or more of the additional elements of social life without having become truly social. Only the termites (insect order: Isoptera), ants, some bees, and some wasps (see Figure 5.1) have come to possess all the factors that are basic to complex social living.

Considered in another way, social life exists wherever there is mutual cooperation of two or more individuals. Such mutual cooperation may come about in two essentially different ways: The first involves two (or more) completely independent creatures abandoning their complete independence and forming ties between themselves where no such ties existed previously. Now the behavior of one is modified by the behavior, or simply the existence, of the other. The second origin of mutual cooperation is for two individuals to separate incompletely, as in the case of a mother and her offspring. Originally the offspring is in a sense part of the mother's body, and although the egg is laid or birth occurs, a behavioral bond holds the two together for a short or longer period of time.

It is generally believed that the latter mechanism is the basis for development of termite, ant, and wasp colonies. Probably each possibility has been followed in some of the several separate origins of social life of the bees (Michener, 1958; Sakagami and Michener, 1962).

Regardless of how the first step was taken, social life in the honeybee colony depends heavily upon six factors: (1) maternal care, (2) maternal survival, (3) division of labor, (4) use of predigested or secreted food, (5) storage of food, and (6) communication. The nature of each of these factors of social life will be reviewed and the extent of its development in some related groups briefly indicated.

Maternal Care   Maternal care in its most primitive form may be nothing more than selection of an oviposition site by the female instead of dropping eggs like rain, as does the walkingstick (Michener and Michener, 1951). Maternal care ranges upward from egg placement in a protected place or on a food plant (cabbage butterfly or corn borer), through guarding the eggs for a short time (several species of stink bugs), or making a nest (earwig) and stocking it with food (most bees), to the elaborate waxen-comb, temperature-regulated, guarded nurseries of the honeybee. Among the honeybees, however, the reproducing female no longer provides the maternal care, since by division of labor maternal care has been shifted to the worker caste.

The female bumblebee, when she establishes her nest, provides the complete care of the eggs and brood, but later when young workers emerge, they relieve the mother of the necessity for foraging and of some of the brood care.

Various degrees of complexity in nest construction exist among bees, as has been described by Michener (1964) and by Kerr and Laidlaw (1956). All members of the genus *Apis* build vertical combs; two species build a multiple-combed nest in a cavity, and the others build a single-combed nest in the open air. Most stingless bees build horizontal combs, several in a cavity, and nest organization is highly complex. One African genus of Meliponini (*Dactylurina*) builds vertical combs with horizontally elongate cells like *Apis*. Bumblebees (*Bombus*) make use of cavities or some such protection in which a crude nest of brood cells and a few honey pots are built.

Nest construction tells part of the story of maternal care, but another part is revealed in the feeding of larvae. There are two main methods. In the case of stingless bees, the entire food requirement of the larva is placed in the cell, the egg is deposited, and the cell sealed. Such feeding is called mass provisioning. Its alternative, progressive provisioning, demands a continual supplying of food to the brood throughout the larval period and is the type of feeding practiced by honeybees. Bumblebees, not as highly evolved socially as the stingless bees, nevertheless practice progressive provisioning to an extent. Some of the Xylocopinae also practice progressive provisioning. Progressive provisioning is considered to be the more highly evolved because it leads to greater contact between parent and offspring and more easily permits food to be varied throughout the developmental period. Even so, only about 400 of some 20,000 species of bees are known to practice it (Michener, 1964).

A clue to how progressive provisioning may arise from mass provisioning is provided by the wasp *Synagris spiniventris* (superfamily:

Vespoidea). When caterpillars are abundant—easily and quickly obtained—mass provisioning is followed. When caterpillars are scarce and providing them requires much time, progressive provisioning is the rule. In this case, one can speculate that the time of oviposition is not rigidly tied to a certain state of provisions in the nest. This unlinking of distinct units of behavior, which are often tied into a behavioral chain, opens the way for further evolution.

Maternal Survival   Maternal survival is just as necessary as maternal care to the development of a family-type social unit. Length of life of females may range from that of the Ephemeridae (Mayflies) whose adult has vestigial mouthparts and lives only long enough to reproduce (a day to a few days) to that of queen termites which may survive for many years (Michener and Michener, 1951). Queen honeybees normally survive for 2 or 3 years. Workers, on the other hand, have a much shorter life. Queen bumblebees live approximately a year, whereas workers survive only until the end of the summer at most. In social species, a lengthy survival of the reproducing female is the rule, and this permits the development of a family-type colony.

Division of Labor   Division of labor is often considered to be the true mark of a social insect. Division of labor not by the sexes but within a sex is the kind of division of particular importance.

One of the most primitive divisions of labor occurs among some species of South African bees of the genus *Allodape* (carpenter bees: Xylocopinae). The mother, by herself, cares for the first larvae, and when these daughters emerge, they look exactly like their mother. The daughters remain in the nest for a while, helping to feed their brothers and sisters in the larval stage. After fertilization, they set up nests of their own, but they have shared the nursery labor with their mother (Wheeler, 1928). Such sharing (rather than division) of labor is on a behavioral level only. In two genera of Australian Xylocopinae (*Allodapula* and *Exoneura*), among which colonies of two to several individuals are found, Michener (1963) noted a primitive division of labor. Some individuals were foragers, and others were egg layers. No external morphological differences between egg layers and workers were found, but a high percentage of workers had no sperms in the spermatheca and their ovaries were slender. Thus a difference in mating behavior has been evolved and is accompanied by an internal difference in ovarian development.

The same lack of external morphological differences is found in *Lasioglossum marginatum* (Brullé) (formerly called *Halictus marginatus* Brullé). Both a queen and a worker caste are found in this species, as evidenced by differences in mating behavior, longevity, and ovarian size, but, according to Plateaux-Quénu (1962), such caste differentiation occurs after emergence of the adult. Factors involved in the caste differentiation have not been identified.

The wasp *Polistes gallicus* L. (superfamily: Vespoidea) has a queen and a worker caste whose differentiation begins with factors acting in the larval stage. Nevertheless, further differentiation can occur among the individuals of the queen caste. A group of such individuals, all potential queens, after mating and overwintering, may together start a nest in the spring. A socially dominant individual becomes the true queen, and the other individuals become workers with a short life, small ovaries, and some worker behavioral characteristics; this is so even though they had been members of the queen caste. Ovarectomy experiments have shown that position in the social hierarchy has an effect on ovarian development and not vice versa (reviewed by Michener, 1961). This example is particularly interesting because of the likelihood that behavior itself is initiating or contributing a morphological and functional differentiation of the individuals in a social group.

Even though many of the Halictidae show no external morphological differentiation of castes, in *Lasioglossum malachurum* (Kirby) (formerly called *Halictus malachurus*), the evolution of labor division has gone beyond the behavioral stage and has entered a morphological stage. Two distinctly different kinds of daughters are produced. The longulus daughters (formerly called *Halictus longulus*) are produced until late in the season; these daughters establish no colonies and lay no eggs. Instead, they take over the food-collecting activities formerly done by the mother. At the end of the season, females of the queen type are produced, and they mate and overwinter (Wheeler, 1928).

Queen bumblebees are usually, but not always, larger than workers. Other external, morphological, caste differences are usually not to be seen, but great physiological and behavioral differences exist. The queen alone mates, overwinters, and founds a new colony in the spring. She is capable of performing all the colony's necessary work, but as soon as the first workers emerge and mature sufficiently to forage, the queen remains in the nest. A real division of labor exists henceforth among the members of the colony. There is a further division of labor among the workers, based on their size, the larger ones doing more of the foraging and the smaller ones more of the work in the nest (Free and Butler, 1959).

Division of labor reaches a peak among stingless bees (Kerr, 1950a; Kerr and Laidlaw, 1956) and honeybees. Queens and workers engage in completely different tasks. Since neither caste can do the work of the other, the life of either one alone is biologically meaningless, like the life of an organ separated from the body of a mammal. Like an organ also, neither queen nor worker can survive in nature apart from the colony for more than several hours or a few days at most. There is a further division of labor within the worker caste which is based on age, but this division is flexible and may be modified according to colony need. Extensive morphological differences support the behavioral division of labor on the caste level, and, at least in the honeybee, several glands wax and wane in synchrony with the successive tasks assumed by the workers.

Some of the caste specializations of the queen are mating behavior, well-developed reproductive system, stinging of other queens only, and production of queen substances. Some of the specializations of the workers are foraging behavior, poorly developed reproductive system, stinging in defense of the colony, production of wax and brood food, presence of honey stomach, and the presence of several modifications of the legs which adapt them for special tasks, most prominent of which is pollen collection. The evolutionary development of the pollen-collecting apparatus is considered in detail by Grinfel'd (1962).

Predigested or Secreted Food   An individual is usually responsible for digestion of the food it eats. Among highly social insects, much predigested or secreted food is utilized and a further division of labor brought about. The queen and larvae receive such food, and it has so profound an effect upon social development that use of predigested and secreted food is considered to be a fourth major factor in social evolution. A queen honeybee may produce eggs in one day equal to one-third to one-half her body weight. She is able to do this as a result of having received concentrated, ready-for-assimilation food produced by the glands of worker bees. Consequently bigger colonies are possible.

Larvae also receive glandularly secreted food for a part of the larval life. They receive honey and pollen during other parts of their life, but honey is a predigested food. Such food leads to rapid larval growth and emergence of adults in a minimum time. Consequently brood-nest space is available for another generation of bees, which again leads to the development of large populous colonies.

Honeybee colonies are not at their maximum population throughout the year but, instead, range from 10,000 or so up to perhaps 70,000 individuals. A maximum population is needed during the short period of blooming of plants from which nectar is obtained. Extensive brood rearing occurs just prior to this nectar flow, and a large population is developed. Use of highly nutritious food facilitates the development of such a population at the exact time to be of greatest benefit to the colony and of the least cost to it in total amount of food consumed.

Even though stingless bees practice mass provisioning of brood cells, glandularly produced food is placed in the cells. This food plus pollen and honey is placed in stratified layers so that food of the larva, even in a sealed cell, varies throughout the developmental period (Kerr and Laidlaw, 1956).

Bumblebees do not feed glandularly produced food, nor is there any feeding of one adult, queen or worker, by another (Free and Butler, 1959).

Food Storage (Hoarding)   A constant food supply throughout the year is necessary to the most highly developed social structure. For an animal that long ago became specialized on food available for only a few weeks in the summer, some problems are apparent. Food storage of a hoarding nature is essential as another bit of behavior—another factor in social life.

Nonsocial bees initiate each new generation during the blooming period of their food plants. Bumblebees maintain their nests all summer but store only enough food to carry the colony through a rainy or otherwise dearth period. *Apis mellifera* stores surplus food in amounts ranging from 50 or 75 pounds up to 500 pounds or more with commercial methods in some especially good beekeeping regions. *Apis indica* stores somewhat less honey than *A. mellifera*. *Apis florea* and *Apis dorsata* store a small surplus: an ounce or two and up to 30 or 40 pounds, respectively. The latter two are restricted to the tropics, as are the stingless bees also. Stingless-bee colonies store an appreciable amount of honey and live the year around.

Communication   Absolutely basic to social activity on every level is communication between individuals. At the most fundamental level, the facts of membership in the same species and readiness to mate must be communicated before reproduction can occur. Guarding a nest, whether by bumblebees, birds, or wolves, involves both intraspecific and interspecific communication. Defending a territory, whether by birds or by man in the suburbs, similarly involves communication. Cohesion of flock, herd, or pack requires communication. Alerting other members of a social unit about the presence of danger or the availability of food requires communication.

Systems of communication may be chemical, optical, auditory or mechanical in nature. Chemical communication, as well as being the most primitive, is also the most limited in variety of possible messages, since in general a separate substance is required for each message. One substance, however, may carry a different message to each of two different castes, two sexes, or two genotypes. Nevertheless, after optical and particularly after auditory and mechanical means of communication are evolved by a line of descent, the floodgates are open for elaborate developments.

The honeybee colony and its individuals, for efficient existence, employ several involved systems of communication. Such communicatory systems have evolved concomitantly with the evolutionary development of the previously discussed social-life factors. Some communication among bees depends on odor. The sex-attractant complex of the queen has been discussed under the section on mating behavior. Scent of the queen is effective in three other ways in the normal bee colony: It is an attractant of worker bees (Gary, 1961; Velthuis and van Es, 1964); it inhibits queen rearing in normal queen-right colonies (Butler, 1961a); and it inhibits oogenesis in worker bees (Butler and Fairey, 1963).

Scents, however, are only partially responsible for the inhibition of oogenesis in workers and the inhibition of queen-cell construction by them. Queen pheromone complex, the most active component of which is 9-oxodec-trans-2-enoic acid, produced in the mandibular glands of the queen and ingested by workers, is important in both phenomena (Butler, 1959; Butler, Callow, and Johnston, 1959; Butler and Fairey, 1963; and Callow, Chapman, and Paton, 1964). (The term "pheromone" designates a substance secreted by an animal which transmits information to another animal.)

Scents produced by worker bees are also effective in communication. The product of Nassanoff's gland located on the seventh abdominal tergum of the worker produces an odor which calls other worker bees to the location of its source. Such a pheromone functions in the clustering of dispersed bees and in recruiting other workers to a rich source of food. Most of the activity in the Nassanoff pheromone is due to the presence of geraniol and both nerolic and geranic acids (Boch and Shearer, 1962, 1964; Free, 1962). Boch and Shearer (1963) have also shown that newly emerged bees produce no geraniol but production increases strikingly at about the time bees begin field activities.

Many years ago von Frisch and Rösch (1926) reported that foraging bees preferentially attracted other bees from their own colony to the foraging area. Therefore a specific colony odor must be present in addition to Nassanoff's pheromone, which is not specific to colony or race. Kalmus and Ribbands (1952) confirmed and amplified the earlier work and then concluded that food sharing among workers (makeup of a colony's food is probably different from any other) and metabolic differences between colonies led to specific colony odors being carried by individual bees.

Such specific colony odors become even more important in recognition of hive mates by guard bees stationed at the entrance (Butler and Free, 1952). If stinging of an intruder of any kind (insect or man) occurs, another odoriferous pheromone comes into play. The sting carries volatile substances which incite other bees to sting. Boch, Shearer, and Stone (1962) have identified iso-amyl acetate as one of the active components of the sting pheromone. The area of chemical communication by bees is under active investigation. The whole field of chemical communication among animals has been carefully and creatively reviewed by Wilson and Bossert (1963).

In addition to recognition of foreigners by odor at the hive entrance, guard bees utilize optical signals to recognize out-and-out robber bees, according to the conclusions of Butler and Free (1952) and Lecomte (1951). It is generally recognized that robber bees display characteristic behavior when approaching a colony. They fly back and forth until an unguarded opening is perceived among the guards at the entrance. A returning forager entering the incorrect hive does so directly with no hesitation whatsoever. The large eyes of the drones may reasonably be supposed to function (along with the sense of smell) in locating mates.

Mechanical means of communication are prominent in the dances of bees, which are too well known to justify extensive review here (von Frisch, 1950, 1955; Lindauer, 1961). It may, however, be stated briefly that a bee returning to the colony from a foraging area with a load of food performs a dance that conveys the information necessary to enable a hive mate to go to the foraging area that supplied the food. If the food is very near the colony, the round dance is performed; if far from the colony, the waggle dance is done (Figure 5.6). The round dance conveys no information on direction, whereas the waggle dance translates the angle between the food source and the sun into an angle on the comb surface between the direction of the waggle run and the vertical. Straight up on the comb surface means straight toward the sun, and any angle to the right or left of straight up means the corresponding angle to the right or left of the sun. The duration of the waggle run has been shown by von Frisch and Jander (1957) to indicate the distance to the food source. Both round- and waggle-dancing bees supply information about the odor of the food source by the odor clinging to their bodies or by the odor of the food in the honey stomach, a sample of which is given to bees after the dance.

It has been discovered recently and independently by Esch (1961) and Wenner (1962a) that sound production also conveys information during the bee's dance. Other types of information are also communicated by sound (Wenner, 1962b, 1964). It should be pointed out that sounds in bee communication apparently are transmitted as vibrations through the substrate; there is as yet no evidence that bees are sensitive to airborne vibrations, but further research may develop new information.

Bee dances are also utilized by scout bees, which hunt for a new nesting site in the process of swarming, to reveal the location of such sites. The vigor of the dance and the perseverance of the dancer reveal something about the quality of the potential nesting site, just as the dance concerning a nectar or pollen source reveals something about its quality. The bees of the swarm engage in a decision-making process when scout bees report several possible nesting sites of differing quality (Lindauer, 1961).

Evolutionary development of this almost fantastic informational dance of the Western honeybee may perhaps be glimpsed by noting which of its component parts are present in contemporary bee species or what bases for its development may lie full of potential but uncalled for in other species that do not now and never have needed such a system of communication.

Looking first at *Apis mellifera's* closest relative, *Apis indica*, one finds a dance that differs no more from that of *mellifera* than one race of *mellifera* differs from that of another (Lindauer, 1961). (Racial differences will be discussed on page 94.)

Apis dorsata, although needing further investigation, seems to dance on the vertical side of its single comb only where the sky can be seen (Lindauer, 1957a, 1961). *A. dorsata* engages in both round and waggle dances.

Apis florea, a more primitive bee, performs a more primitive dance. The workers of this species do not dance on the vertical side of the comb but go instead to a special little horizontal dance platform on the top of the comb. Here a dance is performed the waggle run of which points directly toward the food source. No translation of solar direction to gravitational direction is made by *Apis florea*. This step has not been developed and apparently is not needed since a horizontal dance platform is always available on the top of the comb, which is built around a small limb where the bees can see the sky. *A. dorsata* often attaches its comb to an overhanging rock or other such place which excludes the possibility of having a horizontal dance platform. Even though both *A. mellifera* and *A. indica* dance on the vertical sides of their colony's many combs, they can dance on a horizontal surface such as the entrance board or one of their combs held horizontally if it is exposed to the sky.

The stingless bees, tribe Meliponini, are the nearest relatives of the tribe Apini, the honeybees. Their communication about food sources is less elaborate than that of honeybees. The most complex system discovered in the Meliponini involves use of scent from the food source, an alerting buzz when the forager enters the nest, zigzag runs in the nest which attract attention, scent marking of a trail between the food source and the nest by a substance from the mandibular glands, and guide bees which lead recruit bees to the food source. It seems that some Meliponini may release a great quantity of the mandibular scent at the food source and dispense with trail marking and guidance (Kerr, 1960). Furthermore, in 1965 Esch, Esch, and Kerr discovered that stingless bees use sound to convey information about distance to the food source. Still more primitive bees of this tribe use only scent of the nectar, alerting buzzes, and zigzag runs (Kerr, 1960) and give no information as to the location of the food source (*Trigona droryana;* Lindauer, 1961). Use of these three measures alone amounts only to alerting hive mates. The success of this alerting is not very great when compared with the success achieved by the guidance of recruits. Lindauer and Kerr (1960, or see Lindauer, 1961) provide comparative data which show that there are great differences in recruitment success by the various species (provided the colony populations, and thus the number of potential recruits, were about the same in the various colonies tested). The most primitive member of the Meliponini known, *Trigona silvestrii,* cannot alert its mates to look for an artificial source of food at all unless a scent has been added to the food (Kerr, 1960). The honeybee, which is able to communicate so much information, can get along without the scent.

Communication regarding food sources is nonexistent or almost so among bumblebees, the Bombini. Free and Butler (1959) state that several observers have watched returning foragers and have seen nothing to indicate any sort of communication about the site of a food source. Furthermore, a bumblebee feeding on sirup from a dish does not recruit any of its hive mates. At best, a forager, just returned, searching eagerly for a receptacle into which it can place its collection may excite other colony members to search for food.

Bumblebees in the field show a tendency to alight on those flowers on which other bees are already feeding. Possibly either scent or optical mechanisms are involved in this attraction.

Colonial life of bumblebees being what it is (a small colony which never overwinters in temperate zones), communication about food sources might actually work to the disadvantage of the colony. Its few foragers would concentrate on one plant species, which would soon finish blooming, instead of distributing themselves among a number of species and gaining the security of several food sources. There may have been selection against such communication in bumblebees.

They are not, however, without communication mechanisms. Males engage in flying circuitous routes which they scent-mark along the way.

In all probability such scent attracts virgin queens which are out of the nest and ready for mating (Free and Butler, 1959). Bumblebees are able to differentiate between colony mates and intruders although perhaps not as consistently as honeybees.

Anyone who has disturbed a bumblebee nest knows that disturbed bees immediately engage in an unusual buzzing. In addition to alerting the trespasser as to his whereabouts, this buzzing could also serve to alert and mobilize all bees in the nest for defense. Whether it does so or not is problematical, but discoveries of the significance of sounds and vibrations are appearing increasingly.

One can wonder about the phylogenetic origin of the movements involved in the dances of bees. What movements in a primitive ancestor were appropriated and enhanced by selection to become filled with meaning? Dethier (1957) describes some remarkable gyratory searching movements of the fly, *Phormia regina*, following stimulation by a drop of sugar sirup. After ingestion of the sugar sirup the fly, under constant illumination, engages in a series of clockwise and counterclockwise turnings as it searches around the area where the sirup was ingested. If the illumination is all from one side, the fly "dance" becomes oriented lengthwise, parallel to the light rays. If the fly performs in the dark on a vertical surface the dance becomes oriented to gravity, showing an up and down lengthening. The duration and intensity of the dance are positively correlated to the concentration of the sirup stimulus. These and other facts were compared with the facts of honeybee dances, and amazing parallelisms were brought to light by Dethier. It would seem that this fly, a solitary insect, has within its behavior patterns a sufficient basis upon which natural selection can act to produce the highly evolved dances of the honeybee. That this dance has become ritualized beyond a response to a mere physiological stimulus is a necessary conclusion if one is to explain dances reporting a possible nesting location.

Loss of the Factors of Sociality and the Development of Social Parasitism

Just as individual animals may be parasitized, a social unit itself is sometimes subjected to parasitic attack. Thousands of insect species, ranging from roaches to beetles, live on the social economy (rather than on an individual) in the nests of termite, ant, bee, and wasp colonies. The sort of social parasitism in which the parasite is not closely related taxonomically to its host (e.g., a parasitic beetle in an ant nest) will be passed over in this discussion. We are here concerned with cases in which a host colony is parasitized by a close relative. Such a phenomenon occurs among some ants, bees, and wasps. Michener and Michener (1951) state that in no group is such a wealth of closely related social parasites found as among the ants. There are no such parasitic cousins among termites, stingless bees, or honeybees, but among the bumblebees they are particularly prominent.

Parasitic bumblebees belong to the genus *Psithyrus*, and it is generally agreed that the several *Psithyrus* species originated from the genus *Bombus*. Both an enhancement of certain characteristics and a loss of others have occurred. *Psithyrus* species have lost pollen baskets as well as the tendencies to collect food, build a nest, or incubate brood. There is no worker caste among their progeny. They have gained a thicker, tougher cuticle (protective body covering), more powerful sting, and more pointed mandibles with heavier musculature. Their body color closely resembles that of the host they parasitize.

The various *Psithyrus* species parasitize either one or a few *Bombus* species. Behavior of the parasitic female varies upon entry of the *Bombus* nest. She may be quiet and retiring or outright aggressive. The Bombus female may be killed or permitted to live, but in the latter case it seems that the *Psithyrus* female eats the *Bombus* eggs, not allowing any of them to hatch. In either case, the workers already present are sufficient to ensure the maturation of a number of parasitic males and females.

One of the interesting means of defense against *Psithyrus* invaders involves nonviolent tactics. When a *Bombus fervidus* host colony is invaded by *Psithyrus laboriosus,* workers proceed cautiously to daub droplets of honey on the stranger instead of promptly stinging it as they do most other insect invaders (Plath, 1934). The honey technique works since the invader, getting wetter and wetter, soon leaves the nest.

It is of some interest that a considerable amount of attempted nest invasion occurs within bumblebee species. Queens that emerge late from hibernation may attempt to take over an established nest. Such a fact is pertinent to the origin of parasitism. These *Psithyrus-Bombus* relationships provide some extremely fascinating problems in social behavior and its evolution. Unfortunately there have been few recent studies in this area.

LEARNING

The normal, natural life of the honeybee is interlaced with both opportunities and necessities for learning. These include learning the location of the colony and its odor; the location of food from information given by a dancing bee; the color, odor, and shape of flowers; the location of nectar in a flower; and so on. This section will review most cases of learning that have been demonstrated in honeybees.

At the beginning of the present century, there were confusion and disagreement as to whether or not bees could recognize different colors. In a 1915 paper, von Frisch (1950) reported that they could distinguish certain colors. He conditioned bees to specific colors by putting a food dish, from which the bees collected sugar sirup, on colored cardboard. Later, when the food dish and the training cards were removed and replaced by fresh cards and empty glass dishes, the bees returned to the training color. From this and later work by others (reviewed by Ribbands, 1953; Daumer, 1956), it is known that bees do not see the color red but do see ultraviolet. In some regions of the spectrum visible to them, they are sensitive to small changes in wavelength, but in other regions they are not sensitive to much larger changes.

By similar techniques von Frisch and later workers demonstrated the bees' ability to see certain form or pattern differences, particularly the degree of "brokenness" of a pattern (von Frisch, 1950). This early work provided information about the sensory and learning capabilities of bees, which in turn provided a basis for subsequent, more naturalistic investigations.

In a colony possessing bees of all ages, and during a period of favorable flying weather, young bees between the ages of about 1 and 2 weeks can be expected to take their first flights. Unless they learn the location of their colony on this first flight, their only possibility for continued survival hinges upon their chance discovery of a colony that will accept drifters. In *Apis mellifera*, consequently, there is a premium on knowing the way to go home. If a colony of bees is taken several miles from its home location to a new location and if before any flight has occurred some bees are taken from the colony and released some hundreds of yards from the hive, experience indicates that none will return (Ribbands, 1953, based on Wolf). As the length of time bees are permitted to fly before removal from the hive increases, so does the bees' success in returning to their hive.

A considerable amount of investigation reviewed by Ribbands (1953) and by Lindauer (1961) reveals that bees learn the location of prominent landmarks relative to the location of their hive, they learn the appearance of their hive—its color and form—and the position of the entrance in the hive, and they learn the odor of their home colony. All these factors play a role in homecoming.

It is generally thought that displacement of the hive by a few feet from its original position (the one known by the bees) always results in the same phenomenon: Returning bees gather in the air around the old location. Some may alight on the ground or some other convenient object. Others continue flying around, obviously searching. And this search continues until the old hive has been located, or the bees drift into a foreign colony, or they settle near the old location and perish outside any colony.

Table 5.1
Return of bees to a new colony location at given distances from the previous location (modified from Free, 1958)

Distance moved	Whether empty hive at old site	No. of marked bees flew	Percent returned to new site
5 yards	Yes	17	100
5 yards	Yes	40	90
5 yards	No	46	98
5 yards	No	56	87
15 yards	Yes	176	69
15 yards	No	126	78
15 yards	No	257	96
¼ mile	Yes	174	76
¼ mile	No	145	85
¼ mile	No	55	85
1 mile	Yes	191	75
1 mile	No	126	92

Free (1958) investigated the above phenomenon experimentally. In a series of experiments he moved colonies containing some marked forager bees to new sites 5 yards, 15 yards, ¼ mile, and 1 mile from the original site. He noted the marked bees that flew from the new site and the proportion that successfully returned to it. The results are presented in Table 5.1. No other colonies were close to the original location in these experiments, but, as noted, an empty hive was placed on the original site in some instances.

Obviously most bees found their way home. Further data revealed that many bees visited the old site first and then went on to the new location of their colony. In the more distant moves, however, the presence of an empty hive in the old location appeared to be correlated with the return of a smaller percentage of bees to the new location.

Ribbands (1953) reviewed experiments by Kathariner (1903), von Frisch (1914), and Wolf (1926) designed to assess the influence of color on homing by bees. Colors were affixed to hive fronts for a time. Then colors were changed about and the effect on the bees noted. Wolf extended such experiments to include odor. Without question, both of these aspects of the environment play a role in the bee's return to its colony.

Ribbands and Speirs (1953) carried out a very clever investigation which determined how quickly bees would learn to return to the hive by way of a new entrance location, the old having been closed and the new one displaced in direction by 90 degrees and in height by several inches. In general, the bees reoriented quickly, and age had no effect on their ability to do so. The fact that they reoriented more quickly on the second displacement of the entrance than on the first suggests that bees may be capable of learning to learn (learning set: Harlow, 1949) which, if demonstrated to be true, opens up incredible possibilities. When the bees were confused by new circumstances, memory of a previous entrance location was sometimes demonstrated.

A different kind of displacement experiment demonstrates that bees utilize for homing other factors in the environment besides landmarks, the hive, and its odor. Wolf (1927) trained bees to a feeding place 150 meters north of the colony. These bees at the feeding place were then displaced to other locations as diagramed in Figure 5.5; 50 were taken 150 meters west of the colony, 50 to a point 150 meters east, and 50 to a point 150 meters south of the colony. Bees so displaced were observed to fly south, which was no longer the correct direction to the hive. The time required for each bee to return to its colony was recorded and the average obtained as follows: west, 102 seconds; east, 88 seconds; south, 168 seconds. The controls from the feeding place to the north required only 32 seconds to return to the colony. Wolf's results suggest that the bees remembered the direction of their outward flight relative to the sun and simply reversed their direction on the flight homeward. After having flown the remembered distance, no hive was found, a fact which threw them back on other resources. Only then did memory of landmarks or sight of the colony come into play.

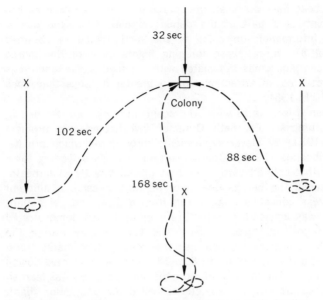

Figure 5.5 Homing by bees displaced from their feeding place (top) to left, right, and south of their colony. X marks the release spot, solid arrows the direction in which they were observed to fly, and broken arrows their supposed routes of return. Average number of seconds required for 50 bees to return is given.

Life for a young bee inside the hive includes a great deal of movement over the combs and around the inside of the nest. In the course of such movement, diverse stimuli are encountered, and various responses are made: Brood is fed, comb is built, foragers are relieved of nectar loads, etc. For a time the view prevailed that in a normal colony there was a somewhat rigid time sequence for the performance of each of these duties in the life of each bee. Presently the predominant interpretation holds that colony need is the primary determinant of individual bee activity. Wenner (1961) suggested that division of labor in the colony may be understood as a Markov process.

At any rate, after the period of hive duties, most bees engage in foraging. Before going to the field, however, most bees follow dancers who report the distance, direction, and scent of such foods. This information is received and acted upon, and the new forager finds the source of food reported by the dance and recognizes it by the scent.

Lindauer (1953) showed that bees who had never seen a dance nor associated with experienced foragers could perform accurate dances in the colony. The dances, consequently, are not learned in any usual sense of the word. Ability to follow a dancer closely and accurately, however, appears to be learned. This must be practiced before a high degree of skill is developed and the complete message received.

As the bee engages in its first foraging flight, other learning occurs.

Although the bees have not seen the source of food, they have experienced its scent as a part of the dance complex. From the dance, however, no information on color of the food source is obtained (Ribbands, 1953). During these foraging flights color of the source (as well as scent) becomes associated with the food. Color, once it is learned, operates as an attractant from a greater distance than does scent (von Frisch, 1950).

The question arises as to when the scent and color are learned in the foraging process. Elizabeth Opfinger investigated this problem carefully (in 1931) by clever experimental procedures which are described by Ribbands (1953). Opfinger arranged a glass feeding table supported so that colored cards could be placed directly underneath. She could expose the bee to one color on its approach, a different one while it was collecting sugar sirup from a dish, and yet a third color while it was engaged in departure circling. It was demonstrated that the color of the feeding place was learned only during the approach flight. She demonstrated also that other optical marks (form and pattern) at a distance of up to 8 to 14 inches were learned during the approach flight. The location of more distant landmarks was learned during the departure circling, sometimes called the orientation flight. Other similar experiments (Opfinger, 1949) indicated that the bees learned the scent of the feeding place also on the approach flight.

As already discussed, bees are sensitive to form differences. The recent work of Leppik (1953, 1964) indicates to what lengths this ability may extend. Bees seem able to distinguish between certain differences in petal numbers on flowers.

Learning is not finished when the food-yielding flower has been found. Von Frisch (1955) describes the process by which a bee learns, when first alighting on a flower, where the nectar is located. She probes here and there with her proboscis until she locates the nectar droplets. After several visits to the same species of flower a bee puts her proboscis directly to the region of the nectaries.

Reinhardt's (1952) study provides excellent data on honeybee learning in a natural situation: foraging on alfalfa flowers. The alfalfa flower is a somewhat complicated mechanism. Bees can obtain nectar by working the flowers from any of three frontal positions or by working through several variations of a side position. The frontal positions frequently lead to tripping of the flower which usually results in the flower parts striking the bee's head and closing the space occupied by the proboscis, trapping the bee momentarily. The side positions are more difficult initially for the bee, because the flower is not directly open from the side, but by use of the side position the bee is not trapped by a tripped flower.

Reinhardt observed individually marked bees visiting alfalfa flowers in 144-square-foot cages. Fragmentary records only were obtained on some bees, but Reinhardt was able to observe 23 bees change from a frontal approach to the flower, with consequent tripping, to the side approach. Several individual records show the bee's performance on a long series of consecutive flower visits and support the hypothesis that learning is involved in the bee's collecting from the alfalfa flower.

If a bee is foraging for pollen the flower must be tripped to make the pollen available. Reinhardt says, "Frequently a pollen bee is seen to struggle momentarily, pushing on the standard petal and punching into the corolla as if in attempts to trip the flower." Not many bees were observed in pollen-gathering activity, and only fragmentary records were obtained on most. Two bees observed extensively showed significant increase in tripping efficiency on the second day.

In summary, inexperienced bees become more successful as experience is obtained in working alfalfa flowers. If working for nectar, many of them learn the side approach and avoid being trapped when the flower is tripped. If working for pollen, they learn to trip more frequently and perhaps learn to extricate themselves from the flower trap more easily. Much other material of great interest is included in Reinhardt's paper.

Weaver (1956, 1957a) investigated bee activity on hairy vetch (*Vicia villosa* Roth.) which also presents some difficulties to the bee. Here some bees are base workers and some are trippers. Weaver states, "The foraging method is learned and becomes fixed through success at foraging from a very few blossoms in one manner during almost random attempts to reach the nectar." When foraging conditions were such that large nectar loads were obtained quickly, more trippers than base workers were found in the field. When the nectar flow became poor, the ratio of trippers to base workers decreased sharply. The state of the nectar flow, therefore, seems to affect the two kinds of foragers differently.

It was demonstrated years ago that bees could learn at what time of day food was available and would visit the food source at this time and no other. Ribbands (1953) reviews time perception in the bee. Time perception seems tied to a 24-hour rhythm, for Beling was unable to train bees to a 19-hour rhythm and Wahl was unsuccessful in the attempt to train bees to a 48-hour rhythm. Wahl was able to train bees to come to two different feeding places each at a different time during the day. Renner (1960) presents recent experimental results (including trans-Atlantic displacement) on the basis of their time sense. Very recently, Taber (1964) considered the factors that influence the flight rhythm of drone honeybees.

Lastly there are somewhat artificial tests of bees' learning ability. Butler (1954) states that bees can be trained to follow a visual trail, such as a tape or a series of colored cards, to a food dish. If the trail is curved so that the food dish lies closer to the hive than the farthest point on the trail, "the bees continue to follow the trail for perhaps half an hour or so. Then they begin to fly home from the dish by the shortest possible route, ignoring the trail, but still continue to follow it on the outward journey" for another hour. By the end of this time, they will ignore the trail for the more direct route. If, now, the food dish is removed, many of the bees searching for it will begin to follow the trail again, another example of their recalling a former successful method when faced by new confusing circumstances.

In a series of experiments designed to locate the olfactory sense organs in honeybees, Frings (1944) obtained information on their learning to associate the odor of coumarin with food. Untrained bees did not extend their proboscises in the presence of this odor, but by additionally exposing their antennal and tarsal contact chemoreceptors to sugar solution and allowing them to feed briefly, an association between coumarin odor and proboscis extension was established. In one group of 28 bees, the association became established in from 2 to 15 trials, with the average at the seventh trial.

Essentially similar techniques were used by Takeda (1961) in an investigation conducted within the theoretical framework of classical conditioning. A generalization was made by the bees between two similar odors but not to a third dissimilar one. Experimental extinction, differentiation, and conditioned inhibition were demonstrated. All these were temporary and not retained until the following day. Thus, spontaneous recovery was observed. Takeda (1961) stated:

☐ At all events, the present experiments on olfactory conditioned responses showed the function of the CNS in the honey bee and also showed that the various phases of the conditioned reflex are very similar to those shown by higher mammals. . . . The question of homology of the underlying mechanisms of the conditioned reflex in mammal and insect is a subject for future research.

Also, there are four other studies that report research on conditioning in bees (Wenner and Johnson, 1966; Johnson and Wenner, 1966; Pessotti, 1963, 1964).

Only a little work has been done with bees in mazes. Kalmus (1937) investigated a relatively simple maze. Weiss (1953) taught bees and wasps successfully to traverse mazes with several choice points provided with color differences. Lopatina and Chesnokova (1959) attached a simple, glass maze to a small hive. One of the alleys was illuminated with light passing through a variety of different filters arranged along the runway, and only this alley was supplied with food. The bees learned to find the food but could not find their way back to the hive if the color chain was removed. There is some difference of opinion as to the meaning of this and other breakdowns in behavior brought about by the removal of the colored lights (Schneirla, 1962). Considerable work has been done with ants in mazes, and ants' accomplishments are compared with those of rats in Schneirla's paper.

It has been shown that bees dancing in the hive for some hours (marathon dancers) change the angle that the straight run of the waggle dance makes with the vertical, in accordance with the passage of time (Lindauer, 1961). To do this correctly the apparent direction of movement of the sun in the sky must be known, and this apparent movement differs in the Northern and Southern Hemispheres, being clockwise from a vantage point in the north and counterclockwise in the south. Lindauer (1960, 1961) concluded that this "knowledge of the sun's direction of movement" is learned.

Only a small amount of attention has been given to the duration of a learned response, or length of memory, in bees. Butler (1949) describes a case in which memory seemed to be retained during the winter. The fountain from which bees collected water during the summer was taken down in each autumn of several years and stored during the winter. Each of several springs, bees were seen flying around the location of the fountain during the previous summer. As soon as the fountain was reestablished, these bees began to collect water from it. Butler concluded that these were bees that had survived the winter and remembered the position of the fountain where they had collected water during the previous autumn.

In another experiment, when colonies were returned to the former general area but not precisely to the previous colony location (after a 2-week stay in an area several miles distant), I have observed a large number of bees return precisely to the previous location of their colony. Unless colony odor somehow persisted on the ground in the former location, one would be inclined to hypothesize that memory for colony location persisted at least over this period of time.

Both of the above are observations, rather than well-designed experiments, and as such are highly suggestive but not definitive. The cases of the bees' return to an old entrance location and their return to a visual trail that they once followed to food when they were faced in each instance by confusing circumstances have already been mentioned. Some further information can be gleaned from conditioning experiments. Schneirla (1953) reviews an experiment by von Frisch in 1920 in which bees were trained to get sugar sirup from a blue box scented with tuberose. In test trials, conducted immediately after training, a tuberose-scented gray box was entered 146 times against 81 times for an unscented blue box. In a second experiment 5 days later, the unscented blue box was entered more frequently than the tuberose-scented gray box, which suggested that the memory for color was greater than the memory for scent.

BEHAVIORAL VARIATION
Differences in Learning Ability

Although intraspecific differences in learning ability are well known in some forms of animal life, it may be surprising to find such differences among honeybees. Lubbock (in 1875, according to Ribbands and Speirs, 1953) reported finding differences between the learning abilities of various individuals. In her investigation of the bee's learning of the scent associated with a foraging place, Opfinger (1949) had occasion to retrain bees to a different scent. There were great differences between individual bees in their retrainability to a scent different from the one they had learned initially. Ribbands and Speirs (1953) presented bees with a similar sort of learning problem insofar as retraining was concerned—in this instance retraining to a new location of the entrance to their hive. Some bees learned the new location by using it once, "most" by the end of several uses during the first day, "nearly all" by the end of the second day, but one bee was still returning to the location of the old entrance after 3 days. Age was found to have no effect upon the quickness with which a bee learned to return to a new entrance. The authors pointed out, "Several extrinsic factors are involved in any problem of orientation and there may be differences in the order and magnitude with which they impinge upon each individual." Such differences in impingement of extrinsic factors might very well be due to differences in heredity of the bee as well as to differences in its prior experience.

In a study of the ability of bees to learn a new location of their hive, Free (1958) noted, "The behavior of the individual bees varied greatly." Frings (1944) found individual differences in the rate of learning to extend the proboscis in the presence of the odor of coumarin (classical conditioning).

In another study involving bees carrying on their normal activities, Reinhardt found some evidence for variation in learning ability. Some bees quickly found and used a successful means of avoiding momentary entrapment by the alfalfa flower, but others failed. Reinhardt points out that the experimental conditions probably contributed to this but feels that these conditions are not the whole explanation. Even though the nature of the investigation precluded carefully controlled identical experiences for each bee, "experience appears not to be the only factor." One bee, for instance, was caught repeatedly by the trap. "Judging by her record, her capacity to learn or retain, doesn't compare well with that of A-4 or E-7" (two other bees).

In none of these cases of individual differences is anything known about hereditary variation in the population studied, and with the exception of age in one instance, nothing is known about the possible differences in environmental factors during the bee's life prior to testing. A rich field lies here awaiting investigation.

Differences in Other Behavior Patterns

It is common knowledge among those acquainted with honeybees that a tremendous range of variation in a wide variety of behavioral characteristics may be found within the species. Unfortunately very few quantitative data have been recorded to substantiate the observations. This does not apply, however, to all characteristics. Consequently this section will be variable in the extent to which its statements are supported.

Within the species *Apis mellifera* L., there are a large number of races. Ruttner (1963) divides these into three groups: European, Oriental, and African races, according to their geographical origin. The most popular races from a commercial standpoint are the Italian (from Italy), the Caucasian (from the Caucasus mountains), and the Carniolan (from the southeastern Alps and the northern Balkans). Although considerable variation in morphology distinguishes these and other races, the conspicuously outstanding variation is in behavior. Pure races are difficult to find outside the limited localities of their origin, but some of the variation resident in the whole group of races can be found in the more or less mixed populations in domestic use.

Brother Adam of England has observed the behavioral characteristics of the various races to a far greater extent than anyone else. His descriptions and commentaries are presented in three papers (Brother Adam, 1951, 1954b, 1964).

Flight Activity   Ribbands (1953) reviews the evidence that certain environmental factors such as nectar supply in the field, temperature, light intensity, wind, rain, and colony size influence flight activity. Hutson (1930) and Filmer (1931) obtained results indicating that their Caucasian bees flew at lower temperatures than their Italians. Hassanein and El-Banby (1960c) studied flight activity throughout the entire day during nectar flows from several different plants. They marked 70 individual bees of the Italian, Caucasian, and Carniolan races and introduced them when very young into a single colony. Records of the exits and returns of the individual bees were made. The results appear to show "that the Caucasian bees started foraging later in the morning, while the Carniolans were the earliest in foraging during the three flowering seasons"—citrus, clover, and cotton. During the citrus and cotton nectar flows, Caucasian and Carniolan races were more active than the Italian in the forenoon and less active in the afternoon. During the clover flowering period the three races seemed to be more nearly equal in activity.

Propolis Collection and Deposition   Propolis is a resinous substance collected from certain trees, mixed with more or less wax, and used to seal joints, cracks, and crevices in a beehive. Caucasians use propolis in quantities far beyond that of the other well-known races (Park, 1938; Ruttner, 1963). Brother Adam (1954b) states that the Egyptian bee uses no propolis and at least some strains of Carniolans use wax instead of propolis.

Nectar Collection and Honey Production   That there are tremendous differences in honey storing (hoarding behavior) cannot be doubted. Years ago, Dr. Otto Mackensen drew my attention to the differences between two of his inbred lines. Colonies of these two lines appeared identical in all obvious ways except that the combs of one were filled with honey and newly gathered nectar whereas the combs of the other were "dry." Similar observations have since been made repeatedly with inbred lines as well as other bees in our own laboratory.

In a 5-year test of three races, Park (1938) found average annual honey production of Caucasians to be 117 pounds; Italians, 146 pounds; and Carniolans, 173 pounds under the conditions of management followed. This result is in contrast to that of Corkins and Gilbert (1932) who found the particular strain of Caucasians with which they worked to produce 71 percent more honey than the average of Italians from several leading strains available in the United States. It must be kept in mind that honey production in such a test involves a complex resultant of many variable factors, many of which have nothing to do with simple industry in hoarding.

Hassanein and El-Banby (1960a) calculated from their results of honey production in relation to the number of bees in the colony that the Caucasian was the most industrious worker, the Italian least so, and the Carniolan between the other two. The important point is that much variation exists, probably within the so-called races under test, and it is largely uninvestigated for either theoretical or practical objectives.

Pollen Collection   As early as 1940 Todd and Bishop pointed out that "Pollen samples from colonies side by side may come from predominantly different sources. . . ." In a comparison of the number of pollen loads collected by two colonies from various sources, Synge (1947) presented striking quantitative evidence of the differences in pollen-collecting behavior. One colony collected twice as much from white Dutch clover but only two-thirds as much from red clover as the other one. There were far greater differences in amounts of pollen collected by the two colonies from less abundant plants. Whether or not these differences reflected pollen preferences or were merely chance attachments of bees of a colony to particular plant species in a particular location is an open question. Nevertheless, Colin G. Butler stated in a personal communication (Rothenbuhler, Gowen, and Park, 1953), "For instance there is no doubt that some of the strains with which we work tend to collect a higher proportion of red clover pollen."

Races of the honeybee also vary in the total amount of pollen collected (Butler, in above citation; Brother Adam, 1954a). Some bees collect far more pollen than others, disregarding the origin of the pollen.

The problem of pollen collection has at last been approached experimentally. Nye and Mackensen (1965) measured the alfalfa-pollen-collecting activity of a number of colonies in 1962. The three highest and the three lowest were selected for breeding, and from these, three high lines and three low lines were established in 1963. Colonies of each of these six lines were tested in five locations. A letter from Nye, considering also their recent results, states that the tests showed significant differences between some of the lines. Colonies headed by sister queens tended to collect alfalfa pollen to a similar degree as compared with those headed by unrelated queens. Certainly there is a strong suggestion of heritability of the tendency for high or low collection of alfalfa pollen.

Further studies of another generation in 1964 and retesting of some of the previously tested colonies lend further strong support to the concept of hereditary components. The importance of this work, both for behavior genetics of bees and for practical apiculture, can hardly be overestimated. Taken with other work on behavior genetics of honeybees, one can envision some of the possibilities with this species.

Brood Rearing   There is a great deal of observational evidence to indicate that some races begin brood rearing earlier in the spring than others or continue it later in the fall.

Corkins and Gilbert (1932) reported a 3-year comparison of Italians and Caucasians in Wyoming; it shows that Caucasians reared more brood than Italians before the nectar flow, but Italians reared more during and after the flow.

Hassanein and El-Banby (1960b) compared Caucasians, Carniolans, and Italians with respect to brood rearing under their conditions in Egypt. There the year can be divided into a fall and winter period, largely with a nectar dearth during which there is a very small amount of brood rearing by an average colony, and a spring and summer period with several nectar flows during which colonies engage in a great deal of brood rearing. The Caucasians tested reared less brood than the other races during each of the periods. Compared with the Italians, the Carniolans reared less brood during the blooming period but perhaps slightly more during the fall and winter period.

Swarming   Swarming is a part of reproduction by bees, and just as reproductive behavior is variable among other animals, it might be expected to be variable among bees. Such seems to be the case, although, again, the evidence is not all that is desired. A great mass of experience has established a reputation of much swarming for Carniolans and little swarming for Italians. A considerable amount of apicultural literature could be cited in support of these reputations. Park's data (1938) cover a 5-year record on 12 colonies of each race for each year. At all times efforts were made to control swarming (reduce or eliminate by reasonable efforts). Nevertheless the Carniolans averaged 3 swarms per year, the Caucasians 2, and the Italians 0.6. There was no question of a strong tendency to swarming in Carniolans.

Defensive Behavior   One thing, perhaps above all others, is known about honeybees: They sting. They sting as a means of defending the nest against intruders ranging from man to robber bees. There is tremendous variation, nevertheless, in the frequency with which a person may be stung by two colonies of bees under practically identical circumstances. In the writings of almost everyone who has written about races of bees, differences in temper are mentioned. Cyprians are notoriously cross, so much so that they are not kept commercially in the United States. Carniolans and Caucasians are gentle. Kerr and Araujo (1958) studied the African races of bees and found *Apis mellifera adansonii* Latr. to be very cross. These bees go beyond defense of the nest and actually drive bees of the Italian race away from flowers in the field.

Rothenbuhler (1964b) has measured defensive activity of several groups of bee colonies, which will be discussed on page 103. Briefly, only one sting was received from a group of seven Van Scoy colonies, whereas a total of 143 stings were received from seven Brown colonies under similar conditions. This experiment presented evidence of wide variation in defensive behavior by two genetically different lines.

Nest-cleaning Behavior   Although most larvae in a normal colony survive the developmental stages and emerge as adults from their cells in the combs, an occasional larva may die. In certain cases of disease—American foulbrood, European foulbrood, and sacbrood, for instance—many larvae may die. During investigations of resistance to

American foulbrood (causative agent: *Bacillus larvae* White) it became clear that bees differ in their responses to dead brood.

In this resistance investigation, Park tested colonies by inserting a piece of comb containing many larvae killed by American foulbrood into their brood nests. The various colonies reacted differently (Park, 1936, 1937) to this inoculation. Some colonies merely cleaned out the dead larvae, some removed the dead larvae and cut the cell walls down to the comb's midrib which separated the two sides of the comb, and a third actually removed a part of the midrib, presumably after doing the above. The basis of this variation was not determined although Park attempted to relate it to levels of disease resistance. Another of his experiments compared the bees' reaction, in eight different colonies, to comb inserts with empty cells and inserts with American-foulbrood-killed larvae. Pieces of comb used as inserts came from colonies other than the ones tested so that all comb was foreign to the colony under test. In every case the insert containing dead larvae was extensively torn down and removed, but the other piece of comb with no dead larvae was simply fastened into place and the torn cells were repaired.

An important advance was made by Woodrow and Holst (1942) who compared the behavior of one resistant and one susceptible colony toward brood individually inoculated with spores of *Bacillus larvae*. About 200 inoculated larvae were tested in each colony. About two-thirds of the inoculated larvae were killed in each case. In the case of the resistant colony, all the dead larvae were removed from their cells several days before the end of the normal larval developmental period. The susceptible colony, on the other hand, although it removed one or more larvae on every day of observation, retained 39 dead larvae in its cells at the end of the experiment. A difference in rate of removal of dead larvae in these two colonies was demonstrated.

Many years later evidence for similar differences in behavior associated with genetically distinct inbred lines was presented (Rothenbuhler, 1964a). Two lines had been selected for resistance and two lines for susceptibility to American foulbrood. Both resistant lines removed foulbrood-killed brood quickly, and both susceptible lines allowed most of it to remain in the combs until the end of the experiment. Further investigation of nest-cleaning behavior will be considered on page 98.

Dancing Behavior　　　Lastly and best known is variation in dancing behavior. Most of von Frisch's early work was with the Carniolan race. If the food source is near the hive, this race performs the round dance, which does not indicate the direction of a food source but simply that it is in the vicinity of the hive. When the food source is at a distance of about 85 meters, Carniolans engage in clear waggle dances, which indicate the direction of the food and also its distance.

In 1950, Tschumi, working with the Italian race, observed a different sort of dance. If the food source was between approximately 10 and 100 meters, this race engaged in a different dance, called the sickle dance. Hein (1950) observed sickle dances also by Dutch bees. That

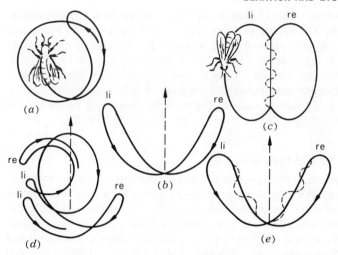

Figure 5.6 Types of dances. (*a*) Round dance; (*b*) sickle dance; (*c*) waggle dance; (*d*) and (*e*) transitional dances. (*From Boch, 1957.*)

there were racial differences in dances was further established by Baltzer (1952) in a study of Italian and German bees. The form of these dances as well as the form of transitional dances between them are shown in Figure 5.6.

In a detailed extensive study Boch (1957) examined the dances of six different races of *Apis mellifera*: Carniolan (*carnica*), German (*mellifica*), Punic (*intermissa*), Caucasian (*caucasica*), Italian (*ligustica*), and Egyptian (*fasciata*). Figure 5.7 presents his findings. It can be seen that all races studied except the Carniolan perform sickle dances. The transition from the round dance to some other type occurs at a different distance in each case. There are also differences among the various races with respect to the distance of the food source reported by the waggle dance. The races differ in the number of complete dance cycles

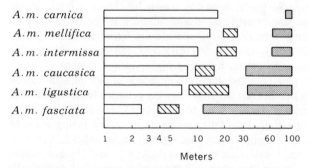

Figure 5.7 Racial differences in bee dances. White = round dance; crosshatched = sickle dance; gray = waggle dance. Space between the bars indicates transitional dances. (*From Boch, 1957.*)

they perform per unit of time to report a source at a given distance. For example, Carniolans perform 9.8 ± 0.19 cycles per 15 seconds if the food is 100 yards distant, whereas the Egyptian bee performs only 7.95 ± 0.13 cycles (Boch, 1957). Because of these racial differences in dancing behavior, bees of the various races misunderstand part of the information received from bees of different races.

Reviews in English of some of the variation in dancing behavior have been presented by Lindauer (1961) and von Frisch (1962).

METHODS OF GENETIC ANALYSIS

Efficient, well-controlled genetic experimentation with bees is absolutely dependent upon artificial insemination of queen bees. The first demonstrated successful technique was that of L. R. Watson (1928), but modifications and additions, principally by Nolan, Laidlaw, Mackensen, and Roberts, were made before the present high degree of success was obtained (Mackensen and Roberts, 1948; Laidlaw, 1949). Artificial insemination of bees is performed by the use of a microsyringe, a dissecting type of microscope, a queen-holding apparatus, and a carbon dioxide anesthetic. Usually 75 to 100 percent of the queens operated upon are successfully inseminated.

The second requirement for genetic analysis of honeybee colony behavior is genetic uniformity of the colony's worker bees. Unless all worker bees in each colony are genetically similar (with respect to the behavioral characteristic under study), one cannot classify colony phenotype in a genetically meaningful way. Major sources of heterogeneity in workers of a natural colony are (1) the heterozygosity of the queen (mother of the workers) and (2) her multiple mating with heterogeneous drones (fathers of the workers). Indeed, whether or not these same sources of heterogeneity exist is a question prerequisite to genetic investigation of any colony of individuals—bees, wasps, ants, or termites. Additionally one must know (3) whether one or several mothers are contributing to the colony's population.

Elimination of the above sources of heterogeneity in honeybees is comparatively easy. Normally, only one queen functions in each honeybee colony. By the use of artificial insemination, a single drone may be used for each insemination, and his sperms are expected to be genetically identical. Heterozygosity of the queen is the last problem; this can be reduced to a minimum by using highly inbred queens.

Inbred queens can be used for certain types of matings in colony behavior genetics but not for all that may be genetically desirable. For instance, inbred queens can be used to produce F_1 worker bees. A backcross of the F_1 to either parental line can be made by taking a drone (a gamete) from an F_1 queen and mating him to an inbred, parental-line queen. The reciprocal cross is not useful, i.e., the F_1 queen mated to a parental-line drone, because it produces a colony of genetically heterogeneous worker bees. Nevertheless the possible matings are sufficient to permit extensive genetic analysis. Haploidy of the drones is a fortunate circumstance, and since haploidy of males exists throughout the Hymenoptera, colony behavior genetics of many social Hymenoptera may be developed. Among termites, by contrast, both the male and female parent are diploid. In such a case, genetic analysis of whole-colony behavior cannot proceed beyond the F_1 generation, because the segregation in F_1 hybrid parents produces colonies of genetically heterogeneous individuals, impossible to classify phenotypically as a colony unit. The fact that more than one pair of reproductives is present in some colonies is a further complicating factor. If genetic analysis of termite-colony behavior is to be completely successful, some new approaches must be designed.

This *inbred-queen* X *single-drone* technique designed for genetic analysis of behavior in honeybee colonies (Rothenbuhler, 1960) has proved to be extremely powerful (Rothenbuhler, 1964b). Never in my experience, nor in that of my associate Victor C. Thompson, has such a variety of behavior patterns been seen in a group of bee colonies as that seen in a group of backcross colonies studied in 1958. The variation included, in addition to hygienic and stinging behavior to be discussed presently, nectar gathering, burr comb building, calmness of bees on combs, tendency to fly or drop to the ground when shaken from combs, and other characteristics. The technique seems certain to be as useful in the analysis of all the other variations in behavior discussed as it has been in two instances. It can provide colonies of genetically uniform individuals for many sorts of nongenetic behavioral experiments. For bee breeding, this technique provides a potent form of gamete selection. Its use in bee breeding would seem to have no immediately visible limits.

In addition to artificial insemination and the *inbred-queen* X *single-drone* technique, a third development of tremendous promise for analysis of behavior and behavior genetics of bees is appearing. This is the small-colony technique. Beekeepers have long used nucleus colonies to care for young queens until they have mated and are ovipositing. Scientists have also used small colonies to reduce expenditures of materials and time necessary for colony manipulation. The use of very small colonies to reduce the complexity of the study of events in a colony is a different thing. Success of the latter will depend upon how well small colonies simulate normal colonies with respect to the characteristic under study (Butler, 1961b). They may be very useful, nevertheless, in the initial development of theory which can then be tested in larger units.

Notably successful utilization of small colonies for research has been demonstrated by Lindauer (1952, 1953) and by Sakagami (1953) (see also Lindauer, 1961). Both these investigators studied division of labor. Sakagami used from 800 to 1,200 bees in his colonies; Lindauer, judging from his illustration, must have used a much smaller number, but this is not certain. In our laboratory, we have often used 200- to 400-bee colonies. Nelson and Jay (1964) are currently involved in studies of bee behavior in 100- to 300-bee colonies. An effort is under way by them to reduce the number further.

One of the problems encountered with small colonies or indeed any observation colony (with glass walls) is the inability of the bees to keep the comb warm, or their heavy clustering to do so. Our laboratory has designed a shelter which holds four observation hives. Temperature within the shelter is regulated by thermostatic control of electrical heating units and of exhaust fans (for cooling in the summer). These have been found to be completely satisfactory.

Flight rooms in which greater control of environmental conditions is obtained are being used to investigate behavior of honeybees (Renner, 1955; Nye, 1962; S. C. Jay, 1964) as well as that of other bees. How helpful such rooms really are remains to be seen.

Confinement of bees' flight space to the interior of a screen cage of from approximately 40,000 square feet down to a few square feet and up to 7 or 8 feet in height has been tried (Farrar, 1963). Like flight rooms, they will probably be useful for some experiments, but to what extent is as yet undetermined. We are currently testing cages approximately 8 by 8 by 16 feet in connection with our observation hive shelters.

GENETIC BASES OF CERTAIN BEHAVIOR DIFFERENCES

The first genetic analysis of a behavior difference among honeybees known to me involved nest-cleaning behavior (Rothenbuhler, 1958, 1964b). Two inbred lines, one selected for resistance to American foulbrood and one selected for susceptibility to American foulbrood, differed strikingly in behavior. The resistant line removed foulbrood-killed larvae from the brood nest mostly on the eighth, ninth, tenth, and eleventh days of larval-pupal life, whereas the susceptible line allowed most of such larvae to remain in the brood nest until the end of the experiments, usually on the fifteenth or sixteenth day. The difference between the responses of these two lines toward diseased brood is clearly shown in Figures 5.8 and 5.9. There is also a great difference in the response of

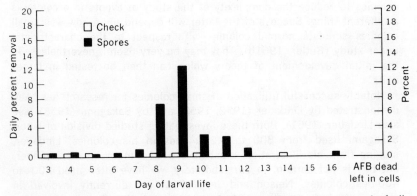

Figure 5.8   Behavior of three Brown colonies resulted in the removal of all American-foulbrood-killed (AFB) individuals before the end of the experiment. (*From Rothenbuhler, 1964b.*)

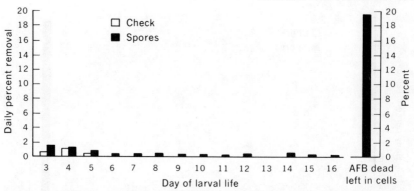

Figure 5.9 Behavior of four Van Scoy colonies left most foulbrood-killed individuals in the comb until the end of the experiment on the sixteenth day. (*From Rothenbuhler, 1964b.*)

these two lines toward hydrogen-cyanide-killed brood (Jones and Rothenbuhler, 1964). The differences in final result with cyanide-killed brood would seem to be a difference in rate of response by the two lines. Whether the difference between the two lines with respect to foulbrood-killed brood is more than a rate difference is unknown at present. The knowledge that a large clear difference existed seemed sufficient to justify genetic analysis, which was carried out in 1957 and 1958.

Before the results of the genetic analysis were published in detail, a considerable amount of time was devoted during the past few years to answering questions regarding the effects of certain environmental factors on nest cleaning. Does the number of dead individuals to be removed substantially alter the time required for removal? Experiments with numbers ranging from about 100 to 2,000 put into the same colony at different times suggests that variation in numbers, within limits imposed by colony population, does not lead to variation in the time required for removal (Jones and Rothenbuhler, 1964). Do differences in the average age of the bees comprising a colony lead to differences in removal time? Experiments over 2 years and involving 18 colonies, having bees of the same age to within 4 days, tested and retested as they progressed from about 5 days of age to about 66 days of age suggest that all bees up to about 28 days of age display normal nest-cleaning behavior (V. C. Thompson, 1964). Inasmuch as any colony in our usual tests would have a large number of young bees, age does not seem a likely complicating factor. Will a colony composed of approximately one-half resistant bees and one-half susceptible bees display nest-cleaning behavior? Such mixed colonies of bees have regularly performed like the control colonies composed entirely of hygienic bees (Trump, 1961). The same experiments indicate also that nonhygienic bees do not learn to be hygienic nor vice versa. What effect on removal does location of the dead brood in the combs of the colony have? Removal is slow or nonexistent in combs of the honey-store region (supers) but very fast in the brood-nest region (Borchers, 1964). What

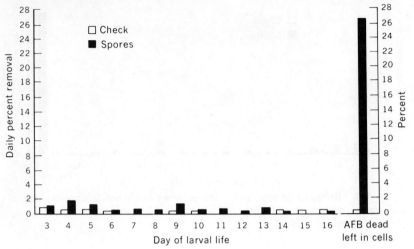

Figure 5.10   Behavior of five F_1 colonies resembled the behavior of the Van Scoys. (*From Rothenbuhler, 1964b.*)

effect does nectar flow have on hygienic behavior? A collection of observations and experiments suggests that a nectar flow (from flowers in the field or sugar sirup from an artificial feeder) has some effect. Incoming liquid food seems prerequisite to hygienic behavior (Rothenbuhler, 1959; V. C. Thompson, 1964; Borchers, 1964; Mourer, 1964).

These results increased our confidence that some ordinary environmental factor had not seriously affected our 1957–1958 results, which were published in 1964. Figure 5.10, which can be compared with Figures 5.8 and 5.9, shows that the gene or genes responsible for hygienic behavior are recessive. Figure 5.11 summarizes part of the

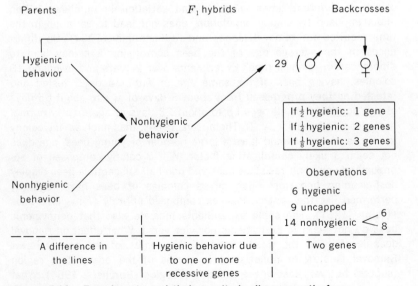

Figure 5.11   Experiments and their results in diagrammatic form.

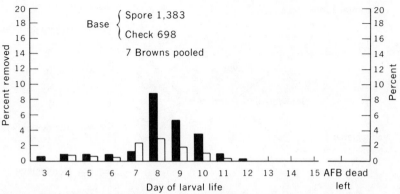

Figure 5.12   Data pooled from seven Brown-line colonies tested at the same time as the backcrosses.

experiment and shows that 29 drones from the F_1 generation of queens were backcrossed to queens from the hygienic line. The proportion of the 29 colonies resulting from these matings was expected to indicate the number of loci affecting hygienic behavior that were segregating in the F_1 queens. Along with this test of the 29 backcrosses, 8 backcrosses to the nonhygienic line, 7 hygienic-line colonies, and 7 nonhygienic-line colonies were tested. Every colony of the pure lines performed as expected, as can be seen in the pooled data presented in Figures 5.12 and 5.13. One of the backcrosses to the nonhygienic line showed hygienic behavior, and we are at a loss to explain this result, especially when no F_1 colony has shown hygienic behavior. The data from each backcross are presented in detail in the original paper (Rothenbuhler, 1964b).

The most striking and wholly unexpected results came from the 29 backcross colonies to the hygienic line. Six of them displayed typical or near-typical hygienic behavior. Nine of them uncapped but did not remove dead brood, leaving 14 as completely nonhygienic. By this time it became obvious that two loci were suggested by the 6-out-of-29

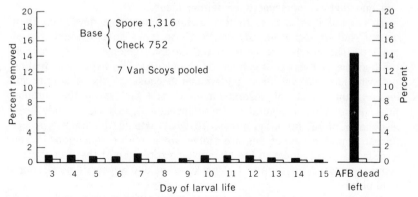

Figure 5.13   Data pooled from seven Van Scoy-line colonies tested at the same time as the backcrosses.

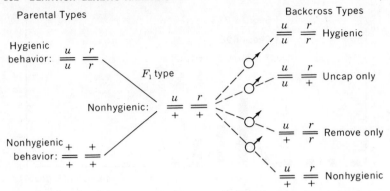

Figure 5.14   Genetic hypothesis offered in explanation of different responses to American-foulbrood-killed brood observed in 63 colonies of
bees. (*From Rothenbuhler, 1964b.*)

result and furthermore that one of these loci concerned uncapping.
Therefore the other must have been concerned with removal, and the
way to test this was to uncap the dead brood and return it to the
colony. Here the results were not decisive but ranged from 0 percent
removal in three cases to a high of 92 percent removal accomplished
in 2 days, when the experiment had to be terminated. Nevertheless,
the interpretation is that six can be classified as removers and eight
as nonremovers.

One emerges with the hypothesis in Figure 5.14. The difference
between the two lines in response to foulbrood-killed brood is due to
genetic differences at two loci, an uncapping locus and a removing locus.
Homozygosity for the recessive allele for uncapping, designated u, and
for the recessive allele for removing, designated r, characterizes the
hygienic line. The nonhygienic line is homozygous for dominant wild-
type alleles. From a backcross of the F_1 generation (by way of a drone
from an F_1 queen) any one of four kinds of colonies may develop:
one with worker bees homozygous for both the uncapping and removing
alleles, two that are homozygous at one or the other of the two loci,
and one that is homozygous at neither locus.

This hypothesis is easy to test. Future generations developed from
queens and drones from any one of these types should breed true at
the loci that are homozygous in the parental colony. If they do not,
the hypothesis must be modified or abandoned. The hypothesis-testing
program has not yet been undertaken because of the necessity for
investigating effects of several environmental factors on the parental
lines. Results from several of these investigations have already been
given and more are in progress. To what extents the phenotypes of
ancestral and F_1 genotypes are stable are important questions. When
knowledge with respect to the phenotypic expression of these genotypes
is reasonably adequate, a test of backcrosses and advanced generations
is in order.

Along with the genetic analysis of hygienic behavior, a partial analysis of the genetic basis of stinging behavior was possible. Counts were made of the number of stings received by the colony operator as he removed (and returned) each comb for inspection of brood present or lost each day in each colony. Such counts were made on the 7 hygienic control colonies (Brown), the 7 nonhygienic controls (Van Scoy), the 8 backcrosses to the nonhygienic line, and the 29 backcrosses to the hygienic line. Results are presented in Table 5.2. It can be seen that the Van Scoy colonies were nonstingers and the Brown colonies stung many times. The backcrosses to the Van Scoy line were gentle. The 29 backcrosses to the Brown line ranged from no stings to 23 stings per colony. Since only one of the six hygienic colonies among these backcrosses stung more than once, there is no evidence that stinging and hygienic behavior have a common genetic basis. (It has been thought that both these characteristics, which have often seemed to be associated, might be due simply to increased vigor in bees.)

That stinging behavior is determined in part by genetic differences seems clear, but the nature of the genetic differences between these two lines with respect to this characteristic is not clear. No separation into classes, as occurred with respect to hygienic behavior, was apparent.

The work on selection for alfalfa-pollen collection was discussed on page 92.

With so much intraspecific variation present in honeybees and with the availability of the necessary techniques for its genetic analysis, it seems that many more behavior-genetic analyses of this species will appear shortly.

Table 5.2
Frequency distribution of six groups of colonies according to the total number of times the beekeeper was stung while engaged in the same operations with each under similar conditions. Fourteen visits were made to each colony

Type of colony	Total number of stings																
	0	1	2	3	4	5	6	7	8	11	15	19	20	21	23	26	31
Seven Van Scoy colonies—none hygienic	6	1															
Seven Brown colonies—all hygienic										1	1	1	1	1		1	1
Eight colonies from backcrosses of F_1 to Van Scoy line—one hygienic	7	1															
Twenty-nine colonies from backcrosses of F_1 to Brown line—six hygienic	9	9	2	3	1	2		1	1						1		
The six hygienic colonies from the 29 backcrosses to the Brown line	2	3					1										
The one hygienic colony from the eight backcrosses to the Van Scoy line	1																

CONCLUDING STATEMENTS

Social life of an advanced complex kind has evolved in only two insect orders: the Isoptera and the Hymenoptera. All the Isoptera, the termites, are social. Most of the Hymenoptera live solitary lives, but all the ants, some wasps, and some bees live highly social lives. Of the six families of bees only the Apidae and Halictidae contain social species. Social life arose repeatedly in Halictidae and at least twice in Apidae (Xylocopinae and Apinae).

Such repeated occurrences (i.e., the rise of social life) in the Hymenoptera and the complete absence of complex social life in all but one of some two dozen other insect orders are striking facts. Hamilton (1964) has pointed out that haplodiploidy, which exists throughout the Hymenoptera, brings about a closer than usual genetic relationship among sisters, and that this situation may be a basic factor in the repeated evolution of social life in the order. In this connection, it is perhaps of basic importance that nearly all the division Aculeata prepare nests of some kind and stock them with food for the larvae. Very early, then, nest building as an item of maternal care was evolved by

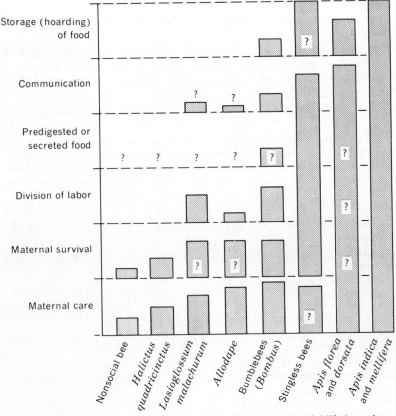

Figure 5.15   Comparative development of six factors of social life in various kinds of bees.

an ancestor of a large part of the Hymenoptera. The first nests may have been built at the food source. Maternal care was extended when behavior concerned with food placement in the larval cell was evolved. Taken together, these two aspects of maternal care went a long way toward providing the base from which further developments could occur.

From this primitive beginning of social life found in essentially non-social bees, it can be seen in Figure 5.15 that various degrees of social life have been evolved. *Halictus quadricinctus* live slightly longer —some days beyond the time required to build a few cells, stock them with food, and oviposit. This additional time is spent in guarding the nest. Whether or not the mother adds enzymes to the stored food or partially digests it is unknown. The *Lasioglossum malachurum* mother cares for the young bees until their emergence, survives presumably until the end of the summer, divides the labor of rearing further progeny with the first-emerged daughters who are morphologically distinct from the mother, and possibly engages in primitive communication with them. The only advance made by *Allodape* is progressive rather than mass feeding of larvae. There are no caste differences in *Allodape.* Bumblebees keep the brood cells warm, have extensive division of labor, communicate to some extent, although not about food sources, and store a few cells of excess food. Honey itself might be considered a predigested food, but nothing beyond this in the way of predigested or secreted food is used by bumblebees, unless they add enzymes to the larval food. Compared with *Apis mellifera* and *Apis indica,* both stingless bees and the other species of *Apis* use simpler mechanisms of communication. Stingless bees utilize mass provisioning. *A. dorsata* and *A. florea* store only a small amount of food, and at least *A. dorsata* moves from one ecological region to another throughout the year.

In the graph of Figure 5.15 such division of labor as is brought about by sex differences and the communication concerned with mating have been ignored. From the graph, two facts seem to be apparent:

1   A remarkable stepwise and gradual acquisition of these factors can be illustrated by the bees even though, as has been pointed out, this does not represent a phylogenetic line of descent. These are modern species which have achieved different levels of social organization.

2   The various factors of social life have not been emphasized equally in the various lines of descent. To illustrate this, stingless bees still use mass provisioning, even though their colonies are much larger social units, and store much more food than *Apis florea.* Although maternal care is not as extensive in *L. malachurum* as among *Allodape* (mass provisioning versus progressive), *L. malachurum* has much more advanced division of labor.

The social Hymenoptera as illustrated by honeybees learn a variety of things. Learning is part of their behavioral capacity and indeed necessary to their existence. When survival depends upon returning to the nest, one has to learn its location. This is only one instance of the necessity of learning.

There are strong indications that differences in learning ability exist in honeybees. These may be both genetically and environmentally induced. There are data and/or impressive observational evidence for differences in many other behavior patterns.

In spite of all the behavioral variations, only the barest start has been made in behavior genetics of this species. Investigation of great differences in nest-cleaning behavior has resulted in a two-locus hypothesis which is subject to much further testing. Stinging behavior was shown to be genetically independent of nest-cleaning behavior, and it seems to be genetically more complex. Selection for alfalfa-pollen collection has produced striking results, and further analyses are under way.

Development of behavior genetics of honeybees is certain to be of great economic value to agriculture and of equally great theoretical value to biology of behavior.

PART II GENES AND MECHANISMS

INTRODUCTION
Benson E. Ginsburg

No one seriously questions the proposition that there are hereditary contributions to behavioral potentials. In an evolutionary series one sees this with respect to behavioral adaptations. Within a species, even one such as the honeybee, where responses are largely built in and modulated by selection to rather narrow tolerances, important heritable variations in behavior occur on a simple genetic basis (see Rothenbuhler, this volume). Among mammals, where natural selection has favored greater flexibility of response to environmental situations, many experimental psychologists (and educators) have entirely ignored hereditary variation and have related performance exclusively to experiential history. Others have taken a more biological attack and demonstrated that the very flexibility of response is genotype-dependent and that there are genetic limits on styles and degrees of nature-nurture interactions.

Genetic effects on morphological features do not involve the serious complexities of defining a phenotype that often occurs with behavior. Whatever else a pigment gene may do, it has a clearly visible and consistent effect, so that at least this aspect of the phenotype of an animal of a given genotype is naturally defined. If the effect is modified in the presence of other genes (e.g., dilution genes) or under certain environmental conditions (e.g., as by cold in the Himalayan rabbit), this, too, is susceptible to straightforward analysis. Grüneberg, in *Animal Genetics and Medicine* (1957), listed a great many such structural mutants in mammals, some of which are of particular interest in behavior genetics, as they involve anomalies of the sense organs and the nervous system, varying from gross teratology to relatively discrete effects. These include pseudencephaly in the mouse, caused by a simple recessive gene as well as by chromosomal translocations in the offspring of irradiated animals; spina bifida in mice and rabbits; syringomyelia in rabbits; absence of corpus callosum in mice; rodless retina; hydrocephalus; microphthalmia; and many others. Grüneberg attempted to erect a pedigree of causes based on sequences of morphological events during differentiation as a model for a gene-action analysis. The inferential reasoning from such structural anomalies to behavior is uncomplicated. Obviously, mice with rodless retinas are visually defective.

The problem of what is a natural behavioral phenotype remains relatively uncomplicated when one leaves the above categories of genetic abnormalities to consider those in which the primary effect is obviously behavioral, occurring with little or no manipulation of the external environment, as in the case of waltzing mice and various ataxic mutants. Here again there is an overt aspect of phenotype that is well defined by nature. The distinction between genotype and phenotype is also fairly obvious, as in the case of pseudencephaly, which may result from a single genetic substitution or, alternatively, from a chromosomal translocation. Grüneberg lists four distinguishable genotypes for hydrocephaly and demonstrates that they operate on different portions of his pedigree of causes. In short, a population has various genetic ways of arriving at the same phenotype, even when it is a grossly abnormal one.

In the subtler and more interesting aspects of behavior, the natural phenotype is not so apparent. Very often its unity is provided by the concept in the mind of the investigator when he creates definitions of "intelligence," "emotionality," "aggressiveness," or "schizophrenia" and constructs instruments of measurement and/or diagnosis that achieve a level of general acceptance. Individuals vary in all these attributes, and it is easy to see how each can have positive or negative survival value, and therefore be subject to selective pressures, or show familial tendencies in pedigree analyses, thereby attesting to a genetic basis. There is, however, no assurance from this kind of analysis that we have matched the building blocks of nature (here conceived of as atomistic) with those of our conceptions. Among psychologists, only the ethologists have asked themselves this question: Is it any wonder, then, that by contrast with the traits that are clearly "given," such as those collated by Grüneberg, where relatively simple genetic models suffice, those contrived by the minds of the psychologists and the clinician are complex and polygenic? Whatever the correspondence between ideal natural units of behavior and entities such as intelligence or categories of mental disease, the latter have utility for us; even if they represent complexes rather than units, or mixtures rather than compounds, the biological analysis and refinement of these contrived categories are both important and possible to achieve.

In Part II, the first chapter, by Caspari, points out that behavioral capacities are as much an aspect of phenotype as are morphological variations and that both depend on processes which, though environmentally labile to varying degrees, are rooted firmly in the genotype. Current concepts of gene structure and gene action are reviewed and evaluated in relation to developmental processes in higher organisms. Relations between genotype and phenotype are discussed and illustrated by means of examples in which behavioral characters are presented as, in principle, not uniquely different from others, except that the environmental effects upon their expression may be broader. Finally, after a detailed review of current concepts of genetic coding and the relationship between genes and characters, the use of the genetic coding model as the biological basis of learning and memory is rigorously evaluated in the context of experimental evidence. Caspari's evaluation provides a basis for conceptualizing and understanding this approach, as well as a model of critical analysis from which the conclusion emerges that there is no evidence for memory molecules at present. The chapter points out that the coding and transfer of information in the nervous system have been conceptually visualized in several ways, which, in turn, can be combined to form more complex models. On the simplest hypothesis the critical events may occur either within the neurons or at the synapses.

In the last chapter of Part II, Eugene Roberts postulates a model in which the critical events occur at the synapses. These synaptic events are controlled by chemical transmitters that serve specialized excitatory and inhibitory functions. The synthesis, release, and catabolism of these substances are the critical components of the model. As these are under enzymatic control, the most likely mechanisms through which hereditary variability can be expressed would be through the genetic determination of the efficiency and balance of these enzymatic processes. Genetic, biochemical, and behavioral data compatible with this hypothesis are presented.

The second chapter deals with the identification and control of genetic variables through the use of (1) selection to phenotypic criteria, (2) Mendelian analyses, and (3) the use of isogenic stocks (including inbred strains and identical siblings). A gene-action approach is illustrated in which biological variabilty is simplified, and equivocal interpretations of the meaning of correlations are avoided through the use of both segregation and selection experiments in which correlated events that are not causally associated may be separated from each other. Population analyses of nature-nurture interactions in relation to early stress are presented. These indicate that thresholds of susceptibility as well as temporal, qualitative, and quantitative responses to early stress are all genotype-dependent.

The chapter by Hamburg reviews the mechanisms involved in psychological-stress reactions and relates them to data in which the genetic determination of various aspects of adrenal cortical hormone metabolism is, in a few selected cases, demonstrably under genetic control. In others, the biosynthetic and catabolic pathways are analyzed in a manner that will permit behavioral inferences, should genetic variability be found to be operable in these reactions.

The chapter on hereditary metabolic diseases, by Hsia, points out that behavioral phenotypes can occur as natural units in the metabolic realm on the basis of simple genetic determination, as in the case of phenylketonuria. It is of interest to note that here, again, simple genetic determination is associated with a unit that is a natural "given." Experimental phenylketonuria, which may be considered a phenocopy of the genetic condition, can be induced in animals and adds to the possibility of studying the mechanisms by which the behavioral effects of phenylketonuria are produced.

Genes affecting metabolic events, including localization of response thresholds to hormones, are cryptic, in the sense that they produce no grossly visible effects, nor do they usually have a one-to-one relationship to a given behavioral test. It is very likely that such metabolic events are, in themselves, the primary phenotypes and that the genetic variability of these metabolic mechanisms can be analyzed in the Mendelian fashion and related to secondary and tertiary events, including behavior.

As more of these hidden genes are discovered and analyzed in terms of their primary phenotypic effects at the molecular level, further analyses dealing with more complex aspects of phenotype can be simplified by putting the genes in question on constant genetic backgrounds, thus eliminating variability due to modifier genes. This restriction of variability can be achieved by repeated backcrossing to an inbred strain. In selected cases, the hidden genes can be followed after crosses without recourse to biochemical analyses by coupling them with obvious physical markers that are closely linked on the same chromosome. The opportunities are just beginning at this level, and this type of biological simplification and analysis may be the key to the neurobiology of the future. It provides the surest presently known method for controlling the otherwise confounding biological variability and for avoiding specious correlations, thereby making possible experimental analyses relating molecular events to organismal behavior.

The concepts and information presented in Part II are, therefore, at the hub of the wheel whose axis consists of the primary molecular events that represent the activities of the genes, and whose spokes represent supervening levels of organization whose rotation may be visualized as depicting interactions with environmental events ranging from the immediate cellular substrates at the hub to the more complex external environmental factors at the periphery. Finally, the entire wheel seen as an organism and its rotation seen as representing genotype-environment interactions trace a path, by means of behavior, toward an adaptive goal, in the company of other organisms with which they may be connected through social hierarchies, food chains, or other aspects of behavioral interdependence.

CHAPTER SIX
GENE ACTION AS APPLIED TO BEHAVIOR
Ernst Caspari

GENERAL THEORY OF GENE ACTION

The phenomena which we understand by the term behavior are phenotypic phenomena. As all phenotypes, they arise during the development of the individual, which is controlled by the action of the genes. It is therefore necessary in the context of this book to review briefly the theory of gene action as it has developed during the past decade.

It is generally agreed that the genic material is deoxyribose nucleic acid (DNA). The chemical structure of DNA has been clarified by Watson and Crick (1953), and the Watson-Crick model has become so generally known and accepted that it is necessary here only to allude to some of its most important features. DNA is a large chain molecule consisting of a helically coiled double strand. The skeleton of each strand consists of a regular alternation of five-carbon sugars (deoxyribose) and phosphates. The two strands are linked to each other by bases, each of which is attached to one of the sugar molecules. Four bases are found in animal DNA: two purines, adenine and guanine, and two pyrimidines, thymine and cytosine. For stereochemical reasons, the bases can be bound to each other only in specific ways: Adenine is combined with thymine, and guanine with cytosine. In this way, the arrangement of bases on one of the backbone chains strictly specifies the arrangement of the bases on the complementary chain. This complementarity of the two chains is assumed to account for the ability of a DNA molecule to replicate itself, one of the main characteristics postulated for genetic material: Each one of the two chains serves as a template for the synthesis of its complement and therefore of a new chain.

A second characteristic to be demanded of the genetic material is its ability to specify a phenotype; it must be assumed to carry in coded form the information necessary for the production of a phenotype. Since the pattern of the sugar-phosphate backbone of the DNA molecule is regular, the information must be assumed to reside in the pattern of purine and pyrimidine bases. In other words, the four bases constitute a four-letter alphabet in which all the information necessary for the development of an organism is coded.

It is well established that one of the important functions of DNA is the determination of the structure of specific protein molecules. Proteins are single-chain molecules consisting of large numbers of amino acids linked to each other. About 20 amino acids are known to occur in proteins. Each species of protein, as defined by a specific function, is characterized by a unique pattern of amino acid sequence. Actually, the total amino acid sequence is known for only a small number of proteins (e.g., human hemoglobin) since the methods used for its analysis are laborious and difficult. Nevertheless, it has been shown in a large number of cases that mutations in the genic material may lead to the replacement of a particular amino acid at a particular position in the protein chain by another amino acid. It is therefore generally accepted that the sequence of purine and pyrimidine bases in the DNA is the code for the sequence of amino acids in the protein.

Since there are only four purines and pyrimidines in the DNA molecule and about 20 amino acids, each amino acid must be represented by a combination of bases in the DNA. The question of the base representation of amino acids is known as the coding problem and has been one of the most active fields in molecular biology in recent years (reviewed by Lanni, 1964). In general, it is accepted that the code for every amino acid consists of a sequential and nonoverlapping triplet of purine and pyrimidine bases. Since triplets of four bases would result in 64 rather than 20 combinations, the code is said to be degenerate, many amino acids being represented by more than one triplet. Much progress has been made in identifying the triplets coding for particular amino acids, even though many problems still remain to be solved. The experimental work whereby the code has been broken consists in the use of artificially synthesized ribonucleotides in the place of messenger RNA and therefore represents primarily the code for messenger RNA (see also Cairns, 1966).

Since DNA is primarily involved in the determination of the amino acid sequence of proteins, the question of the transfer of information from DNA to the protein becomes the basic question of genic action. It is known that, in the process of information transfer, another nucleic acid, ribosenucleic acid (RNA), is involved. Ribosenucleic acid consists of nucleic acid chains, similar to DNA, in which, however, the sugar ribose takes the place of deoxyribose, and the pyrimidine thymine is replaced by uracil. By contrast with DNA, RNA is not a unitary type of molecule but exists in three types of compounds with differing structures and functions.

Transfer RNA

Transfer RNA consists of relatively small molecules, comprising about 75 to 100 nucleotide[1] residues. In ground-up centrifuged cells, it is found in the supernatant, nonparticulate fraction and is therefore known

---

[1] Purine or pyrimidine base bound to a sugar-phosphate group.

also as soluble RNA (s-RNA). It consists of a single chain of nucleotides which is, however, bent upon itself so that it is in part single-stranded and in part double-stranded like DNA. Complete determination of the nucleotide sequence of some specific transfer RNAs (e.g., alanine transfer RNA of yeast) has recently been achieved (Holley et al., 1965). Since it is known that the base sequence of s-RNA is dependent on genic DNA, this implies that the DNA nucleotide pattern for this particular gene has been established.

Soluble RNA has the particular function of carrying amino acids to the site of protein synthesis. It can, therefore, be combined with amino acids by means of specific enzymes; the energy for this process is supplied by the usual energy-supplying substance in biological reactions, adenosine triphosphate (ATP). The s-RNAs, as well as the enzymes involved in these reactions, are specific for the amino acid in question, so that at least 20 different s-RNAs must exist; actually the number is probably greater, so that a certain amount of degeneracy exists also at the transfer-RNA level. However, all transfer-RNAs possess the same terminal sequence of three nucleotides: adenine-cytosine-cytosine; it is the terminal adenine that reacts with the carboxyl (COOH) group of the amino acid. Further on in the s-RNA molecule there must be sequences that enable it to recognize the specific amino acid with which it can react.

Ribosomal RNA

Over 80 percent of the RNA found in the cytoplasm of cells is incorporated into small particles called ribosomes. They consist of about 60 percent RNA and 40 percent protein and can be clearly seen in electronmicrographs of animal cells as particles about 150 angstroms in diameter. They are frequently, though not always, arranged in regular rows on the surface of lipoprotein membranes, the so-called endoplasmic reticulum. The ribosomal RNA is a large molecule, consisting of about 1,000 nucleotides, and differs from s-RNA also in its base composition. By ultracentrifugation, several types of ribosomes of different size can be distinguished but there appear to be only two types of fundamental ribosomal particles; the others are apparently associations of these two.

Messenger RNA

Messenger RNA (m-RNA) is a large single-stranded RNA molecule which may be found in the nucleus and in the cytoplasm and is frequently combined with ribosomes. It is synthesized with great speed and has, at least in microbial organisms, a very short lifetime. Its pattern of nucleotides is complementary to that of one of the two DNA strands of a gene. This can be shown by separating the two strands of a DNA molecule from each other, by means of high temperature, and combining the resulting single-stranded DNA with m-RNA. After slow cooling, hybrid molecules are obtained, containing one strand of DNA and one of RNA. This is a very sensitive test for complementarity of base patterns in nucleic acids, since the pairing of single nucleotides with others is highly specific. It is therefore accepted that one of the DNA strands of a gene acts as a template for the formation of messenger RNA and that m-RNA is synthesized with the aid of an enzyme, RNA-polymerase, on the DNA template, using uracil (instead of thymine) to pair with adenine. RNA-polymerase can be purified and used in a reaction to synthesize single-stranded RNA *in vitro*.

Information transfer from DNA to protein proceeds, then, in two steps: *transcription* of the message from DNA to m-RNA, and *translation* from messenger RNA to protein. The transcription process proceeds in the nucleus and results in the formation of m-RNA molecules complementary to one of the DNA strands. In the process of translation, m-RNA, ribosomes, and s-RNA-amino acid complexes are involved. In this process, the ribosomes are present in the form of compounds of two particles each, one of which is smaller than the other. The m-RNA strand is attached to these ribosomal particles, probably at the border where the larger and the smaller particles adjoin. Since m-RNA is a long strand, it is usually attached to a number of ribosomal particles. Such ribosome—m-RNA complexes, looking somewhat like a string of beads, have been isolated from cells and seen under the electron microscope. The s-RNA—amino acid complexes become also attached to the ribosome—m-RNA complex. The m-RNA apparently selects the particular amino acid—s-RNA compound to be attached. It is usually assumed that the s-RNA "recognizes" a particular triplet of the m-RNA. The amino acids attached to the s-RNA are in this way brought into orderly alignment with each other and are linked to each other by peptide bonds. In this way the protein is synthesized, amino acid by amino acid, starting at the N-terminal end. The synthesis of one protein molecule of human hemoglobin takes about 90 seconds.

Many proteins function as enzymes, catalyzing particular biochemical reactions. If a gene specifying a particular enzyme molecule is changed, the protein chain constituting the enzyme will be accordingly changed. Changes in the genetic material, the DNA, are known as *mutations.* Mutations may be of a number of different types, but the simplest and least extensive is the replacement of one base in the DNA by another. This mutational change will be transmitted, by the usual mechanism of DNA replication, to the offspring of the mutated cell. Furthermore, the change in one nucleotide may change the meaning of the code word of which it forms a part; consequently, a wrong amino acid may become incorporated into the enzyme molecule. The result is frequently a protein molecule that cannot carry out its enzymatic function. If the product of the reaction catalyzed by the enzyme is necessary for the life of the organism, the organism will die unless the missing enzymatic product is supplied by the environment. In this case, the presence of an inactive protein related to the normal enzyme can frequently be demonstrated by serological techniques. Geneticists, for reasons of technical expediency, like to work with gene mutations that result in a completely inactive protein, leading to a complete block of an enzymatic reaction. It should not be assumed, however, that a completely inactive enzyme is the only possible consequence of an amino acid substitution. Cases exist in which the mutant enzyme has a lower activity than the normal enzyme (so-called "leaky" mutants) or in which the mutant enzyme is active only under specified environmental conditions, e.g., temperature-sensitive mutants. Finally, it may be assumed that there are mutations leading to altered proteins that have full enzymatic activity. This class of mutations would not be found experimentally, since the methods for finding mutations depend on the observation of phenotypic differences. Their existence is, however, indicated by the fact that homologous proteins in different species, e.g., the insulins of different mammals, usually show differences in their amino acid constitution, even though they carry out the same function.

It appears that a particular protein chain is determined by a single sequence of nucleotides in the DNA which is situated in a particular position in the genetic material, its *locus*. In a bacterial chromosome, the locus of a gene can be specified by its spatial relationship to other genes, whereas in an organism with several chromosomes the locus of a particular gene is specified by a specific chromosome or linkage group, and a particular position within this chromosome. A number of loci controlling particular enzymatic steps are known in microorganisms. It is a general rule that all mutations affecting a particular enzymatic step will be found to be situated at one locus only and that all mutations at a specified locus will affect the same enzymatic step, though frequently to different degrees. This principle is best documented for the bacteria *E. coli* and *S. typhimurium* and for the fungus *Neurospora*, in which a number of mutations controlling specific enzymatic steps are known. Its general validity for multicellular organisms is usually assumed but not strictly demonstrated. For in the genetically best investigated organism, the fly Drosophila, only about twelve cases of genic control of specific enzymes are known, while in man, where a number of genetic diseases due to protein alterations have been described, the linkage groups have not been worked out, so that the position of the genes cannot be determined.

Finally, it should be mentioned that not all the DNA is concerned with the production of m-RNA and the coding of proteins. Ribosomal RNA and the transfer RNAs are also determined by specific nucleotide sequences in the DNA. This has been demonstrated by means of DNA-RNA hybridization experiments, similar to those described earlier for messenger RNA. Other DNA sequences are concerned with the regulation of the activity of genes responsible for protein structure, a phenomenon which will be discussed in the next chapter (see also Jacob and Monod, 1963). Finally, it has been suggested from experiments with bacteriophages that some genes may be involved in the organization of cytoplasmic particles (Edgar and Epstein, 1965).

GENE ACTION IN HIGHER ORGANISMS

The principles of gene action described in the first part of this chapter have been mostly discovered by work on microorganisms. But they apply also to multicellular animals and plants. In many cases gene mutations have been found to result in abnormal enzymes and other proteins, and at least in human hemoglobins, in which a number of genetically altered forms have been found, many mutant hemoglobins are characterized by single amino acid substitutions. The process of information transfer in multicellular organisms is apparently similar to that in microorganisms. Ribosomes and transfer RNA are similar to those of bacteria; m-RNA arises in the nucleus, is transferred to the cytoplasm, but may have a longer life-span than that of microorganisms. DNA is accumulated in the chromosomes of the nucleus, but the structure of the chromosomes is certainly more complex than the simple DNA double strand found in bacteria and many viruses. The DNA of chromosomes is associated with chromosomal proteins, the most plentiful of which are the histones, a strongly basic type of protein. Furthermore, the chromosomes contain a certain amount of RNA, and the nucleolus, a structure formed at a particular chromosome, is very rich in RNA. But even though the structure of the chromosomes is more complex in higher organisms, all the evidence available indicates that the processes of transcription and translation are essentially identical with those in microorganisms.

One consequence of mutations should be mentioned; it is not peculiar to higher organisms but is more conspicuous and important the more complex the system. If an enzyme that catalyzes a reaction $A \rightarrow B$ is replaced by a mutant enzyme unable to catalyze the reaction, the reaction will be blocked. This block will have two consequences: The product B will be missing, and the substrate A will accumulate. Although the primary action of the gene is unitary, i.e., the determination of the structure of a protein, it will produce secondary phenotypic consequences. The phenomenon of a single gene's having multiple phenotypic effects is known as *pleiotropism* (see Caspari, 1952). The above example shows that a metabolic block will immediately result in two pleiotropic effects. Examples of pleiotropism of this type have been described for eye-color mutants in insects, but some of the best-known examples are phenylketonuria and galactosemia in man (see Chapter 9 by Hsia). These cases also demonstrate that the pleiotropic effects of these genes are not restricted to lack of product and accumulation of precursor but that both of them may have further biochemical and physiological effects, including effects on intelligence, a behavioral character.

The analysis of the mechanisms for pleiotropic gene action at the biochemical and developmental levels constitutes one of the most complicated tasks of developmental genetics. Interference by the stored precursor of a blocked enzymatic reaction with a second enzymatic reaction not directly influenced by the specific mutation involved constitutes a relatively simple biochemical model. The more generalized a character is, the more it is likely to be influenced by the pleiotropic effects of several genes. Some general physiological characters, such as viability and fertility, seem to be affected by the majority of genes. A single-gene mutation at a specific locus will, then, usually result in a number of secondary pleiotropic effects; conversely, any character of an organism, except the structure of specific proteins, will usually be affected by the action of numerous genes.

One further complication of gene action in higher organisms should be mentioned: Most multicellular animals and plants contain two complete sets of chromosomes, one derived from the father and one from the mother; they are "diploid." The possibility exists, therefore, that the genes derived from the father and from the mother at a particular locus, the two *alleles* at this locus, may be different from each other. An organism containing two different alleles at one locus is called "heterozygous" for this locus. For the phenotype of a heterozygous organism a number of terms are in use; they will now be described, since they are frequently used in behavior genetics.

It is possible that the heterozygote shows the characters appertaining to both alleles in the same cell. This phenomenon, called codominance, is frequently found for proteins and antigenic carbohydrates and seems to be characteristic for biochemical characters close to the gene. In hemoglobin abnormalities, for instance, the heterozygote often contains both the normal and the abnormal hemoglobin, though usually not in equal amounts. A second possibility is that in individual cells either one or the other allele is expressed. This situation, which was regarded as unusual in the past, may be more frequent than originally assumed. Thirdly, the heterozygote may be intermediate between the two homozygotes. The most frequent phenotypic interaction in heterozygotes consists, however, in the situation in which the heterozygote closely resembles, or may be phenotypically identical to, one of the homozygotes. This situation is known as dominance, and it is important to realize that the terms "dominant" and "recessive," which played a large role in the work of earlier geneticists and are still emphasized in some elementary genetics texts, are only terms describing one of the possible phenotypes of the heterozygote. Finally, overdominance or heterosis is sometimes observed, i.e., cases in which a heterozygote has a more extreme phenotype than either of the homozygotes.

The appearance of a heterozygote with respect to the different pleiotropic effects of the same pair of alleles may be different. Let us assume, for instance, that a gene blocking an enzymatic reaction leading to pigment formation is observed. The heterozygote may show codominance at the protein level, both active and inactive enzymes being present. If these proteins are studied by means of tests for enzymatic activity, intermediacy will be observed, simply because active and inactive proteins are present. At the level of pigment formation, dominance may be the result, since the reduced enzymatic activity may still be sufficient to form the full amount of normal pigment (see Egelhaaf and Caspari, 1960). Finally, the same alleles may have effects on viability or mating behavior, and for these characters dominance of one or the other allele, or even heterosis, may be observed (Caspari, 1950). It will be shown later that, even though the designation of the phenotypic appearance of the heterozygote is not very instructive from the point of view of gene action and developmental mechanisms, it becomes of primary importance when the evolutionary fate of a pair of alleles is considered.

THE ACTION OF GENES IN DEVELOPMENT

The phenomenon of development consists, in principle, in the fact that the cells of a multicellular organism, originally similar in composition and structure and presumably identical in their genetic constitution, become different from each other with regard to chemical constitution, morphological appearance, and function. The term "differentiation" is unfortunately used in two different senses: In classical embryology, it is an antonym to "determination" and designates the morphological changes undergone by a cell in its development. In modern cellular biology, the term designates the arising of chemical differences between originally identical cells during development. For example, an embryonic cell irreversibly determined to become a nerve cell in the later course of its development would be regarded as "undifferentiated" under the first definition, whereas it would be "differentiated" under the second, since a cell determined to become a nerve cell must have become in some way different from a cell determined to become an epidermal cell. In this chapter, the term "differentiation" will be used in the second sense.

The problem of gene action in development can be stated in the following terms: Different types of cells derived from the same egg cell produce different proteins; e.g., red blood cells contain mainly hemoglobin. The proteins are coded in the DNA. It must therefore be assumed that in some way different genes are active in different cells. The word "active" means, in this context, the production of a gene-dependent specific protein. In any particular cell, some genes must be turned on and others turned off; i.e., the proteins depending on these genes may or may not be formed.

Regulatory mechanisms which determine the activity of genes have been intensively studied in bacteria. The work of Jacob and Monod (1961) on the inductive enzyme β-galactosidase in $E.$ $coli$ is the principal case on which the theory is based; other cases of enzyme regulation found in bacteria can be accounted for by similar models. In wild-type $E.$ $coli$ the enzymes involved in the breakdown of lactose are "inducible"; i.e., they are formed only in the presence of an inducer substance (e.g., lactose) in the medium. The structure of the enzyme molecules is dependent on so-called *structural genes* which determine the amino acid sequence of the protein, as described in the previous part of this chapter. The inducibility, on the other hand, is determined by *regulatory genes*, which determine under which conditions the enzyme is formed. Mutations of regulatory genes may result in the protein's being formed in the absence of the inducer substance or in the inability of the cell to form any enzyme at all. Thorough investigation of this system has revealed the presence of a dual genetic control of enzyme production. One locus, the operator, is closely linked to the structural gene; it controls and probably initiates the transcription process. The second locus involved in the control is called the regulator; it may be located at a position removed from the structural gene and produces a *repressor substance* of unknown nature (it has been suggested that it may be a protein or a nucleic acid) which is released into the cytoplasm and blocks the action of the operator locus. The inducer substance removes the repressor substance by some reaction with it and, in this way, permits the production of m-RNA by the structural gene.

In this system, therefore, the coordinated activity of two loci, operator and regulator, controls the activity or inactivity of a structural-gene locus. Many aspects of cell differentiation can be described by invoking variations of similar control mechanisms. Models of this kind have been constructed by Jacob and Monod (1963) for a number of situations which may be assumed to occur in differentiation. However, it is difficult to substantiate any particular hypothesis for a case in a higher organism. One of the main difficulties appears to be the definition of criteria by which regulatory genes can be recognized and distinguished from structural genes in higher organisms.

Nevertheless, some cases exist in which regulatory-gene systems in multicellular organisms have been demonstrated. The controlling genes in corn, described by McClintock (1961), are the most thoroughly investigated examples. In these control systems, affecting well-investigated structural genes for pigment quality and quantity, starch composition, etc., two chromosomal elements are involved, one of which has to be located close to the structural gene while the other one may be situated on a different chromosome. The effects of these controlling elements are observed as patterns of pigment or starch formation on a developing kernel or plant. The controllers determine whether a particular cell and its descendants will or will not produce the substance (pigment or carbohydrate) which is formed under the influence of the structural gene, in other words, whether the gene is turned on or off. The similarity of this system, particularly with respect to the control by two loci, with the operator-regulator system of *E. coli* is striking. Evidence for the existence of regulatory genes has also been obtained for hemoglobin in man (Motulsky, 1962b and 1964).

Regulation of gene action at the cellular level has been the subject of several lines of research in the past few years. Some of them will be briefly mentioned here. It has been noted that certain regions in the giant salivary-gland chromosomes of Diptera show enlargements, so-called "puffs," in particular regions at different times of development. It is generally assumed that these regions correspond to particular genes and that the puffs indicate activity of these genes. This assumption has been proved for one special case (Beermann, 1961).

Some factors initiating this gene activation (puffing) have been studied by a number of authors (Clever, 1961). It has turned out that the appearance and disappearance of some of these puffs may be controlled by the hormones inducing metamorphosis. Clever observed that a few minutes after the injection of the metamorphosis hormone, ecdysone, a puff at a specific location appears; its time of persistence is dependent on the concentration of the hormone. Subsequently, puffs at other locations in the giant chromosomes appear at specific times after hormone injection. From these results it may be conjectured that the pupation hormone initiates the activation of one particular gene in the chromosome and that this gene, possibly by means of one of its reaction products, causes other genes to become activated.

The possibility that some of the hormones are involved in the activation of certain genes may also be indicated for mammals. Adrenal cortical hormones, for instance, may affect the activity of numerous enzymes in the liver of mammals; some become increased in their activity, while others become less active, with the same concentration of the hormone. These data are hard to evaluate, since changes in "activity" of an enzyme may involve changes in the rate of synthesis or in the rate of breakdown of the enzyme or, alternatively, an influence on the state of activity (activation or blocking) of an enzyme which does not change in its relative concentration. There are only a few cases in which a decision between these alternatives has been possible. The same difficulty applies to cases of higher animals in which the activity of an enzyme is increased as a result of increased supply of its substrate, e.g., the relationship reported between liver alcohol dehydrogenase and alcohol consumption (see McClearn, 1965, p. 802). These cases, which seem to resemble inductive enzyme formation in bacteria, may or may not be due to similar mechanisms. Different mechanisms for different cases seem to be indicated by the investigations that have been carried out.

At all events, the material on insects strongly suggests an effect of hormones on gene activation, and similar effects appear likely in vertebrates. Hormones are released into the circulation by ductless glands and frequently affect different types of cells in different ways. From a genetic point of view, they have an unspecific activity and may represent a regulatory mechanism which is superimposed on the intracellular regulation of genic activity. The hormones themselves are the products of biosynthetic processes and are, for this reason, dependent on the presence of gene-controlled enzymes (see Chapter 8 by Hamburg). Structural genes controlling an enzyme involved in hormone synthesis may therefore appear as regulatory from the point of view of structural genes controlled by the hormone. The difference between regulatory and structural genes in higher organisms is, therefore, to a certain degree a matter of definition. The controlling elements in corn, however, show certain characters that differ strongly from those of orthodox structural genes (McClintock, 1961) and must therefore be regarded as a different category of genes. In higher organisms, apparently there are several categories of genetic controlling mechanisms: first, controlling elements which determine the state of activity of genes directly and, secondly, structural genes which control the activity of other genes indirectly, by means of their reaction products, e.g., hormones. This picture agrees with Clever's conclusions on the activation of genes in the salivary glands, where the first puff, induced directly by the pupation hormone, is assumed to induce the later-appearing puffs on the chromosome.

In a recent discussion on the activity of genes in development, Paigen (Paigen and Ganschow, 1965) has postulated that, in addition to structural genes and regulatory genes, two more types of genes must be assumed to play a role: architectural genes, which are concerned with the organization of cellular organelles, and time-controlling genes (temporal genes). The evidence for the existence of architectural genes is scanty but suggestive. It is known that genes affecting eye pigments in insects affect not the biosynthetic steps in pigment formation but the presence or quality of the ribonucleoprotein granules on which the pigments are deposited. The production of bacterial viruses after infection may be regarded as the synthesis of rather complex cytoplasmic particles in bacteria. Besides the involvement of structural genes specifying the proteins present in the virus, it has been demonstrated that genes are necessary for assembling the different parts of the virus into a complete particle; for instance, genes are apparently needed to attach the head and tail of the bacteriophage to each other (Edgar and Epstein, 1965). In some instances of cytoplasmic particles the situation is more complex. The plastids of plants, and probably also the mitochondria, contain their own DNA and possess the ability for identical reproduction. These particles would be expected to show a certain degree of genetic autonomy but will not be completely independent of chromosomal genes, because some of their building blocks result from the activity of chromosomal genes. Such a complex interaction of cytoplasmic particles and nuclear genes has been clearly established for the inheritance of plastids and is suggested for mitochondria.

The problem of temporal genes is most difficult, since it is hard to imagine how a time-controlling mechanism, a genetic clock, could be coded in a nucleic acid chain. Nevertheless, evidence for the existence of such genes has been found in the experiments of McClintock (1961) on controlling elements in corn. The time at which the structural genes are turned on and off is controlled by the quantity and the qualitative state of the regulator elements ("activator"); one activator gene may cause an early switch of gene activity, two activators induce switches in the middle of development, while three activator elements (the tissue in question is triploid) induce very late switches.

These examples show that, in the course of differentiation, specific genes become activated and that this activation is under the influence of regulatory genes of different types. The mechanisms whereby regulatory genes control the action of structural genes have been investigated only for a few cases, and there is at present no reason to assume that all conform to the same pattern. It may very well be that different levels of control of gene activity have become superimposed on each other in the course of evolution.

The second set of problems with which developmental genetics is concerned centers on the question of how it is "decided" that a particular gene is turned on in particular cells but not in others. Generally the decision is made for groups of cells rather than for individual cells. The problem may therefore be designated as the problem of pattern formation. Kühn (1965) decided that three fundamental mechanisms in the determination of pattern development must be distinguished: unequal cell division, interaction between cells and cell groups of the same organism, and environmental influences.

Unequal cell division has already been mentioned in the references to McClintock's work in corn. Frequently, in the course of gene activation processes, there arise twin spots that consist of two adjacent regions of different genic expression. These must be interpreted as the results of an unequal cell division: The two products of a mitotic division assume different states of gene activation and transmit them to their daughter cells. Besides this chromosomal mechanism of unequal cell division, there is a second mechanism in which the two daughter nuclei of a division, though originally identical, come to lie in different cytoplasms. It may then be assumed that the cytoplasmic differences in the two cells give rise to different gene activities. This type of mechanism is assumed to play a role in the cleavage divisions in early development, particularly in some lower animals, such as snails and Tunicates. It has been thoroughly investigated for the development of bristles in flies, scales in moths, and stomata on the leaves of plants (Stebbins, 1965). It should be pointed out that, in all known cases where unequal cell division has been established as a means of differentiation of cells, a limited number of cells are involved in the process.

In classical embryology, the greatest emphasis has been put on the interactions of cells in development. These processes of pattern formation include directed growth processes, which result in differences between cells with respect to their position in the cell continuum; the interactions of groups of cells with each other, frequently by chemical means, as in the process of induction; and the self-organizing ability of groups of cells, known as "morphogenetic fields." It would be impossible at the present time to describe these phenomena in terms of the action of individual genes. The generalized picture may be formulated by stating that the interrelationships of the cells of an organism to each other, chemical as well as mechanical, lead to differentiation of the cells, presumably by the activation and inactivation of specific parts of the genome.

It is important to realize that the description of developmental processes due to cell interactions in genetic and molecular terms becomes very complex. There is no doubt that in many of these processes numerous proteins as well as smaller molecules are involved. Even though it is known that many genes interfere with specific developmental processes of this type, results of the analysis of such cases have usually been disappointing. In a case of genetically conditioned anophthalmia in the mouse, for example, Chase and Chase (1941) found that the primary effect of this condition is a delay in the rate of outgrowth of the optic cup from the brain, so that it approaches the epidermis at a time when the competence for lens formation has already passed. Numerous processes are involved in the organization of a normal eye, which are designated by the following terms: growth rate of the eye primordium, inductive activity of the eye cup on the epidermis, competence of the epidermis to react on the inductive stimulus of the eye cup with lens formation, and influence of the lens on the organization of the eye. The genetic condition described here interferes with one of these processes in a quantitative way. Every one of these processes is under the influence of a number of genes, as shown in Chase and Chase's example: The genetic condition investigated is not simple; one major gene and several modifiers seem to be involved. Cell interaction processes may therefore be assumed always to involve the action of a large number of genes, the coordination of their activities having been established in the evolutionary history of the species.

The third mechanism indicated by Kühn is the influence of environmental factors on differentiation and determination. This factor is usually neglected in textbooks of embryology, since it does not appear to be an important mechanism in normal animal development. Indeed, Kühn's own examples for this phenomenon are all taken from the plant kingdom. But it will be necessary here to keep this possibility in mind, since the behavior of animals shows rather striking examples of environmental determination of development.

GENE AND CHARACTER

In the preceding section we dealt mostly with processes under the control of individual genes. Only in the discussion of pattern formation due to interaction of cells was the influence of several genes on the same developmental process mentioned. If, instead of starting with individual genes obtained through spontaneous or induced mutations, one starts a genetic investigation with a character, defined by some measurement of the organisms in a population, one frequently finds individual variations and can investigate them by crossing experiments. In these crosses, one finds usually that the character in question is under the control of several genes; it shows *polygenic inheritance*. The only exception is variation with respect to characters that are in a one-to-one relationship with structural genes, such as specific proteins and blood-group antigens. This principle is true for animal populations as well as for human populations. In the latter case, rare hereditary diseases that are kept in the population by mutation frequently turn out to be due to single genes, while variation in other characters is usually due to polygenic systems.

Most investigations in behavior genetics have turned up polygenic differences. The reason is that they are based either on selection experiments, a method uniquely designed for the analysis of polygenes, or on the finding of behavioral differences in previously selected strains of rodents or breeds of dogs, in which a number of genes for which the original population was polymorphic had become homozygous through inbreeding. It should be emphasized that the frequent finding of polygenic differences with respect to behavioral characters is due to this method of investigation and is in no way restricted to behavioral characters. Single-gene mutations influencing behavioral characters are well known in man, e.g., phenylketonuria, amaurotic idiocy, etc., and are also known to occur in animals.

The principles of polygenic inheritance and the methods used in its analysis are discussed in other chapters (by Broadhurst, Bruell, and R. C. Roberts). Here it will only be mentioned that, while there are methodological differences in the investigation of unifactorial and polygenic inheritance, there are no differences in principle. It had been suspected that polygenes constitute a particular type of gene, different from the orthodox structural genes and located at particular regions on the chromosomes. There is no reason at present to uphold this difference: It has been possible to analyze the components of relatively simple polygenic systems; Rothenbuhler's (1964b) investigation of the hygienic behavior of the honeybee, which can be divided into two behavioral components each controlled by a single Mendelian gene, is a classical example from behavior genetics. Furthermore, many orthodox Mendelian genes have been shown to have pleiotropic effects on generalized physiological and behavioral characters; effects of mutants in insects on mating behavior have been particularly thoroughly investigated. In other words, whether a particular gene appears as a single gene or as a component of a polygenic system depends on the level of phenotypic character being investigated and on the method used for the isolation of the genes.

While there is, therefore, no difference between orthodox genes and polygenes in principle, there are certainly differences not only in the method of their investigation but also in the information that can be obtained from them. Particularly if information is desired concerning primary gene action, individual mutant genes offer the best chances for meaningful results, and actually all our knowledge of gene action at the molecular level is derived from mutations at individual loci. In polygenic systems, it is most likely that a large number of biochemical differences will be found which may interact with each other in complex ways. Furthermore, the same behavior phenotype may be produced on different genetic bases and may therefore be due to different biochemical conditions. It is no accident that in many cases the investigation of the biochemical basis of behavioral characters, which had been originally found as strain differences and were therefore due to a polygenic system, led to unclear and even contradictory results.

An example may explain this reasoning. Phenylketonuria in man leads to a defect in intelligence. If it is desired to investigate the biochemical basis of intelligence, this method has the advantage that the whole problem of the definition of intelligence can be avoided and that a biochemical mechanism interfering with normal intellectual functioning can be isolated. It would not be expected that the results would give a complete picture of human intellectual functioning, expressed in biochemical terms. On the other hand, we can select from a human population individuals with different IQs and try to find biochemical differences between them. It appears unlikely that a simple relationship would be found; whatever is found can be, at most, a statistical correlation of uncertain meaning.

While polygenic inheritance is, therefore, not particularly suitable for the investigation of gene action, it is of fundamental importance for our understanding of natural selection and the process of evolution. Natural selection does not usually favor isolated genes over their alleles but, rather, coadaptive gene complexes (Dobzhansky, 1955), i.e., systems of polygenes which in combination lead to superior adaptedness of their carriers. A relatively simple example is again the system of hygienic behavior in the bee (Rothenbuhler, 1964a and 1964b). Neither one of the two genes involved would by itself increase the probability of survival of the population, but in combination they are of obvious survival value in case of an infection with foulbrood.

Some remarks on heterosis with respect to behavioral characters should be made. The term "heterosis" is used for two different phenomena, both of which may result in superiority of heterozygotes over homozygotes. In inbreeding, frequently a loss of desirable characteristics is observed, due to recessive genes which were present in the original population and have become homozygous accidentally during the process of inbreeding. In two different inbred strains, different recessive genes become fixed, and in crosses between different strains the dominant alleles at these loci are reintroduced, resulting in an overall superiority of the hybrid over both original strains. On the other hand, heterozygote superiority with respect to survival value occurs frequently in wild populations without prior inbreeding, where it leads to genetic polymorphism (Dobzhansky, 1955). In most of these cases, coadapted gene complexes are involved; they are held together by chromosome aberrations inhibiting recombination between the genes constituting the coadaptive gene complex.

But some cases of heterosis due to overdominance in a single pair of alleles are known (Caspari, 1950). In the moth *Ephestia*, a pair of alleles determines the presence or absence of a particular component of the pigment in the testis, a biochemical character which by itself seems to be indifferent. But the two alleles have pleiotropic effects on viability and mating behavior, one allele causing increased survival, while the other in homozygous condition increases the mating probability of the males. In the heterozygote, both favorable pleiotropic characters are dominant, so that the selective value of the heterozygote is superior to that of either homozygote. It is in agreement with this finding that wild populations in America and Europe are polymorphic for this pair of alleles.

One aspect of heterosis is increased "developmental homeostasis" (Lerner, 1954). The fundamental observation is the following: For many characters, the phenotypic variability is higher in homozygotes than in heterozygotes. This is explained by the statement that the developmental system in heterozygotes is better buffered, i.e., less influenced by environmental factors, than in the homozygotes. This superior buffering of the development is regarded as another expression of heterosis. It should be mentioned that even though a reduction in phenotypic variability is frequent with respect to morphological and physiological characters, it is by no means general. With respect to behavioral characters, an increase in variability in the heterozygote has frequently been found. It has been explained by the assumption that for mammals at least an increased variability of behavior may be an expression of heterozygote superiority (Caspari, 1958). It should be pointed out, however, that greater variability in heterozygotes with respect to behavioral characters is by no means universal (Winston, 1964).

It appears, therefore, that no general rules for the variability in heterozygotes can be made, particularly with respect to behavioral characters. If the reasoning mentioned above is applied, the behavior of heterozygotes should depend not so much on whether the character is morphological, physiological, or behavioral but rather on the influence of variability in this character for the fitness of the organism. It is important to realize that many behavioral characters are themselves mechanisms tending to adapt the organism to its environment and that the adaptational and evolutionary aspects of behavior must be kept in mind when genetic differences in behavioral characters are studied (Ginsburg, 1963a).

It has been mentioned in the previous paragraphs that developmental processes in general may be influenced by the environment in which they proceed. Since the heredity-environment problem has been widely discussed with respect to behavior, some general discussion of this topic seems needed in this context. It was mentioned in the first section that even in microorganisms there exist mutants in which the production of an enzyme is dependent on environmental factors, such as temperature. Genes whose expression depends on environmental factors are by no means rare. This fact has been expressed by geneticists by the statement that the genes control the norm of reaction of the organism on the environment. Nevertheless, little work has been done on this aspect of genetics, because workers in the field have intentionally chosen to work with mutations, whose expression is *not* influenced by the environment, for experiments on gene localization and gene action can be carried out more efficiently in these mutants.

Generally speaking, the development of animals is not strongly influenced by environmental variation within normal limits. Waddington has introduced the term *canalization* for the limitation of the influence of environment on developmental processes once they have been initiated. In plants, particularly in the development of stem and leaves, environmental influences such as temperature, moisture, and light are known to affect growth and development profoundly. Bradshaw (1965) has pointed out that this difference between plant and animal development is related to the fact that growth and development are the only means whereby an individual plant can adapt to variations in environmental conditions, whereas animals accomplish the same adaptive task by means of behavior. It is therefore not astonishing that in some respects the development of behavioral characters is more like plant than animal development.

In plant breeding, it is generally taken for granted that a particular character, e.g., yield, is strongly dependent on an interaction of genes and environment. By selective breeding a strain may be obtained which will produce a superior yield under one set of environmental conditions. But this strain will not be expected to produce a superior yield under all possible environmental conditions, and for every set of environmental conditions a particular set of genes is needed to ensure superior yield. In other words, yield—and the same applies to other characters—cannot be said to depend on genes or to depend on environment; rather, it depends on an interaction of these two components, each combination leading to a specified result. The same principle applies to animal development, except that frequently the variations due to environmental factors are not as obvious as in plants, because the developmental system is more strongly canalized.

Animal development can, however, be altered by extreme environmental influences. In insects, near-lethal heat shocks have been used in most experiments; in mice, because of the fact that development proceeds in the uterus, drugs or ionizing radiation is more effective. In all these experiments it was found that the nature of the abnormalities induced by the environmental factor is dependent on the time at which the environmental agent acts. The period of time during which a particular phenotypic character can be influenced by an environmental agent is designated as its *sensitive period*. The sensitive period for a particular character has a beginning and an end; neither before nor after the sensitive period can the character be influenced by environmental conditions. For example, a sensitive period for the wing pattern of butterflies and moths exists in the early pupal period; it can even be subdivided into sensitive periods for different pattern elements which, however, may overlap each other in time. It should be noted that the term "sensitive period" must be distinguished from the term "critical period." The critical period divides a developmental process into two parts; before the critical period the developmental process may be inhibited by certain operations or influences, while after the critical period it is completely determined. An example is the metamorphosis of insects: If the glands producing the metamorphosis hormone are removed or inhibited before the critical period, metamorphosis will not proceed, and the animal will remain a larva. Similar operations after the critical period have no effect on metamorphosis. The terms "sensitive period" and "critical period" are well-established embryological terms which refer to related but not identical phenomena. Recently, the terms have come into use in the study of the development of behavior, particularly with reference to early learning, imprinting, and the influence of early handling on emotional development. Unfortunately, the two terms are used interchangeably by workers in the field of behavior; careful distinction between them might be of value in the interpretation of the phenomena involved.

The abnormalities induced by environmental influences frequently are phenotypically indistinguishable from the phenotypes induced by certain mutations but are distinguished from them by the fact that they are not transmitted through the chromosomes to the offspring. Such nonheritable phenotypic changes induced by the environment have been called "phenocopies" of the mutant genes by Goldschmidt. Phenocopies were thoroughly investigated in the 1930s because it was believed that the sensitive period during which a phenocopy can be induced corresponds to the time when the mutant gene and its wild-type allele are "active." The concept of the activity of a gene was not clear at the time of these investigations, and it is now generally accepted that the time of production of a specific protein by a structural gene has no relationship to the sensitive period in which a phenocopy of the gene is produced. The sensitive period is rather a period in which certain developmental processes then going on can be disturbed either by an environmental shock or by certain mutant genes.

The occurrence of phenocopies gives rise to the phenomenon of *genetic assimilation* (Waddington, 1961). If Drosophila are subjected to environmental shocks, e.g., high temperature, at a particular sensitive period, flies with a missing or reduced cross vein on the wing may be induced, phenocopies of the sex-linked gene crossveinless. If these changed flies are mated to each other and treated again, strains may be obtained by selection which give, on stimulation in the sensitive period, a higher percentage and a greater expression of the abnormal phenotype. Finally, after prolonged selection for formation of the phenocopy, crossveinless flies may arise even without the heat shock. The genetic changes that have gone on in this experiment are due to selection from the original strain. The genes that have been selected for were already present in the original populations, and they influence the developmental process (in this case, formation of the cross vein) in such a way that it becomes less well canalized. The function of the environmental stimulus in this experiment is that it serves to uncover genes affecting the developmental process; they could not be observed in the original population because of lack of variability. The experiment is, therefore, a selection experiment for genes lowering the level of canalization. The mutant locus copied by the modification is sometimes, but not always, involved in the selection; in the case of "crossveinless" the *cv* locus was not among the genes selected. This depends not so much on the action of the gene as on the genetic constitution of the strain from which the selection started, since only genes that were present originally in at least two alleles can respond to selective pressure.

The development of behavioral characters is in many respects similar to the development of morphological characters. Particular behavior patterns appear at different times, mature, and become more differentiated. The basis of the development of behavioral characters is not understood at the biochemical level. This is not astonishing, since even simple stimulus-reaction relationships in lower organisms cannot be analyzed in molecular terms and since developmental processes in general present difficulties in this respect. However, it should be emphasized that the development of the stereotyped behavior patterns of lower organisms does not seem to involve problems beyond those posed by any developmental system. Nor does the influence of environmental factors on the development of behavior pose any new problems; in higher animals behavioral characters seem to be less thoroughly canalized than many morphological characters.

However, one aspect of environmental influences on behavior deserves special mention. When the question of heredity versus environment is raised, the "environment" is not temperature, light, or any other form of stimulating energy. Environment in this case refers to the fact that in many organisms the development of behavioral patterns is strongly and irreversibly influenced by previous stimuli and by previously acquired behavior patterns. Particularly mammals and birds have the ability to modify behavioral patterns according to previous experiences; they possess a memory in which previous experiences can be stored, and they are able to modify their behavior accordingly; i.e., they can learn. Even among the highest mammals, such as man, it is by no means the case that all behavior patterns are learned or even subject to learning; some reflexes remain at the completely automatic level, and many aspects of food uptake are modified by learning but are not dependent on it. On the other hand, there exist many behavior patterns, particularly among higher animals and man, which are not only modified by learning but are also dependent on learning for their normal development. If the organism does not have the opportunity to learn, the behavior patterns either will not develop at all or will develop in an abnormal way. In these cases, the ability to learn particular behavior patterns, the necessity to learn them, the time when they can and must be learned seem to be themselves under genetic control. This seems to be indicated by the fact that in related species similar behavior patterns may be more or less dependent on learning. The copulatory pattern of rodents, for instance, seems to be under hormonal and genetic control, similar to the mating patterns of insects, while in the great apes the development of the copulatory pattern is completely dependent on learning by mating with an experienced animal.

THE BIOLOGICAL BASIS OF MEMORY

The problem of the biological basis of memory will be treated briefly, because certain similarities between heredity and memory have impressed investigators throughout the history of biology. Richard Semon (1904) made this analogy the basis of a Lamarckian theory of heredity and evolution. Attempts to treat both phenomena under a common aspect have continued since that time. The basis of the analogy is the fact that in both cases information is stored in coded form. Since the mechanism of the storage of information in the genetic material is well understood, it is tempting to ask the question whether the information storage involved in memory might have a similar biochemical basis.

Fuller (1964b) has pointed out that on the basis of our present knowledge two different models of information storage in the nervous system can be suggested. One of them involves the interaction of nerve cells with each other, e.g., facilitation of impulse conduction between certain neurons or the establishment of new associations between neurons. The other possibility is a permanent change in the biochemical activity of certain nerve cells; because nucleic acids are known to be able to store biological information and RNA has been shown to be involved in nerve activity, particular attention has been paid to RNA metabolism. There is no reason why both mechanisms might not be involved in memory. Nevertheless, the question is frequently posed as an either-or question (Barondes, 1965).

The hypothesis that, in the process of learning and information storage in the nervous system, changes in the circuitry of the brain are involved was accepted without question up to about 1958. There was actually not much evidence for changes in nervous connections, although changes in action currents in particular parts of the brain incidental to learning have been observed. With increasing knowledge about information storage and information transfer in the cell, mentioned in the first section of this chapter, the attention of many workers turned to the biochemical processes that accompany learning, particularly to the metabolism of RNA.

There exists abundant evidence that nervous activity is usually accompanied by an increase in the rate of RNA metabolism. In most of these investigations, however, no attempt has been made to specify the type of RNA involved. Any increase in the rate of protein turnover and protein synthesis is accompanied by an increase in RNA synthesis, particularly affecting ribosomal RNA. Such observations do not permit the distinction between effects on the circuitry of the brain and information storage at the cellular level, since it would be expected that the outgrowth of new nerve connections would proceed under considerable protein and RNA synthesis.

A relationship between changes in organization of the brain and increase in RNA synthesis is suggested by an experiment carried out by Morrell (1961). If a region of the cerebral cortex is made epileptogenic by the use of ethylchloride, a secondary focus of epileptogenic activity will gradually form in a corresponding position in the other hemisphere. If this secondary focus is isolated by severing its nervous connections, it becomes inactive but continues to show increased electrical activity on stimulation, even months after it has been isolated. This indicates that there occurred a permanent change in the nervous tissue which was initiated by the injured tissue of the primary focus. Morrell furthermore states that both the nerve cells in the peripheral regions of the primary focus, its most active part as well as the secondary focus region, show in cytological preparations an increased RNA content, particularly in the region of the dendrites. The important fact of this experiment is the evidence it supplies that changes in the nerve cells may persist even in the absence of their normal connections with other nerve cells. On the other hand, the connections may be assumed to be important in the establishment of the secondary focus, since it develops as a consequence of the injury to the primary focus by the drug.

The involvement of RNA activity in learning and in the retention of memory has ben repeatedly studied by the use of drugs whose mode of action in the information-transfer processes is relatively well understood. Two examples of this approach will be discussed here. Dingman and Sporn (1961) injected 8-azaguanine into the brain of rats; 8-

azaguanine is an analog of the base guanine and is incorporated into RNA. Dingman and Sporn could demonstrate that much of the injected 8-azaguanine was incorporated into the RNA of the brain cells, but it disappeared relatively fast. Injected rats were tested for retention of a previously learned task and for the speed with which they could learn a new maze. Recall of a previously learned task is not affected by the treatment, while the rate of learning a new task is reduced. Actually, the learning curves of treated and control animals appear to run parallel, so that the main difference seems to be restricted to the first trial. The meaning of this experiment is highly dubious because it has not been established which type of RNA is affected and whether the effect is direct effect of the drug on brain RNA or one of its secondary effects on metabolism.

Similar experiments have been carried out by Flexner and collaborators (1963, 1964). They used the antibiotic puromycin, a drug which is known to inhibit protein synthesis on the level of translation from m-RNA to protein; it is assumed that it inhibits the release of the protein chain from the ribosome–m-RNA complex. They injected puromycin into various parts of the brains of mice and tested for retention of a previously learned avoidance discrimination task. Injection blocked the memory if the drug was applied 1 day after the task had been learned, but not if administered 11 to 38 days after the task had been learned. This result demonstrates a different biochemical basis for short-term and long-term memory. It also shows that the effect on short-term memory was obtained only after injection of puromycin into the temporal region of the brain, which results in a block of protein synthesis in the temporal cortex and in the hippocampus. Injection into other parts of the brain was ineffective.

Although these results give some insight into several aspects of memory, they cannot give direct information on the problem of biochemical mechanisms involved in information storage in the brain. Even though the primary action of a drug on the mechanism of genetic information transfer may be well understood, it must be considered that blocks in a biochemical reaction have secondary effects on other reactions, analogous to the situation outlined in our discussion of pleiotropic gene action. In the experiments of Flexner et al. (1964) chloramphenicol, another drug blocking protein synthesis at the translation level, possibly by inhibiting the attachment of m-RNA to the ribosomes, was also used. Injection of chloramphenicol into the temporal area of the brain produces convulsions but does not affect memory. The difference between the action of puromycin and chloramphenicol, both of which are supposed to block protein synthesis at the translation level, points to the difficulty in relating a drug effect on memory to an action of the drug at the molecular level.

A more direct approach to the problem has been taken by Hydén and Egyházi. In order to determine the amount and base composition of RNA in individual cells, they applied a sensitive technique, which had been worked out by Edström for this purpose, to specific types of nerve cells involved in a learning process. The technique (Egyházi and

Hydén, 1961) consists in dissecting individual neurons from some region of the brain under a stereomicroscope. After fixation, the total RNA of the cell can be extracted by treatment with the enzyme ribonuclease, and the amount of RNA determined by light absorption at 2,570 angstroms, the peak absorption of nucleic acids. For the determination of base ratios, the RNA is hydrolyzed, and the resulting free bases are separated from each other by electrophoresis on a cellulose fiber 25 to 35 microns thick. A similar technique may also be used to determine amounts and base compositions of isolated RNA from individual nuclei.

This method was applied by Hydén and Egyházi (1962) to the neurons of rats in a learning experiment. Rats learned to balance on a wire to reach their food. Since this learning process concerned a balancing performance, it was suspected that the cells of Deiter's nucleus, which are favorable for the use of this technique because they are very large, would be affected. The RNA from neurons from Deiter's nucleus of trained rats was compared with that of normal rats and of rats in which the organs of equilibrium had received an unspecific stimulation. The amount of total RNA increased after learning and also after unspecific stimulation but, in both cases, the base ratio of the total RNA remained the same. The total RNA values are an expression of the activity of the cytoplasm, since only 4 percent of the total RNA is found in the nucleus. When the base composition of nuclear RNA was determined, it was found that it was changed considerably after learning, showing an increase in adenine and a decrease in uracil; unspecific stimulation had no effect on the base composition of nuclear RNA. The increase in total, probably cytoplasmic, RNA appears to be an unspecific indication of neuronal activity, while the storage of information is connected with an alteration in the base composition of nuclear RNA.

Similar results were obtained with cortical cells of rats which had been trained to use the left paw instead of the originally preferred right paw for obtaining food. Neurons in an area of the right cortex known to be concerned with the control of handedness were investigated, the corresponding area of the right side of the same animal being used as a control. No difference in the amount of total RNA per cell was found, but there was a significant difference in the base composition of nuclear RNA. In normal control rats, no difference in base composition in the RNA of nuclei from the right and left side was found (Hydén and Egyházi, 1964).

Both experiments agree in the finding that in the learning process a permanent change in the base composition of nuclear RNA takes place. Hydén and Egyházi argue that this change is probably not due to the RNA of the nucleolus, since nucleolar RNA has a base composition very similar to cytoplasmic RNA and is regarded as the precursor of ribosomal RNA. It is therefore concluded that the learning situation induces the synthesis of an RNA with specific base composition in the chromosomes and, since the base ratio of RNA is determined by genic DNA, it may be assumed that in the learning situation specific gene loci in specific cells are induced to produce an RNA which is regarded as messenger RNA. The formation of a memory trace would, then, be analogous to the induction process in *E. coli* described earlier in this chapter. The cell specificity would be dependent on the stimulus involved in learning, as in the first case, or on the activity controlled by the learning process, as in the second case, in other words, by the circuitry of the brain. The question may be raised whether individual cells can give rise to only one messenger RNA, i.e, whether only one gene locus can be turned on in a particular cell or whether several genes may be involved.

The theory of memory proposed by Hydén and Egyházi describes learning as a process which is in principle not different from other developmental processes. It involves first the establishment of the circuitry of the brain, which is a very complex developmental process involving a specific arrangement of cells and their differentiation, and secondly a turning on of specific genes in the nerve cells by environmental stimuli. Once these genes are turned on, they may apparently persist in this state permanently; similar persistence is suspected to take place in many other processes of differentiation, and Jacob and Monod (1963) have presented models in which an inducible system can stay locked in a particular state. The model proposed by Hydén and Egyházi reduces the difference between learned and nonlearned behavior; it may simply depend on whether a stimulus turning on particular genes in certain neurons is environmental or endogenous.

CHAPTER SEVEN
GENETIC PARAMETERS IN BEHAVIORAL RESEARCH[1]
Benson E. Ginsburg

Although the title of this chapter implies a breadth that is coextensive with the scope of the entire volume, the substance will emphasize those aspects of the methodology of our own laboratory that appear to have general applicability, illustrated in the context of the particular researches in which these approaches have been most fruitful.

SELECTION AND INBRED LINES

One of the major interests of our group, which may be used to illustrate a number of generic problems, is the assessment of the hereditary contribution to audiogenic-seizure susceptibility in mice. This is a phenomenon that has been described by some workers as genetically complex (Fuller, Easler, and Smith, 1950; Frings and Frings, 1953) and by others as genetically simple (Witt and Hall, 1949; M. L. Watson, 1939). It has been considered to be a unitary trait and, therefore, a behavioral phenotype in its own right. It has also been treated as one aspect of a more generalized phenotype—susceptibility to stress—induced in a variety of ways (Ginsburg, 1963b). It is clearly influenced by a multiplicity of environmental factors, but it is also associated with particular inbred strains or genotypes which do not interact in equivalent fashion with environmental variables.

If one accepts the broadest description of the phenotype as including a spectrum of behavior that goes far beyond convulsive activity, together with the hypothesis of a complex polygenic basis, one may legitimately wonder whether there has not been confusion between contents and container (Ginsburg and Laughlin, 1966) in the sense that a series of correlated behavioral attributes may have been fortuitously associated in an inbred strain or other selectively derived stock without any necessary causal relationships. Further genetic experiments would

[1] The work reported in this chapter was supported by Grants B-1216, MY-3361, and MH-03361 from the National Institutes of Mental Health; Contract N6 ori 02037, NR 110-500 from the Office of Naval Research; and a grant from the Dr. Wallace C. and Clara A. Abbott Memorial Fund of the University of Chicago.

then be expected to reveal that the phenotype is pluralistic, involving genetically independent and dissociable components. Were such a genetic dissection to be performed with a wild population, the meaning of "phenotype" would, in a sense, be further confused, as it is certainly possible that the genetically dissociable components became associated through the action of selective factors to produce a multifactorial phenotype having survival value. In this sense, the test of genetic parsing would not be critical in terms of providing a definition of the phenotype, as it would then be the coadapted complex that forms the significant evolutionary entity. Carried to its logical extension, this point of view leads to the conclusion that it is the total genome that is relevant, since genes typically interact, and that, further, the genetic constitution of a population, including its breeding structure and its multiple genetic pathways to overlapping phenotypes, constitutes the relevant entity.

While this is patently true, and the population matrix provides the encompassing configuration from which components isolable by experimental techniques (including those of Mendelian analysis) derive their biological significance, it is also true that the whole, which is surely more than the sum of its parts, is not amenable to understanding unless we discover its components and explore their interrelationships. Nowhere are the latter more complex than in the behavioral realm, where the control of biological variability through genetic means is essential to even the most elementary understanding of the nature-nurture problem.

It is for this reason that studies of twins have achieved so prominent a position in human behavior genetics and that inbred strains and isogenic stocks are so often used in corresponding animal research. However, the use of inbred strains or isogenic stocks is also subject to constraints. Each such stock, for example, represents only a single biological situation. Therefore a sampling of X such stocks, each involving N individuals, has dealt with X biological situations, each replicated N times, rather than with a sample of NX individuals. Moreover, an inbred stock represents a highly artificial genetic condition—virtual lack of heterozygosity—seldom, if ever, encountered in nature, and the sampling of X such conditions has an unknown and often indeterminable relationship to the distribution of genotypes and phenotypes in the natural population from which the stocks were originally derived. There are also limitations on the reproducibility of results. Since mutations occur, some heterozygosity eventually creeps in, so that genetic change is not precluded. Inbred stocks may not always be more uniform phenotypically than random-bred animals. Some inbred strains have a very low threshold of interaction with environmental variables and are, as a consequence, highly variable despite their relative genetic uniformity (Russell, 1941).

In spite of these desiderata, isogenic stocks and inbred strains afford excellent and even unique opportunities to investigate the effects of genotype on behavioral characteristics. Examples of such studies abound in the work of Scott and his coworkers on agonistic behavior in mice (Scott, 1942; Scott and Fredericson, 1951); the work of Fuller et al. (1950), Vicari (1951), Abood and Gerard (1955), and others on audiogenic seizures; the work of Bruell (1962) on some half-dozen quantitative behavior differences in mice; the work of C. S. Hall (1938), Lindzey (1951), and Broadhurst and Jinks (1961) on emotionality; of Rodgers and McClearn (1962) on alcohol preference; and many others too numerous to mention.

Considering the wide spectrum of behaviors affected by genetic variables in the studies mentioned, and the additional fine-grain demonstrations of the effects of particular chromosomes on behavior afforded by the work of Hirsch (1962)—and the strain, species, and subspecies differences represented in the work of King and Shea (1959)—one gets the impression that there are likely to be genetic influences affecting virtually all aspects of measurable behavior and that the usual tests in which only environmental variables are intentionally manipulated owe a significant part of their variability to the nature-nurture interaction. In spite of this, the usual methodology assumes the constancy of the former and so, in fact, assigns the product entirely to the latter. Oftentimes the natural or genetic contribution has an indirect effect on the outcome of standard experiments. The C3H agouti mouse, for example, is often visually deficient. Yet it has been used in experiments where visual cues are important and where it has been assumed to be normal. With the occurrence of such relatively gross errors, it is not only imaginable but demonstrable that subtler differences can be involved in many other stocks without the experimenter's knowledge, even after these are documented in the literature. This is hardly remarkable in view of the fact that the journals in which genetic results are described belong to a different culture from those in which behavioral experiments are reported, and the twain do not meet often enough.

It must also be remembered that the genotype and phenotype have no one-to-one relationship. The genotype is a potential for development that may vary with varying environmental circumstances, and it is precisely this point that should prove interesting to the psychologist. Moreover, a phenotype, especially an adaptive one, may rest on a multiplicity of genetic bases, each having a somewhat different underlying mechanism. Teleologically speaking, it is reasonable to expect that a population should find a variety of genetic ways of adapting and should maintain genetic flexibility as insurance against the biological inability to adapt to a change in conditions. It is, therefore, of more than academic interest to resolve a population into its genetic elements in relation to behavioral capacities under varying environmental conditions and to analyze the differences in underlying gene-controlled mechanisms that will yield an understanding of the actual biological bases of behavior. Hirsch (1963 and 1967) has demonstrated that these genetic determinants are of importance in the ontogeny of individual differences with which psychology is, in the last analysis, concerned in a very major way. This chapter will provide additional illustrations of the importance of genetic parameters in behavior research and of the utility of using a gene-action approach in this field.

The C57BL/10 strain of mouse has proved highly variable on a series of behavioral tests ranging from fighting capacity, through the fecal-bolus emotionality test, to susceptibility to audiogenic seizures. In a series of selection experiments, progeny from the extremes of each distribution were assortatively mated with each other for seven successive generations, in an attempt to derive high and low lines for each measure. Every experiment was independently done; that is, there was no attempt to select high and low lines on several measures simultaneously. Progeny tests involving the eighth generation were then made. As Figure 7.1a to e illustrates, there were no significant differences between the high and low lines on any of the tests. There were likewise no significant differences between the eighth generation and the original generation preceding the selection, thus affording proof that the variability was nongenetic.

Selection for Aggressiveness

Figure 7.1a and b illustrates the lack of effect of seven generations of selection on latency to fighting and success in fighting among males isolated from weaning to 75 to 80 days of age. Ten males were used to establish the initial latency values and success levels. These were fought daily, using the test procedure described by Ginsburg and Allee (1942). Initial encounters were between naïve mice. The subsequent eight sets of encounters were round robin, in which each mouse met every other mouse once according to a predetermined order. Each pair of mice was observed for 20 minutes, unless a fight that could

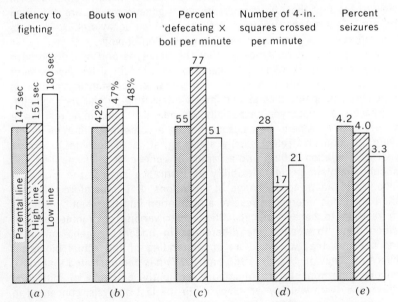

Figure 7.1   Mean scores of parental generation of C57BL/10 mice compared with those of the eighth generation after selection for high and low scores.

be scored as a victory for one mouse or the other developed sooner. Latency to fighting was recorded for the mouse that initiated a given attack. If the attacked mouse fought back, this was scored as a "contingent latency." Sires used in the selection experiment for fighting latency were chosen in each generation on the basis of cumulative latency scores recorded in seconds. Contingent latencies were arbitrarily valued at one-half the test period (600 seconds) for selection purposes and were used only when a mouse did not initiate an attack during a 20-minute test period but did fight back if attacked by the other mouse. If he initiated an attack later in the bout, the latency to this attack was used as his score. If a mouse responded to an attack by submitting or running away (see Ginsburg and Allee, 1942), this was scored as a 20-minute latency. If no fighting occurred during a test, this was scored as a 20-minute latency for both mice. In the first generation of selection, the three highest-scoring (longest latency) and lowest-scoring mice were selected as sires in an attempt to establish high- and low-latency sublines. In no instance were the scores of the two groups of sires overlapping in any generation of selection. The data in Figure 7.1a are expressed in terms of average latency to initiate a fight. Thus, although contingent latencies and defeats were used in the selection scores, only the latencies to the initiation of an attack are represented in the final data.

As female mice do not fight readily, they were not tested for aggression. Dams were, therefore, selected on the basis of the performance of their male offspring. The criteria were that every female used in the selection experiment must have had two or more litters, each containing two or more tested males whose averaged scores placed them in the upper (high latency) or lower 40 percent of the distribution. In a few instances, females were selected as dams during the first three generations on the basis of male progeny whose scores were closer to the mean. All these were replaced in later generations of selection by females who met the criteria. In each subsequent generation of selection, ten males from the presumptive high-latency subline were tested as described above, and the three highest scorers picked as sires. Ten males from the presumptive low-latency subline were separately tested, and the three lowest scorers were used as sires. Females were selected as dams in the same manner; that is, only females with high-scoring male progeny from the presumptive high line were eligible as dams for long-latency mice, and only those with low-scoring male progeny from the presumptive low line were eligible as dams for the short-latency line.

Selection for success in winning fights (Figure 7.1b) overlapped that for high and low latency to fighting, in that the same populations and the same experiments were used to score the mice, although the matings for each generation of selection were not identical for the two behaviors. Selection for sires and dams was based on criteria analogous to those used in the latency experiments, except that the proportion of bouts won was used to select sires, and the proportion of bouts won by at least two sons in each of two litters was used to select dams. The rank-order correlation between short-latency fighting and proportion of fights won over all eight generations of selection (number of animals $= 150$, each tested nine times) was $+.36$. Scores for generation VIII were obtained by matching two arbitrarily chosen groups each consisting of five low-line males with five males from the high line, round robin fashion ($N = 20$). Scores were separately tabulated for each line and expressed as the average proportion of wins per mouse per line.

Selection for Emotionality

Two criteria were used for emotionality measures. One was a 1-minute fecal-bolus test (Figure 7.1c). The second was an activity test carried out at the same time (Figure 7.1d). Both tests were administered at 28 days of age in a white porcelain-enameled box 17 by 15 by 17 inches, the bottom of which was marked into 4-inch squares. The defecation scores were low, and only about one-half the mice defecated during the test period. Parents of the presumptive low subline were chosen from among the animals having zero defecation scores. Those for the high subline were the three top-scoring males and females selected from a group of 30 of each sex in each generation. After the first generation, only highs derived from highs and lows derived from lows in each previous generation were selected to carry on the "lines." One hundred animals, representing the first 50 produced by the highs and the first 50 produced by the lows, were then compared in the eighth generation. No significant differences were found between these two populations or between the means and distribution of scores in the two populations in generation VIII and the original populations from which selection was started. Figure 7.1c compares the means of the products of the number defecating by the fecal-bolus count for each line, where the N is adjusted to 100. Figure 7.1d shows the number of 4-inch squares crossed after the mouse, which was placed in the center of the test area, had positioned itself.

Selection for Susceptibility to Audiogenic Seizures

Mice were tested for seizures by the method described by Ginsburg and Miller (1963). Those selected as parents for the presumptive low seizure line met a criterion of total absence of seizures on four consecutive one-per-day trials beginning at 28 days of age. Those selected as "highs" exhibited at least one clonic-tonic convulsion on the same test schedule. After the first generation, highs were selected from highs and lows from lows, as in the preceding selection experiments. Two hundred animals, representing the first 150 produced by the lows and the first 50 produced by the highs (which were rare) were compared in the eighth generation (Figure 7.1e). No statistically significant differences in seizure incidence, severity, or latency were found between the high and low "lines" in generation VIII or between these and the original population.

By contrast, similar selection for high and low seizure lines in the

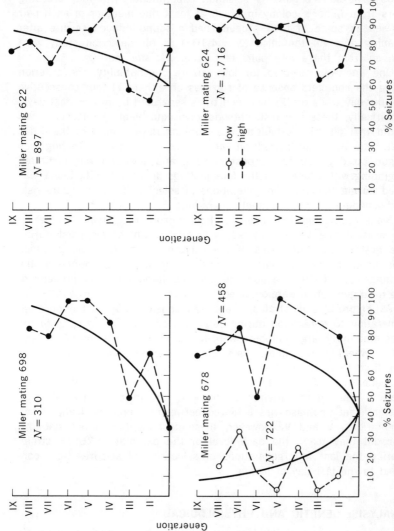

Figure 7.2 Selection for high and low seizure susceptibility in C57BL/6 mutant mice.

progeny of irradiated C57BL/6 mice carried out in our laboratory by S. Maxson (1966) resulted in significant differences between the progeny of the selectively bred animals. Normally, the parent strain is highly resistant to seizures at weaning age. The stock from which selection produced high seizure lines originated in the colony of D. S. Miller and was attributed by her to radiation-induced mutation (Miller, 1963). Maxson[2] began his selection experiments with four breeding pairs from Miller's colony in which at least one member of each pair as well as some of their progeny had a sound-induced seizure after weaning age. All the mice ($N = 4{,}103$) in his selection study were derived from these four pairs of brother X sister matings.

One line was selected for low seizure susceptibility, the criterion being the complete absence of seizures on either (1) four consecutive one-per-day trials or (2) two such trials separated by two nontest days. Empirically, these two test methods gave equivalent results, as it was found that all mice exhibiting a seizure on days 2 or 3 of the 4-day test sequence also seized on either day 1 or day 4. Testing was begun on days 30 to 32 post partum, and selection was instituted beginning with progeny of the four matings described. Schedule 1 was used during the first five generations of selection for low seizure risk, and schedule 2 for subsequent generations.

Maxson selected four lines for high seizure risk, using a criterion of at least one seizure on test schedule 1, which was used during the first seven generations of the experiment, or schedule 2, which was used in the remaining generations of selection. In one of the matings (no. 678) selection was not instituted until the progeny of the third generation were obtained.

As Figure 7.2 illustrates, there was considerable fluctuation from generation to generation during selection for low risk, in spite of the fact that fairly large numbers were involved in generations IV, VI, and VII. In the high-risk lines, the fluctuations were most extreme during the first three generations, and the maximum effect was obtained by generation IV in all but one instance, where it occurred one generation later. In the attempt to lower seizure incidence by selection, a significant decrease was achieved between generations I and II and generations IV and V. However, these decrements did not stabilize, nor did they reach the base level of the parental C57BL/6 strain. During the last four generations of selection, all seizures were confined to the last day of testing.

ANALYSIS: GENETIC AND PHYSIOLOGICAL

Thus far only one of the high reactor lines has served as the basis for a Mendelian analysis and has been studied physiologically. This is the one derived from mating 624. The significant changes occurred between generations III and IV and between generations VI and VII.

[2] The author is indebted to S. C. Maxson for permission to use his unpublished data.

The latter increase has been maintained during three further generations of selection.

The data suggest that reactor-line 624 differs from the parent stock by a single-gene mutation which was assorted and fixed in the homozygous state by selection (Table 7.1). Investigations are now in progress to determine whether any locus previously known to affect seizure susceptibility is involved. Ginsburg and Miller (1963) have presented evidence that two independently segregating autosomal genes interact to produce various levels of seizure susceptibility in DBA X C57BL mice. One of these genes (designated A) is associated with a discrete enzymatic effect on nucleoside triphosphatase, found by J. S. Cowen (1966) to be localized in the granular cell layer of the dentate fascia of the hippocampus (see also Ginsburg, 1965b). The other (designated B) has been associated with the peculiar responses of these strains to glutamic acid, which has a marked palliative effect on seizures in particular genotypes (Ginsburg, 1954, 1965b). In addition, A. Lehmann (personal communication) has demonstrated that the seizure-prone phenotype occurs on still another genetic basis, which she designates ss, in albino mice originally selected by Frings and Frings (1953) for high and low seizure susceptibility. In a cross between these high- and low-seizure albinos, no seizures occur in F_1 (Ss), and the segregation ratios in F_2 and backcrosses conform to expectation for an autosomal recessive gene predisposing to convulsive seizures. These high-seizure mice (HS) are not affected by glutamic acid (Ginsburg, 1954), and their hippocampal histochemistry is distinguishable from the reaction obtained in the presence of the AA genotype

Table 7.1
Monogenic assortment of C57BL/6 X C57BL/6s crosses tested for seizure incidence beginning 30 to 32 days post partum

Generation	Observed incidence, %	Expected, %	χ^2	P
P_1 $\left(\dfrac{\text{1 or more seizures per 4 trials}}{\text{number of animals tested}}\right)$				
C57BL/6	$\dfrac{1}{59}$ 2	0		
C57BL/6s	$\dfrac{56}{56}$ 100	100		
F_1	$\dfrac{118}{414}$ 28	28		
F_2	$\dfrac{116}{278}$ 42	39	.97	.325
Bx to C57BL/6	$\dfrac{23}{179}$ 13	14	.185	.67
Bx to C57BL/6s	$\dfrac{107}{181}$ 59	64	1.944	.168

There were no differences between the results of reciprocal crosses.
F_1 males $(N = 206)$ had 13 percent fewer seizures than F_1 females $(N = 208)$. This difference is significant at the .003 level. No other significant sex differences occurred.

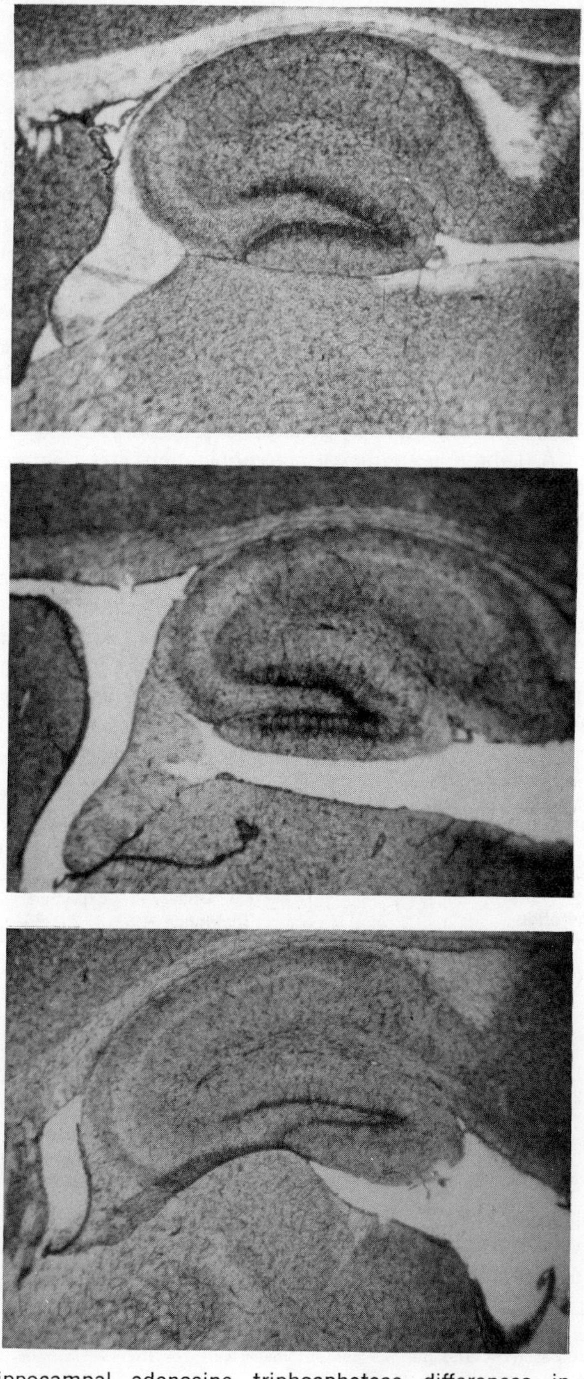

DBA/1

HS

C57BL/10

Figure 7.3 Hippocampal adenosine triphosphatase differences in three strains of mice.

(see Figure 7.3).[3] Interestingly enough, however, the effect is in the same direction as that obtained with the A gene and appears also to involve a difference in morphology, in which the granular cell layer appears more blunted, as though it had an extra segment and, therefore, a larger mass than the other strains. It had not previously been feasible to determine, either on the basis of phenotypic expression or on the basis of the hippocampal effects, whether the same gene is involved in DBA and HS seizure susceptibility. The morphological and histochemical aspects of the phenotype will now make it possible to find out.

The high-seizure mice derived by Maxson from the Miller stocks are indistinguishable from the original C57BL/6 mice in terms of brain histochemistry and, therefore, have presumably not mutated at the A locus (Figure 7.3). Experiments are in progress to determine whether or not they respond to glutamic acid in the manner to be expected if the B locus is involved. Should they not respond, then crosses with HS mice from Lehmann's colony will be made in order to determine whether or not the mutation has occurred at the S locus and whether S is really A or an allele of A.

The results of these experiments indicate that seizure susceptibility, which is only one dramatic aspect of a broader behavioral phenotype, occurs in laboratory mice on several rather simple genetic bases, each of which produces morphological and/or biochemical effects in the brain, some of which are highly local. The discovery of more such genes and the identification of their primary effects may be expected to reveal the variety of ways in which convulsive activity can occur on a natural basis. The identification of the site and nature of the primary gene effects will suggest that other behaviors may be involved and will thus help to complete the description of the behavioral phenotype. The genetic analyses are made more certain as a result of being able to use morphological and biochemical markers, rather than a behavior that is modifiable environmentally and occurs on an actuarial probability basis. In addition, the correlation of presumed causal mechanisms with given genes on identifiable backgrounds provides the basis for more certain reasoning at the level of causation than mere correlation with phenotype. With the latter approach, it would be difficult to establish the relationship between the ATPase activity of the granular cell layer and seizures, since the latter can occur when the enzymatic activity is high (DBA) or low (C57BL/6s), and since even in DBA X C57BL crosses the correlation is not between the enzymatic activity and the seizures but between the total genotype and the seizures, as the behavioral expression of the A gene depends upon the particular B alleles with which it is paired, as well as the total genetic background (Ginsburg and Miller, 1963). $AAbb$, for example, has a "seizure-prone" granular cell layer (high activity) but is phenotypically seizure-resistant when obtained on a DBA/1 X C57BL/10 background, whereas $AaBb$

---

[3] The author is indebted to J. S. Cowen for permission to reproduce the figures illustrating the hippocampal ATPase reaction.

(intermediate granular-cell-layer activity) and $aaBB$ (low activity) are seizure-prone. $A\_\_B\_\_$ and $\_\_ \_\_BB$ are also susceptible to seizures. The same DBA/1 crosses on a C57BL/6 background produce quite different results, in which the expression of A and B are incrementally equivalent, and only $aabb$ is seizure-resistant (see Ginsburg and Miller, 1963).

CHOICE OF BEHAVIORAL PHENOTYPES

With reference to the glutamate effect on seizures, collaborative work with S. Ross has demonstrated that in those genotypes where glutamate is supportive in helping withstand the effects of the sound stress it also provides an improvement in learning ability on a series of elevated Hebb-Williams mazes, but only when the patterns are of an order of complexity that appears to be at the threshold of the animal's ability to learn. By contrast, genotypes that are not affected by glutamate in the seizure situation are not helped on these mazes by this substrate. This work is closely comparable to that of Hughes and coworkers (Hughes and Zubeck, 1956, 1957; Hughes et al., 1957), who also found that the effects of glutamate on behavior were strain-specific.

Although most of our inbred strains gave consistent results on the open-field fecal-bolus test developed by Hall as a presumed emotionality measure, several samples of C57BL/10 mice did not. Selection experiments failed to confirm the hypothesis of possible genetic differences among "emotional" and "nonemotional" mice of this strain (Figure 7.2). We therefore moved to another measure and used the conditionability of the fecal-bolus response as our indicator (Figure 7.4).[4] Since we were using the mice for experiments on audiogenic seizures, the tests were made in the seizure chamber and involved contrasting naïve with previously stressed animals, using three time periods:

1   From the time the animal could be seen and observed in its home cage to the time it was placed in the chamber
2   For 1 minute prior to the onset of the sound stimulus
3   During the sound-stimulus phase

Replicated experiments have established that the conditionability of the fecal-bolus response is a more sensitive and reliable measure of all our strains than the open-field test performed conventionally.

GENOTYPE-ENVIRONMENT INTERACTIONS

As Figure 7.4 shows, the behavioral profiles are different for each of the strains tested. The genotype is, therefore, an important determinant of the effect of experience on this emotional response.

We have also found (Ginsburg and Allee, 1942), in common with Scott and coworkers (Scott, 1942; Scott and Fredericson, 1951), that

4 The author gratefully acknowledges the technical assistance of Dale Lang Messner.

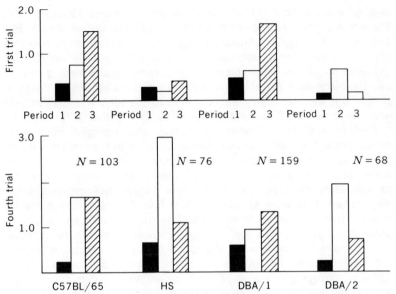

Figure 7.4   Boli/period/mouse, before and after exposure to bell.

there are profound differences in fighting behavior among isolated males of various strains of mice and that these differences are strain-specific. In the initial experiments, there was an apparent discrepancy between our results and those of Scott with respect to the rank order of the three strains that we tested in common [C-3H agouti, C (Bagg) albino, and C57BL/10]. In our experiments, the C57BL/10 mice were the most successful fighters in paired interstrain encounters, whereas in Scott's work the reverse was true. One possible explanation of this difference is that in one of the studies the fights were stopped as soon as they were initiated, whereas in the other they were permitted to

Table 7.2
Initial* versus later bouts won by C57BL/10 mice in encounters with C3H agouti and C–albino males ($N = 20$ males per strain†)

	Number of encounters	Number of fights	Initial bouts won	Later bouts won	P (Initial versus later wins)
C57 X C3H + C–albino	200	111	$\frac{7}{20}$	$\frac{60}{91}$	.01
C57 X C3H	100	74	$\frac{3}{10}$	$\frac{39}{64}$	.07
C57 X C–albino	100	37	$\frac{4}{10}$	$\frac{21}{27}$	.03
P (strain difference)		56×10^{-8}		.12	

* An initial bout represents the first actual fight engaged in by the C57BL/10 mouse, regardless of which animal initiated the aggression.
† Ten males of each strain were fought round robin fashion with ten males from each of the other two strains.

continue until a clear outcome was reached (Ginsburg, 1965a, p. 59). The simplest interpretation of this finding, however, is that the C57BL/ 10s are more easily conditioned to fight vigorously, although they may not do so initially. In attempting to repeat these experiments with descendants of the original strains, the same results were achieved; i.e., the black mice did not do as well on initial encounters as the other strains but became highly aggressive with experience and eventually won the majority of the interstrain encounters (Table 7.2).

Another and more interesting finding also emerged from these experiments. The methods for handling the mice in the two laboratories in question differed. In one, the mice were picked up by the tail wtih a forceps, while in the other they were transferred by being permitted to walk from one cage to another with an apposed entrance, or, if they balked, they were scooped up in a box. Recent evidence demonstrating that differences in early handling produce differences in later behavior suggested that the fighting experiments should be repeated in order to determine whether this could have been a factor in producing the differences in behavioral outcome mentioned above (Ginsburg, 1965a). Accordingly, males of each strain were matched as closely as possible for age and weight and were assigned to either a forceps-handled or a nonhandled group. (Weighings and cage transfers were carried out on a once-a-week schedule.) Litter mates were represented in each. Round robin encounters were then instituted on the same basis and schedule as in the original experiments (Ginsburg and Allee, 1942). Half of the animals of each group were scored according to the method used by Scott (1942); i.e., the fights were stopped as quickly as possible after initiation; the other half were scored according to the method of Ginsburg and Allee, where the fights were permitted to proceed to a definite outcome.

With either method of evaluation, the albino and agouti strains were not affected by the differences in once-a-week handling; however, the C57BL/10s exhibited a significant difference in outcome in both test situations, the forceps-handled animals being less aggressive than their counterparts, according to the measures used (Table 7.3). As previously reported, this strain appears unusually labile in its susceptibility to environmental influences (Ginsburg, 1958, 1963a, 1965a). In inter-

Table 7.3
Effects of handling on aggressive behavior of mice

Strain	Number of fights initiated per round robin*		P
	Forceps-handled	Nonhandled	
C57BL/10	27	55	.01
C3H agouti	22	26	
C–albino	19	18	

Data on interrupted and completed fights have been combined, as differences between these categories were not significant at the .05 level.
* $N = 10$ animals per group per strain (5 on interrupted fight schedules + 5 on completed fight schedules). Scores represent fights initiated out of 100 opportunities to fight (= 1 round robin).

strain contacts run exactly as the intrastrain encounters described in Table 7.3, the forceps-handled C57 males ranked below both the handled and nonhandled males of the other two strains in number of fights won, whereas the nonhandled C57 males ranked above them on the same measure. The difference was highly significant.

There is, therefore, an interaction between genotype and early experience such that aspects of aggressive behavior that are profoundly affected by early-handling variables in one genotype may be totally unaffected in another.

Denenberg and his coworkers (1964) have demonstrated a profound effect of foster rearing by rats in diminishing fighting among C57BL/10 mice.

It is of interest to note that seizures have an effect upon aggressive behavior of male mice (Ginsburg, 1963b, 1965a). One or more convulsions experienced by DBA males at weaning age profoundly reduce the incidence of fighting after sexual maturity.

That genotypic variables influence aggressive behavior has long been known (see Ginsburg, 1958, pp. 412–413; Scott, 1958; Fuller and Thompson, 1960). What is of greater interest to the student of behavior genetics is that the genotype demonstrably and predictably refracts experiential input according to its own biological properties. Early handling, consisting of removing each immature mouse individually from the nest and placing it in a small dish for 2 minutes, has little effect on the later aggressive behavior of HS mice. It does, however, affect the fighting activity and latency to fighting of other strains (see Ginsburg, 1963b, p. 236). The direction and magnitude of the effect are strain-specific, as are the time and duration of the experiential input that is most effective in producing a later behavioral departure from the control. Thus (see Ginsburg, 1963b, Table 7) the critical period in which early handling produces the greatest change in latency to fighting among adult males is the same for DBA/2 and C57BL/6 mice. However, the change produced is not the same. In the DBA/2 mice, fighting latency is reduced, whereas in the C57BL/6 animals it is markedly increased over normal control values, although the handling procedures applied to both strains are exactly the same, and the control latencies are almost identical. A comparable increase in latency to that produced in C57BL/6 mice by early handling is produced in C57BL/10 animals; however, among the 10s, the critical period for producing the maximum effect occurs earlier. In DBA/1 mice, the maximum effect is produced by cumulative experience during the entire preweaning period. Here, as in DBA/2, latency to fighting is reduced by the handling procedure.

Aggressiveness was tested by paired encounters between mature male mice isolated from weaning until 75 days of age.[5] They were then tested in a circular chamber 9 inches in diameter divided by a removable wooden partition, which separated each member of the pair initially and after the fight was stopped. Observations were made by

[5] The author wishes to acknowledge the technical assistance of Mrs. J. Jumonville.

means of a 10- by 12-inch mirror placed 12 inches above the top of the cylinder and positioned at a 45-degree angle. Tests were made once a day for 5 days, using at least 10 pairs of mice in each group. Latencies were measured from the time the partition separating the animals was lifted until the beginning of an actual attack.

In addition to the latency measure used to illustrate the interaction between genotype and experiential variables, observations have been extended to include other behavioral measures in a time distribution study by J. Jumonville (1967). These were also analyzed according to strain-experience interactions. Significant components of aggressive behavior included grooming, tail rattling, attacking, wrestling, submissive posture, jumping, and nosing. Jumonville also measured the fecal boli deposited per minute during the aggression tests and found that, for the DBA strains, animals that had not seized when exposed to sound stimulation at weaning age defecated less than those that had. The effect of the seizures exhibited at weaning age, as already mentioned, was to reduce fighting after sexual maturity. These findings are of particular interest in view of the effects of various types of shock (insulin shock, electroshock) on the subsequent behavior of disturbed human patients, where aggressive (including self-destructive) tendencies are also reduced.

Early handling had no detectable effect on seizure susceptibility, although the DBA/1 mice that had been handled daily during their entire preweaning period exhibited a markedly reduced fecal-bolus count during the 1-minute period that they were in the chamber prior to the onset of the sound stimulation. By using this measure as well as the percent of animals defecating during this pretest period on the first two test days, it was found that the response increased on the second day, whether or not a seizure occurred on the first day. For DBA/1 mice, there was no correlation between the fecal-bolus response and seizure susceptibility on the same day. Some of the other strains exhibited significant correlations on these tests.

So far as the aggressiveness measures are concerned, the strains tested assort in the following order when ranked according to the proportion of 5-minute trials resulting in fights ($N = 100$ to 150 for each group):

HS, more aggressive than A/Jax, C57BL/10, C57BL/6, DBA/1, DBA/2

DBA/2, more aggressive than A/Jax, C57BL/10, C57BL/6, DBA/1

DBA/1, more aggressive than A/Jax, C57BL/10, C57BL/6

None of the other differences were statistically significant. On the basis of the same measure, significant differences in the effects of handling occurred as follows:

Effective critical period	Increased aggression	Decreased aggression
2–14 days	C57BL/10	HS, DBA/1
15–27 days	C57BL/10, A/Jax	
2–27 days	C57BL/10, A/Jax	DBA/1, DBA/2

As may be seen, the results obtained from the latency measures do not correspond exactly to those based on incidence of fights. Other behaviors noted during the aggression tests were also significantly affected by early handling (Jumonville, 1967).

These same strains interact differentially with other forms of early stress, including low temperature (Ginsburg, 1965a), electroshock, and shaking (Jumonville, 1967). Genotypic differences in response to early stress can, therefore, determine very different outcomes in later life on a wide variety of behavioral measures. Moreover, the time of the critical period at which a given manipulation is most effective is both genotype- and behavior-specific. The direction and magnitude of the response to the same external stress may be quite different, depending on the genotype, which, on this basis, makes a determining contribution to the rank of an individual in a statistical distribution as a result not only of early stress but also with respect to manipulation by conditioning, drugs and antimetabolites, and even surgical insult to the brain (Ginsburg, 1954, 1958, 1963b, 1965a, 1965b; Maxson, 1966). Unless the genetic contributions to these interactions are taken into account, the investigator is in danger of "statisticking" away important biological variables contributing to the very behavior he is attempting to understand. The modal response of a species, breed, or stock is the measure usually investigated by psychologists (e.g., rat learning) and ethologists (e.g., species-specific behavior). However, the biological variability in potential for behavioral response is equally important in three respects:

1   As a clue to the evolutionary history of the species

2   As a potential for further evolution, especially if conditions of life change

3   As a means of understanding individual variability and of dealing effectively with it in a given situation

FUTURE BEHAVIOR-GENETIC ANALYSIS

As illustrated in the sections of this chapter dealing with the physiological genetics of susceptibility to audiogenic seizures, a multiplicity of genotypes can predispose to a common behavioral phenotype. The genetically mixed materials belonging to such a phenotype encompass a variety of underlying morphological, physiological, and biochemical mechanisms, each of which is consistently associated with a given genotype within the phenotype. Hence, attempts to deal with the analysis of causal mechanisms underlying behavior are often confused and confounded by lack of appropriate genetic controls. Selection experiments, the use of inbred strains, multiple backcrossing of mutants to an inbred-strain background, and the usual Mendelian analyses provide the methodological tools for instituting such controls and thereby of identifying and separating the components of what is otherwise a complex mixture of variables. Once this has been done, it is a far simpler matter to distinguish biologically meaningful from causally unrelated correlations by means of a gene-action approach.

At the population level, the necessity of distinguishing biologically based individual differences in capacity to respond to controlled environmental input is equally important. Gertrude Stein's "rose is a rose is a rose" only if obtained by self-pollination for a long-enough series of generations. Otherwise, $\sum\limits_{2}^{\infty}$ (rose + rose + rose) is the more descriptive statement. Analogously, the Gaussian distribution, if applied to aspects of the aggressive behavior of the laboratory mouse, represents an amalgam of genome-environment interactions. The stability of the distribution depends upon the extent to which the Hardy-Weinberg law is applicable under the particular circumstances affecting the population in question, and equally upon the reproducibility of the range of environmental variables from generation to generation. Animals taken from such a population and tested for interaction effects with normally impinging environmental variables would be expected to constitute a sample of every part of the population curve. On this view, the latter does not owe its variability primarily to chance deviations but rather to the binomial expansion interpreted to mean that the location of the point opposite each coefficient represents a discrete sampling of genotypes whose numerical representation is given by that coefficient and is determined by the phenotypic outcome of genome-environment interactions. It may be dissociated into a number of subpopulations by selection experiments, the mean of each of which will constitute a point on the original distribution. Only where the environmental variables are sufficiently overriding so as to produce a single outcome against the range of usual biological variability would the conventional interpretation of the normal curve be expected to apply.

Elsewhere in this volume, Hamburg has provided a physiological model for genetic-environment interactions in relation to applied stress. Beginning with stimulus input, there can, on our combined conceptions, be genetic variability in the degree to which a given applied stress can be reflected in neural events that are capable of stimulating particular secretory cells in the median eminence of the hypothalamus. There can also be a genetic influence on the rate at which the structural elements involved in detection of the stimulus and those concerned with the transmission of neural impulses achieve their full ability to respond. The secretory cells may also vary genetically in threshold stimulation required to initiate activity, in the relation between the intensity of stimulation and the degree of response, and in the developmental rate at which their full capacities are reached. If there are no genetic anomalies gross enough to interfere with the synthesis of the appropriate polypeptides in these secretory cells, the response mechanism of the target cells in the pituitary to these specialized substances appearing in the hypothalamic portal circulation may also vary genetically, as is strongly suggested by data on individual, strain, and species differences in adrenal steroid responses to stress. These may be equally validly considered to depend upon biological variability in adrenal cortical responses. Even if there were no genetic variability in any of these mechanisms, the ontogeny and style of end organ response could provide ample latitude for genetic variability. The availability of pure strains for which the most sensitive periods for behavioral change resulting from early stress are strain-behavior-specific, as are the directions, magnitudes, and characteristics of the effects, provides tools by means of which such problems may now be meaningfully subjected to experimental investigation.

The methods and concepts of behavior genetics constitute an in-dispensable tool for the meaningful investigation of problems of be-havior at every level, from the molecular, through the organismic, to the population. Without them, the organism remains a black box; nature-nurture interactions will remain mysterious; fruitless contro-versies relating to biogenesis versus psychogenesis of clinical syn-dromes will remain as unsettled as heretofore; drug therapy and psy-chotherapy will proceed on a wholly empirical basis; and countless correlational studies will provide the bases for theories, with no non-circular way of testing their foundations. Once behavior is taken to be as characteristic a component of the total phenotype as immuno-chemical specificity or gross morphology, the same biological principles that have been applied so effectively in these less labile realms may be applied to the study of behavior, with a much enhanced probability of achieving a similar breakthrough.

CHAPTER EIGHT
GENETICS OF ADRENOCORTICAL HORMONE METABOLISM IN RELATION TO PSYCHOLOGICAL STRESS
David A. Hamburg

PSYCHOLOGICAL STRESS AND THE ADRENAL CORTEX
CNS Regulation of Adrenocortical Function

The past decade has seen the emergence of a dynamic research area concerned with the coordinated functioning of nervous and endocrine systems in the adaptation of the whole organism to environmental conditions (Weitzman, 1964; Nalbandov, 1963; Scharrer and Scharrer, 1963; Reichlin, 1963; G. W. Harris, 1962; Tepperman, 1962; Green, 1962; Greer, 1962; Bajusz, 1960, 1961, 1962, 1963; Strom, 1961; Mason, 1959a; Fortier and de Groot, 1959; Harris, 1957).

For many years the brain and the endocrine glands have been viewed largely as separate entities, mediating quite different functions. The brain has been viewed mainly as mediating the organism's relation to the environment through behavior, i.e., oriented to the external environment. The endocrine system has been viewed mainly as regulating reproduction, growth, and metabolism, i.e., oriented to the internal environment. In both directions, recent research has achieved an increasingly refined analysis at cellular, subcellular, and molecular levels. Concomitantly, research has increasingly clarified the interdependence of the parts of the organism and the coordination of the whole in relation to changing environmental conditions. In this perspective, the brain and the endocrine system may profitably be viewed as a functional unit in adaptation. The integrative functions of the nervous and endocrine systems are no longer viewed largely as separate realms; rather, the superordinate integration of these two systems in managing the affairs of the whole organism has become a major object of research.

Experimentally, a great deal of evidence has accumulated in support of the concept that the endocrine system functions largely under CNS control; at the same time, there is a growing body of evidence that circulating hormones exert feedback controls on the brain. Thus, information from both internal and external environments is integrated in the brain (particularly in the brain stem), and the function of the various endocrine glands is adjusted accordingly (Scharrer and Scharrer, 1963; Tepperman, 1962).

Within this framework, much work has centered on the hypothalamus and its relation to the pituitary, the "master gland" of classical endocrinology (Harris, 1957; Nauta, 1963; Guillemin and Schally, 1963; Ganong, 1963; Sayers, 1962; Saffran, 1962; Sloper, 1962; E. Anderson, 1961; Vogt, 1960; Saffran and Saffran, 1959; Sayers et al., 1958; Donovan and Harris, 1957). It has become increasingly clear that neurosecretory cells in the median eminence of the hypothalamus secrete several substances that selectively elicit the secretion of the various tropic hormones of the anterior pituitary gland. These substances reach the pituitary in vertebrates through a special system of vessels, the portal veins. These vessels link the median eminence of the hypothalamus with the anterior pituitary and thus provide the principal bridge between nervous and endocrine systems. This portal system provides a final common pathway of neuroendocrine integration, in the sense that a variety of neural circuits may transmit information through this pathway to the anterior pituitary and hence to the gonads, thyroid, and adrenal cortex, via their respective tropic hormones. For the purposes of this discussion only one of these functional subdivisions is immediately relevant: CNS regulation of the function of the adrenal cortex via its tropic hormone, ACTH (adrenocorticotropin).

There is abundant evidence from neuroendocrine research of the past decade that structures in the hypothalamus and limbic system are involved in the regulation of ACTH secretion and hence in the secretion of cortisol (hydrocortisone) (Nauta, 1963; Matsuda et al., 1964a, 1964b; Brodish, 1963; Kendall et al., 1964; Davidson and Feldman, 1962, 1963; Leeman et al., 1962; Smelik and Sawyer, 1962; Taylor and Farrell, 1962; Setekleiv et al., 1961; Slusher and Hyde, 1961a, 1961b; Royce and Sayers, 1959; Schally et al., 1958; Slusher, 1958; McCann et al., 1958; Mason, 1958a, 1958b; Nauta and Kuypers, 1958; Smelik, 1958; Newman et al., 1958; Saffran and Schally, 1955). The sequence of events precipitated by stressful stimulation may be roughly outlined as follows: Impulses from the cerebral cortex impinge upon hypothalamic centers in and around the median eminence. Large neurosecretory cells in this region then secrete corticotropin-releasing factor (CRF). CRF is a polypeptide (very similar to vasopressin) which is secreted by the neurosecretory cells into the portal veins, which descend around the pituitary stalk into the anterior pituitary gland.

The arrival of the stimulus at the anterior pituitary from the hypothalamus does not produce an indiscriminate discharge of the various tropic substances. There appear to be specific controls that permit secretion of each tropic hormone selectively, based upon neurosecretion of a distinct releasing factor for each tropic hormone. Evidently, these releasing factors are closely related polypeptides.

A variety of approaches to the study of brain-pituitary relations have been undertaken (Tepperman, 1962; E. Anderson, 1961). One major approach to the analysis of hypothalamic control of the anterior pituitary has been to interrupt all neurovascular connections which normally connect them. Another major approach, growing in utilization in recent years, has involved electrical stimulation, with subsequent measurement of adrenocortical function, sometimes by indirect indices but lately more often by direct measurements of corticosteroids. These two major types of approach may be further specified in the following methods:

1   Studies of animals in which the anterior pituitary has been surgically removed and pituitary grafts implanted in sites far removed from the hypothalamus
2   Stimulation of various brain regions chiefly in the hypothalamus by means of stereotactically placed, chronically implanted electrodes
3   Blocking of certain anterior pituitary responses, such as ACTH release, by stereotactic placement of electrolytic lesions in hypothalamic areas
4   Inhibition of release of ACTH by placement in the hypothalamus of tiny fragments of adrenocortical tissue or of hydrocortisone
5   Pituitary-stalk section experiments, sometimes including a mechanical barrier to the regeneration of the portal vessels from above
6   Studies of pituitary hormone content after placement of hypothalamic lesions
7   Extraction and purification of hypothalamic substances which selectively stimulate the release of anterior pituitary hormones

The level of output of cortisol in adrenal vein blood is very largely determined by the intensity of the ACTH stimulus to the adrenal cortex. The secreted cortisol feeds back information to the hypothalamic-pituitary cells. Thus, an increase in cortisol concentration leads to a decrease in ACTH release. Conversely, ACTH release is enhanced by a lowering of cortisol in the systemic circulation. However, it is abundantly clear that ACTH may also be increased, sometimes quite strikingly, in the absence of any such lowering of the circulating concentration of cortisol; the latter increases occur through strong CNS stimulation of the anterior pituitary-adrenocortical system.

In general, the pituitary transplantation studies have indicated the importance of the portal system for normal functioning of the anterior pituitary. However, these studies also indicate that some ACTH may be released independently of hypothalamic control. Similarly, studies utilizing interruption of the hypophysial stalk and portal system demonstrate serious impairment but not abolition of the ACTH-releasing mechanism. The hypothalamic-lesion studies have also contributed substantially to the evidence that the hypothalamus exerts important controlling influences on ACTH release. Destruction of the tuberoinfundibular nucleus and of the median eminence limits the capacity of the pituitary to release ACTH under stress but does not prevent a basal secretion of ACTH. Of all the hypothalamic and limbic areas that have been studied through lesion experiments, the region most consistently associated with significant impairment of ACTH release is the median eminence.

A particularly interesting direction of neuroendocrine research in recent years has involved the elucidation of influences from higher centers playing upon the hypothalamus and through it upon the anterior-pituitary–adrenocortical system (Nauta, 1963; Ganong, 1963; Setekleiv et al., 1961; Slusher and Hyde, 1961b; Mason, 1958a, 1958b; Nauta and Kuypers, 1958). In this work, attention has chiefly been centered on the limbic system. This is the great limbic lobe of Broca, completely enveloping the brain stem; it consists of the old cerebral cortex, from an evolutionary viewpoint, as well as subcortical structures. Nauta's work has substantially clarified limbic-hypothalamic-midbrain relations. On the basis of this research, Nauta conceives of the hypothalamus as a major relay station intercalated between neuron pools of the limbic system on one side and the midbrain on the other. Several workers have shown that stimulation of various limbic areas through chronically implanted electrodes in waking mammals, including primates, can produce substantial increases or decreases in the circulating concentration of cortisol. The amygdala and hippocampus have been particularly important in these experiments; both areas appear to have a major role in the regulation of ACTH release and hence in the regulation of adrenocortical activity. However, several other limbic structures appear also to be significantly involved in this regulation. Altogether, there is a growing body of information on the neural circuitry underlying the regulation of endocrine responses in accordance with varying environmental conditions.

Adrenocortical Function in Naturally
Occurring Psychological Stress

Psychological-stress research in man has developed a considerable body of evidence in recent years indicating that the anticipation of personal injury may lead to important changes not only in thought, feeling, and action but also in endocrine and autonomic processes and hence in a wide variety of visceral functions (Hamburg, 1959, 1962; Michael and Gibbons, 1963; Lloyd, 1963; Friedman et al., 1963; Fishman et al., 1962; Elmadjian, 1962; Suwa et al., 1962; Horigan, 1960; Mason, 1959b; Pace et al., 1956; Bliss et al., 1956). Much work in this field has centered on the changes in adrenocortical functioning that occur in association with emotional distress. Investigators have generally found the adrenal cortex to be stimulated via the CNS under environmental conditions perceived by a person as threatening to him. Usually such personally threatening conditions precipitate clearly detectable emotional distress. In some studies, it has been possible systematically to correlate the extent of emotional distress with the plasma or urinary corticosteroid level, each assessed independently. Early work in this field relied heavily on bioassays and 17-ketosteroid measurement. The results of those earlier studies have been greatly strengthened by more recent ones depending upon precise, reliable biochemical methods for measuring 17-hydroxycorticosteroids. Several hundred persons have been studied in various laboratories under conditions of moderately intense distress. The results are quite consistent, showing significant elevation in both plasma and urinary 17-hydroxycorticosteroids over the observations made under nondistress conditions. Moreover, many of the persons in the distress groups have been studied on repeated occasions, and the elevated 17-hydroxycorticosteroid levels have been found to be quite persistent or recurrent when the distress remained unabated. With relief of distress, a tendency toward a substantial decrease in corticosteroids has been observed. In some instances, it has been possible to study individuals undergoing extremely intense distress; under these conditions, exceedingly high corticosteroid levels have been observed. Although the data are generally less adequate, similar studies relying upon newer biochemical methods for measurement of epinephrine, norepinephrine, and aldosterone, under conditions of emotional distress, have yielded similar results. Thus, it now appears likely that emotional distress in man is associated with elevated blood and urinary levels of four adrenal hormones. These elevated levels probably reflect increased secretory activity by the gland.

Experimental Data on Adrenocortical Function in Psychological Stress

During the past decade experiments have generated substantial data indicating a linkage between emotional responses and adrenocortical function in man (Bliss et al., 1956; Wadeson et al., 1963; Handlon et al., 1962; David et al., 1962; Euler et al., 1959; Persky et al., 1958, 1959; Hamburg et al., 1958). In the early work, the stress interview was the most widely employed tool. For example, Bliss used stress interviews as well as several other experimental techniques as a means of producing mild-to-moderate emotional perturbation in human subjects. Emotional responses were usually of short duration and modest intensity. A significant rise in plasma 17-hydroxycorticosteroids could be demonstrated only when subjects were separated into those who showed clear-cut responses of emotional disturbance and those who were relatively undisturbed. This type of differentiation was carried further by the Michael Reese group who worked out reliable rating scales for three kinds of emotional response (anxiety, anger, and depression) and applied these scales systematically in a stress-interview situation, with each subject being used as his own control for a period of 4 consecutive days. The extent of increase in anxiety, anger, and depression and a combined distress rating were found to be significantly and linearly related to the plasma 17-hydroxycorticosteroid level. In general, the greater the emotional distress, the greater tendency toward plasma 17-hydroxycorticosteroid elevation. This same group later used a perceptual distortion situation as an experimental stress technique. The situation was set up in such a way that the subject's basic ability to perceive his environment accurately was challenged. Under these conditions, a tendency toward an elevation in plasma 17-hydroxycorticosteroid was again observed, particularly in those who felt personally threatened by the procedure.

Striking elevation in plasma 17-hydroxycorticosteroid levels has been produced in rhesus macaque by conditioning procedures associated with anticipation of noxious stimulation (Mason et al., 1957). Both a conditioned fear situation (anticipation of pain) and a conditioned avoidance situation (avoidance of anticipated pain) were associated with sharp increases of plasma 17-hydroxycorticosteroid levels. The changes produced in these experiments were as great as those produced by a large dose of ACTH, thus suggesting the possibility that adrenocortical secretory activity might approach its maximal rate under acute stress conditions. In control studies, animals in a familiar situation undergoing no experimental procedures did not show plasma 17-hydroxycorticosteroid elevation during sessions of similar duration; neither did monkeys which were pressing levers for food rewards on several reinforcement schedules. The plasma 17-hydroxycorticosteroid elevations observed in the conditioned fear and conditioned avoidance situations could be elicited repeatedly over a period of several months when the monkeys were studied once weekly.

The reverse has also been reported, i.e., conditioned suppression of stress response (adrenal ascorbic acid depletion) (Komaromi and Donhoffer, 1963). Rats were accustomed to handling. A group given physiological saline intravenously showed adrenal ascorbic acid depletion 90 minutes after the injection. When the rats were accustomed to the injection (20 days), the ascorbic acid response was diminished, though not abolished. However, when habituation was associated with feeding immediately following the injection, the ascorbic acid effect was abolished completely.

In the past few years, motion pictures have been used in an experimental technique for eliciting emotional and adrenal responses in man. Von Euler and associates (1959) used cuttings from fiction and documentary motion pictures, showing a variety of scenes involving physical injury and human suffering. One hour's exposure of healthy young normal subjects to this material elicited a significant increase in excretion of epinephrine and a slight increase in urinary excretion of 17-ketosteroids, pregnanediol and norepinephrine, but no increase in plasma 17-hydroxycorticosteroids. As in previous studies (Bliss et al., 1956; Persky et al., 1959) those subjects who became most personally involved in the stress procedure and who experienced the most intense emotional reactions also showed the greatest tendency to elevation of adrenal hormones.

Another recent study (Wadeson et al., 1963; Handlon et al., 1962) calls attention to the psychological conditions under which the circulating concentration of plasma 17-hydroxycorticosteroids may be lowered rather than raised. When 19 normal young adult male subjects viewed Disney nature-study films, it was found that the plasma 17-

hydroxycorticosteroid levels were lowered to a significant degree in comparison with (1) a control period in which no films were shown and (2) a showing of emotionally arousing films. The clear difference in plasma 17-hydroxycorticosteroid response to arousing and bland films suggests that the adrenal cortex may respond to events of emotional significance within the range of mildly stressful, ordinary experience and that the CNS regulation of adrenocortical function involves lowering as well as raising plasma 17-hydroxycorticosteroid concentrations. It is important to note that in all these experiments with man, the experimenters have necessarily been limited for the most part to distress responses of low intensity and short duration, since ethical considerations in human experimentation preclude more radical interventions.

Altogether, the extensive research of the past decade on adreno- cortical function in man in naturally occurring and experimental stress situations has provided substantial evidence of significantly increased secretory activity of the adrenal cortex under these conditions. In- creasingly, a variety of workers in different disciplines have become interested in the potential physiological and clinical significance of this phenomenon. Bush (1962), in a comprehensive review of chemical and biological factors in the activity of adrenocortical steroids, makes the following statement.

☐ Another important development in recent years has been the steady growth of the idea that emotional factors of various kinds probably play the major part in causing fluctuations of adrenocortical secretion rate in man, monkeys, and possibly other species. It is probable that very severe burns, and large doses of certain agents such as bacterial pyrogens, his- tamine, and peptones, cause a brisk release of ACTH, that is independent of any emotional concomitants; but it is extremely doubtful whether any of the physical stimuli which are commonly supposed to be "stresses" are effective in causing the increased secretion of ACTH at all. Thus, severe exercise, cold, and fasting produce little or no effect on the secretion and metabolism of cortisol in man unless they are part of a situation that pro- vokes emotion. On the other hand, strong emotion in the absence of any recognizable physical stimuli or "stresses" regularly causes maximal in- creases in the secretion rate of cortisol and its concentration in peripheral blood. . . . It is difficult to overemphasize the importance of these findings for adrenocortical physiology. . . . The increases in concentration of corti- sol in plasma and in the excretion rate of metabolites of cortisol that are seen during such relatively common and mild periods of anxiety as being given an oral examination or interview for a senior position are such as are being seen otherwise only in patients with severe Cushing's syndrome. Much further work is needed to elucidate the question of what part this phenomenon might play in psychosomatic disease.

Individual Differences in Adrenocortical Responses to Psychological Stress

Up to this point, we have been considering general tendencies: for example, that distress is associated with elevations in 17-hydroxy- corticosteroids and that relief of distress is associated with a fall in corticosteroid levels. Indeed, we may summarize several major trends of the evidence on general tendencies of adrenocortical function under conditions of psychological stress: (1) there is an important set of CNS regulatory functions acting upon the anterior-pituitary–adreno-cortical system, particularly through brain structures in the hypo-thalamus and limbic system; (2) elevations in plasma and urinary 17-hydroxycorticosteroids are regularly observed under difficult circum-stances perceived by an individual as threatening to him; (3) there is a positive correlation between the degree of distress experienced by the individual and the tendency toward corticosteroid elevation. It is im-portant to note that we have also observed individual differences in 17-hydroxycorticosteroid excretion, consistent over several months and through several stressful experiences (Mason and Hamburg, unpub-lished observations). These consistent individual differences have been observed in relation to both (1) the range within which the person's 17-hydroxycorticosteroid excretion fluctuates under ordinary circum-stances for him and (2) the responses in 17-hydroxycorticosteroid excretion to a difficult, disturbing experience. This finding of consistent individual differences in adrenocortical response to environmental con-ditions, including those of a stressful character, touches on the im-portant problem of differential susceptibility to psychological stress. For many years, clinicians in a variety of fields have been impressed with the vulnerability of some individuals to stressful experience and the striking resistance of others under what appear to be similar conditions of personal threat and emotional distress. The precipitation and exacerbation of a variety of illnesses have been associated with emotional crisis—not only psychiatric disorders but a rather wide range of medical problems, prominently including but not limited to the classical psychosomatic disorders. Yet it is abundantly clear that many individuals undergo the ubiquitous stressful experiences of living without developing such disorders. In principle, there are good reasons for anticipating that a great variety of genetic and environ-mental factors might contribute to the formation of consistent in-dividual difference in stress response and hence to differential sus-ceptibility. In the present context, we are considering only one source of such individual differences in stress response but a potentially important one: genetically determined enzymatic differences in syn-thesis or disposal of adrenocortical hormones.

HUMAN BIOCHEMICAL GENETICS OF ADRENOCORTICAL HORMONES

Normal Biosynthesis and Catabolism of Corticosteroids

The following is a brief account of the principal metabolic pathways of corticosteroids (Dorfman, 1961, 1962; Forsham, 1962; Fortier, 1962; Lieberman, 1961; Berliner and Dougherty, 1961; Soffer et al., 1961; Bush, 1960; Heftmann and Mosettig, 1960; Tomkins and McGuire, 1960). Cholesterol is the most important precursor of the steroid hormones. For this and other reasons, one important line of inquiry in recent years has centered on the biosynthesis of cholesterol. The current biochemical consensus on the sequence of events in cholesterol biosynthesis has been arrived at through carbon-by-carbon degradations of isotopically labeled cholesterol and its synthetic intermediates. After it was demonstrated that mammals have the ability to synthesize cholesterol, isotope experiments indicated that this synthesis *in vivo* is accomplished from small molecules rather than from large precursors. A variety of experiments with mammalian organs, notably the liver, pointed to acetate as the primary sterol precursor. Later experiments involving the stepwise degradation of cholesterol biosynthesized from labeled acetate-C-14 showed that each carbon in the sterol molecule arises either from the carboxyl or methyl carbon of acetate. It appears that almost all mammalian tissues can accomplish the biosynthesis of cholesterol from acetate, notably including the adrenal gland. Important steps in this biosynthetic pathway include the following: activation of the free acetate ion by condensation with coenzyme A to form acetylcoenzyme A; the formation of mevalonic acid (involving the rate-limiting step in sterol biogenesis); the formation of squalene (a 30-carbon intermediate); the cyclization of squalene, leading to the first sterol intermediate, lanosterol; and the conversion of lanosterol (C-30) to cholesterol (C-27).

The next phase involves the biosynthesis of adrenocortical steroid hormones from cholesterol. Some years ago, Bloch first showed that cholesterol is a precursor of steroid hormones by administering deuterium-labeled cholesterol to a pregnant woman and isolating labeled pregnanediol from her urine. The conversion of cholesterol to corticosteroids has been demonstrated in a variety of experiments, both *in vivo* and *in vitro*. Important steps in this biosynthetic pathway include the following: cleavage of the side chain of the cholesterol molecule, yielding pregnenolone (a 21-carbon steroid) via 20-hydroxycholesterol (ACTH is the main regulator of reactions from cholesterol to pregnenolone); the oxidation of pregnenolone to progesterone, an important hormone in its own right and a crucial intermediate in the biosynthesis of other steroid hormones; and the hydroxylation of progesterone in three positions to form cortisol. These final hydroxylation steps are of great interest biochemically, genetically, and clinically. The progesterone molecule is quite similar to the cortisol molecule but lacks three hydroxyl groups, at the C-11, C-17, and C-21 positions. One striking characteristic of these hydroxylation reactions is their dependence upon the availability of TPNH. Known defects in these hydroxylation reactions will be described below. Before doing so, it is worth noting that most of the biosynthetic pathway outlined above also applies to formation of aldosterone. However, aldosterone is not hydroxylated at the C-17 position. It is formed through corticosterone, which is hydroxylated at the C-11 and C-21 positions; the biosynthesis of aldosterone is accomplished through the 18-hydroxylation of corticosterone followed by oxidation of the alcohol to an aldehyde function at the C-18 position.

The main features of cortisol catabolism will now be briefly sketched. The half-life of free cortisol in plasma, as measured with C-14-labeled cortisol, is about 2 hours. There is virtually no destruction of the steroid ring in cortisol catabolism. The principal site for catabolism of the hormone is the liver. The first step is the reduction of the double bond between C-4 and C-5, hence the incorporation of two hydrogens in the A ring and formation of dihydrocortisol, which is biologically inactive. Next, the α-3 ketonic group is reduced, forming the tetrahydro derivative. The tetrahydro derivative is then conjugated with an acid, mainly glucuronic acid, but also to a lesser extent with sulfuric and phosphoric acids. The enzyme glucuronosyl transferase is crucial in the conjugation with glucuronic acid. This conjugate is quite water-soluble and is rapidly excreted in the urine through glomerular filtration with no appreciable reabsorption by the tubules. In addition, some cortisol (5 to 10 percent) is further degraded in the liver to 17-ketosteroids, principally the etiocholanolones. Cortisol and cortisone are freely interchangeable by enzymatic action in the liver, hence the finding of substantial amounts of tetrahydrocortisone in the urine, even though no measurable cortisone is present in the blood. Both tetrahydrocortisone and tetrahydrocortisol may be reduced further to the 20-hydroxy derivatives, the cortolones and cortoles. Mean values for 24-hour urinary excretion of major end products of cortisol in adults are as follows: tetrahydrocortisone, 5 mg; tetrahydrocortisol, 3 mg; tetrahydrocortol and -cortolone, 3 mg; free cortisol, 30 μg; and 11-hydroxy 17-ketosteroids, 1 mg.

Genetically Controlled Abnormalities of Adrenocortical Hormone Metabolism

Much interesting work in recent years has gone into the discovery and elucidation of several defects in hydroxylation during biosynthesis of cortisol; these defects have now had considerable biochemical and clinical clarification, with some genetic evidence as well (Tomkins and McGuire, 1960; Brooks, 1962; Motulsky, 1962a; Wilkins, 1962; Fukushima et al., 1962). Together they comprise the clinical syndrome, congenital adrenal hyperplasia. They will be discussed in some detail below. In addition, a small amount of work has been done on enzymatic defect in the conjugation of corticosteroids with glucuronic acid (Childs et al., 1959). Further, one study has presented some evidence suggesting the possibility of a shift in the excretory pattern of corticosteroid metabolites in some individuals when under stress (Persky, 1957).

In general, the genetics of hormone metabolism has not been extensively studied (Motulsky, 1962a). It now appears that there are a number of intriguing possibilities waiting for exploration in this field; perhaps it will become an important branch of human biochemical genetics within the next decade. For example, several defects in the synthesis of thyroid hormone have been identified, and the biochemical nature of the defect has been elucidated (Motulsky, 1962a; Williams and Bakke, 1962). Each of these leads to hypothyroidism.

Similarly, one form of diabetes insipidus appears to be due to a genetically determined deficiency in secretion of antidiuretic hormone (Levinger and Escamilla, 1955; Cannon, 1955; Forssman, 1945, 1955). Interestingly, this deficiency appears to be not an enzymatic defect in synthesis of the hormone but rather a quantitative reduction of nerve cells in the supraopticohypophysial system, that is, a genetic defect in the CNS control mechanism pertinent to a set of endocrine functions.

The following are clinical syndromes of abnormal adrenocortical hormone metabolism (Forsham, 1962; Tomkins and McGuire, 1960; Brooks, 1962; Motulsky, 1962a; Wilkins, 1962; Fukushima et al., 1962).

Defective Hydroxylation at C-21   This is the most common and most adequately studied of the congenital adrenal hyperplasias (Tomkins and McGuire, 1960; Brooks, 1962; Motulsky, 1962a; Wilkins, 1962; Fukushima et al., 1962; Eberlein and Bongiovanni, 1960; Birke et al., 1958; Dyrenfurth et al., 1958; Childs et al., 1956; Cleveland et al., 1962; Jacobsohn, 1962). Since the predominant symptoms are those associated with excessive androgens (e.g., hirsutism in girls and precocious puberty in boys), attention might be drawn to the gonads. However, work of the past decade has shown clearly that the difficulty is due to excessive secretion of adrenal androgens rather than gonadal androgens. Important links in this chain of evidence may be cited as follows:

1   Such patients usually have low levels of circulating corticosteroids.

2   High levels of ACTH are found in the blood.

3   Administration of ACTH in large doses yields no appreciable rise in plasma corticosteroid levels.

4   Large quantities of androgenic 17-ketosteroids are found in the urine.

5   Pathological evidence reveals diffuse, bilateral adrenocortical hyperplasia.

The sequence of events in pathogenesis is itself an interesting commentary on neuroendocrine relations. Cortisol cannot be synthesized or can be synthesized only in very small quantities, owing to a genetically determined defect in a hydroxylating enzyme. This information is fed back to the hypothalamus by means of the very low circulating concentration of cortisol. In turn, information is conveyed via the hypothalamic-pituitary portal system, leading to a substantial increase in ACTH secretion. The high ACTH concentration would normally have the effect of greatly increasing cortisol secretion, but, in the presence of the enzymatic defect crucial to cortisol synthesis, the principal effect is merely to increase greatly the secretion of adrenal androgens, which normally are produced only in very small amounts. This excessive adrenal androgen production in turn leads to the symptoms of masculinization.

In the C-21 hydroxylation defect, the block is in the 21-hydroxylation reaction from 17-hydroxyprogesterone to 11-desoxycortisol (compound S), the immediate precursor of cortisol. There is then an accumulation of compounds behind the block as well as metabolites of these compounds. Thus there is a notable accumulation of 17-hydroxyprogesterone and of its main metabolites, pregnanetriol and 17-hydroxypregnenolone. In addition, a variety of unusual steroids appear in the urine; these steroids can be largely explained by a defect in 21-hydroxylation. They are products of alternate routes of metabolism of compounds whose major route is blocked by the enzyme deficiency. The quantities of pregnanetriol and 17-ketosteroids appearing in the urine in this syndrome are exceedingly large; on the other hand, the quantities of tetrahydrocortisone and tetrahydrocortisol are usually well below normal. In most cases, the block is not complete; although the production of cortisol and aldosterone is seriously deficient, enough cortisol is usually produced so that there is masculinization without adrenal insufficiency, and enough aldosterone is produced so that sodium loss is not dangerously excessive.

However, within this category there is a subgroup called the salt-losing syndrome. In these cases, no tetrahydrocortisol is excreted, and there is even more urinary pregnanetriol; this indicates that the block in 21-hydroxylation is virtually complete. The adrenal, even though highly stimulated by the large quantities of ACTH reaching it, is unable to produce enough 21-hydroxylated steroids to prevent salt loss and adrenocortical insufficiency. Patients are seriously lacking in both cortisol and aldosterone; they die early unless treated promptly with large doses of both hormones. There is a highly interesting intermediate group of patients who are able to produce minimally adequate quantities of cortisol under ordinary conditions but who become "salt losers" in the presence of superimposed stress, such as infections (Eberlein, 1958). The implications of this type of situation for psychological-stress research will be discussed below.

Defective Hydroxylation at C-11   This is a rare syndrome characterized by the lack of the normal enzymatic hydroxylating system at the C-11 position (Tomkins and McGuire, 1960; Brooks, 1962; Motulsky, 1962a; Wilkins, 1962; Fukushima et al., 1962; Reynolds and Ulstrom, 1963; Bergman et al., 1962; Eberlein and Bongiovanni, 1955, 1956; Bongiovanni and Eberlein, 1955). The block here is in the final step of cortisol synthesis, i.e., from 11-desoxycortisol to cortisol. Thus 11-desoxycortisol (compound S) accumulates behind the block; moreover, the next to last step in the pathway of aldosterone synthesis is also blocked, i.e., the step from 11-desoxycorticosterone to corticosterone, thus leading to an accumulation of 11-desoxycorticosterone behind the block. The latter compound is a potent mineralocorticoid and is instrumental in the pathogenesis of the hypertension which is characteristic of this C-11 syndrome. A principal excretion product here is tetrahydro S, which appears in large quantities in the urine. Both compound S and tetrahydro S appear in the blood in relatively high concentrations. In contrast to the C-21 syndrome, no 11-oxy-genated 17-ketosteroids are found in the urine.

5-Pregnenolone Block   This too is a rare syndrome, quite recently discovered, in which synthesis of cortisol is interrupted and early precursors accumulate, leading to formation of androgens (Forsham, 1962). Profound adrenocortical insufficiency ensues.

3β-hydroxysteroid Dehydrogenase Defect   This is another rare syndrome recently discovered (Bongiovanni, 1961). It is found in infants with the salt-losing syndrome whose defect seems to involve the 3β-hydroxysteroid dehydrogenase system, since 3β-hydroxy-Δ^5-steroids predominate in the urine. There is an absence of cortisol metabolites and those C-21 steroids such as pregnanetriol that are usually found in congenital adrenal hyperplasia.

Etiocholanolone Excess   In this syndrome, there is also deficient production of cortisol, with the distinctive feature of etiocholanolone accumulation (Bondy et al., 1958; Gardner and Migeon, 1959). Since etiocholanolone is capable of causing periodic fever in man (probably via a central-nervous-system action), recurrent episodes of fever are a part of the clinical picture. Both the fever episodes and the high circulating levels of etiocholanolone may be eliminated by corticosteroid treatment.

Intermediate Form of Congenital Adrenal Hyperplasia in Women This syndrome is characterized by oligomenorrhea or amenorrhea, anovulatory temperature curves, and elevated urinary 17-ketosteroids, pregnanetriol and pregnanediol (Gold and Frank, 1958; DeAlvarez and Smith, 1957; Greenblatt et al., 1956; Jones et al., 1953). Treatment with cortisol corrects the condition, leading to menstruation, ovulatory temperature cycles, and decrease in output of 17-ketosteroids, pregnanetriol and pregnanediol.

Lipoid Hyperplasia of the Adrenals with Generalized Steroid Hormone Insufficiency   Here an enzymatic defect early in the pathway of steroid hormone biosynthesis precludes the formation of steroid hormones from cholesterol (Prader and Siebenmann, 1957). The adrenals become enlarged and filled with lipides. The ovaries and testes are also involved. The synthetic incapacity includes gonadal as well as adrenal steroids. These patients die from adrenal insufficiency in early infancy. The nature of this syndrome is not yet well established.

Addison's Disease   While the adrenal insufficiency of Addison's disease may be produced in a variety of ways, one subgroup appears to have a genetic basis (Stempfel and Engel, 1960; Meakin et al., 1959; Mitchell and Rhaney, 1959; Shepard et al., 1959; Brochner-Mortensen, 1956; Mosier, 1956; Berlin, 1952). A few cases of un-complicated familial Addison's disease have been reported. This familial condition may be detected in infancy and is associated with adrenocortical hypoplasia. It has also been described in older children, in whom it was characterized by inadequate cortisol secretion but no clinically significant deficiency in aldosterone production.

Defect in Conjugation with Glucuronic Acid   In general, there is exceedingly scant information on genetically determined enzymatic defects involving catabolism of adrenocortical or any other hormones. However, in principle, the possible accumulation of toxic metabolites of one or another adrenal hormone under stress is a matter of much interest. One study has a direct bearing on this problem (Childs et al., 1959). The investigators were basically concerned with bilirubin metabolism in nonhemolytic familial jaundice (Crigler-Najjar syndrome). This is the result of an autosomal recessive gene; the key defect is the absence of the enzyme glucuronyl transferase. This enzyme in liver microsomes normally conjugates various substances, e.g., bilirubin, salicylates, and cortisol, thereby facilitating their excretion. Childs and associates administered C-14-labeled cortisol and tetrahydrocortisone (as well as other test substances) to one patient with this disease. They demonstrated a partial deficiency in conjugation of both steroids with glucuronic acid. However, the patient was able to dispose of both labeled steroids by other means; evidently alternate pathways were used. (For example, perhaps more steroid was conjugated as sulfate or phosphate.) However, the amount of cortisol used was quite small (175 μg, which would give little elevation of plasma 17-hydroxycorticosteroids); moreover, it was a single-injection, short-term experiment. It is possible that the significance of the partial deficiency would be more striking in a long-term situation.

Genetic Analysis of Human
Adrenocortical Metabolic Abnormalities

Several systematic studies have been carried out on families of patients with congenital adrenal hyperplasia, and the evidence has been subjected to formal genetic analysis (Childs et al., 1956). The distribution of cases in families is generally quite consistent with an autosomal recessive gene which, in the homozygous condition, leads to the clinical disorder. The disease is not so rare as had been thought. The systematic studies have lead to estimates of the frequency of heterozygote carriers ranging from 1 in 35 (Switzerland) (Prader, 1958) to 1 in 125 (Maryland) (Childs et al., 1956). Consanguinity in parents of affected patients is only slightly higher, as would be expected with a relatively frequent recessive gene rather than a very rare one. Since the metabolic defects differ in the various subtypes of the disease, it is not surprising that most of the evidence indicates they are inherited separately. Males suffering from the disease are more difficult to recognize, since virilization is less obvious in males than in females. This is the probable reason for the excess of females in the major studies to date. Efforts have been made to detect the carrier state in the heterozygote parents of patients with congenital adrenal hyperplasia. Childs and associates (1956) administered ACTH to 20 such parents to stimulate adrenocortical activity. The response of these parents in pregnanetriol excretion (derived from 17-hydroxy-progesterone) was significantly higher than in 18 control subjects, though there was considerable overlap. Further work is needed on detection of heterozygotes.

Patients with the most severe enzymatic defects have genital abnormalities at birth. Another relatively large group of patients is recognized at about 3 years of age; this group consists mainly of boys in whom the diagnosis is not made earlier because of a recognition difficulty mentioned above. Another interesting subgroup has emerged recently, consisting of virilization occurring shortly after puberty, with high levels of urinary pregnanetriol and 11-oxopregnanetriol (Brooks, 1962). The urinary steroid-excretion pattern is quite similar to that observed in infants with congenital adrenal hyperplasia. These patients apparently have a partial defect in hydroxylation at C-21 which does not result in high androgen secretion until puberty. The evidence to date suggests a familial occurrence of this delayed-onset syndrome. There is also some genetic evidence in the lipoid hyperplasia syndrome. In view of the high incidence of consanguinity and the familial occurrence, recessive inheritance is probable (Prader and Siebenmann, 1957). In passing, it is worthwhile to note that there is also some genetic evidence regarding a disorder of the adrenal medulla (Carman and Brashear, 1960). The tendency to develop pheochromocytoma with consequent hypersecretion of catechol amines appears to be inherited as a dominant; 10 kindreds including 25 cases of familial pheochromocytoma have so far been reported.

Experimental Approaches to Genetic Factors in Adrenocortical Function, Utilizing Laboratory Animals

The application of endocrinological and biochemical techniques to inbred mouse strains has lately provided another view of genetic factors in adrenocortical function.

Thiessen and Nealey (1962) tested five inbred mouse strains, mainly utilizing eosinophil counts as an indirect estimate of adrenocortical function. (In general, eosinophil levels tend to fall as the circulating concentration of corticosteroids increases. However, the eosinophils are also affected by other factors, e.g., epinephrine, and therefore represent only a rough assessment of adrenocortical response.) They found that the five strains differed in (1) resting eosinophil levels; (2) the eosinophil decrease resulting from handling and blood sampling, this decrease presumably reflecting in part the adrenocortical response to stress; (3) adrenal weights; (4) behavior; (5) adrenal-behavior relations. In this study, as in several others, the C57BL strain appeared particularly responsive to stressful conditions.

In the Department of Psychiatry at Stanford University, Levine and Treiman (1964) have made direct measurements of plasma corticosterone (the principal corticosteroid secreted by the adrenal of mice and rats) at several time intervals after the exposure of four inbred mouse strains to (1) noxious stimulation (electric shock); (2) a novel environment; (3) control conditions. Significant differences in the temporal pattern of plasma corticosterone response to stress (noxious stimulation and novel environment) were observed among the four strains. The C57BL/10 and A/Jax showed significantly higher plasma steroid levels, which also remained elevated longer than those of the DBA/2 and AKR mice. Such strain differences might well provide the starting point for investigation of the underlying genetic and bio-chemical mechanisms.

Christian (1955) has reported that wild house mice responded to group living with a greater degree of adrenal hypertrophy than did albino laboratory mice. This is reminiscent of Richter's earlier finding of great differences in adrenal size between wild and domestic rats (Richter, 1954). He trapped numerous wild Norway rats and compared them with domesticated Norway rats (Wistar strain). The adrenals of the domesticated strain were one-third to one-fifth as large as those of the wild rats; the difference was in the size of the adrenal cortex.

A recent interspecies observation is interesting in the context of experimental pharmacogenetics (Meier, 1963). Since corticosteroids generally have a well-documented protein catabolic effect, it is plausible that large doses of such hormones in pregnancy might produce damage in the offspring. In fact, it has been possible experimentally to produce abortions, macerated fetus, and congenital abnormalities in rabbits and mice by administration of cortisone to the pregnant mother. However, in rats (Wistar strain), cortisone does not produce such effects, even in high dosage. The resistance of pregnant rats to cortisone may provide a useful experimental model for the analysis of a similar resistance of pregnant women to cortisone.

RESEARCH OPPORTUNITIES

In the two preceding sections, we have placed in juxtaposition two bodies of information which in the past have not been related to each other: (1) CNS regulation of adrenocortical function, especially the effects of psychological stress on adrenocortical function, and (2) genetics of adrenocortical hormone metabolism, especially as elucidated by biochemical defects leading to clinical disorders of adrenocortical function. It now remains to consider how the juxtaposition of these two bodies of information may bear upon the general field of behavior genetics, particularly in relation to psychiatric and psychosomatic problems.

To pursue this line of inquiry, it will be useful to consider further some pertinent aspects of human biochemical genetics as they apply to the problem area under discussion. The important research of recent years on disorders such as phenylketonuria and galactosemia has made fundamental contributions in genetics and medicine, has developed powerful techniques for human biochemical genetic analysis of a wide range of clinical problems, and has pointed up sharply the relevance of such approaches to brain function and behavior (H. Harris, 1959; Hsia, 1960). Among other things, these studies have made clear that a genetic change affecting one particular biochemical reaction may indirectly bring about much more extensive biochemical change than the loss or diminution of a single reaction. When certain cells lose the normal ability to carry out a particular reaction, the compound behind the block may accumulate in abnormally large quantities or may be metabolized to other compounds in abnormally large quantities, or both (Sutton, 1961). Either event can have a profound metabolic effect, and so a single-gene mutation may have wide ramifications within the organism, some of which are quite significant for brain function and behavior. Even if the primary defect is unknown, the secondary effects crucial in pathogenesis may be manageable, as illustrated by the treatment of diabetes mellitus with insulin. If the primary site of the deficient biochemical reaction can be identified, effective methods of treatment or prevention may be worked out, basically in two ways at the present time: (1) by compensating for the deficiency of a hormone or (2) by preventing the damaging effects of accumulated intermediates (Sutton, 1961).

For a variety of reasons, it would be desirable to be able to detect the presence of potentially damaging genes in an apparently healthy person. To detect such genes, it is essential that they show some functional difference from the normal genes. Ordinarily this difference consists of a decrease in the primary gene product. Increasingly in recent years, it has been possible to develop biochemical methods for discovering the small difference in function between homozygous and heterozygous individuals (Childs and Young, 1963). In clinical disorders mediated by recessive inheritance, such as those discussed above, it is possible to study both homozygotes and heterozygotes by directing attention not only to patients but to their families. A number of instances are now known in human biochemical genetics in which detailed study of the heterozygotes has revealed minor abnormalities which are evidently caused by the abnormal gene in single dose. The abnormality of the heterozygote in such instances is usually qualitatively similar to that of the homozygote but quantitatively much less. Consequently, the individual so affected usually does not show any obvious pathology and appears to function adequately under ordinary conditions. However, as we shall see, there is an important possibility that these individuals may be vulnerable in specific ways when a heavy load is placed upon the partially deficient system. The point deserving emphasis is that the rapid development of research in this field is making clear that detailed investigation of heterozygotes with disorders inherited recessively increasingly makes possible the detection of minor biochemical abnormalities in apparently healthy individuals, some of which are likely to take on clinical significance over the long term.

Lederberg (1963) has recently called attention to the importance for human welfare of the full diagnosis of heterozygotes. While a variety of useful techniques for such diagnosis have emerged in recent years, and additional ones are evolving in promising ways at the time of this writing, it will suffice here to call attention to one established, valuable technique by way of illustration. The tolerance test or loading experiment, long useful clinically in the diagnosis of diabetes mellitus, has proved a valuable tool in detecting latent biochemical abnormality (Sutton, 1961). An enzymatic deficiency that may not be visible at ordinary substrate levels in the body can become fully visible if its substrate is supplied exogenously at greatly increased levels. In this way, environmental fluctuations in substrate level are effectively overcome. Such loading experiments have been notably successful not only in the area of carbohydrate metabolism but also in the elucidation of phenylketonuria. They have proved effective in revealing variations between homozygous normal and heterozygous individuals who are otherwise phenotypically normal. Such tests are now increasingly employed as a means of detecting minor variation in many biological substances.

There is reason to believe that this type of approach is an important frontier in human biology. In presenting principles of human genetics, Stern (1960) makes the following comment pertinent to the present discussion:

☐ Most biochemical analyses of gene defects have been concerned with rare defects. The biochemistry and the genetics of many rather common defects is less well understood, largely because most rare defects are caused by practically complete absence of a normal metabolic process, while the more common defects may be correlated, not with the absence of the biochemical reaction but only with a quantitative abnormality. Obviously, it is easier to discover which genes result in a reaction's absence than to find which genes lead to less striking quantitative variations in reactions. Yet there is every reason to assume that genes, by regulating biochemical processes, determine not only striking abnormalities but also less striking ones and, above all, much of the variability among normal human beings.

Similarly, in Tepperman's perspective (1962) on endocrine and metabolic physiology he makes the following comment:

☐ It is a mistake to think of genetics and the endocrine system only in relation to inborn errors which manifest themselves in overt disabilities and dysfunctions. Endocrine strength and resiliency is probably no less hereditary than is endocrine weakness and susceptibility to disease. Even those people who have no obvious endocrine disease doubtless have varying amounts of what might be called "endocrine reserve." Whether or not they eventually succumb to an endocrine disorder may depend largely upon environmental factors. For example, the man with a "weak" beta cell in his islets of Langerhans may avoid overt diabetes mellitus if he does not become obese.

It is for this reason that a frank acknowledgment of the importance of heredity in metabolic and endocrine diseases is not defeatist as some observers have asserted. In the future, increased methodological sophistication may make it possible to detect not merely malfunction of the endocrine glands but "borderline compensation" as well. Measures may then be devised which will make possible the prevention of manifest disease.

A variety of "borderline compensation" situations can be imagined which would be quite susceptible to decompensation under conditions that placed a heavy load upon the partially defective system. One interesting example of this type of situation has been briefly mentioned above: the intermediate group of patients with a defect in C-21 hydroxylation who are able to produce minimally adequate amounts of cortisol under ordinary conditions but who become "salt losers" in the presence of superimposed stress such as infection (Eberlein, 1958). It is very much worth investigating whether intense emotional distress would be capable of precipitating decompensation in a situation of this kind. In principle, it seems entirely reasonable that such instances can be found.

Another interesting example of endocrine "borderline compensation" has been found in the area of thyroid function and disease (Motulsky, 1962a; Williams and Bakke, 1962). The compensatory thyroid enlargement known as simple goiter is most commonly caused by an environmental factor, iodine deficiency. The pathological changes in the thyroid reflect the gland's tendency to compensate for the interference with normal hormone production. However, some facts indicate that genetic susceptibility plays a role as well. For example, many individuals who live in hyperendemic goiter areas do not develop goiter; moreover, familial occurrence of simple goiter in nonendemic areas has been observed. There is now considerable evidence that incomplete defects in thyroid hormone synthesis occur in adult heterozygotes who develop simple goiter without clinical hypothyroidism. For example, a partial deficiency in the enzyme dehalogenase, which is essential in thyroid hormone synthesis, has been discovered in relatives of goitrous cretins, some but not all of whom had goiters. Motulsky (1962a), in reviewing the area of genetics and endocrinology, makes the following comment on this situation:

☐ It is not unlikely that heterozygotes, having a partial defect, are able to maintain normal thyroid function under usual conditions but develop goiters when faced with such stresses as iodine deficiency, pubescence, and pregnancy, or exogenous goitrogens. Since there are always many more heterozygotes (for example, in a population with an incidence of homozygotes of one in ten thousand, there would be 2% heterozygotes), a significant proportion of patients with goiter may be heterozygous for these genetic defects. An individual without any of the cited genetic susceptibility factors presumably would develop a goiter if the exogenous stress, such as marked iodine deficiency or high goitrogen intake, or both, were severe enough. Recent findings on genetic factors in goiter make it likely that simple goiter usually has no single etiology, but is a striking example, of the interaction of multiple hereditary determinants with various environmental and endogenous factors.

This type of situation calls attention to several potentially important points bearing on the relations of psychological stress and partial defects in metabolic pathways of adrenocortical hormones:

1   Such partial defects may be fairly common, reflecting heterozygote carriers of recessive genes, even though the gross clinical disorder (homozygous condition) is rare.
2   Such partial defects might make no gross difference in behavior except under severe stress.
3   A variety of clinically significant interactions between genic and environmental factors may reasonably be expected.

The probable diversity of genetic factors relevant to several major clinical syndromes in psychiatry and psychosomatic medicine deserves emphasis here. (The same might well be said of the probable diversity of relevant environmental factors and of the complex interactions between genic and environmental factors pertinent to the major clinical syndromes.) Disorders such as schizophrenia, depression, alcoholism, and hypertension cover quite a broad range of clinical territory and may well subsume any number of subtypes within each category. At present, investigators differ in their judgment as to whether each of these diagnostic categories reflects a single clinical entity or a variety of subtle entities, though opinion seems to be moving in the latter direction in recent years. In any event, it is entirely conceivable that there might be a variety of genetic and environmental pathways through which an individual could develop a clinical disorder that would presently fall under the diagnostic rubric of, say, schizophrenia. Research efforts toward differentiation within each of these categories on whatever grounds—genetic, biochemical, behavioral, clinical—are very much worthwhile. In my judgment, broad diagnostic categories such as schizophrenia and depression are in fact quite heterogeneous, and in the foreseeable future it will probably be increasingly meaningful to speak about the group of schizophrenias and the group of depressions. These are, of course, empirical questions; some of the psychiatric-psychosomatic categories will no doubt turn out to be more heterogeneous than others. In discussing this general problem in medical genetics, Motulsky (1962a) makes this important observation:

☐ Recent work in genetic diseases has resulted in the almost constant finding that an entity that appears clinically uniform may represent the end result of different gene mutations (genetic heterogeneity). Considering the complex enzymatic control of metabolism, it is not too surprising that interference with different metabolic steps necessary for a given function may lead to a similar end result. In order to demonstrate that a given disease is the result of different mutations, one should ultimately demonstrate biochemical differences between the various forms of the disease; such heterogeneity can be strongly suspected if different modes of inheritance (i.e., recessive or X-chromosome-linked recessive) are detected in different families with a similar disease. This approach, with careful correlation of detailed genetic, clinical, and biochemical data, has been extremely helpful in the definition of disease entities and for an understanding of the pathophysiology of disease.

A fine example of the principle involved here has been provided by research on mental retardation (H. Harris, 1962; Polani, 1962). Not so very long ago, it seemed reasonable to speak of this as a single broad diagnostic entity, at least as a kind of clinical shorthand. However else the individuals might differ, they had in common a serious, sustained difficulty in learning, leading to a progressive lag in intellectual development as compared with their peers. However, important differentiation has occurred in recent years through some of the most notable advances in human genetics. Several of the syndromes that have been clarified by a biochemical approach, such as phenylketonuria and galactosemia, have been associated with mental retardation. Similarly, the cytogenetic approach has been fruitful in clarifying some syndromes associated with mental retardation, such as Down's syndrome. Each of these constitutes quite a small fraction of the total number of cases of mental retardation. From the present pace of events, it is altogether likely that a number of other syndromes will be clarified and differentiated out of the broad category in the next decade. Thus, in the field of mental retardation, we can concretely point to a number of pathways through which an individual may acquire the central behavioral disorder: serious, sustained difficulty in learning. This provides one useful model for future research on other major mental disorders. Kety (1959) and Caspari (1962) have called attention to the probable genetic and biochemical diversity in schizophrenia. Similarly, Garmezy and Rodnick (1959) have presented evidence of psychological heterogeneity among schizophenic patients.

It is well known that behavioral characteristics of adaptive significance tend to be polygenic in inheritance. A complex behavior pattern is very likely to involve many genes in its anatomic and physiological substrate; therefore, comprehensive analysis of all genetically contributing factors would be exceedingly difficult and indeed can be envisioned only as a long-range possibility. However, from the viewpoint of the present approach, this need not be discouraging to investigators. Many single-gene defects may adversely affect the behavior pattern and therefore take on clinical significance. The field of intelligence, so much studied in behavior genetics because of its obvious adaptive and clinical significance, provides a good example. It is perfectly clear that genetic factors in intelligence are polygenic; that the anatomic and physiological substrate of learning and problem solving is exceedingly complex, involving as it does a large part of the primate brain. So far, our understanding of this wide gamut of genetic factors pertinent to intelligence is quite limited. Yet we have already begun to see in recent years, as noted above, a number of single-gene defects which are highly relevant to intelligence, which indeed produce profound impairments of great clinical significance. Similarly, a great variety of neural, endocrine, and cardiovascular structures and functions are relevant to stress responses. Therefore, a comprehensive understanding of genetic factors relevant to stress responses is a long-term program whose fulfillment will be slow in coming; yet it is altogether possible that the next decade will see important advances in our understanding of a variety of single-gene defects which have a bearing on stress responses. Some of these may not be compatible with a viable organism. Others may be trivial in consequences. In between lies a broad area of potential significance for psychiatric and psychosomatic problems.

CHAPTER NINE
THE HEREDITARY METABOLIC DISEASES[1]
David Yi-Yung Hsia

The purpose of this chapter is to emphasize the relationship between behavior and the metabolic diseases that are clearly genetic in origin. It is hoped that study of the effects of a single-gene mutation upon intelligence and personality will provide clues to the patterns of behavior caused by multiple-gene changes. This will be done in three parts: The first will deal with the present-day concept of "inborn errors of metabolism." The second will describe in detail the work that has been done with one of these hereditary metabolic diseases, namely, phenylketonuria. The third will summarize some of the studies that have been initiated in a number of other similar conditions.

INTRODUCTION

The concept of "inborn errors of metabolism" was first suggested in 1908 by Sir Archibald Garrod. In the Croonian Lectures delivered at the Royal College of Physicians (Garrod, 1909), he suggested that four metabolic disorders, albinism, alkaptonuria, cystinuria, and pentosuria, had certain features in common. First, in all four conditions, the onset of the particular abnormality could be dated to the first day or weeks of life, especially when a special effort was made to do so. A second characteristic was their familial occurrence in a considerable number of cases. A third feature was that the conditions were relatively benign and compatible with a normal life expectancy. A fourth feature noted by other clinicians of his day was the frequency with which these disorders occurred among the offspring of consanguineous marriages. Although Garrod provided all the concepts upon which modern biochemical genetics is based, his work was largely ignored by the classical geneticists of his day.

Instead, evidence for the chemical translation of genetic influence

[1] These studies were aided by grants from the Chicago Community Trust, the Kettering Foundation, the Illinois Mental Health Fund, and the U.S. Public Health Service (HD 330) (T1-AM-5186).

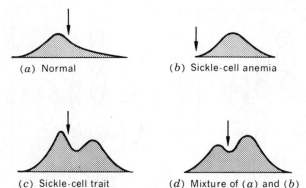

(a) Normal   (b) Sickle-cell anemia

(c) Sickle-cell trait   (d) Mixture of (a) and (b)

Figure 9.1   The electrophoretic patterns of normal and sickle cell hemoglobins. The single pecks $a + b$ correspond to an electrophoretically hemogenous material whereas the heterozygote (c) carries a mixture of normal adult and sickle cell hemoglobin. (*From Pauling, 1955.*)

had to be provided by workers in other fields. In the 1930s, Scott-Moncrieff (1939) showed that the structure of the anthocyans, the principal water-soluble plant pigments, was genetically determined and that changes at single loci would cause great differences in pigment color. Slightly later, Beadle and Tatum (1941) working with induced mutants of *Neurospora crassa* showed that the acquisition of single essential growth factors could be traced to single chemical reactions, each dependent upon a different enzyme. From this emerged the concept of "one-gene–one-enzyme" control of chemical reactions.

In 1949, Pauling and his coworkers made the important observation that the hemoglobin (Hb) from a patient with sickle cell anemia had a distinct electrophoretic pattern. It was found that Hb S migrates as a positive ion whereas Hb A migrates as a negative ion in a 0.1 M phosphate buffer, pH 6.9 (Figure 9.1). They suggested that this represented an abnormality in the structure of the hemoglobin molecule and proposed that these conditions should be regarded as "molecular" diseases. In 1956, Ingram showed that, if the globin portion of the hemoglobin molecule is broken down by tryptic digests, some 28 peptides, each with an average chain length of 9 to 10 amino acids, result. These can be separated on paper, first by electrophoresis and then by chromatography. The peptides prepared from Hb S differ from those of Hb A only in the location of one peptide group (number 4) (Figure 9.2).

Perhaps the most striking progress has been made in the realm of enzyme defects of a hereditary nature. Although so far it has not been possible to study in detail the molecular structure of each enzyme, it has been feasible to measure enzyme function. Detailed studies of the role of enzymes have provided considerable information on the genetic and biochemical mechanisms responsible for a number of hereditary diseases. At the present time, there are well over 100 con-

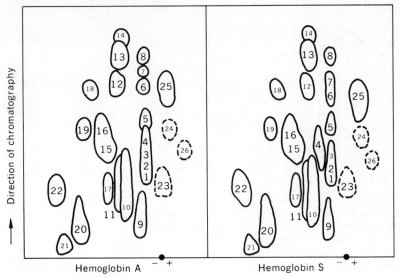

Figure 9.2   Fingerprints of trypsin digest of hemoglobins A + S, showing the number of peptide pairs. Dotted lines indicate peptides that become visible only after heating the chromatogram. (*From Ingram, 1958.*)

ditions which could fit into Garrod's concept of an "inborn error of metabolism." We shall discuss in detail the one that has probably been most extensively studied.

PHENYLKETONURIA

Phenylketonuria is a hereditary disease characterized by mental retardation and the presence of phenylpyruvic acid in the urine. It holds a unique position among the hereditary metabolic diseases because it is easily detectable and, when diagnosed early, the mental deficiency can be prevented with the use of a special diet.

Clinical Features

Phenylketonuria can frequently be diagnosed by a careful history and physical examination before the mental defect becomes evident. In a survey, Partington (1961) showed that 17 out of 36 patients exhibited vomiting during the first few weeks of life; in most instances, this was severe enough to warrant the parents' seeking medical advice. Seven children had projectile vomiting, and three were operated on for pyloric stenosis. He also noted that 12 of the patients had a history of unusual irritability and 13 showed epileptic seizures. Six of the children showed generalized infantile eczema between the first and fourth month, and this persisted with remissions and exacerbations for several months. In addition, five children had dry skin with repeated minor inflammatory lesions or nonspecific rashes.

The mental defect becomes evident between the fourth and sixth month of life. In our experience, the earliest case was referred at 5 months of age because the child did not behave quite up to the norm for that age and was found to have a strongly positive ferric chloride test. Unfortunately, the majority of our cases are referred between 10 months and 3 years, the average age of diagnosis being about 15 months, when a successful result cannot be expected from dietary therapy.

Paine (1957) has described very completely the clinical manifestations of a group of 106 untreated phenylketonuric patients. Although they are frequently of average stature, the majority of these patients tend to be small in terms of height and weight. This is especially true of the very retarded children in whom a food intake adequate for maintenance of nutrition is often difficult to achieve. Although there are many exceptions, phenylketonurics tend to have blue eyes, blond hair, and fair skin. These observations have been confirmed by comparing affected individuals and their unaffected siblings, using reflectograms for hair color and comparing iris colors, using the Martin scale. The decreased pigmentation is particularly striking among Japanese phenylketonurics who have brown hair instead of the usual black hair.

Neurological examination has shown that about one-half of the phenylketonurics show from a mild to a marked microcephaly. Hand posturing is a very striking characteristic, especially among those of low intelligence. These purposeless movements include rhythmic pill-rolling movements of the hand, irregular tic-like motions, aimless to-and-fro movements of the fingers, and frequent habitual fiddling of the fingers close before the eyes. The movements are accompanied by rhythmic rocking back and forth which may continue for hours. Phenylketonurics frequently show tremor of the hands and increased reflexes. It is believed that enhancement of this reflex activity is related to damage of developing pathways within the central nervous system. Phenylketonurics are prone to severe temper tantrums which are hard to control.

About three-quarters of the patients with phenylketonuria show abnormal electroencephalogram patterns. The most frequently seen abnormalities are spike and wave complexes of the petit mal variant type which are found even in the absence of convulsive seizures. Other show abnormalities of voltage or rhythm only.

Intelligence quotients of phenylketonuric patients are to be interpreted with caution. In the first place, test values are inaccurate when the majority of patients are in the imbecile or idiot range and cannot communicate satisfactorily. Furthermore, most of the published data is based on institutionalized cases where there has been little stimulus on an individual basis. This leads us to believe that the IQ of phenylketonurics may, in fact, be somewhat higher than currently believed. However, despite this, there seems to be little doubt that the overwhelming majority of untreated phenylketonurics have an IQ of less than 50; probably most of these are in the range of 25 or less (Figure 9.3).

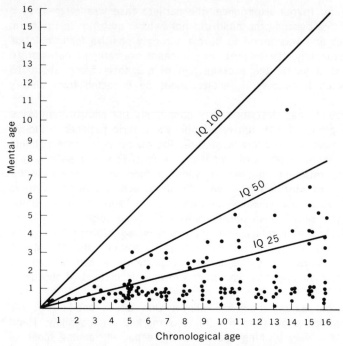

Figure 9.3   The mental ages of 106 untreated phenylketonuric children tabulated against their chronologic ages. (*From Paine, 1957.*)

The 25 or so known atypical phenylketonurics who have normal or near-normal intelligence without having been placed on a special diet are of considerable interest (Hsia and Knox, 1957). Most of these children appear to be superficially normal both neurologically and mentally and were discovered either because of a ferric chloride test performed in a routine survey or because they were siblings of typical cases. We have had the unusual opportunity of working with three such children. The first patient was a 13-year-old school girl who appeared to be superficially quite normal. She was accidentally discovered because we were carrying out a series of studies among all the siblings of known cases. A more careful evaluation of her school record showed that although she was in the sixth grade she was not doing well and remarked, "Why am I so dumb when everyone else is so smart?" IQ tests showed that she had achieved a mental age of 9 years and 7 months on the Stanford-Binet and an IQ of 78 on the Wechsler Intelligence Scale. The other two patients were siblings aged 9 and 12 who were doing entirely normal work in the third and sixth grades in school. Their IQ tests showed values of 98 and 94 on the Stanford-Binet Intelligence Test. Their serum phenylalanine levels were 26.6 and 24.3 mg %, and their urine phenylpyruvic tests were strongly positive.

Atypical phenylketonurics of normal intelligence undoubtedly occur for a variety of reasons. If one considers the distribution of intelligence among all phenylketonurics, one would expect between 1 and 2 percent to have a level in excess of 60. The description of 25 such cases among about 1,000 known phenylketonurics (2.5 percent) is probably not excessive. It has also been suggested that some of these patients may represent a *forme fruste* of phenylketonuria. Woolf et al. (1961) have reported on two sisters with near-normal intelligence who showed fasting serum phenylalanine levels of 8.5 and 9.9 mg/100 ml and in whom the phenylpyruvic acid could be detected only by means of paper chromatography. Phenylalanine tolerance tests of their parents and one offspring showed a characteristic heterozygote range. Undoubtedly, there must be phenylketonurics somewhere between homozygotes and heterozygotes in terms of mental and chemical findings, who either represent another allele for the abnormal gene or reflect the influence of a modifying gene.

The pathological findings in the central nervous systems of phenylketonurics have not been striking. A number of workers have described some demyelination (or perhaps late formation of myelin) in this disease. This has not been confirmed by other workers. Recently it has been shown that the cerebroside content of the brain in phenylketonurics is decreased, and this may be of importance in consideration of the etiology of mental defect (Crome et al., 1962).

Heredity

In a large survey covering institutions for mental defectives from all over the world, Jervis (1954) reported that 312 out of 48,536 patients (0.64 percent) were phenylketonurics. If one accepts this as representative of the incidence of phenylketonuria in the defective population and 1 percent the incidence of defectives in the general population, then the incidence of phenylketonuria in the general population may be assumed to be of the order of 6 per 100,000.

The genetic mechanism for phenylketonuria is precisely known. In

Table 9.1
Plasma *L*-phenylalanine levels after ingestion of 0.1 gram of *L*-phenylalanine per kilogram of body weight*

| | | Hours after dose | | | |
		1	2	4	Sum of hourly levels
Controls	Mean	0.55	0.55	0.30	1.41
	S. D.	±0.186	±0.168	±0.076	±0.366
	Range	0.30–0.90	0.29–1.02	0.21–0.50	0.87–2.19
Heterozygotes	Mean	1.14	1.03	0.76	2.93
	S. D.	±0.187	±0.187	±0.292	±0.458
	Range	0.84–1.43	0.72–1.41	0.45–1.62	2.10–4.03
t		9.8	8.4	6.7	11.7

*Values given are the means, standard deviations, and ranges (expressed as micromoles per milliliter of plasma) in 19 adult controls and 19 heterozygotes (parents). (From D. Y.-Y. Hsia et al., *Nature*, 178:1239, 1956.)

a definitive study involving 266 sibships, Jervis has shown that the disease is transmitted by a single autosomal recessive gene. This was done by showing an equal sex distribution and determining the ratio of affected to normal children, using the Weinberg "sib method," which gave a percentage and standard error of 27.37 ± 2.57, and the proband method, which gave values of 22.38 ± 2.66. Since both are within one-third of the standard error of the hypothetical 25 percent by the Lenz a priori method, the hypothesis that this is an autosomal recessive gene is corroborated.

If an individual is heterozygous for a given trait, he carries one "normal" and one "mutant" gene, and the molecules derived from the "mutant" gene may be expected to differ from those derived from the "normal" gene within the same individual. On superficial examination, such individuals often appear to be normal. However, more careful physical and biochemical studies might reveal minor departures from the norm, and these individuals can then be identified in a population.

The detection of heterozygous carriers for phenylketonuria was first successfully accomplished by Hsia and coworkers (1956) by means of a phenylalanine tolerance test. A load of 0.1 gram of L-phenylalanine per kilogram of body weight was administered by mouth, and samples of blood were taken at 1, 2, and 4 hours for the determination of phenylalanine. It was found that the heterozygotes had plasma phenylalanine levels averaging twice those in normal controls (Table 9.1).

The ability to detect heterozygous carriers for phenylketonuria from normal controls with a minimum of overlap has permitted biochemical confirmation of the recessive mode of inheritance of the mutant gene among parents, grandparents, uncles, aunts, first cousins, etc. It has also been possible to investigate the incidence of mental disease in heterozygotes. In the past, several workers have suggested that there was an increased incidence of psychosis in later life among relatives of phenylketonuric patients, particularly of the senile melancholic type. Phenylalanine tolerance tests have been performed for a group of older patients of the latter category, and there has been no finding of an increased incidence of heterozygotes for phenylketonuria in the group (Pratt et al., 1963).

Pathogenesis

Phenylketonuria is caused by a deficiency of the liver enzyme, phenylalanine hydroxylase, which converts phenylalanine to tyrosine (Jervis, 1953; Wallace et al., 1957; Mitoma et al., 1957) (Figure 9.4). This results in the accumulation of L-phenylalanine in the blood and spinal fluid and causes three main effects:

1 The excessive phenylalanine is converted by its transaminase to phenylpyruvic acid. This in turn is converted to phenyllactic acid, phenylacetic acid, and phenylacetylglutamine which is excreted in the urine (Woolf, 1951).

2 The excessive phenylalanine inhibits the hydroxylation of tryptophan (Nadler et al., unpublished), and this results in a decrease of 5-

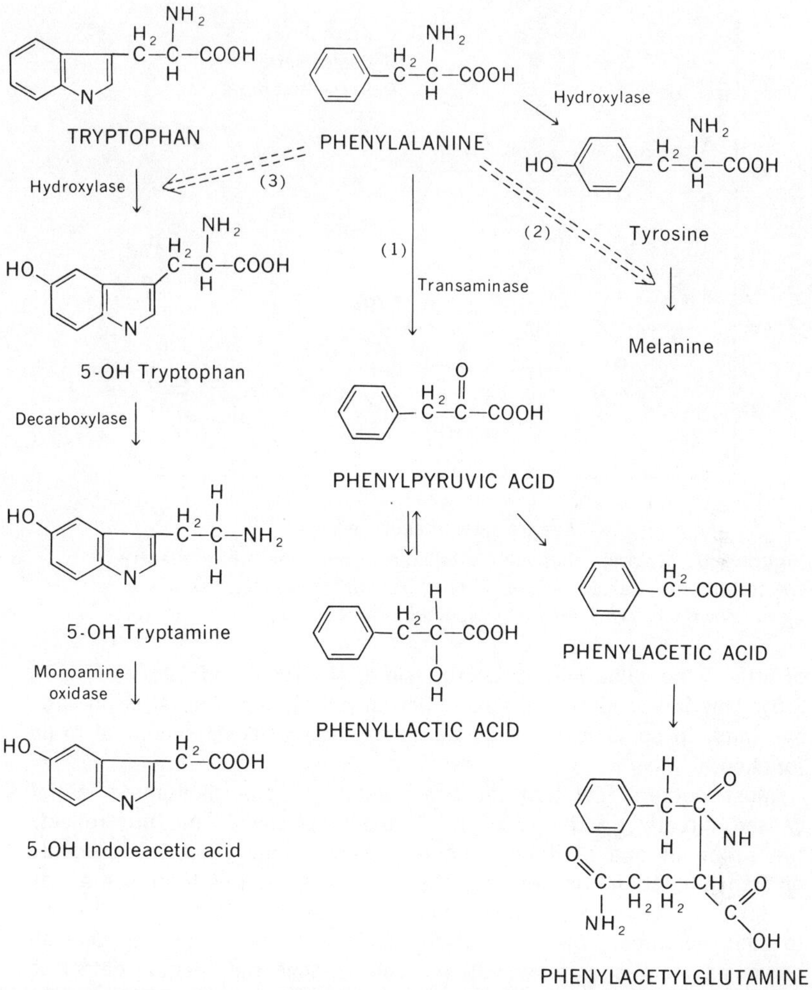

Figure 9.4 The site and main consequences of the metabolic block in phenylketonuria. The metabolites that are increased are given in capital letters, and the metabolites that are decreased are given in lower-case letters.

hydroxytryptamine in the blood and 5-hydroxyindoleacetic acid in the urine (Pare et al., 1957).

3 The excessive phenylalanine level inhibits the normal pathways of tyrosine, causing a decreased production of melanin, and this is responsible for the light pigment in the skin and hair of such patients (Dancis and Balis, 1955).

Although the etiology of the mental defect remains unknown, there is evidence that, if the affected children are treated with a low-phenylalanine diet at an early age, near-normal mental development will result. A similar course of treatment in an older affected child is

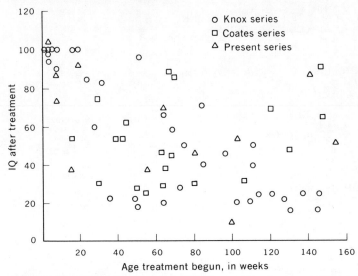

Figure 9.5   Relation between intelligence and age when dietary treatment was started. (*From Hsia et al., 1962, with the permission of the Northwestern University Medical School.*)

of little or no value (Knox, 1960; Hsia et al., 1962 and 1966a) (Figure 9.5). This indicates that the reduction of phenylalanine level at an early age must in some way be related to the normal development of brain function.

Most workers feel that the brain defect in phenylketonuria is not caused directly by the excessive levels of phenylalanine but reflects the action of one of its metabolites. Some believe that the products of transamination, particularly phenylacetic acid, might in some way be toxic to the brain. However, massive doses of this compound given to adult volunteers produced discomfort but no truly alarming central-nervous-system symptoms. Others believe that the mental defect is related to the decrease of 5-hydroxytryptamine in the brain. It now appears likely that the mental defect is the end result of a critical shortage of brain 5-hydroxytryptamine during early infancy, caused in part by the physiological lack of this metabolite at birth and in part by the excessive phenylalanine levels in this disease.

Experimental Phenylketonuria

In 1958, Auerbach and his coworkers introduced a powerful tool for the investigation of phenylketonuria in both biochemical and behavioral studies. It was shown that a supplementary dose of from 5 to 7 percent of L-phenylalanine added to the diet of a rat, guinea pig, or monkey induces all the biochemical manifestations of phenylketonuria. There is an increase of phenylalanine in the serum and of phenylpyruvic acid in the urine of treated animals and a concomitant decrease of 5-hydroxytryptamine in the blood and 5-hydroxyindoleacetic acid in the urine (Huang et al., 1961). More recently, it has been shown that such animals also exhibit a decrease of brain 5-hydroxytryptamine (Hsia et al., 1963; McKean et al., 1962; Yuwiler and Louttit, 1961).

From the beginning, it has been apparent that a high-phenylalanine diet altered the pattern of behavior of experimental animals (Auerbach et al., 1958; Polidora et al., 1963; Woolley and van der Hoeven, 1963, 1964; Yuwiler and Louttit, 1961). Auerbach and his coworkers (1958), using a temporal discrimination test (involving licking a dry dipper and being rewarded by the delivery of water on a 2-minute fixed-interval schedule), found that the frequency of appearance of positive accelera- tion on days 6 to 8, expressed as the percentage of the number of intervals showing any responses, averaged 82.9 percent for normal rats and 46.8 percent for phenylketonuric rats, a difference which was significant at the 0.1 level. In learning to swim a water maze, where escape from water was the only motivation, the same group (Polidora et al., 1963) showed that the phenylketonuric animals consistently performed more poorly (in transit speeds) on maze tests over 3 days when the animals were between 72 and 90 days old.

Yuwiler and Louttit (1961) found that more errors were made on the Hebb-Williams maze and more trials were required to meet a criterion on a successive discrimination problem by animals fed phenylalanine than by their normal controls. Woolley and van der Hoeven (1963, 1964) have carried out a series of experiments with adult mice, using a simple maze. They found that increases in cerebral serotonin or serotonin plus catechol amines resulted in decreases in learning ability. By contrast, decreases in serotonin plus catechol amines brought about an increase in learning ability so that such mice were slightly superior to normal mice in this respect. When these experiments were repeated with infant mice treated within 24 hours after birth, rather different results were obtained. By administering a fine suspension of DL-phenylalanine and L-tyrosine by stomach tube nine times daily to newborn mice, they were able to produce experi- mental phenylketonuria and thereby a situation much more analogous to the human phenylketonuric infant. Control groups included mice treated when 7 to 8 weeks old; untreated mice; and mice given water hourly. Such controls gave an average score of 7.5. Animals given the high-phenylalanine diet from birth had an average score of 6.3. These results were compared with those of animals given reserpine, which reduces serotonin, and those given chlorpromazine, which blocks the receptors for serotonin; these were found to have average scores of 6.6 and 6.5. Thus, it would appear that the reduced learning ability in animals made phenylketonuric on an experimental basis may be related to a combination of reduced 5-hydroxytryptamine caused by immaturity and by excessive phenylalanine levels (Hsia et al., 1963).

Treatment

When the enzyme defect is intracellular, the systemic administration of the missing enzyme is obviously of limited value. Instead, treatment has consisted of eliminating from the diet the compound that is present in excess as a result of the metabolic block. Over a time, this

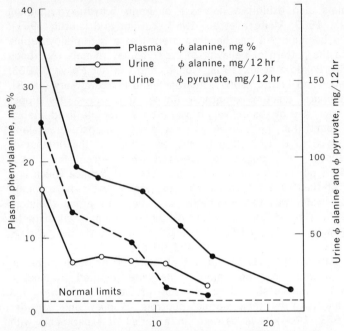

Figure 9.6 The effect of a low-phenylalanine diet on levels of phenylalanine and the phenylpyruvate in the plasma and urine of a phenylketonuric. (*From Hsia.* Postgrad. Med., *22:203–210, 1957, with the permission of the IPMA Publishing Co.*)

treatment reduces the excessive accumulation of this compound within the body, and the patient shows a normal metabolic pattern. In 1953, Bickel and coworkers, at the suggestion of Dr. L. I. Woolf, prepared a special protein hydrolysate low in phenylalanine content. When this was administered to a phenylketonuric child, the plasma phenylalanine level was greatly reduced, and the phenylpyruvic acid disappeared from the urine (Figure 9.6).

The low-phenylalanine diet is made up of four major components (Phenylketonuria, 1958): (1) low-phenylalanine protein hydrolysate available in the form of either Lofenalac or Ketonil; (2) 1 percent protein fruits and vegetables; (3) sugar and butter; and (4) multivitamins and ferrous sulfate supplements.

All patients should be seen at least once every 6 weeks, and the plasma phenylalanine level should be kept below 3 mg/100 ml. The phenylpyruvic acid excretion cannot be used as a measure of biochemical control, since it disappears from the urine when the plasma phenylalanine level falls below 10 mg.

As a general rule, the younger phenylketonurics take the diet quite well. There is more difficulty in inducing the older ones to eat, primarily because they have had previous experience with regular foods. Only two types of complications have been reported with this synthetic diet. During the early phases of the program, a number of workers inadvertently reduced the L-phenylalanine intake to below the minimal requirements for maintenance. As a result, the infants started to lose weight and develop gross amino-aciduria due to tissue protein breakdown. This was corrected by the administration of 2 ounces of whole milk daily, and the infants thrived thereafter. Careful balance studies indicate that the optimal dietary phenylalanine levels for a 6-month-old phenylketonuric infant should be approximately 25 mg/kg/day, which is about one-seventh the amount taken by the nonphenylketonuric infant on a normal diet (Paine and Hsia, 1957). The amounts for an older child or adult should be about 10 mg/kg/day, which is also considerably less than for a normal adult.

In 1959, Dodge and coworkers reported on two phenylketonuric children in whom hypoglycemia developed during the course of treatment with a phenylalanine-restricted diet. Refusal to take an adequate amount of an unpalatable diet over a period of weeks resulted in a state of undernutrition accompanied by fatty metamorphosis of the liver. A relatively short period of fasting caused severe hypoglycemia, convulsions, and coma.

There appears to be little question that the biochemical abnormality in phenylketonuria can be completely corrected and the patients can be maintained in a satisfactory physical condition for many years on the synthetic diet low in phenylalanine content. It is much more difficult to assess the possible beneficial effect of this diet upon the mental development of patients with phenylketonuria. In the first place, it has been shown that there is considerable variability in the mental development of such children without treatment, including a few with near-normal mentality. Secondly, environmental factors undoubtedly have a role in altering this pattern. In our own experience, phenylketonuric children kept at home invariably score better than those from institutions. Despite these limitations, sufficient time has elapsed so that a preliminary appraisal of the value of the diet can now be made.

It appears fairly certain that a diet low in phenylalanine has little or no effect upon phenylketonuric children 6 years of age or older. In a control study involving 24 patients and carried over a test period of 12 to 15 months, it was found that the only patients who showed possible beneficial effects were the four for whom the diet was started before the age of 3 years (Hsia et al., 1958). Approximately 100 phenylketonuric infants under 3 years have now been treated with the diet, and there appears to be no question that the diet has a positive influence in increasing the intelligence of phenylketonuric children, with those being started at the earliest age having the highest intelligence ultimately (correlation coefficient between age and IQ being .42. Of the 17 for whom treatment was initiated before 3 months of age, 14 achieved an eventual IQ in the range for normals (Figure 9.5) (Coates, 1961; Hsia et al., 1962; Knox, 1960).

This represents one of the first examples of mental deficiency which can be prevented by means of medical or dietary therapy and serves as a model for future work in this direction.

Because of the obvious success of the dietary treatment, most workers have been reluctant to terminate the diet in treated cases. In fact, there have been several reports of regression following cessation of the diet in successfully treated cases. Others have reported a normal development pattern when the diet was discontinued after several years of treatment. Since this is still an open question, we have not discontinued the diet for our successfully treated patients.

Prevention

Since a delay in treatment frequently results in subnormal intelligence, it is crucial that phenylketonurics should be diagnosed within the first months of life, before any signs of mental retardation appear. Two diagnostic approaches have been tried. Until recently, all the tests have centered in trying to detect phenylpyruvic acid in the urine. Since there is frequently a delay in the appearance of phenylpyruvic acid in the urine, it has usually been necessary to test the urine of the infant at between 4 and 6 weeks of age. This has made follow-up of neonates somewhat difficult as not all mothers bring their infants for a well-baby examination at that time. In a survey carried on in Cardiff, Wales, only 25 percent of the mothers responded (Gibbs and Woolf, 1959). In other areas, up to 70 percent of the babies born were treated but only after a follow-up program had been instituted (Allen, 1960).

For this reason, increased attention has recently been focused upon screening blood phenylalanine levels by means of a heel prick before the baby is discharged from the newborn nursery. Guthrie (1961) proposed a semiquantitative method for detecting elevated phenylalanine levels in newborn infants before discharge. This test is based on the fact that the inhibition of growth of *Bacillus subtilis* by beta-2-thienyl alanine in a minimal culture medium is specifically prevented by proline, phenylalanine, phenylpyruvic acid, or phenylacetic acid. A small amount of fresh blood obtained by skin puncture is transferred immediately to a piece of thick filter paper. The blood spot is dried, autoclaved, punched out, and placed over the agar surface containing the inoculum and the inhibitor, and interpreted 16 hours later. A halo of growth of the organism surrounding the disk indicates an increased phenylalanine content which is in proportion to the size of the halo. This method has had a frequency of false positive cases of about 0.1 percent in a sample of 157,780 infants.

We (Hsia et al., 1964) have recently developed a quantitative method to determine serum phenylalanine that requires 25 μl of serum from capillary blood. Using this technique, we surveyed 4,000 newborn infants and found the mean and standard deviation for this sample was 2.09 $\pm$ 0.51 mg/100 ml. The values were not appreciably influenced by maternal age and gravida or by the sex, race, birth weight, or age of the infant. The lowest recorded serum phenylalanine for a phenylketonuric infant was about seven standard deviations above the mean for the normal newborn population (Figure 9.7). In this survey, only 8 of the 4,000 infants (0.2 percent) temporarily showed serum phenylalanine levels above 4.0 mg/100 ml.

Thus, practical methods are now available for the screening of new-

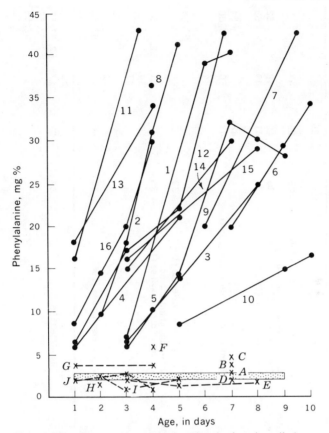

Figure 9.7 Serum phenylalanine levels in phenylketonuric infants (circles) and unaffected siblings (crosses)
compared with mean and one standard deviation of normal newborns (dotted area). (*From Hsia et al., 1964.*)

born infants for this condition. It is strongly recommended that such
tests be routinely performed on every newborn infant so that mental
retardation in those found to be phenylketonuric may be prevented.

OTHER HEREDITARY
METABOLIC DISEASES

Half a century of rapid progress in the basic sciences has given body
and substance to what was, by necessity, supposition in Garrod's time.
The original four inborn errors of metabolism have now been increased to well over 100. Nearly a quarter of these conditions involve
some aspect of mental deficiency and behavior. The important biochemical, genetic, therapeutic, and screening information on these
conditions is summarized in Table 9.2. For further significant advances
that have been made in our understanding of phenylketonuria and other
hereditary metabolic diseases, see Hsia (1966a, 1966b, and 1967).

Table 9.2
List of hereditary metabolic diseases associated with mental deficiency

	Clinical description			Enzyme studies			Hetero-zygote detection	Treat-ment	Screening test
Condition	Year	Refer-ence	Organ	Deficient enzyme	Ref.				
Tay-Sachs disease	1881	1							
Gaucher's disease	1882	2					39		
Thyroid deiodinase defect	1897	3	Thyroid	Iodotyrosine deiodinase (1956)	29		29	47	
Galactosemia	1908	4	Erythrocytes	Galactose-l-phosphate uridyl transferase (1956)	30		40	48	52
Wilson's disease	1912	5	Serum	? Ceruloplasmin (1952)	31		41	49	53†
Niemann-Pick disease	1914	6							
Hurler's syndrome	1917	7					42		54
Phenylketonuria	1934	8	Liver	Phenylalanine hydroxylase (1954)	32		43	50	55, 56
Thyroid iodide organification defect	1950	9						47	
A-beta lipoproteinemia	1950	10					44		
Lowe's syndrome	1952	11							57, 58‡
Crigler-Najjar syndrome	1952	12	Liver	Glucuronyl transferase (1957)	33		45		
Maple syrup urine disease	1954	13	Leukocytes, skin	Branched-chain ketoacid oxidative decarboxylase (1960)	34, 35			51	52, 57, 58
Pyridoxine dependency	1954	14						14	
Thyroid iodotyrosyl coupling defect	1955	15						47†	
Hartnup disease	1956	16							
Leucine-induced hypoglycemia	1956	17						17	
Argininosuccinic aciduria	1958	18	Liver	Argininosuccinase (1961)	36		46		57, 58‡
Cystathionuria	1959	19							57, 58‡
Thyroid iodide trapping defect	1960	20						47	
Histidinemia	1961	21	Skin	Histidase (1963)	37		37	27*	52, 57, 58‡
Hyperglycinemia	1961	22							57, 58‡
Hyperprolinemia	1962	23							57, 58‡
Hydroxyprolinemia	1962	24							57, 58‡

	1962	25	Liver	Homocystathionine synthetase (1964)		
Homocystinuria					38	57, 58‡
Citrullinuria	1963	26				57, 58‡
Pyridoxine deficiency	1963	27			27	
Hydroxykynureninuria	1964	28		28		

* Treatment possible, but not yet tried.

† Method for screening applicable only to older children.

‡ Method for screening available but not confirmed in affected newborn.

1 Tay, W. Symmetrical changes in the region of the yellow spot in each eye of an infant. *Trans. Ophthalmol. Soc. U. K.*, 1:55–57, 1881.

2 Gaucher, P. C. E. De l'epithelioma primitif de la rate: hypertrophie idiopathique de la rate sans leucemie. Paris, 1882.

3 Osler, W. Sporadic cretinism in America. *Trans. Congr. Amer. Physicians & Surgeons*, 4:169–206, 1897.

4 Von Reuss, A. Zuckerausscheidung im Säuglingsalter. *Wien. Med. Wochschr.*, 58:799–803, 1908.

5 Wilson, S. A. K. Progressive lenticular degeneration. A familial nervous disease associated with cirrhosis of the liver. *Brain*, 34:295–509, 1912.

6 Niemann, A. Ein unbekanntes Krankheitsbild. *Jbr. Kinderh.*, 79:1–10, 1914.

7 Hunter, C. A rare disease in two brothers. *Proc. Roy. Soc. Med.*, 10:104–116, 1917.

8 Fölling, A. Über Ausscheidung von Phenylbrenztraubensaure in de Harn als Stoffwechselanomalie in Verbindung mit Imbezillität. *Z. Physiol. Chem.*, 227:169–176, 1934.

9 Stanbury, J. B., and Hedge, A. N. A study of a family of goitrous cretins. *J. Clin. Endocrinol*, 10:1471–1484, 1950.

10 Bassen, F. A., and Kornzweig, A. L. Malformation of the erythrocytes in a case of atypical retinitis pigmentosa. *Blood*, 5:381–387, 1950.

11 Lowe, C. V., Terrey, M., and MacLachlan, E. A. Organic-aciduria, decreased renal ammonia production, hydropthalmos, and mental retardation. *A.M.A. Amer. J. Dis. Child.*, 83:164–184, 1952.

12 Crigler, J. F., Jr., and Najjar, V. A. Congenital familial nonhemolytic jaundice with kernicterus. *Pediatrics*, 10:169–180, 1952.

13 Menkes, J. H., Hurst, P. L., and Craig, J. M. A new syndrome: progressive familial infantile cerebral dysfunction associated with an unusual urinary substance. *Pediatrics*, 14:462–467, 1954.

14 Hunt, Andrew D., Jr., Stokes, Joseph, Jr., McCrory, Wallace, and Stroud, H. H. Pyridoxine dependency. *Pediatrics*, 13:140–145, 1954.

15 Stanbury, J. B., Ohela, K., and Pitt-Rivers, R. The metabolism of iodine in two goitrous cretins compared with that in two patients receiving methimazole. *J. Clin. Endocrinol*, 15:54–72, 1955.

16 Baron, D. N., Dent, C. E., Harris, H., Hart, E. W., and Jepson, J. B. Hereditary pellagra-like skin rash with temporary cerebellar ataxia, constant renal aminoaciduria and other bizarre biochemical features. *Lancet*, 2:421–428, 1956.

17 Cochrane, W. A., Payne, W. W., Simpkiss, M. J., and Woolf, L. I. Familial hypoglycemia precipitated by amino acids. *J. Clin. Invest.*, 35:411–422, 1956.

Table 9.2 (Continued)

18 Allan, J. D., Cusworth, D. C., Dent, C. E., and Wilson, V. K. A disease probably hereditary, characterized by severe mental deficiency and a constant gross abnormality of aminoacid metabolism. *Lancet*, 1:182–187, 1958.

19 Harris, H., Penrose, L. S., and Thomas, D. H. H. Cystathioninuria. *Ann. Hum. Genet.*, 23:442–453, 1959.

20 Stanbury, J. B., and Chapman, E. M. Congenital hypothyroidism with goiter: Absence of an iodid-concentrating mechanism. *Lancet*, 1:1162–1165, 1960.

21 Ghadimi, H., Partington, M. W., and Hunter, A. A familial disturbance of histidine metabolism. *New Engl. J. Med.*, 265:221–224, 1961.

22 Childs, Barton, Nyhan, William L., Borden, Margaret, Bard, Leslie, and Cooke, R. E. Idiopathic hyperglycinemia and hyperglycinuria: A new disorder of aminoacid metabolism. *Pediatrics*, 27:522–538, 1961.

23 Schafer, I. A., Scriver, C. R., and Efron, M. L. Familial hyperprolinemia, cerebral dysfunction, and renal anomalies occurring in a family with hereditary nephropathy and deafness. *New Engl. J. Med.*, 267:51–60, 1962.

24 Efron, M. L., Bixby, E. M., Palattao, L. G., and Pryles, C. V. Hydroxyprolinemia associated with mental deficiency. *New Engl. J. Med.*, 267:1193–1194, 1962.

25 Carson, N. A. J., and Neill, D. W. Metabolic abnormalities detected in a survey of mentally backward individuals in Northern Ireland. *Arch. Dis. Childhood*, 37:505–513, 1962.

26 McMurray, W. C., Rathburn, J. C., Mohyuddin, F., and Koegler, S. J. Citrullinuria. *Pediatrics*, 32:347–357, 1963.

27 Scriver, C. R., and Hutchison, J. H. The vitamin B_6 deficiency syndrome in human infancy. *Pediatrics*, 31:240–250, 1963.

28 Komrower, G. M., Wilson, V., Clamp, J. R., and Westall, R. G. Hydroxykynureninuria: A case of abnormal tryptophan metabolism probably due to a deficiency of kynureninase. *Arch. Dis. Childhood*, 39:250–256, 1964.

29 Querido, A., Stanbury, John B., Kassenaar, A. A. H., and Meijer, J. W. A. The metabolism of iodotyrosines III. Diiodotyrosine deshalogenating activity of human thyroid tissue. *J. Clin. Endocrinol.*, 16:1096–1101, 1956.

30 Kalckar, H. M., Anderson, E. P., and Isselbacher, K. J. Galactosemia, a congenital defect in a nucleotide transferase. *Biochim. Biophys. Acta*, 20:262–268, 1956.

31 Scheinberg, I. H., and Gitlin, D. Deficiency of ceruloplasmin in patients with hepatolenticular degeneration. *Science*, 116:484–485, 1952.

32 Jervis, G. A. Phenylpyruvic oligophrenia. *A. Res. Nerv. Ment. Dis.*, 33:259–282, 1954.

33 Schmid, Rudi, Axelrod, J., Hammaker, L., and Rosenthal, I. M. Congenital defects in bilirubin metabolism. *J. Clin. Invest.*, 36:927, 1957.

34 Dancis, J., Hutzler, J., and Levitz, M. Metabolism of the white blood cells in maple syrup urine disease. *Biochim. Biophys. Acta*, 43:342–343, 1960.

35 Dancis, J., Jansen, V., Hutzler, J., et al. (as listed in *Index Medicus*). The metabolism of leucine in tissue culture of skin fibroblasts of maple syrup urine disease. *Biochim. Biophys. Acta*, 77:523–524, 1963.

36 Westall, R. G., and Tomlinson, S. Argininosuccinase activity in argininosuccinic aciduria. *Proc. 5th Int. Congr. Biochem., Moscow, Abstrs.* 16, 103, 1298, London, 1961.

37 La Du, B. N., Howell, R. R., Jacoby, G. A., Seegmiller, J. E., Sober, E. K., Zannoni, V. G., Canby, J. P., and Ziegler, L. K. Clinical and biochemical studies on two cases of histidinemia. *Pediatrics*, 32:216–227, 1963.

38 Mudd, S. H., Finkelstein, J. D., Irreverre, F., and Laster, L. Homocystinuria: An enzymatic defect. *Science*, 143:1443–1445, 1964.

39 Aronson, S. M., Perle, G., Saifer, A., and Volk, B. W. Biochemical identification of the carrier state in Tay-Sachs disease. *Proc. Soc. Exp. Biol. Med.*, 111:664–667, 1962.

40 Kirkman, H. N., and Bynum, E. Enzymic evidence of a galactosaemic trait in parents of galactosaemic children. *Ann. Hum. Genet.*, 23:117–126, 1959.

41 Sternlieb, I., Morell, A. G., Bauer, C. D., Combes, B., De Bobes-Sternberg, S., and Scheinberg, I. H., with technical assistance of Brousseau, J. C. Detection of the heterozygous carrier of the Wilson's disease gene. *J. Clin. Invest.*, 40:707–715, 1961.

42 Teller, W. M., Rosevear, J. W., and Burke, E. C. Identification of heterozygous carriers of gargoylism. *Proc. Soc. Exp. Biol. Med.*, 108:276–278, 1961.

43 Hsia, D. Y.-Y., Driscoll, K. W., Troll, W., and Knox, W. E. Detection by phenylalanine tolerance tests of heterozygous carriers of phenylketonuria. *Nature*, 178:1239–1240, 1956.

44 Salt, H. B., Woolf, O. H., Lloyd, J. K., Fosbrooke, A. S., Hubble, D. V., and Cameron, A. H. On having no beta lipoprotein. *Lancet*, 2:325–329, 1960.

45 Childs, B., Sidbury, J. B., and Migeon, C. J. Glucuronic acid conjugation by patients with familial nonhemolytic jaundice and their relatives. *Pediatrics*, 23:903–913, 1959.

46 Coryell, M. E., et al. A family study of a human enzyme defect, argininosuccinic aciduria. *Biochem. Biophys. Res. Com.*, 14:307–312, 1964.

47 Stanbury, J. B. The metabolic basis for certain disorders of the thyroid gland. *Am. J. Clin. Nutrition*, 9:669–675, 1961.

48 Komrower, G. M., Schwarz, V., Holzel, A., and Goldberg, L. A clinical and biochemical study of galactosemia. *Arch. Dis. Childhood*, 31:254–264, 1956.

49 Walshe, J. M. Treatment of Wilson's disease with penicillamine. *Lancet*, 1:188–192, 1960.

50 Bickel, H., Gerrard, J., and Hickmans, E. M. Influence of phenylalanine intake on phenylketonuria. *Lancet*, 2:812–813, 1953.

51 Dent, C. E., and Westall, R. G. Studies in maple syrup urine disease. *Arch. Dis. Childhood*, 36:259–268, 1961.

52 Guthrie, R. Proc. Am. Public Health Ass., Kansas City, Mo., November, 1963.

53 Aisen, P., Schorr, J. B., Morell, A. G., Gold, R. Z., and Scheinberg, I. H. A rapid screening test for deficiency of plasma ceruloplasmin and its value in the diagnosis of Wilson's disease. *Am. J. Med.*, 28:550–554, 1960.

54 Berry, H. K. Procedures for testing urine specimens dried on filter paper. *Clin. Chem.*, 5:603–608, 1959.

55 Guthrie, K., and Susi, A. A simple phenylalanine screening test for detecting phenylketonuria in large populations of newborn infants. *Pediatrics*, 32:338–343, 1963.

56 Hsia, D. Y.-Y., Berman, J., and Slatis, H. A. Screening program for phenylketonuria in newborn infants. *J. Amer. Med. Ass.*, 188:203–206, 1964.

57 Efron, M. L., Young, D., Moser, H. W., and MacCready, R. A. A simple chromatographic screening test for the detection of disorders of amino acid metabolism. *New Engl. J. Med.*, 270:1378–1383, 1964.

58 Scriver, C., Davies, E., and Cullen, A. M. Application of a simple micromethod to the screening of plasma for a variety of amino-acidopathies. *Lancet*, 2:230–232, 1964.

CHAPTER TEN
SYNAPTIC NEUROCHEMISTRY: A PROJECTION[1]
Eugene Roberts

INTRODUCTION

This chapter presents in an informal way some of my present thoughts and feelings about the role of neurochemical endeavor in the neurobiological sciences. More extensive, documented articles containing some of these ideas can be found elsewhere (Roberts, 1965; Roberts and Baxter, 1963; Roberts et al., 1964).

One of the chief motivations for engaging in a chemical study of the nervous system is the hope that the knowledge gained eventually might contribute to an understanding of the working of the human mind. The force of this motivation is very great, indeed, as seen from the excitement engendered by each new report of an "abnormal" molecule in tissue fluids of schizophrenics, the report of a new mental "wonder drug," or the enunciation of a new theory of memory. There is no doubt that the structure of society and the course of human events would be greatly altered by real knowledge, at the molecular level, of our mental machinery. However, the intense effort devoted to each new finding and the attendant narrowing of the perspective often have led to the delusion that the one and only solution is on the threshold of being achieved. Frequently this has resulted in a waste of effort on the part of the initial investigators and of those who subsequently have undertaken to test the validity of the results,

[1] This investigation was supported in part by funds from the Max C. Fleischmann Foundation of Nevada, a grant from the National Association for Mental Health, and Grant NB-01615 from the National Institute of Neurological Diseases and Blindness, National Institutes of Health.

because the observations have dealt with isolated chemical findings which were not related to a meaningful model or framework.

The general solutions to the various problems of the functions of the nervous system of various species will come from an integrated understanding of the wiring diagrams (neuroanatomy), the electrical and chemical characteristics of the conducting units and transmitting mechanisms in the circuits (neurophysiology and neurochemistry), and the behavior of the organisms whose activities are regulated by them (psychology). In the case of human beings, the influence of a complex social environment cannot be neglected. The logical role of the neurochemist is to supply information which will describe the structure and organization of the conducting and transducing units in molecular terms and their function at a chemocybernetic level. The units and supporting elements of neural circuits are made up entirely of chemical substances of varying degrees of complexity. The syntheses and degradations of the macromolecular constituents which catalyze the multitude of ongoing intracellular chemical reactions and which form the structural components of the cells are under genetic control. The genetic expressivity of a cell is subject to complex environmental influences so that at any one time the state of the system is a resultant of the interaction of multiple genetic and extragenetic factors. In the circuits of the nervous system, in contrast to ordinary electrical circuits, the greatest proportion of the messages between the units is carried through the intervention of a variety of chemical transducers which must be formed, stored, liberated, bound to membranes, and eventually destroyed.

THE SYNAPSE AS A COMMON GROUND FOR CHEMIST AND BIOLOGIST

It is obvious that the chemist must choose a meaningful functional unit of the neural machinery with which to work. One of the major obstacles to progress in the past has been the absence of a single model to which both the neurobiologist and neurochemist could relate. It appears to me that the detailed examination of the organization of chemically transmitting synapses would give the chemist the best chance of obtaining data that would allow maximal opportunity for the development of correlations with observations from other disciplines.

The minimal functional unit that must be studied is the entire synaptic apparatus of which the essential elements are the pre- and postsynaptic endings, an extraneuronal compartment consisting of glial end feet and extracellular space, and the blood vessels in the synaptic area. Obviously, if understanding is to be gained of what happens in interneuronal circuits in which the nodal points are synapses, a thorough knowledge must be gained about what happens at individual synapses. It is hard to envision how chemical measurements performed on homogenates, slices, or even on individually dissected nerve-cell bodies covered with axo-somatic synapses and containing dendritic stumps could give the required information. However, recent studies of the central nervous system of the leech have indicated the feasibility of combined morphological, physiological, and chemical approaches to the study of neuron–glial–extracellular-space–capillary relationships (Kuffler and Potter, 1964; Nicholls and Kuffler, 1964, 1965).

A SIMPLIFIED SYNAPTIC MODEL:
PRE- AND POSTSYNAPTIC RELATIONS

The purpose of this chapter is to show that a hypothetical model or framework can be constructed which may lead to biochemical experimentation that is germane to the unique function of the nervous system: the generation, conduction, and regulation of nerve impulses. The relationships to be discussed for illustrative purposes, which are only a few of the many possible, include only those at the presynaptic and postsynaptic neuronal sites.

Upon excitation of a neuron, there is presumed to be a release of transmitter (or transmitters) so that there is a vectorial flow from the depolarized axonal presynaptic endings onto the postsynaptic membranes of dendrites or soma or onto presynaptic endings of other axons. The nature of the specific interaction that takes place between transmitter and membrane must be a function both of the chemical nature of the transmitter and of the structure and state of the reacting membrane. Excitation (nerve activity) occurs when the permeability of a membrane is changed in such a fashion that depolarization results. Inhibition results when the liberated substance blocks the depolarizing action of excitatory influences acting upon the same membrane. The ionic mechanisms of excitation and inhibition have been discussed extensively (Eccles, 1964).

In general, it is believed that in the vertebrate central nervous system excitatory transmitter is liberated at axo-dendritic endings and that inhibitory transmitter may be liberated at axo-axonic and axo-somatic connections (Eccles, 1964). It also has been suggested recently (Roberts, 1965; Roberts et al., 1964) that immediately following depolarization a substance (or substances) that could act as a direct synaptic feedback inhibitor might be liberated from a bound or stored form from postsynaptic sites into the extraneuronal synaptic environment in amounts bearing some quantitative relationship to the degree of depolarization. The possible relationships are illustrated schematically in Figure 10.1. The excitation is shown to occur first presynaptically and then postsynaptically. The substance (or substances) released from the presynaptic endings could act synergistically with other facilitatory (depolarizing) influences and antagonistically to the inhibitory (hyperpolarizing) influences. When an excitatory transmitter affects a postsynaptic membrane and depolarization results, it is suggested that an *instantaneous* postsynaptic liberation of an inhibitory substance may take place.

Likewise, somewhat later, as a consequence of nerve activity taking

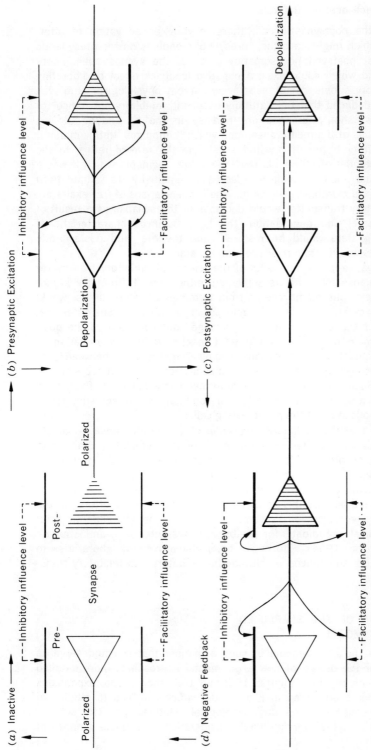

Figure 10.1 The synapse operating as a hypothetical cybernetic physiological unit during one cycle of activity. This representation places emphasis on the role of the pre- and postsynaptically liberated transmitters, which are designated on the diagram by the lines between the pre- and postsynaptic endings. Inhibitory and facilitatory influences at a particular synapse that do not originate within the synapse are indicated separately.

place in the postsynaptic cell, there may be an activation of interneurons which might have their presynaptic endings on the presynaptic
(axonal) or postsynaptic (somatic) sides of the synapse being considered and which could exert a negative feedback effect by liberating
an inhibitory substance (or substances). From physiological studies it
has been deduced that the naturally occurring inhibitory substance (or
substances) should produce an increase in the K^+ and/or Cl^- ion
conductances of the membranes upon which it impinges. In this manner,
free inhibitory transmitter would accelerate the rate of return to the
resting potential of all depolarized membrane segments which it would
contact and would stabilize (decrease sensitivity to stimulation)
undepolarized membrane segments. The concentrations of free excitatory
and inhibitory transmitter would decrease at the synapse as a result of
rapid removal and degradation, the ionic balances characteristic of
the resting state would be restored, and the pre- and postsynaptic
membranes would attain their prestimulus state.

The negative-feedback effects of the release of inhibitory transmitter
onto a synapse following its activation, either directly from the stimulated postsynaptic membrane or from terminals of interneurons, would
prevent excessive presynaptic release of excitatory transmitter per
stimulus and a too extensive and prolonged depolarization of the postsynaptic membrane. This would tend to ensure the accumulation of
sufficient quanta of the various chemical messengers between presynaptic volleys to maintain normal capacity for activity at the synaptic
junction. Such a system would serve to maintain a minimally fluctuating
activity at a synapse under any given condition of afferent stimulation
and metabolic state of the participating cells.

For the maximally effective operation of a synaptic servomechanism
there would be required a coordination of the sequential changes in
properties of pre- and postsynaptic membranes with the formation,
storage, release, and metabolic degradation of the various membrane-
active substances involved. At the present time the two known substances believed possibly to serve as excitatory transmitters in the
vertebrate central nervous system are acetylcholine and glutamic acid;
the only likely candidate for inhibitory transmitter is γ-aminobutyric
acid (Krnjevic, 1964). Efforts are being made in many laboratories to
identify more substances which may be excitatory and inhibitory transmitters.

EXPANSION OF THE SIMPLIFIED
SYNAPTIC MODEL

To employ the above model in such a manner that it would serve as
a basis for the design of new experimental approaches, more detailed
proposals are required. Figures 10.2 and 10.3 show partial expansions
of the scheme presented in Figure 10.1 which can lead to studies of
specific biochemical and pharmacological relationships. The scheme
formalizes a highly oversimplified consideration of a single synapse at

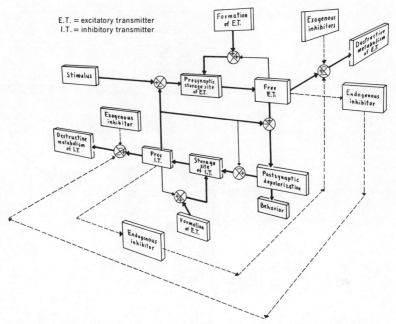

Figure 10.2 Expansion of Figure 10.1 with emphasis on some details of regulation within the excitatory transmitter (E.T.) and inhibitory transmitter (I.T.) systems.

which a single excitatory transmitter (E.T.) and a single inhibitory transmitter (I.T.) are active. Eventually it may be found that several E.T.s and I.T.s are active at individual synapses. Consideration will be given only to gross membrane effects and transmitter storage, release, and metabolism. Elsewhere, a beginning has been made toward a detailed examination of some other pertinent features of a synapse, such as ionic events during activation, possible rate-limiting metabolic steps, glial-neuron-blood relationships, changes in connectivity at a synapse with use and disuse, and neuronal specificity (Roberts, 1965; Roberts et al., 1964).

Occasionally it has been helpful to think of the E.T. system in relation to acetylcholine and the I.T. system in relation to γ-aminobutyric acid (γABA). However, it must be pointed out that in no instance have both of these systems yet been proved to operate at a particular synapse, and it is certain that these substances are not operative at all synapses. The synthesis of acetylcholine, a substance exerting excitatory action on some neural membranes, is catalyzed by choline acetylase and its hydrolysis by cholinesterase. γABA, an inhibitory substance which acts by increasing conductances of membranes to K^+ and Cl^- ions, is made from L-glutamic acid by glutamic decarboxylase and degraded by γABA-α-ketoglutarate transaminase. Both γABA and acetylcholine are found in free and bound forms in the

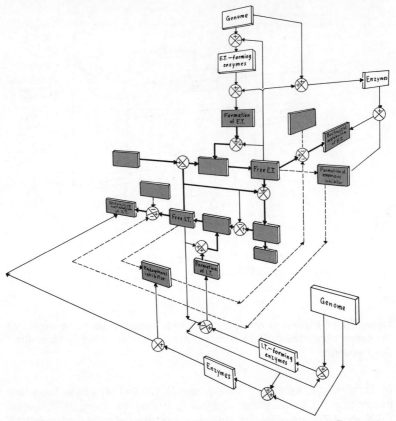

Figure 10.3 Expansion of Figure 10.2 with emphasis on possible genetic regulatory factors in the E.T. and I.T. systems.

vertebrate CNS, and experiments have shown that when these substances are applied to neural tissue they can undergo physical binding to membranes.

REGULATION WITHIN THE
E.T. AND I.T. SYSTEMS

When a particular neuron is effectively stimulated, there is liberated into the synaptic cleft a free form of some of the E.T. that is contained in the presynaptic terminals in a bound form (Figure 10.2). The free form of E.T. is bound to the apposing postsynaptic membrane, producing depolarization, and the largest proportion of the liberated transmitter is most likely destroyed rapidly by enzymes on the postsynaptic membrane. Under conditions of stimulation which are normal for a particular synapse, the formation of E.T. within the presynaptic ending would take place at a rate commensurate with the rate of depletion so that, within relatively narrow limits, a constant level of E.T. would be maintained in the presynaptic storage site. In the absence of all synaptic activity the formation of E.T. would be stopped, possibly by product inhibition, substrate limitation, or both. The latter control mechanism is not shown in Figures 10.2 or 10.3 either for E.T. or I.T.

If acetylcholine is an example of E.T., it might be suggested that when the acetylcholine content reaches high levels in the nerve endings the choline acetylase system might cease to function, either because of inhibition by acetylcholine itself or because the quantity of free choline and acetyl CoA, the precursors of acetylcholine, might become limiting. After stimulation and release of acetylcholine, the acetylcholine level in those portions of the nerve ending from which it was released could fall below the inhibitory level, and some of the choline and acetate liberated by hydrolysis could be transported back into the presynaptic site for reconversion to acetylcholine. Acetyl CoA formation also could be enhanced by increased glycolysis consequent to depolarization (see Roberts, 1965, for discussion).

Under some circumstances (excessive stimulation, overproduction, blockade of destructive enzymes) the rate of release of free E.T. would exceed the rate of removal, and some of the undegraded transmitter might be transported back into the presynaptic site and be available for use again. If the condition of excessive accumulation of E.T. should persist for long enough periods, it might act directly or indirectly at a genetic level (Figure 10.3) as a repressor for the formation of messenger RNA involved in the formation of the enzymes which form E.T. and as a de-repressor for the formation of the RNA messengers for enzymes involved in its destructive metabolism or conversion into other substances (i.e., endogenous inhibitor).

Several of the considerations employed above in the case of E.T. would apply to I.T. concerning the possible coordinations involved in maintaining a steady-state situation with regard to storage, release, synthesis, and destruction and will not be discussed again. I.T. would be released into the synapse only after postsynaptic depolarization. It would decrease the excitability of the presynaptic and postsynaptic membranes to the depolarizing influence of a given afferent stimulus, since presumably in the presence of I.T. fewer quanta of E.T. would be liberated per impulse and the sensitivity to depolarization of the postsynaptic membrane per quantum of E.T. would be less than in the absence of I.T. Under ordinary circumstances either the rate of removal of free I.T. would be great enough to remove all of it from the synaptic region between impulses or the rates of liberation and removal would be so coordinated that a constant amount of free I.T. would be present at all times between stimulations, supplying a modulatory "tone." It would be more likely that at a particular synapse there would be a constant background level of free γABA than of free acetylcholine, since the affinity of substrate for enzyme and turnover number of the γABA-transaminase, the γABA-destroying enzyme system, are much lower than those of cholinesterase. An additional defense against the escalation of free I.T. content would derive from its inhibitory effect on depolarization of membranes; increased amounts of free I.T. at a synapse would decrease its own release from postsynaptic storage sites or from presynaptic terminals of interneurons.

REGULATION BETWEEN THE
E.T. AND I.T. SYSTEMS

Some cross regulations may exist between the E.T. and I.T. systems which may become particularly important when rapid, large increases either in I.T. or E.T. are produced by extensive blockade of the respective degradative enzymes with exogenously administered or endogenously formed (dashed lines, Figure 10.2) inhibitors. Products may then be formed from I.T. and E.T. which ordinarily may not be formed at all or which may exist only in low concentrations because of the relatively low affinity of E.T. and I.T. for the enzymes involved. If, when the breakdown of either E.T. or I.T. is inhibited, products are formed that can inhibit the breakdown of I.T. or E.T., respectively, the enhanced excitability or depression attributable to the gross increases in one or the other of these substances at a synapse would be balanced by an increase in a *physiologically* equivalent quantity of the other.

In contrast to the small rapidly occurring adjustments (millisecond time scale) taking place during normal synaptic activity, changes such as those discussed above would be expected to be relatively great and to persist for long periods of time (minute-hour time scale). In the case of increased E.T., an initial period of increased excitability, as measured by some physiological parameter, would be followed by a gradual increase in I.T. content which would be paralleled by a return of excitability to normal levels. Similarly, when I.T. would be increased by blockade of its destruction, a period of decreased excitability would be followed by an increase in E.T. content and the return of excitability to normal. In both cases discussed above, the eventual result would be the maintenance of essentially normal synaptic activity in the presence of increased levels of both E.T. and I.T. If the latter suggestions are correct, good correlations between estimates of E.T. or I.T., alone, and synaptic excitability would be expected only at the beginning of the induction of changes in contents of E.T. or I.T. However, at all times, changes from the control levels in the *relative* amounts of E.T. and I.T. should be well correlated with changes in excitability.

CHANGES IN REACTIVITY
OF MEMBRANES

Another factor which must be important in the regulation of synaptic activity is the state of the pre- and postsynaptic membranes. It can easily be imagined that the plasma membranes, which contain structural and enzymatic proteins, lipides, glycolipides, and polysaccharides (Emmelot et al., 1964), would be subject to many local influences which might affect their physical state [pH, concentration of small charged molecules (organic and inorganic), hormones, availability of water, exogenously administered drugs, etc.]. Let us suppose that in the environment of a particular synapse a change occurs which suddenly decreases the sensitivity of the presynaptic membranes to stimulation (see Figure 10.2). The same stimulus that had been causing the liberation of a given amount of E.T. would liberate less than before, the effect on the postsynaptic membrane would be less, and less I.T. would be released following postsynaptic depolarization. The decreased liberation of I.T. would result in an increased sensitivity of both pre- and postsynaptic membranes to stimulation and a return toward normal responsivity of the synapse. Likewise, an environmental change producing a decreased sensitivity of the postsynaptic membrane would lead to less liberation of I.T., enabling the synapse to return to normal operation. Conversely, an increased sensitivity of the membranes responding to stimulation would result in an increased release of I.T. and a heightened inhibitory "tone" which would counterbalance the hypersensitivity and keep the resultant synaptic activity within the normal range.

FUNCTION OF NEURONAL CIRCUITS AND SYNAPTIC ACTIVITY

Our eventual goal is to be able to test some of the above hypotheses at intact individual synapses by ultramicro analyses of pertinent chemical variables performed simultaneously with measurements of electrical changes at pre- and postsynaptic endings. To date it has not been possible to do this in vertebrate synapses. The only experimental work which can be discussed with reference to the model presented in this chapter is concerned with correlations between gross chemical measurements in brains of animals and some aspect of behavior. Although far from ideal, it is not entirely unreasonable to make inferences about synaptic activity from some measurable parameters of behavior.

A single external event can produce the activation of circuits of neurons. In any such circuit of synapses there would be expected a variation in thresholds with regard to the amount of stimulation required to activate the individual synapses. Until the synapse with the highest threshold is activated, the circuit, as such, will not fire, and only those synapses in the chain up to the inactive member will be able to reflect the impingement of a single externally applied stimulus. When a whole behaving organism is presented with an effective stimulus, which consists of a perceived pattern of changes in the environment, it would be expected that many neuronal circuits would be set into activity. It may be assumed that the greater the number of pertinent circuits operative under a given pattern of stimulation, the greater would be the probability of a rapid and efficient processing and analysis of the data for use of the perceiving organisms and the more effective would be the response to the stimulus (behavior).

If the above assumption is correct, it should be possible to make reasonable predictions about the effects of various types of treatments (including that with drugs) on complex neuronal functions, if their effects on synaptic function are known. The types of effects of the treatments would be expected to be inhibitory, excitatory (or facilitatory), or pathological. Inhibitory and facilitatory substances or procedures would exert their detectable influences by acting in such a manner that more or less stimulation, respectively, would be required to initiate firing of the synapse with the highest threshold, and, therefore, more or less stimulation would be needed to activate the entire circuit. There are a number of potential sites of these effects at the synapse (pre- and postsynaptic membranes, glial cells, capillaries). The primary effects at any of these sites could be on some aspect of their physical structure, on an enzymatic function that may be related to the metabolic maintenance of the site, or on the synthesis or degradation of a transmitter substance involved in maintaining coordinated function between the synaptic components.

The following substances are known to increase neural excitability and also to facilitate learning when given in moderate doses: strychnine, picrotoxin, amphetamine, 5,7-diphenyl-1,3-diazadamantan-6-ol, thyroxine, nicotine, caffeine, potassium, and physostigmine. Substances known to decrease neural excitability, which also have been shown to inhibit learning, are calcium, various barbiturates, ether, and carbon dioxide (Glickman, 1961; Hudspeth, 1964; McGaugh et al., 1961). Substances or procedures exerting pathological effects at the synapse, which may include high levels of those which in smaller amounts cause excitation or inhibition, would act by producing sufficient disorganization in synaptic function so that at least some synapses in key circuits would cease to function as effective cybernetic units, and, therefore, the circuits of which these synapses are a part also would not function properly. Procedures producing convulsions and spreading depression would be included in the pathological category.

Some excitatory substances, when given in large amounts, produce paroxysmal discharges and seizures. However, it is certain that the modes of action of these substances at a molecular level are widely different, ranging from alterations produced in the physical state of the neural membranes by direct combination with some component in them to the inhibition of cholinesterase, which results in the accumulation of acetylcholine. Whatever their mechanism of action, these substances destroy the effectiveness of the systems of checks and balances that exist between inhibitory and excitatory phenomena at synapses in key neural circuits and allow uncoordinated excitation to take place. Similarly, some inhibitory substances, probably acting by a variety of mechanisms, can cause anesthesia and finally death when given in excess. Indeed, little about the detailed mechanism of action can be inferred from the study of the effects on intact animals of substances that either cause convulsive seizures or prevent them if corresponding measurements of pertinent substances in the CNS are not carried out at the same time.

TYPES OF STUDIES WHICH EVENTUALLY MAY BE RELATABLE TO THE SYNAPTIC MODEL

Elevation of I.T.

In a recent series of experiments (Kuriyama et al., 1965) the time course of changes in brain content of γABA and sensitivity to electro-convulsive shock were studied in mice after administration of ami-nooxyacetic acid (AOAA), an inhibitor of γABA transaminase. The increase in γABA content as a function of time after AOAA administration (25 mg/kg) was biphasic. A plateau in γABA level was attained between 3 and 4 hours, and a subsequent secondary rise took place between 4 and 6 hours. Elevated values were still observed after a 24-hour priod. There was a remarkable decrease in electroshock-seizure incidence (75-ma stimulus) during the first 1½ hours after AOAA. Subsequently, the susceptibility to seizures began to return to normal, attaining the control values at 6 hours, at which time the γABA content was maximal. Figure 10.4 shows a three-dimensional plot in which seizure incidence and whole brain γABA contents (as percent of control values) are plotted as a function of time for 6 hours after the administration of AOAA. Only during the first 1½-hour period was there a correlation of decrease in seizure susceptibility with increase in γABA content. Thereafter, the seizure susceptibility increased while the γABA content continued to rise.

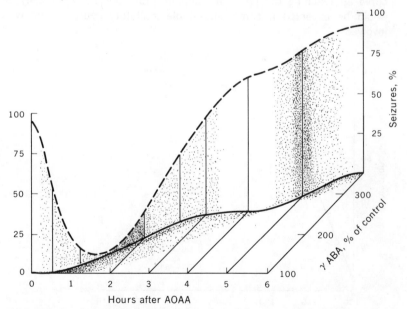

Figure 10.4 Three-dimensional plot of seizure susceptibility and γABA content of brains of mice as a function of time after administration of 25 mg/kg of aminooxyacetic acid. Arrows point away from the inflection points of the curves from which they are drawn.

One of the possibilities suggested by the above data is that the increases in γABA levels and changes in seizure susceptibility after administration of AOAA are completely unrelated. Another possibility is that a relationship exists but is localized to a particular, small region in the brain or can be found only by analysis of those neuronal regions from which neuroactive substances are liberated during function in the CNS. The latter points are now being further explored in our laboratory by study of smaller brain regions and by analysis of isolated nerve endings from normal and AOAA-treated mice with regard to γABA content and the activity of the enzymes involved in its metabolism.

Another way to view the problem is consistent with the formulation in Figure 10.2. γABA may, indeed, be importantly involved in decreasing neuronal excitability soon after γABA transaminase blockade with AOAA, and total brain γABA may be directly related to the physiologically effective amounts of the substance; but compensatory increases in excitatory factors, decreases in inhibitory factors other than γABA, or both, may take place, with the consequent restoration of normal sensitivity to electroshock in spite of the persistence of elevations in γABA content. With the above in mind, it was of interest to examine the type of relationship between γABA content and protection against seizures during the first 1½ hours after AOAA (Figure 10.5). At the lower levels of γABA a linear relationship was found to exist between seizure susceptibility and brain γABA content, with the curve approaching the point of complete protection asymptotically, as would be expected if some saturable inhibitory neuronal site were involved.

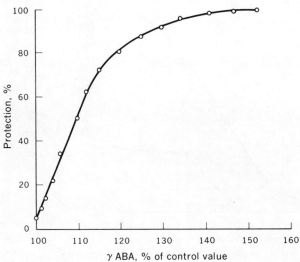

Figure 10.5 Relationships between brain content of γABA and protection against seizures (75-ma current) during the first 1½-hour period after administration of aminooxyacetic acid (25 mg/kg).

A further look at Figure 10.4 reveals the interesting fact that the inflection point of the first portion of the γABA curve occurred at the time when the curve of seizure susceptibility began to rise toward normal, and the inflection point of the latter curve just preceded the beginning of the second rise in γABA content. This would be understandable in terms of a biochemical servomechanism in which a progressive, self-limited increase in amount of effective excitatory transmitter (increased rate of formation, decreased rate of destruction, increased sensitivity of receptor) or a progressive, self-limited decrease in effective amount of one or more inhibitory substances other than γABA would be triggered off at the time when the increase in γABA is occurring at the maximal rate (inflection point); likewise, at the time of maximal rate of change of the compensatory change, a secondary rise in γABA could be induced through an increase in the rate of its formation or a decrease in the rate of utilization. Glutamic acid, a potentially important excitatory factor in the CNS, did not increase after AOAA. Work on acetylcholine content is in progress.

Elevation of E.T.

A well-studied case of a persistent elevation of acetylcholine content in the brain is that of the *ep* strain of mice (Kurokawa et al., 1961, 1963; Naruse et al., 1960), in which susceptibility to convulsions after successive movements causing loss of postural equilibrium is inherited as a Mendelian dominant characteristic. The convulsive threshold to pentamethylenetetrazole is lower in the *ep* strain than in the normal strain studied. Spontaneous convulsions do not occur in the *ep* mice under ordinary cage conditions. A morphological basis for the defect could not be found on histological examination of these mice. Postural stimulation did not induce convulsions in animals younger than 7 weeks and brought on seizures only in an incomplete form in animals between 7 and 10 weeks of age. Full-blown, but not lethal, convulsions could be produced in all animals of both sexes of the susceptible strain at all subsequent ages.

The acetylcholine levels of the brains of *ep* mice were 40 to 60 percent higher than those of the other strains studied. At 4, 5, 6, and 7 weeks of age and in adults the levels for the *ep* strain were 2.09, 2.31, 2.54, 2.66, and 2.86 $\mu g/g$, respectively; for a control strain (*gpc*) the corresponding values were 1.47, 1.65, 1.94, 1.95, and 1.95 $\mu g/g$, respectively. The brain γABA content in the *ep* mice also was 40 to 50 percent higher than in the control strain, while both glutamine and glutamic acid levels were approximately 30 percent lower. Thus, *both acetylcholine and γABA levels* were found to be higher to approximately the same relative extent in the brains of the *ep* mice. The enzyme catalyzing the synthesis of acetylcholine from choline and acetyl CoA, choline acetylase, was found to be 50 percent higher in the *ep* mice than in the other three strains tested. The increased choline acetylase activity probably furnishes the explanation for the higher value of acetylcholine in the brains of the *ep* mice, since the activities of the cholinesterase and the enzyme catalyzing the synthesis of acetyl CoA did not differ significantly between the *ep* mice and the control strains. In a study of the content of particle-bound acetylcholine in homogenates of brains of *ep* and normal mice it was found that the labile component (liberated by osmotic dilution) in the *ep* strain was twice that found in normal mice; the level of stable acetylcholine was approximately the same. The labile fraction of bound acetylcholine may be the fraction liberated during nerve activity, since decreases in this fraction were found to parallel decreases in total acetylcholine in animals that had been convulsed (Kurokawa et al., 1963).

The above data tentatively may be placed into the scheme shown in Figures 10.2 and 10.3 as follows: For some, as yet unknown, genetic reason (failure of repression or too much de-repression?) the activity of the enzyme that forms E.T. is greater than normal in the CNS of the *ep* mouse. This results in an increased level of labile, or physiologically available, E.T. in presynaptic storage sites of E.T. A given external stimulus releases more than normal amounts of E.T. at the synapses of the activated neuronal circuits. This would predispose to indiscriminate spread of impulses and convulsions, a tendency compensated· for by higher levels of I.T. in physiologically available storage sites and an increased release of I.T. at synapses consequent to nerve activity. As a result, the animal could live out its life without visible handicap under ordinary laboratory conditions. The neuronal system poised at the above level would have less stability than the normal one; thus, intense stimulation would lead to incoordination in the least-compensated circuits, which would be reflected behaviorally in a failure to adjust normally to certain types of environmental stresses.

Decreased Sensitivity of Neuronal Membranes

It is not unreasonable to assume from data in the literature that barbiturates may exert their inhibitory action at synapses by decreasing the excitability of both pre- and postsynaptic membranes. Reasoning from our synaptic model, it could easily be suggested how, during the addiction to these drugs, compensatory phenomena might take place which would cause more E.T. and less I.T. to be liberated per impulse, thus enabling essentially normal synaptic transmission to take place in spite of decreased sensitivity of the membranes. It has been reported that barbiturates cause an increase of bound acetylcholine in the cerebral cortex of the cat (McLennan and Elliott, 1951) and a decrease in total γABA content of the cerebral cortex of the rat (Baxter, personal communication). If the barbiturate should be withdrawn suddenly, the sensitivity of the membrane sites would return to normal, probably shortly after the blood level had fallen, while it would be expected that the enzyme systems overproducing E.T. and underproducing I.T. would take longer to return to levels of the predrug period. Therefore, hyperexcitability would be expected to result upon drug withdrawal. Indeed, convulsions and other neural disturbances occur in animals and man during barbiturate withdrawal, indicating a pathological predominance of excitatory influences. If the above general concept has some validity, then an elevation of γABA levels during the withdrawal of barbiturates in addicted organisms should decrease the severity of the symptoms.

A pertinent experiment based on the above suggestion was performed in which dogs were addicted to barbital (Essig, 1963). After complete withdrawal of the drug, a control group was given saline and an experimental group was given AOAA, which elevates γABA levels. There was a dramatic difference in the behavior of the two groups. All the controls showed convulsions, status epilepticus being attained in a number of instances, and a 56 percent mortality occurred in the first 7 days after withdrawal. Far fewer seizures and deaths were found among the treated dogs, some of the animals showing no seizures at all during the 7-day period of administration of AOAA. When administration of AOAA was terminated at 7 days after cessation of the barbiturate, several of the dogs showed convulsions for the first time. The latter observation is in keeping with the suggestion of a slow return to normal of the compensatory changes developed during barbiturate addiction. Although not proved, it seems likely that AOAA exerts at least part of its inhibitory action in this instance through increasing the level of γABA in the CNS. Dilantin, acetazolamide, and pyridoxine failed to show anticonvulsant effects under these conditions.

GENERAL COMMENT

The time is not yet ripe for a complete experimental test of the model presented in this chapter. First must come a much more complete identification of synaptically active excitatory and inhibitory transmitters and the attainment of pertinent information about the mechanisms of their formation, storage, release, mode of action, and degradation. In each case crucial questions must be answered, like those about γABA. Is it released from the presynaptic terminals of all neurons in the CNS or just from the terminals of special neurons which are part of feedback circuitry and which liberate γABA onto presynaptic and/or postsynaptic elements of particular synapses? Or is γABA formed chiefly postsynaptically (in dendrites and perikarya) in neurons throughout the CNS, maintained in a bound or sequestered form, and liberated into the synaptic cleft upon depolarization of the postsynaptic membrane in amounts which bear some quantitative relationship to the degree of depolarization?

We must move step by step, from types of experiments in which correlations are made between a measurable behavioral variable and biochemical measurements in whole brains or grossly dissected portions thereof, to the ultimate goal of a kinetic study of the interplay of pertinent biochemical and physiological factors in a single synapse which is functioning as an integral part of intact neuronal circuitry and whose function is being monitored by measurement of the electrical responses at both pre- and postsynaptic terminals. Recent experiments (Gibson and McIlwain, 1965) in which continuous intracellular recordings have been made from cells of guinea pig cortex slices suitably maintained *in vitro* have given a promising new tool.

Difficult problems must be faced in any attempts to determine whether or not there is a causal relationship between the alteration of a particular biochemical variable and a measurable synaptic or behavioral change in the organism in which the change is induced. The model presented requires that the *balance* between the E.T. and I.T. systems, rather than the absolute amount of either E.T. or I.T., would be important in the regulation of activity at synapses. It is now necessary to turn from more simple and comfortable courses of action and to undertake the multivariant analytical approaches which are worthy of the cybernetic control mechanisms that exist in synapses.

The type of synapse discussed in this and preceding chapters in largely conjectural biochemical and biological terms would be a suitable nodal element for circuits with plastic properties in which there is plasticity because the connectivity of the net as well as that of the nodal elements is use- and time-dependent (Roberts, 1965; Roberts and Baxter, 1963; Roberts et al., 1964). In such networks there could be "memory without record." The actual demonstration that synapses have the general properties suggested in this chapter would make a great contribution to the understanding of many of the integrative functions of the nervous system observed in behaving organisms.

PART III QUANTITATIVE GENETIC ANALYSIS

INTRODUCTION
R. C. Roberts

Just about a hundred years ago, Mendel uncovered the fundamental concepts of genetics. His success in an area where others before him had floundered can probably be attributed to two factors. Firstly, Mendel recorded his data quantitatively and examined them by numerical methods; this approach led him to formulate the cardinal principles of segregation and independent assortment. Secondly, and just as importantly, Mendel carefully chose to work with certain characteristics of his peas that could be sharply distinguished one from the other; one cursory glance at a pea seed would be sufficient to establish whether it was wrinkled or round.

When Mendel's work was rediscovered ("reexamined" would probably be a better word) at the turn of the century, it was just his care in' choosing his characteristics that led to some fierce polemics concerning the generality of his results. Whereas Mendel's tall and short peas, for instance, showed a distinct contrast, the same is not generally true of, say, human stature. Among human beings there is a continuous distribution of stature from the very tall to the very short and, as Galton had shown, the inheritance of such traits appeared to be intermediate; offspring tended to reflect the mean height of their two parents and did not partition themselves into neat ratios of tall to short. This apparent anomaly was reflected in the sharp cleavage of opinion in British genetics during the early years of this century. The major dramatis personae were probably Bateson on the one hand, who was actively multiplying examples of Mendelian inheritance, and on the other hand the biometricians, especially Weldon, who claimed that traits such as height and weight could not involve the Mendelian mechanism. Initially, at least, the biometricians tended to dismiss Mendelism as a trivial exception to the common-blending type of inheritance displayed by continuously varying characters.

Outside Great Britain, these differences of opinion and of approach were not thrown into such sharp relief. In the United States, East, who worked with ear length in corn, had proposed the multiple-factor hypothesis to explain his results by about 1910. Contemporaneously in Sweden, Nilsson-Ehle drew an identical inference from the inheritance of kernel color in wheat. Indeed Mendel, in the second part of his original paper, had hinted at a similar explanation of what he called the "enigmatical results" from the inheritance of flower color in crosses of *Phaseolus* species. However, the precise synthesis of ideas about single- and multiple-factor types of inheritance was not finally established until about 1918–1920. This was achieved independently by R. A. Fisher in Great Britain and Sewall Wright in the United States, who showed that the results on the inheritance of continuously varying characters were a necessary con-

212

sequence of "Mendelian" inheritance as the number of factors increased.

Although this common ground was established almost 50 years ago, the methods and concepts of monogenic and polygenic inheritance have to some extent remained distinct. Basically, they have a common feature. Individuals with observable phenotypes are compared with specified relatives, say parents or sibs. But where the individuals cannot be placed unambiguously in clearly defined categories, i.e., where the variation is continuous, special methods must be invoked, and the genetic questions to ask must be phrased appropriately. Arithmetical summaries of simple ratios must give way to a somewhat more elaborate statistical treatment. For this reason, the genetic analysis of continuously varying characters deploys special techniques that have been developed for the purpose.

Part III of this book describes some of the concepts and methods whereby the genetics of continuously varying characters may be examined. Many laboratory measurements of behavior are of this kind, and where adequate data exist, it also seems that the measurements are, in part, the products of polygenic systems of inheritance. The resolution of the variance of such a measurement therefore demands that the appropriate genetic techniques should be employed. No claim is made that Part III is an exhaustive treatise on quantitative genetics; neither is any particular approach advocated at the expense of others. Rather, it has been our aim to present some general ideas of what quantitative genetics is about and how the genetic investigation of quantitative traits may proceed. If any reader deems such an approach to be relevant to his research requirements, it is our hope that he may at least gain some reading knowledge of the language of quantitative genetics and, to that extent, facilitate his access to further literature on the subject.

Part III comprises four chapters. The first is by Roberts, who outlines some concepts and methods in quantitative genetics, with special regard to outbred populations. In the second chapter, Hirsch describes how the genetic analysis of strains of Drosophila selected for geotaxis may be pursued further by establishing the influences of individual chromosomes and the interactions between them. Bruell, in the third chapter, examines the consequences for behavioral studies of the crossing of inbred strains, with a discussion of the evolutionary implications of genetic findings. If several inbred strains are completely intercrossed, the data can be assembled into what is known as a diallel table. A special genetic analysis of such data is described by Broadhurst in the fourth chapter. Broadhurst also discusses some topics of general interest in quantitative genetics; these are environmental control, maternal effects, scaling, and genotype-environment interactions.

CHAPTER ELEVEN
SOME CONCEPTS AND METHODS IN QUANTITATIVE GENETICS[1]
R. C. Roberts

INTRODUCTION

Quantitative genetics is the study of the inheritance of characters where the individual effects of the genes concerned are obscured from view. Let us consider for a moment a typical example of such a character—body weight in the mouse. We know of several genes that affect body weight; for instance, the *yellow* coat-color gene makes the mouse noticeably heavier. Some mutant genes, such as *pygmy* or *obese*, are most easily identifiable through their effect on weight. However, the effect of such genes is large enough for them to be recognized on a given genetic background, and they are best studied through the established methods of formal genetics; their dominance, pleiotropic, and epistatic relationships can be described, and they can often be located on the linkage map of the mouse. But if we take a closer look at a litter of mice, segregating for, say, the *obese* mutant, we could immediately score the animals by eye into obese ones and "normals." Among the two classes, we could detect further differences in body weight, though perhaps we should require a balance to rank them accurately. We should discover further, if we were to carry out the appropriate experiments, that these smaller differences in weight are also in part heritable. But we can no longer recognize the genes

[1] *Acknowledgment*: Readers familiar with D. S. Falconer's *Introduction to Quantitative Genetics* will recognize the source of much of the material in this chapter. It is a pleasure to acknowledge my indebtedness to this source and to thank Dr. Falconer for reading this manuscript.

responsible for this variation, nor are we sure exactly in what way they affect body weight. Whereas we know that the *obese* gene increases the amount of fat, we could not be sure whether the residual variation in weight is due to slight variation in adiposity, in skeletal size, or in what. Through elaborate experimentation, we could perhaps discover answers to some of these questions, but we should fail to disentangle the individual genes involved because their effect is masked by other genes, perhaps having an identical effect. We are now in the domain of quantitative genetics, where individuals differ because they contain different samples of the alleles that potentially affect the character but where the investigator cannot know which particular alleles affect particular individuals. This kind of variation immediately implies a statistical approach. In practice, these *quantitative characters,* as we shall call them, are also subject to a certain amount of environmental variation. Our example of body weight is an obvious case; differences in nutrition, for instance, will obviously be reflected in the measurements. But environmental variation is not limited to quantitative characters, for it is to be found also in the expression of single genes.

It seems reasonable to presume, at this stage, that many aspects of behavior, such as activity and learning, require a quantitative scale of measurement in their treatment. Certainly, some genes exist whose behavioral effects are clear without refined measurements. As a specific example, the gene causing phenylketonuria in human beings has a well-known and marked effect on intelligence. This gene is therefore a source of variation in intelligence among individuals, but no one would argue that this gene, or genes like it, accounts for all the genetic variation in intelligence. Indeed, such genes do not appear as a cause of variation among the so-called "normal" individuals at all. Yet, these "normal" individuals are known to vary, and such variation is known to be, in part, of genetic origin. Similarly, phenylketonuric individuals vary among themselves, and it may be presumed that this variation also is caused, in part, by some of the genes that are responsible for variation among nonphenylketonuric individuals. Although genes such as the one causing phenylketonuria are dramatic in their effect, their occurrence is relatively infrequent, and they cannot therefore be regarded as an important source of variation in intelligence in a human population. The study of genetics is thus limited in scope if it is confined to genes causing "abnormalities." Quantitative genetics finds its application in the resolution of the "normal" range of variation.

The following, then, are the premises of quantitative genetics. A quantitative character is presumed to be affected by a number of genes, whose individual effects are small. Whereas it is clearly absurd to suppose that all the effects of such genes are interchangeable at the level of the primary gene action, the effects are nevertheless deemed to be interchangeable at the level of the measurement. This says no more than that a given body weight, for instance, can be the end product of several distinct developmental pathways. It is a further premise that the genes affecting a quantitative character can exhibit all the properties associated with the so-called "Mendelian" genes, such as dominance, epistasis, and linkage, but that "quantitative" genes possess no properties that are unknown in formal genetics.

Quantitative genetics differs from formal genetics essentially by virtue of the fact that the effects of all the genes involved cannot be identified individually. It does not differ, as is sometimes supposed, because a character is measured rather than scored nor because such a character often exhibits a large amount of environmental variation. Superimposed on the framework of formal genetics is a statistical approach which has given rise to a specialized methodology. This has inevitably led to specialized concepts and to specialized thinking, but the concern is still with genes and not with mathematical relationships.

THE RECOGNITION OF THE GENETIC
BASIS OF A QUANTITATIVE CHARACTER

Suppose we were interested in a character exhibiting a range of continuous variation—characters such as intelligence in a human population or activity scores in mice. We should need to know whether differences between individuals were in any way related to differences in their genotypes and also whether such genotypic differences occurred at one locus or at more. We should eventually need to know what proportion of the variation was attributable to genetic sources.

The first of these questions is sometimes easy to answer, especially if it is rephrased. Given known genotypic differences between animals, do corresponding differences appear in the character that we are examining? The answer to this question is facilitated in some laboratory species, such as mice or Drosophila, by the use of genetically homogeneous material, most commonly inbred strains. Any variation within such a strain is presumed to be environmental in origin. If, however, an analysis of variance reveals excess variation over this amount between strains kept under uniform conditions, there is clear evidence that the character shows genetic variation. It is preferable that several strains should be employed in the search for genetic variation in this way and that these should furthermore be of diverse origin. Even so, the absence of a difference between strains is not clear proof that the character shows no genetic variation in the species at large, for the strains examined may have been fixed for the same genes controlling this character or for genes whose sum effect is similar. However, this is perhaps unlikely. In species where inbred material is unavailable, it may be possible to substitute geographical races or even subspecies for inbred strains.

Having found genetic variation, we need to know whether it is controlled by one gene or by more than one. In the presence of environmental variation, this may not be as easy as it sounds. The approach is to examine for segregation groups of progeny from parents of known values, employing backcrosses, intercrosses, etc., exactly as we would if the character displayed little or no environmental variation, and where the classes of progeny might then be discrete. Perhaps the most critical type of mating would be between parents of intermediate value, to see whether the full range of expression is obtained in the progeny, as expected on a single-locus model. It is seen that environmental variation would make it difficult to recognize more than one locus and well-nigh impossible to recognize even a single polyallelic locus. A confusing situation can also arise in the case of threshold characters. Such characters may under certain conditions mimic single-gene characters rather closely.

Unless we can establish that the genetic variation of the character is controlled by a single locus, or by a few recognizable loci, we have in effect a quantitative character, and the genetic questions to ask must be phrased accordingly. These questions will not refer particularly to individual animals, whose genetic constitution is unknown, but rather to populations of animals, where the sum effect of the genes is more clearly seen. We need to know how much of the total variation is due to various genetic causes, for it is axiomatic that the importance of a source of variation is proportional to the contribution it makes to the total variation. We need to know the extent to which relatives, on average, resemble one another, for this is what we really mean when we say that a character has a genetic basis. (All characters are genetic in the sense that genes are *sine qua non* for their existence.) We need to know whether we can manipulate the genes to change the level of expression of the character, otherwise our research work may become circumscribed. And we need to know how those genes are most readily manipulated, otherwise our time may be wasted. It is to questions such as these that much of what follows will be devoted.

THE SOURCES OF VARIATION
IN THE POPULATION

Let us first examine what genetic factors affect the average performance of a population of animals, and what factors contribute to differences between individuals. To do this, we need to determine the *population mean* and the *genotypic variance*. An understanding of these parameters is necessary for the further development of ideas about quantitative genetics. The symbolism, terminology, and derivations used throughout this chapter follow those used by Falconer (1960).

The *phenotypic value* (P) of an individual consists of two parts, a *genotypic value* (G), determined by the individual's genetic constitution, and an *environmental deviation* (E), which may be either positive or negative. The environmental deviations are taken to be such that their sum over the whole population is zero.

$$P = G + E \quad \text{where } \Sigma E = 0$$

It is a fundamental feature of the formulation that G and E are uncorrelated, and since some behaviorists may consider this an unwarranted premise, the point should perhaps be elaborated. It seems obvious that an animal's genotype may influence its choice of environ-

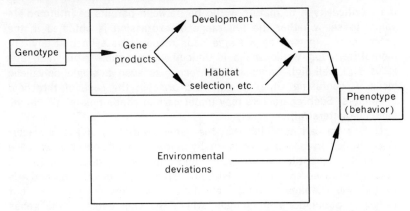

Figure 11.1   Relationship between genotype, environment, and phenotype, in schematic form, whereby the environmental deviations are *defined* as being independent of the genotype.

ment, which would therefore introduce a correlation between the two. This indeed may be so, even under a carefully regulated laboratory regime. However, this phenomenon, where it may exist, should be examined in the context of the pathway from the genotype to the phenotype, in this case behavior. The ramifications in this pathway are manifold and complex; for the present purposes, let us consider the grossly simplified version depicted in Figure 11.1. The genotype may influence the phenotype either by means of biochemical or other processes, labeled for convenience as "development," or by means of influencing the animal's choice of environment. But this second pathway, just as much as the first, is a genetic one; formally it matters not one whit whether the effects of the genes are mediated through the external environment or directly through, say, the ribosomes. In either case, the genotype affects the phenotype, and in this sense, all that comes between the two can be lumped together in a "black box" and treated as various parts of the same genetic process. This has complete operational validity, for the properties of a system can be explored (and often must be explored) without specifying completely the individual components of the system. There are of course powerful precedents for this approach in the physical sciences. But to return to the absence of a correlation between the genotypic values and the environmental deviations, the environment is *defined* as that which affects the phenotype independently of the genotype. If an effect stems from the genes, it is genetic; any other effect is an environmental one.

Since $\Sigma E = 0$, it follows that the average genotypic value of the individuals in a population is equal to their average phenotypic value and that either can be referred to as the population mean (M). Symbolically,

$$\overline{P} = \overline{G} = M$$

To see what genetic factors affect the mean level of performance in a population, we shall begin with a single locus with two alternative alleles, A_1 and A_2. Let the frequency of A_1 be p, and that of A_2 be q, where $p + q = 1$. It follows that, if diploid organisms mate at random, there will appear among the offspring three genotypes, A_1A_1, A_1A_2, and A_2A_2 in the ratio of p^2: $2\,pq$: q^2, respectively. We must now assign scale values to the three genotypes. Let the homozygote A_1A_1 exceed in value the homozygote A_2A_2 by a quantity $2a$ on the scale of measurement. In other words, if the midpoint between the two homozygotes is regarded as the zero level, A_1A_1 has a deviation of $+a$, and A_2A_2 a deviation $-a$. Unlike many models of "Mendelian" dominance, the value of the heterozygote need not coincide with that of either homozygote. Let the deviation of the heterozygote, A_1A_2, from the zero level be d, which can assume either sign and any value.

Genotype

Genotypic value

Thus, if $d = +a$, this represents the complete dominance of the A_1 allele, with respect to this character, while if $d > +a$, it represents an *overdominant* situation. The illustration above depicts partial dominance of the A_1 allele.

We can now derive an expression for the population mean. The model can be expressed in tabular form as in Table 11.1. Multiplying each value by its frequency and summating over the three genotypes, the mean (m) of the population for this particular locus is

$$m = p^2a + 2pqd - q^2a$$
$$= a(p^2 - q^2) + 2dpq$$
$$= a(p + q)(p - q) + 2dpq$$
$$= a(p - q) + 2dpq$$

since $p + q = 1$.

Summating over all relevant loci, the population mean (M) for the character measured is

$$M = \Sigma a(p - q) + 2\Sigma dpq$$

Let us consider briefly the factors that affect the mean genotypic value and thus the mean phenotypic value of the population. The mean depends on (1) the values of a and d and (2) the gene frequency q. For instance, it is easily seen that, at intermediate gene frequencies, the population mean is largely determined by the value of the heterozygotes, e.g., if $p = q = \frac{1}{2}$, $p - q = 0$ and the numerical value of pq is at a maximum. At more extreme gene frequencies, the heterozygotes

Table 11.1

Genotype	A_1A_1	A_1A_2	A_2A_2
Value	$+a$	d	$-a$
Frequency	p^2	$2pq$	q^2

have correspondingly less effect on the mean. It may be mentioned here that, since a and d are fixed values for any given locus under a specified set of conditions, the only genetic way of changing the population mean is by changing the gene frequencies.

The average genotypic value of a population is easily measured, as indicated, by the average phenotypic values. However, the genotypic values of individuals are theoretical values which could be measured only if a genotype could be replicated many times and placed in all the environments to which the population is subjected. In any case, the genotypic value of an individual, considered as a summation over many loci, is not of paramount interest in genetic work, for it refers to a unique combination of genes drawn at random from the population gene pool. Considering again at a single locus of a diploid organism, the two genes at that locus, whether the same or different, are represented in the gametes with equal frequencies. The part of the genotypic value of the parent transmitted to the offspring is the *average effect* of those two genes, considered separately. Take, for instance, the A_1 allele, uniting at random with other alleles of the A locus in the population. In p cases, it unites with another A_1 allele, giving genotypes of value $+a$; in q cases, it unites with A_2, giving genotypes of value d. The mean value of genotypes deriving from the A_1 alleles is therefore $pa + qd$. The average effect (α_1) of the A_1 allele is the deviation of these genotypes from the population mean for that locus:

$$\alpha_1 = (pa + qd) - [a(p - q) + 2dpq]$$
$$= q[a + d(q - p)]$$

Similarly, it may be shown that the average effect of the A_2 allele is

$$\alpha_2 = -p[a + d(q - p)]$$

If we were now to allow one allele to be replaced by its alternative form, what would be the effect on the population mean? Suppose that an A_2 allele were to be replaced by an A_1. This hypothetical procedure enables us to determine the *average effect of gene substitution* at the locus in the population. The A_2 allele is distributed between the genotypes A_1A_2 and A_2A_2 in the ratio p/q. There is therefore a probability of p that the substitution would change the A_1A_2 genotype to A_1A_1 and thus change the value from d to $+a$. Likewise, there is a probability of q that the effect of the substitution would be to change the value from $-a$ to d. Thus, the average effect of gene substitution (α) is

$$\alpha = p(a - d) + q(d + a)$$
$$= a + d(q - p)$$

The same formula can be derived by supposing that the substitution is in the opposite direction, i.e., if an A_1 allele were to be replaced by an A_2, except that the sign of the expression would then be negative. This expression shows that the average effect of a gene is, in fact, the average effect of a gene substitution, weighted by the opportunity for that substitution to occur, i.e., by the frequency of the alternative allele. Thus,

$$\alpha_1 = q\alpha$$
$$\alpha_2 = -p\alpha$$

It is also seen that the average effect of a gene substitution is the difference between the average effect of the two alleles.

$$\alpha_1 - \alpha_2 = (q + p)\alpha = \alpha$$

These algebraic manipulations may appear for the moment to be somewhat irrelevant, but they are the foundations of some of the basic concepts of quantitative genetics. They lead immediately to one such concept, the *breeding value*, which, unlike the genotypic value, can be measured in an individual animal. It refers to the average effect of the individual genes that a parent transmits to its offspring, and we have just seen that this represents only a part of the genotypic value. For instance, in a fully dominant situation, where $d = +a$, the genotypic values of A_1A_1 and A_1A_2 are identical, whereas their breeding values are clearly different, because the latter will produce some A_2A_2 offspring while the former cannot. The breeding values of the three genotypes can be written as the sum of the average effect of their two genes considered separately. As the average effects were derived as deviations from the population mean, the breeding values are also expressed as deviations.

If an individual is mated to a sufficiently large number of animals drawn at random from the population, its breeding value can be measured directly as twice the deviation of the progeny group from the population mean. It has to be twice, because the gametes of the individual concerned unite at random with others, which by definition have a deviation of zero from the population mean. Thus, if a male mouse has progeny whose weight, on average, exceeds the population mean by 2.5 grams, the male's breeding value is +5 grams. If it is mated to a female whose own breeding value is also +5, the expected genotypic value of the offspring is +5. But if the female has a breeding value of −7, the expected genotypic value of the offspring is then −1 gram.

The breeding value is often called the *additive value* of the genotype, and the average effect of a gene may be referred to as its *additive effect*. The variance associated with this source will later be termed the additive variance. Although this terminology is probably self-explanatory, care should be taken not to confuse it with additive gene

Table 11.2

Genotype	Breeding value
A_1A_1	$2q\alpha$
A_1A_2	$(q-p)\alpha$
A_2A_2	$-2p\alpha$

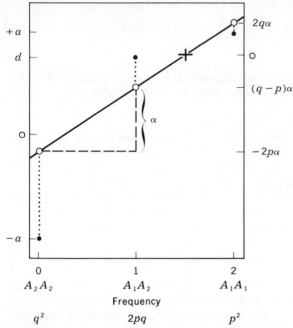

Figure 11.2   Graphical representation of genotypic values (closed circles) and breeding values (open circles) of the genotypes for a locus with two alleles A_1 and A_2, at frequencies p and q. Horizontal scale: number of A_1 genes in the genotype. Vertical scales of value: on left, arbitrary values as explained in the text; on right, deviations from the population mean. The figure is drawn to scale for the values $d = \frac{3}{4}a$ and $q = \frac{1}{4}$. (*From Falconer, 1960. By courtesy of author and publisher.*)

action. Additive gene action implies that the heterozygote value lies midway between that of the two homozygotes. The additive effect of a gene refers specifically to its average effect in the population, which has been shown to depend, *inter alia*, on the dominance. Thus, a fully dominant gene $(d = +a)$ has a perfectly well-defined additive effect, while its action would be described as nonadditive.

The breeding or additive value, then, represents a part of the genotypic value. The difference between the two is represented graphically in Figure 11.2. This is a plot of values against the number of A_1 alleles. The closed circles represent genotypic values, while the line is the least-squares regression fit to these points, weighted by the number of individuals that each point represents. This weighting is therefore determined by the gene frequencies in that population. The value at which the line intersects the position of the genotypes on the gene-dosage axis represents the additive value of those genotypes (open circles). The difference between the additive value of the heterozygote and that of either of the two homozygotes represents the additive effect of the gene substitution.

$$2q\alpha - (q - p)\alpha = (q + p)\alpha = \alpha$$

The deviation of each point from the line represents the part of the genotypic values not accounted for by the additive values. These deviations are known as *dominance deviations* and are always present when the degree of dominance is not zero. If we label the additive effect A and the dominance part D, then

$$G = A + D$$

and

$$P = A + D + E$$

In accordance with the terminology we have adopted, the dominance deviations may be referred to as *nonadditive* effects of the genes. The dominance deviations are the residual effects of combining genes in pairs, over and above the average effects of the genes considered separately.

The algebraic expressions for D for each locus are obtained by subtracting A from G, after first expressing G on the same basis as A, that is, as a deviation. Thus for the A_1A_1 genotype

$$\begin{aligned} G - M &= a - M \\ &= a - [a(p - q) + 2dpq] \\ &= 2qa - 2pqd \end{aligned}$$

Now,

$$\begin{aligned} A &= 2q\alpha \\ &= 2qa + 2q(q - p)d \end{aligned}$$

so that

$$\begin{aligned} D &= (G - M) - A \\ &= -2pqd - 2q(q - p)d \\ &= -2q^2d \end{aligned}$$

Repeating the procedure for the other two genotypes, we can derive Table 11.3.

This shows that only two factors contribute to the dominance deviations, the gene frequency and the heterozygote deviation. Whereas

Table 11.3

Genotype	A_1A_1	A_1A_2	A_2A_2
Frequency	p^2	$2pq$	q^2
Value	a	d	$-a$
$G - M$	$2q(a-pd)$	$a(q-p) + d(1-2pq)$	$-2p(a + qd)$
A	$2q\alpha$	$(q-p)\alpha$	$-2p\alpha$
D	$-2q^2d$	$2pqd$	$-2p^2d$

dominance, in the gene-action sense, contributes to α and thus to A, additively acting genes $(d = 0)$ do not contribute to D. Figure 11.1 shows why the dominance deviations of the homozygotes are always negative if d is positive. It is worth emphasizing also that G, A, and D are all affected by the gene frequencies. This means that the values of these quantities for any locus refer specifically to one population and, strictly speaking, at one point in time. Any change in gene frequency will alter the relative proportion of heterozygotes to homozygotes. This will affect the slope of the line in Figure 11.2, which in turn alters everything.

For a single locus, the relationship $P = G + E = A + D + E$ is exhaustive; it fully describes the system, as far as it goes. But when we consider a genotype as a combination of several or many loci, another term must be added to accommodate the interaction between loci:

$$G = A + D + I$$

where I is the interaction deviation between two or more loci, representing the effect on the genotypic value of an individual over and above the average effects of those loci considered separately. This interaction can be of many different kinds, as in formal genetics. We need not consider the interaction term in any detail here, except to recognize it as a source of variation in the population.

Thus far, we have been concerned to identify the sources of variation in the population and to examine qualitatively the factors associated with each source that gives rise to variation. We must now study the actual variances a little more closely.

Since $P = G + E$, it follows that

$$V_P = V_G + V_E + 2\, cov_{GE}$$

where the Vs represent the variances associated with each source, and cov is the covariance. According to our definition of G and E, there is no covariance between the genotype and the environment. Thus,

$$V_P = V_G + V_E$$

In behavior, there may be psychometric problems to resolve, but once this is done, the phenotypic variance (V_P) is easily measured. However, it was said earlier that G (and therefore E) could not be measured for individual animals unless we could replicate the genotype at will and measure it under a range of environmental conditions. Now a situation that may approximate this, possibly very closely, is the case of a highly inbred line or of a cross between two such lines. In this case there is a replicated genotype, and the variance within such a genotype must be wholly environmental. In practice, it would obviously be preferable to obtain as many such genotypes as possible and to obtain an estimate of the environmental variance from within each one. A snag in this context is that inbred lines have often been found to be more variable than one would expect, on the basis of their supposed genetic uniformity. The problem is discussed by Biggers and Claringbold (1954) and by Grüneberg (1954). The reason for the extraordinary variability of many inbred strains is probably related to their unusual degree of homozygosity, which, according to Lerner's (1954) postulate of genetic homeostasis, renders them more sensitive to environmental sources of variation. By the same token, F_1s may be too insensitive, again relative to outbred stocks.

Any attempt to estimate V_E, in practice, should therefore not only utilize as many inbred lines as practicable but also the crosses between them. If the two provide different estimates of V_E, they should probably be averaged, although in some cases it may be preferable to use the F_1s only. It is impossible to provide objective guidance on this point; the decision must be a subjective one, based on the experimenter's knowledge of his material. The estimate of the environmental variance so obtained refers only to the population that the inbreds and their crosses represent. Its extrapolation to other populations may lead to wrong conclusions. But within such a population the component within strains or crosses estimates V_E, while the component between genotypes estimates V_G.

This partitioning of the total variance indicates the importance of the genotype in determining the phenotype. The ratio V_G/V_P indicates *the degree of genetic determination*. This ratio is sometimes referred to as "heritability in the broad sense," which is clumsy usage. Some writers have referred to it simply as "heritability," without a qualification, which is confusing. The term "heritability," as used in the genetic literature and throughout the remainder of this chapter, will be defined in a narrower, but more useful, sense below.

In the same way as G was split into $A + D$, so can V_G be partitioned:

$$V_G = V_A + V_D$$

It can be shown from Table 11.3 that A and D are not correlated, so that there is no covariance term. The term V_A is the variance of breeding values, associated with variation in the average (additive) effect of the genes. It is thus termed the *additive variance*. Its importance in the theory and practice of quantitative genetics is paramount. It is furthermore a useful concept, as it can be estimated directly in a population, in a way that will be described shortly. The ratio V_A/V_P is termed the *heritability*, which is therefore defined as the proportion of the phenotypic variance due to additive genetic sources. It is the main cause of resemblance between relatives, and it also indicates the reliability of the phenotype as a guide to breeding value.

Although V_A is termed the additive variance, it is worth stressing again that this does not imply additive gene action. It measures the variation in breeding value, which we have seen to depend upon, among other things, dominance.

The remainder of the genotypic variance, with respect to a single locus, is the variance of the dominance deviations. The algebraic expressions for the additive and dominance variances are obtained as follows:

"Variance" is defined as the mean-squared deviation from the population mean. As we have already determined both A and D in terms of deviations, in Table 11.3, all that is required is to square the values obtained, multiply by the frequency, and summate over the three genotypes.

Thus,

$$V_A = 4p^2q^2\alpha^2 + 2pq(q - p)^2\alpha^2 + 4p^2q^2\alpha^2$$
$$= 2pq\alpha^2(p + q)^2$$
$$= 2pq\alpha^2$$
$$= 2pq[a + d(q - p)]^2$$
$$V_D = 4p^2q^4d^2 + 8p^3q^3d^2 + 4p^4q^2d^2$$
$$= 4p^2q^2d^2(q^2 + 2pq + p^2)$$
$$= (2pqd)^2$$

What, then, do these expressions tell us about the additive and dominance variances? Firstly, let us examine their relative magnitudes. Since

$$V_A = 2pq[a + d(q - p)]^2$$

and

$$V_D = 2pq(2pqd^2)$$

it is seen that $V_D > V_A$ only when

$$2pqd^2 > [a + d(q - p)]^2$$

Substitution of values here will show that this will hardly ever occur except at intermediate gene frequencies when $d > a$, that is, in overdominant situations. The first conclusion is therefore that the additive effect of the genes usually contributes more to the variance than the dominance deviations. This conclusion is depicted graphically in Figure 11.3. Both expressions reveal another important conclusion, namely, that genes contribute more variance at intermediate than at extreme frequencies.

Extending the treatment to more than one locus, there is the additional complication of variance due to interaction effects. The theoretical consequences of interaction variance have not been developed very extensively, but the general conclusion seems to be that its contribution to genotypic variance is not very great. Interaction may occur between two loci or between more, but the more loci concerned, the less their relative contribution to the interaction variance. Although the partitioning is purely a theoretical one, several types of interaction variances are recognized, as this helps in the formulation of some problems. The interaction may be between the breeding values of two loci (V_{AA}), between the breeding value of one and the dominance deviations of the other (V_{AD}), between two dominance deviations (V_{DD}), and so forth, to multilocular situations (V_{AAA}, V_{AAD}, etc.). The full formula for the partitioning of phenotypic variance then becomes

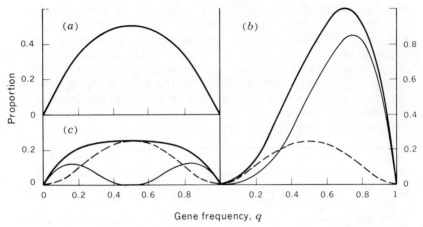

Figure 11.3   Magnitude of the genetic components of variance arising from a single locus with two alleles, in relation to the gene frequency. Genotypic variance, thick lines; additive variance, thin lines; dominance variance, broken lines. The gene frequency q is that of the recessive allele. The degrees of dominance are in (a) no dominance ($d = 0$); in (b) complete dominance ($d = a$); and in (c) "pure" overdominance ($a = 0$). The figures on the vertical scale, showing the amount of variance, are to be multiplied by a^2 in graphs (a) and (b), and by d^2 in graph (c). (*From Falconer, 1960, by courtesy of author and publisher.*)

$$V_P = V_A + V_D + V_I + V_E$$
$$= V_A + V_D + V_{AA} + V_{AD} + V_{DD} + \cdots + V_E$$

In practice, the dominance and interaction variances are grouped together into *nonadditive* genetic variance. As mentioned earlier, the really important division is usually into additive genetic variance and the remainder, nonadditive and environmental together. Occasionally, it may be possible to obtain an independent estimate of the environmental variance, in which case a realistic partitioning may be

$$V_P = V_A + V_{NA} + V_E$$

where V_{NA} is the nonadditive genetic variance. Beyond this we can seldom go. The additive variance is of prime importance. The nonadditive part sometimes contributes to the resemblance between relatives, as will be shown shortly, and it is also very pertinent to the study of inbreeding, to be discussed later.

Before we leave the subject of variance components, a word must be said about *environmental variance*. By definition, it refers to all variation not attributable to genetic causes, and the sources of it may be many. Where possible, genetic experiments should be designed to reduce the environmental variance, where likely sources of such variance are known. Nutrition, climatic, and housing factors are obvious examples; in mammals, maternal effects constitute another source of environmental variance. But over and above all the identifiable sources, there is usually still some residual environmental variance, often a considerable amount. It is convenient in this context to recognize two kinds of environmental variance; one refers to factors that cause individuals to differ, the general *environmental variance* (V_{Eg}), while the other refers to variation within an individual, the *special environmental variance* (V_{Es}), for instance, variation from day to day or even from minute to minute. The latter is the reason for replicating measurements in order to assess the phenotypic value of an individual. Obviously, repeated measurements will reduce the V_{Es} component of V_P. If we now rewrite our formula as

$$V_P = V_G + V_{Eg} + V_{Es}$$

the variance of the mean of n repeated measurements becomes

$$V_{P(n)} = V_G + V_{Eg} + 1/n\, V_{Es}$$

We should perhaps reflect momentarily on the biological validity of repeated measurements. Behavioral scientists are well aware of the difficulties here. The implicit assumption is that the second and subsequent measurements do, in fact, measure the same effect as the first measurement. But it is possible that the very nature of the first measurement may render this assumption untenable.

These, then, are the sources of variance recognized either in practice or in theory. Variance components are the building blocks of quantitative genetics. Before discussing some aspects of methodology in quantitative genetics, we must first examine in more detail the causes of resemblance between relatives.

THE CAUSES OF RESEMBLANCE BETWEEN RELATIVES

In general, resemblance between groups, be they relatives or otherwise, is dependent upon the covariance of members of a group. This factor can be described in a variety of ways, the choice being usually one of convenience. For instance, where the groups concerned are groups of two, a product-moment correlation or a regression coefficient is the simplest expression to employ. But where groups consist of more than two individuals, then components of variance are extracted from an analysis of variance; they can be conveniently expressed as an intraclass correlation (t).

$$t = \frac{\sigma_B^2}{\sigma_W^2 + \sigma_B^2}$$

where σ_B^2 is the component of variance between groups and σ_W^2 is the component within groups.[2] To say that groups differ is another way of saying that members of a group are alike. The greater the

[2] Some readers may be unfamiliar with components of variance and intraclass correlations. An exposition of them may be found in many statistical textbooks, and a monograph by Haggard (1958) deals exclusively with the subject. It should be noted also that the symbol t for the intraclass correlation, used extensively in the genetic literature, is not to be confused with Student's t.

component between groups, the greater the relative similarity of members within a group. Thus the component of variance between groups can equally well be regarded, qualitatively at least, as the covariance of members of a group.

Just as variance was split up into genotypic and environmental components, so now must covariance be treated in the same way. We shall begin again with a single locus and later introduce the complication of interaction when we summate over loci. Two examples will be worked out, to illustrate the two ways of looking at covariance mentioned above. We shall examine first the genetic covariance, i.e., how the genotype of an individual is related to the genotype of a specified relative. Environmental covariance between relatives should be avoided by experimental design but, as we shall see shortly, this is not always possible.

Let us determine first the genetic covariance of an offspring with *one* of its parents, i.e., the covariance of their genotypic values. The genotypic value of, say, genotype A_1A_1, expressed as a deviation from the population mean, was given in Table 11.3 as $2q(a - pd)$. The algebra is simplified if we now express this as $2q(\alpha - qd)$ by substituting $a = \alpha - d(q - p)$. The mean genotypic value of the offspring of A_1A_1 is one-half of the breeding value of that genotype, i.e., $\frac{1}{2}$ of $2q\alpha$. By extending this reasoning to the other two genotypes, we arrive at Table 11.4.

With the genotypic values now expressed as deviations, all that remains is to multiply the two together and also by the frequency, to obtain the covariance of an offspring with one of its parents (cov_{OP}). The expression simplifies to

$$cov_{OP} = pq\alpha^2(p + q)^2 + 2p^2q^2\alpha d(-q+q - p+p)$$
$$= pq\alpha^2$$

Since V_A was seen earlier to be $2pq\alpha^2$,

$$cov_{OP} = \frac{1}{2}V_A$$

This is intuitively reasonable; a parent's genetic contribution to an offspring is one-half the average effect of its genes. Over the whole population, therefore, the covariance of offspring with one parent is one-half the additive variance.

The other kind of relationship that we shall examine concerns the genetic covariance of half-sib groups, i.e., groups of progeny having one parent in common, usually the sire. The covariance among members of a half-sib group is, as explained earlier, the variance of the

Table 11.4

Genotype	Frequency	Genotypic value of parents	Genotypic-value of offspring
A_1A_1	p^2	$2q(a - qd)$	qa
A_1A_2	$2pq$	$(q-p)a + 2pqd$	$\frac{1}{2}(q-p)a$
A_2A_2	q^2	$-2p(a + pd)$	$-pa$

means of such groups. Given a parent of specified genotype, the mean value of its progeny, expressed as a deviation, is shown in Table 11.4. Squaring the deviations and multiplying by the frequency therefore give the variance of means of half-sib groups, this being the covariance (cov_{HS}) that we require:

$$cov_{HS} = p^2q^2\alpha^2 + \tfrac{1}{2}pq(q - p)^2\alpha^2 + q^2p^2\alpha^2$$
$$= pq\alpha^2[\tfrac{1}{2}(p + q)^2]$$
$$= \tfrac{1}{2}pq\alpha^2$$

Since $V_A = 2pq\alpha^2$,

$$cov_{HS} = \tfrac{1}{4}V_A$$

This again is intuitively reasonable, since two half sibs have, on an average, one-quarter of their genes in common. The additive effect of these genes is represented by $\tfrac{1}{4}$ V_A.

By exactly the same principles, although a trifle more laboriously at times, genetic covariances among other types of relationships can be derived. For instance, the covariance between an offspring's value and the mean value of its two parents ($cov_{O\bar{P}}$) can be shown to be, again, $\tfrac{1}{2}$ V_A. Although this covariance, in absolute terms, is the same as that with one parent, it is now of greater relative importance if it is expressed as a correlation or regression coefficient, since the variance of midparental values is only one-half the variance of the parents considered singly.

Another covariance of some importance is that among full-sib groups (cov_{FS}), which can be shown to be $\tfrac{1}{2}$ $V_A + \tfrac{1}{4}$ V_D, an example where the nonadditive genetic variance contributes to the resemblance between relatives. This is because full sibs share not only the additive effects of half their genes, accounting for the $\tfrac{1}{2}$ V_A term, but also a quarter of the dominance deviations, as a result of being identical for one-quarter of all the loci segregating in that mating.

Extending the treatment now to multilocular situations, the expressions for covariance must be expanded to accommodate the interaction terms. The reason for partitioning the variance into V_{AA}, V_{AD}, V_{DD}, etc., will now become apparent. Qualitatively, we may approach the question as follows: If the breeding values of two genes interact, giving rise to V_{AA}, the coefficient of V_{AA} in the expression for covariance is the probability of those two genes being present in the two relatives. For instance, if the genotype A_1-B_1- shows an interaction deviation, then in the covariance of an offspring with one parent, the probability of offspring and parent both containing A_1 or both containing B_1 is $\tfrac{1}{2}$, giving rise to the term $\tfrac{1}{2}$ V_A, as before. The probability of both containing A_1 and also B_1 is $\tfrac{1}{4}$, giving a term of $\tfrac{1}{4}$ V_{AA}. Thus,

$$cov_{OP} = \tfrac{1}{2}V_A + \tfrac{1}{4}V_{AA}$$

Similarly,

$$cov_{HS} = \tfrac{1}{4}V_A + \tfrac{1}{16}V_{AA}$$
$$cov_{FS} = \tfrac{1}{2}V_A + \tfrac{1}{4}V_D + \tfrac{1}{4}V_{AA} + \tfrac{1}{8}V_{AD} + \tfrac{1}{16}V_{DD} + \cdots$$

In general, if, in the expression for covariance, V_A takes the coefficient x, and V_D takes y, then the full formula to allow for interaction becomes

$$cov = xV_A + yV_D + x^2V_{AA} + xyV_{AD} + y^2V_{DD} + x^3V_{AAA} + \cdots$$

The general conclusion is that as V_I is usually rather small compared with V_A or even V_D, the interaction components do not figure prominently as a cause of resemblance between relatives. But in some situations, their effects may not always be negligible.

The last cause of resemblance between relatives is any similarity of the environment to which they may be subjected. As we depend upon the degree of resemblance between relatives to arrive at genetic conclusions, it is obvious that environmental causes of similarity must be excluded or else accommodated. Experimental design can often be employed to reduce the environmental covariance but, for mammals especially, total exclusion cannot always be ensured.

The component of environmental variance previously symbolized V_{Es} will not, by definition, contribute to the similarity of relatives; it refers to variation within an individual with respect to repeated measurements. But the general environmental variance V_{Eg} may cause trouble. It is convenient in this context to repartition the environmental variance into a component called *common environment* (V_{Ec}), tending to make members of a family more alike and therefore different families more different, and a second component, symbolized V_{Ew}, the *within-group component*, which causes individuals to differ irrespective of whether they are related or not. To avoid confusion, it should be noted that V_{Ew} contains all of V_{Es} and some of V_{Eg}, on the previous partitioning. We are concerned here only with V_{Ec}, the environmental influences common to members of a family or group. Possible sources of V_{Ec} come through housing, e.g., if related Drosophila are stored in the same bottle or mice in the same cage. In wild populations, relatives may tend to occupy the same habitat. Sources such as these may be allowed for, if recognized, and under laboratory conditions the component of variance due to common environment can often be assessed directly by setting up replicate bottles of flies, etc. In mammals, however, one cause of V_{Ec} that little can be done about comes through maternal effects, where full sibs, in particular, resemble each other not only because of their genetic covariance but also because the sibs occupied the same uterus in the foetal stage and suckled the same dam postnatally. Where maternal effects exist, the expression for the covariance of full sibs should be rewritten

$$cov_{FS} = \tfrac{1}{2}V_A + \tfrac{1}{4}V_D + \tfrac{1}{4}V_{AA} + \cdots + V_{Ec}$$

It is largely because of the common-environment component that, in genetic studies, full sibs tend to be of less use than the more distantly related half sibs. Full sibs have been singled out for comment, but other relationships should be closely examined in every experiment for possible sources of V_{Ec}, such as maternal effects or managemental factors. The common environment component may produce effects that deserve study in their own right; maternal effects, for instance, as revealed by differences between reciprocal crosses, are often of considerable interest. But in genetic work, the common environment factor is usually best excluded where possible, for it tends to be confounded with genetic components in such a way that it may not be separable.

The covariance may be estimated from the measurable parameters of the population. Consider, firstly, the covariance of offspring with one parent. The variance of the parents is, by definition, the phenotypic variance of the population. Knowing V_P, then, we could assess cov_{OP} directly from the expression

$$b_{OP} = \frac{cov_{OP}}{V_P}$$

Similarly, the total variance in a population of sib groups is likewise V_P, and the component between groups has been shown to be the covariance of members of a group. The covariance of half sibs and of full sibs can therefore be estimated directly from the intraclass correlation (t). Lastly, the variance of midparental values ($V_{\bar{P}}$) is

$$V_{\bar{P}} = V(\tfrac{1}{2}X + \tfrac{1}{2}Y)$$
$$= \tfrac{1}{4}V_X + \tfrac{1}{4}V_Y$$

where X and Y are the phenotypic values of the two parents. If

$$V_X = V_Y = V_P$$

then

$$V_{\bar{P}} = \tfrac{1}{2}V_P$$

Thus the covariance of offspring with the midparental value is obtained:

$$b_{O\bar{P}} = \frac{cov_{O\bar{P}}}{\tfrac{1}{2}V_P}$$

In practice, however, such covariances are not usually estimated, any more than covariances are normally assessed directly in statistical work. Genetic covariances, in absolute terms, are of little direct interest; their magnitudes vary, as we have seen, according to the closeness of the relationship. They are therefore reduced to a common base;

Table 11.5

Relationship	Covariance	Regression (b) or intraclass correlation (t)	
Offspring: one parent	$\tfrac{1}{2}V_A$	$b = \dfrac{\tfrac{1}{2}V_A}{V_P}$	$= \tfrac{1}{2}h^2$
Offspring: midparent	$\tfrac{1}{2}V_A$	$b = \dfrac{\tfrac{1}{2}V_A}{\tfrac{1}{2}V_P}$	$= h^2$
Half sibs	$\tfrac{1}{4}V_A$	$t = \dfrac{\tfrac{1}{4}V_A}{V_P}$	$= \tfrac{1}{4}h^2$
Full sibs	$\tfrac{1}{2}V_A + \tfrac{1}{4}V_D + \cdots + V_{Ec}$	$t = \dfrac{\tfrac{1}{2}V_A + \tfrac{1}{4}V_D + \cdots + V_{Ec}}{V_P}$	$> \tfrac{1}{2}h^2$

since V_A figures prominently in them all, the practice usually is to express V_A as a fraction of V_P. This ratio was described earlier as the heritability. Table 11.5 shows how the covariance is related to the heritability (symbolized h^2).

This, then, shows in principle how estimates of heritability are derived. It shows also the final product of our examination of variances and covariances, and how different relationships can be analyzed and unified into one ratio that refers to the whole population. Before we examine in more detail how heritabilities are determined in practice, we should examine the importance and usefulness of heritability as a concept and its role in quantitative methodology.

HERITABILITY

We have just seen how the formulation of the resemblance between relatives leans heavily on the heritability, as the additive genetic variance is the main cause of the resemblance. The greater the heritability, the greater the covariance and therefore the similarity between relatives. Secondly, the magnitude of the heritability, ranging from 0 to 1, determines the reliability of the phenotype as a guide to breeding value, something which confers a predictive role on the heritability. To see this, let us determine the regression of breeding value on phenotype. Let

$$P = A + R$$

where R is the remainder, nonadditive genetic and environmental. Then

$$cov_{AP} = cov_{A(A+R)}$$
$$= cov_{AA} = V_A$$

since $cov_{AR} = 0$.

So

$$b_{AP} = \frac{cov_{AP}}{V_P} = \frac{V_A}{V_P} = h^2$$

The higher the heritability, the more accurate is the phenotype as a guide to breeding value. Given the phenotypic value, the breeding value can be predicted from the heritability according to the formula

$$Exp\ A = b_{AP}P = h^2P$$

This leads directly to another predictive role, that of response to selection. Expressing both A and P now as deviations, if S (selection differential) is the deviation of the parents from the population mean, the expected deviation of the progeny, R (response to selection), is given by

$$Exp\ R = h^2S$$

since the breeding value is, by definition, the mean genotypic value of the offspring.

The concept of heritability has, therefore, two connotations, one descriptive and one predictive. The first describes the proportion of the total phenotypic variance that is additive genetic, and it leads to a causal description of the resemblance between relatives. The second indicates the reliability of the phenotype as a guide to breeding value, and it leads to the prediction of the results of certain manipulative processes. Quantitative genetics in one of its most pragmatic aspects, namely, animal breeding, makes extensive use of the concept of heritability in the development of its theory. But it is no less important in a more "fundamental" aspect, the investigation of the genetics of a quantitative character. Without the assessment, in some form, of the additive genetic variance—the variance due to the average effect of the genes in the population—the genetic study of such characters can scarcely begin.

In the context of heritability, it is important to remember the basic raw materials that went into the process, namely, a, d, q, and V_E. This means that any estimate of heritability refers to one population only and will not necessarily hold if, for instance, the gene frequencies are changed. The extent to which the results from one population can be extrapolated to another depends on the similarity, both genetic and environmental, of the two populations. Nevertheless, as far as laboratory animals and also domestic livestock are concerned, there is generally wide agreement on the relative magnitude of the heritability for a given character in a given organism, with the following general conclusion. Characters that have been subjected to natural selection, i.e., components of natural fitness, such as fertility or maternal performance, tend to have low heritabilities. The suggestion is that the additive variance has been largely exhausted through the action of natural selection in the past. On the other hand, characters such as the fat content of milk in cattle or bristle number in Drosophila, which do not seem to be such direct components of natural fitness, have in general much higher heritabilities. It will be interesting to see, as more information accumulates, how some behavioral characters fit into the picture.

THE ESTIMATION OF HERITABILITIES

Let us examine a little more closely how heritabilities are estimated in practice, although the principles on which the methods rest have already been given.

The first step involves the practical decision of the type of relationship to employ. The overriding concern at this stage is to avoid environmental sources of covariance that would lead to the wrong answer by inflating the estimate of the heritability. If maternal effects, for instance, are known to affect the character, the use of offspring and dams or of full sibs is vitiated, and other types of relationships, such as sire and offspring or half sibs, must be chosen for the estimation. Such procedure can be indicated only by experience of the biological nature of the character. Other sources of environmental covariance, which may result from feeding, housing, handling, etc., should be excluded as far as possible. Their effect will differ for different characters, and there is no substitute for common sense in avoiding the pitfalls in this respect.

The first method of heritability estimation implicit in Table 11.5 is the regression of offspring on parent. Whatever the type of relationship involved, it is of course bound to include parents and offspring and with the possible biases mentioned above in mind, and provided that the parents have been measured at the right age, etc., it is usually worthwhile to obtain one estimate of the heritability by the regression method, even though the structure has been designed for a sib analysis or some other type of relationship. Little more need be said about the regression method, as the regression coefficient of the mean value of the offspring on the value of one parent measures one-half the heritability, while the regression of offspring on the mean value of the two parents measures the heritability itself. The regressions are calculated from paired observations in the usual way. However, three possible modifications of the regression method should be mentioned.

The first modification is applicable if the mean values of the offspring are based on families differing in size. These can be weighted, so that rather more attention is given to the larger families, according to methods suggested by Kempthorne and Tandon (1953) or by Reeve (1955). Usually such adjustments do not greatly alter the estimate of the heritability, but they do increase the precision of the estimate if the families differ widely in size.

The second modification concerns the variances of the two sexes. It was assumed in our treatment of covariance that these were equal. If they are not, two separate regressions should be taken, that of sons on sires and that of daughters on dams. The average of the two will then provide an overall estimate of heritability, though this will be rather a meaningless figure unless the two do not differ.

The third possible modification becomes imperative if a sire is mated to more than one dam, as would be the case, for instance, in a half-sib structure. The mean of the offspring of one mating could not then be regressed on the midparental value, as one of the parents would be common to several paired observations. The mean of all the sire's offspring could still be regressed on the sire's own value, but the number of sires is seldom sufficient to make this profitable. Under the circumstances, the procedure is to regress the offspring on the dam within each sire group, pooling the degrees of freedom and the sums of squares and of products to obtain a weighted average regression. Such a regression, involving basically only one parent, estimates half the heritability, as before, and is known as the *intrasire regression of offspring on dam.*

The regression method of estimating heritability is therefore conceptually straightforward. We shall now examine the other method of heritability estimation suggested by Table 11.5. This involves the derivation of the intraclass correlation from the analysis of variance of a sib structure, and its theory is somewhat more complicated.

Table 11.6

Source	Degrees of freedom	Composition of mean square
Between sires	$s-1$	$\sigma_W^2 + k\sigma_D^2 + dk\sigma_S^2$
Within sires		
Between dams	$s(d-1)$	$\sigma_W^2 + k\sigma_D^2$
Within dams	$sd(k-1)$	σ_W^2

In practice, the breeding structure usually involves a mixture of full sibs and half sibs. If s sires are each mated to d dams, each of which produce k offspring for measurement, the structure lends itself to a standard hierarchical analysis of variance, as shown in Table 11.6. If the ds and ks are unequal, then there are standard statistical techniques to adjust the coefficients of the components as they affect the mean squares. Note also that the symbol σ^2 is employed here to denote observational components of variance, to distinguish them from causal components, symbolized V.

From the analysis, we can now derive three components of variance, one between sires (σ_S^2), one between dams (σ_D^2), and a third one within groups of full sibs (σ_W^2). The next step is to relate these observational components to the causal components that we have derived previously and thereby derive the ratio V_A/V_P, the heritability.

Firstly, it follows by definition that the sum of the observational components estimates the total phenotypic variance σ_T^2, though the sum may not tally exactly with the total variance as observed.

$$\sigma_T^2 = \sigma_S^2 + \sigma_D^2 + \sigma_W^2 = V_P$$

Secondly, σ_S^2 measures, in fact, the component of variance of half-sib groups, since it is in the nature of the analysis that the deviations of full sibs from their own family means are removed separately. Thus, σ_S^2, the component between half sibs, is equal to the covariance of half sibs, which we saw earlier to be $\frac{1}{4}V_A$. Thirdly, σ_W^2, the component of variance within full-sib groups, is the complement of the component between full-sib groups, $\sigma_{B(FS)}^2$, and the two must add up to the phenotypic variance:

$$V_P = \sigma_W^2 + \sigma_{B(FS)}^2 = \sigma_W^2 + cov_{FS}$$

Thus

$$\begin{aligned}\sigma_W^2 &= V_P - cov_{FS} \\ &= V_P - \tfrac{1}{2}V_A - \tfrac{1}{4}V_D - V_{Ec} \\ &= \tfrac{1}{2}V_A + \tfrac{3}{4}V_D + V_{Ew}\end{aligned}$$

Table 11.7

Observational component	Causal component
$\sigma_S^2 = cov_{HS}$	$\frac{1}{4}V_A$
$\sigma_D^2 = cov_{FS} - cov_{HS}$	$\frac{1}{4}V_A + \frac{1}{4}V_D + V_{Ec}$
$\sigma_W^2 = V_P - cov_{FS}$	$\frac{1}{2}V_A + \frac{3}{4}V_D \qquad + V_{Ew}$
$\sigma_T^2 = V_P$	$V_A + \quad V_D + V_{Ec} + V_{Ew} = V_P$

Lastly, $\sigma_D{}^2$ may be obtained by subtraction:

$$\begin{aligned}
\sigma_D{}^2 &= V_P - \sigma_W{}^2 - \sigma_S{}^2 \\
&= V_P - (V_P - cov_{FS}) - cov_{HS} \\
&= cov_{FS} - cov_{HS}
\end{aligned}$$

These derivations are summarized in Table 11.7.

The interaction variance could easily be included among the causal components by reference to the genetic covariances given previously; it has been neglected here to simplify the presentation, and its effect, in any case, is not usually very great.

We thus see how the total variance can be partitioned into observational components which in turn can be equated to causal components. The heritability can now be obtained from the analysis. Since

$$\frac{\sigma_S{}^2}{\sigma_T{}^2} = \frac{1}{4}\frac{V_A}{V_P} = \frac{1}{4}h^2$$

then,

$$h^2 = \frac{V_A}{V_P} = \frac{4\sigma_S{}^2}{\sigma_T{}^2}$$

Unfortunately, the other causal components are not deduced so easily, since there are basically only three equations for four unknowns. $\sigma_D{}^2$ could be utilized in exactly the same way to obtain, in this case, an upper limit to the heritability. If, in fact, the estimate thereby obtained does not greatly exceed that obtained from the between-sire components, it indicates that neither V_D nor V_{Ec} is very important. But V_D, V_{Ec}, and V_{Ew} can be estimated from this analysis only if they are good reasons for believing that one of them can be safely neglected and equated to zero.

SOME CONSIDERATIONS OF EXPERIMENTAL DESIGN IN THE ESTIMATION OF HERITABILITIES

Having obtained an estimate of a heritability, one requires some indication of its reliability. This becomes a statistical question of attaching standard errors to regression coefficients or to intraclass correlations, as the case may be. Formulae for these standard errors are given in the statistical literature, and without going into detail, we must examine what they tell us about experimental design. Much of the material in this section is discussed in more detail by Robertson (1959).

Sheer physical considerations limit the size of any experiment, since the number of animals that can be measured is restricted either by space or by the time and labor involved in the measurement. However, the restrictions so imposed still leave room for the manipulation of family size, and a choice has to be made between measuring a few families accurately, i.e., by the use of a large number of offspring per family, or measuring more families less accurately. The two must be balanced to give the optimal design, which is the design that minimizes the sampling variance of the heritability estimate. We shall deal with the regression and intraclass methods, in turn, as before. We shall employ the following symbols in this section:

Y = mean value of offspring
X = value of one parent or mean value of two parents
σ_Y^2 and σ_X^2 = respective variances
n = number of offspring per parent
N = number of families
T = total number measured

Thus, if n offspring and one parent are measured per family, $T = N(n + 1)$; if both parents are measured, $T = N(n + 2)$. In sib structures, where the parents need not be measured, $T = Nn$. It is supposed that T will be limited by the total facilities available, and different methods can therefore be compared for efficiency, given the same total facilities.

Let us consider first the regression technique, as utilized to determine the heritability from the relationships:

$b_{YX} = \frac{1}{2}h^2$ (for one parent) or h^2 (for midparent)

Let the sampling variance of b be denoted by σ_b^2. It is well known that

$$\sigma_b^2 = \frac{1}{N - 2}\left(\frac{\sigma_Y^2}{\sigma_X^2} - b^2\right)$$

The subsequent algebra is simplified if an approximation is derived on the basis that N is fairly large and the b^2 is fairly small, as both will tend to be. Thus,

$$\sigma_b^2 \approx \frac{1}{N}\frac{\sigma_Y^2}{\sigma_X^2}$$

Now, σ_X^2 is V_P in the case of one parent, and $\frac{1}{2}V_P$ for the midparental values. But σ_Y^2, the variance of the mean of offspring groups, depends on the number in the group and the intraclass correlation (t) between members of the group. It can be shown to be

$$\sigma_Y^2 = \frac{1 + (n - 1)t}{n}V_P$$

(That this formula is reasonable can be appreciated by substituting $n = 1$ or $t = 0$. The values for σ_Y^2 then become V_P or V_P/n, as expected.) We can now substitute in the formula for σ_b^2:

$$\sigma_b^2 \approx \frac{1 + (n - 1)t}{nN}$$

for the regression on one parent. It is twice as great for the regression on midparent. Assuming T to be fixed, it can be shown that σ_b^2 is minimal when $n = \sqrt{(1 - t)/t}$ for one parent, or $\sqrt{2(1 - t)/t}$ for midparent.

This, then, gives the optimal number of progeny to measure per family and is now seen to depend on t. The intraclass correlation is of course closely related to the heritability, which is not known when the experiment is designed. The optimal design can therefore be derived only a posteriori, but limits to the optimal value of n can be obtained a priori, since t for full-sib families can vary only between 0 and $\frac{1}{2}$, in the absence of complications. Substitution of possible values of t will show that, if the offspring are regressed on one parent, about 10 offspring per family should be measured if the heritability is around 2 percent, while the number drops to 2 or so for a heritability around 50 percent. If midparental values are employed, the numbers are 14 and, still, about 2, respectively. In complete ignorance of what to expect, a number of 5 or so is a reasonable one to employ in order to minimize the sampling variance.

The optimum structure that we have derived is seen to be independent of the total facilities available. We can now derive the standard error of the heritability estimate, in terms of T. For illustration, we shall consider a character of 20 percent heritability, so that $t = 0.1$. The optimum n then is 3 for single-parent regression, and 4 for midparent. Bearing in mind that, since $h^2 = 2b$, $\sigma_{h^2}^2 = 4\sigma_b^2$, the formulas already given can be manipulated to show that

For single-parent regression: $\sigma_{h^2}^2 = \dfrac{6.4}{T}$

For midparent regression: $\sigma_{h^2}^2 = \dfrac{3.9}{T}$

Two intuitively obvious conclusions emerge: First, the larger the value of T, the smaller the sampling variance, and, second, estimates derived from midparental values are more precise, for given total facilities.

The final point concerning the regression method is what precision is required to make the experiment worthwhile. Continuing with our example of a heritability of 20 percent, suppose we require a standard error of not greater than 10 percent. This is not very ambitious—just sufficient to demonstrate that the heritability is not zero. Then, if σ_{h^2} is 0.1, $\sigma_{h^2}^2$ becomes 0.01. We can now determine T, and since $T = N(n + 1)$ or $N(n + 2)$, where n is 3 or 4, respectively, we can calculate N to be 160 for single-parent regressions, and 65 for midparent regressions.

These results are summarized in Table 11.8, which shows the number of animals required to estimate a heritability of 20 percent with a standard error of 10 percent.

Table 11.8

	Single parent	Midparent
Number of parents measured per family	1	2
Number of offspring measured per family (n)	3	4
Number of families required (N)	160	65
Total number of animals measured (T)	640	390

These, then, are the minimal requirements, in terms of animals, in order to estimate a heritability of 20 percent within the broadest acceptable limits. Any increase in the precision of the estimate would require more facilities.

Turning now to sib analyses, utilizing the intraclass correlation for estimating the heritability, we can deal with two simplified structures within the one framework. These are either half-sib or full-sib families but not a mixture of both; this last case will be mentioned briefly later. A simple half-sib structure involves mating a sire to several dams, each of which provides one offspring for measurement. In a full-sib structure, each sire is mated to one dam only, and the mating provides several offspring. The number of families (N) is then equivalent to the number of sires in each case. If each family consists of n offspring, the total number of animals measured (T) is Nn, as the parents for these analyses do not need to be measured. From the analysis of variances between sires, we derive the intraclass correlation. The use of full sibs implicitly assumes that dominance and common environment are unimportant. Granted this assumption, then, the intraclass correlation in the case of full sibs estimates $\frac{1}{2}h^2$. In the case of half sibs, $t = \frac{1}{4}h^2$. Now the variance of the intraclass correlation (σ_t^2) is found in the statistical literature to be

$$\sigma_t{}^2 = \frac{2[1 + (n-1)t]^2 (1-t)^2}{n(n-1)(N-1)}$$

If both N and n are fairly large, as they will tend to be, this formula can be approximated without much loss of accuracy to

$$\sigma_t{}^2 \approx \frac{2(1+nt)^2 (1-t)^2}{nT}$$

Expressed in this way, $\sigma_t{}^2$ can be shown to be at a minimum when $nt = 1$, or $n = 1/t$.

As in the case of regression methods, the optimum structure is again seen to depend on the intraclass correlation. It is a small step now to express n, the optimum number of offspring to measure per family, in terms of the heritability; in the case of full-sib families, the optimum n is $2/h^2$, and $4/h^2$ in the case of half sibs. Thus, for a heritability of 20 percent, 10 offspring per family should be measured if full sibs are employed, and 20 offspring where half sibs must be used. But as the heritability is unknown at this stage, the optimum structure can be achieved only fortuitously, as before. In the complete absence of any knowledge of what to expect, experiments should be based on 10 to 15 full sibs or 20 to 30 half sibs, as the case may be. This, on average, seems to lead to the least loss of information. It should be recognized, however, that these optimal structures are not always easy to attain, and the size of sib groups is often dictated more by the reproductive capacity of the organisms than by statistical considerations.

The sampling variance of the heritability estimate can be deduced as follows: Given the optimum structure of $nt = 1$, the above formula for σ_t^2 can be simplified further:

$$\sigma_t^2 \approx \frac{8(1-t)^2}{nT} \approx \frac{8}{nT} = \frac{8t}{T}$$

since we can neglect $(t^2 - 2t)$ with little loss of accuracy.

This leads directly to the sampling variance of the heritability estimate:

For full sibs: $\sigma_{h^2}^2 = 4\sigma_t^2 = \dfrac{16h^2}{T}$

For half sibs: $\sigma_{h^2}^2 = 16\sigma_t^2 = \dfrac{32h^2}{T}$

We can now determine the scale of experimentation required to achieve a standard error of a given magnitude. Continuing with our previous example of a heritability of 2 percent and presuming that we are aiming, as before, at the modest objective of reducing σ_{h^2} to 0.10, or $\sigma_{h^2}^2$ to 0.01, we can substitute this value in the formulae and solve for T. From the relationship $T = nN$, we can further calculate N, the number of families that should be measured. The results are summarized in Table 11.9.

Full sibs, where they can be employed, are therefore twice as efficient as half sibs. It will be noticed also, referring back to Table 11.8, that, for the specific example chosen, a half-sib structure is of the same efficiency as the regression on one parent. A choice has often to be made between these two methods, and by substituting values in the formulae we have derived, it can be established that the following general rule applies. For a given total number of animals measured, a half-sib structure with optimal design gives a more accurate estimate of the heritability than the regression of offspring on one parent if the heritability is below 20 percent; for higher heritabilities, the opposite holds.

The half-sib structure just discussed, namely, one offspring per dam, becomes the most efficient one to use if only the component of variance between sires can be employed to estimate the heritability. If it is desired to use the between-dam component as well, the situation becomes more complicated, as the family group will then consist of a mixture of half and full sibs. Under these conditions, it can be shown that the optimal design is to mate three or four dams per sire, with $2/h^2$ offspring measured per dam. In the absence of any prior estimate of the heritability, about 10 offspring per dam should be measured.

Table 11.9

	Full sibs	Half sibs
Number of offspring measured per family (n)	10	20
Number of families required (N)	32	32
Total number of animals measured (T)	320	640

This sketchy consideration of experimental design, with rather crude algebra at times, leads to one unambiguous conclusion. It is that estimates of heritability become meaningless if they are based on small numbers. Even when the number of animals rises into the hundreds, the estimates are still not very precise. It may not always be possible to collect all the required data in one generation; in these circumstances, it may be expedient to pool information from more than one generation, on the assumption that the sources of variance do not alter in the meantime.

The implications of this section on experimental design, with specific regard to the estimation of heritabilities, can be summarized as follows:

1 Sheer physical considerations impose a limit on the total facilities available for any experiment. But it is possible to manipulate the variables, especially the breeding structure, to maximize the information that may be gained from these facilities. The concern is to reduce the sampling variance of the estimate of the heritability to a minimum.

2 Given the optimum structure for an experiment, it is possible to predict, within limits, the standard error of the estimate of the heritability. This gives the investigator a realistic idea of the scale of experimentation necessary, lest he should embark on a program doomed to futility from the start.

THE PROVISION OF MATERIAL
FOR RESEARCH

So far in this chapter, we have been concerned with the genetics of quantitative characters in a static situation; we have examined ways of describing the genetics of a population as we find it, with respect to any character in which we may be interested. This descriptive approach is calculated to shed light on the inheritance of the character and to discover how the genetic variables affect the level of its expression. Paramount among the genetic variables are the gene frequencies. We must now examine ways in which gene frequencies can be manipulated to alter the level of expression of a character and thereby extend our understanding of the genetic control of particular biological compositions.

In the remainder of this chapter, we shall concern ourselves mostly with two agencies that can change the gene frequencies rather rapidly. The first is inbreeding, the effects of which are dispersive, resulting from random changes in gene frequencies within lines, though the overall frequencies in a population of such lines do not change. The second is selection, resulting in directional changes in gene frequencies, brought about by the differential fertility of individuals. We shall not consider mutation, as its effect on a quantitative character in the absence of selection is unimportant.

It should perhaps be said here that studies of inbreeding and of selection are not, on their own, potent genetic methodologies. The additional conclusions that they permit about the genetics of a population are seldom rigorous, for the reason that the changes that they bring about defy close scrutiny. Similar results are often the products of different situations, and repeated experiments do not always provide the same answers. This is a reflection of the fact that sampling errors play a large part: first, in the determination of the genetic composition of the base population and, second, in any subsequent manipulation. Nevertheless, information from inbreeding and from selection programs is of value, and the results are of cumulative importance. But no doubt the chief use of inbreeding and of selection is in the provision of special strains for specific research requirements. The particular kind of research for which these strains are employed depends, of course, upon the organism and upon the character involved. For instance, the usefulness of strains of mice susceptible to cancer needs no elaboration; they have been employed in a multitude of ways in cancer research. Again, in their book, Fuller and Thompson (1960) refer several times, and in different contexts, to Tryon's maze-bright and maze-dull rats, illustrating the usefulness of this kind of material in research programs. Instances such as these could be multiplied to illustrate that the end products of inbreeding or selection are often of more value than any information about the route whereby the end product was obtained. Because of this, we shall deal with inbreeding and selection, in turn, from the point of view of developing special strains, as well as deriving information about the genetics of the population.

INBREEDING

Inbreeding can be defined as the union of gametes containing alleles identical by descent. Alleles are said to be identical by descent when they are the division products of one such allele that occurred in the past. The *coefficient of inbreeding* (symbolized F) is the probability that the uniting alleles are identical by descent and can be regarded as a correlation (ranging from 0 to 1) between the uniting gametes. This definition of inbreeding is, however, a theoretical one, and in practice it must be modified. For who knows whether or not any two alleles that are alike are identical by descent, if traced sufficiently far back? Therefore a base line must be fixed arbitrarily, and beyond this line no ancestries will be traced. Any specified degree of inbreeding then becomes one relative to this base population, which by definition has an inbreeding coefficient of zero. Likewise, the practical definition of inbreeding becomes the mating of individuals that are more closely related than the average relationship between all the individuals of that population.

The effects of inbreeding are widely known and do not require detailed comment here. Inbreeding is a dispersive process, in the sense that homozygotes are increased at the expense of heterozygotes. It tends to fix the alleles at a particular locus; i.e., they all become alike, and further heterozygosity can arise only through mutation, which will initially occur in one individual only. The chance of fixation of any one allele is proportional to its initial gene frequency. Thus, if a number of lines from a base population are inbred simultaneously, a proportion p will become fixed for the A_1 allele, and q, or $1 - p$, for allele A_2, where p and q are the respective gene frequencies. If we consider a second locus, B_1B_2, with initial frequencies r and s, respectively, a proportion pr of lines will become fixed for A_1 and B_1, ps for A_1B_2, etc. As the number of loci increases, the probability of two lines being fixed for all the same alleles soon becomes negligibly small; this, perhaps, is the way in which the dispersive process is most clearly visualized.

Although the overall probability of fixation is determined by the initial gene frequency, it is purely a matter of chance which particular lines are fixed for A_1, which for B_2, etc. Because of this random dispersion, the final genetic composition of a line is unpredictable. It is presumably on this account that some inbreeding programs are subjected to simultaneous selection toward some given phenotype. In the light of present-day knowledge, however, this practice seems to have little to commend it; it is roundly condemned, with specific reference to behavior, by Broadhurst (1960). For one thing, selection is much more potent on its own when unaccompanied by the opposing dispersive effects of inbreeding. This is because a favorable gene may begin to become associated with an unfavorable one, and under a system of close inbreeding, they tend to remain so, despite selection. And even with only a few loci involved, this is a likely occurrence. For the establishment of strains for further research, inbreeding as such has little to contribute, unless homozygosity or random dispersion become ends in themselves, as indeed they are in special cases. And quite apart from all this, inbreeding has other ill effects, to be mentioned shortly.

Inbreeding, in the sense of increasing the homozygotes at the expense of heterozygotes, occurs in small populations, for it is intuitively obvious that small numbers make it more likely that uniting alleles are identical by descent. Theoretically, the change in inbreeding coefficient (ΔF) from one generation to the next can be shown to be, approximately,

$$\Delta F = \frac{1}{8N_M} + \frac{1}{8N_F}$$

where N_M and N_F are the numbers of male and female parents, respectively. Thus, five breeding pairs increase the inbreeding coefficient by 5 percent per generation, so that, with even so small a number, the increase in the inbreeding coefficient is not alarming. We must, however, distinguish here between rapid inbreeding, such as occurs when sibs or other close relatives are mated, and slow inbreeding through, say, a restriction of the population size. Selection, particularly natural selection favoring heterozygotes, has much more scope when the inbreeding is slow. Thus, if heterozygotes, which in this context include heterozygous segments of chromosome, have any advantage in fitness, the ensuing natural selection will retard the approach to homozygosity. In fact, Hayman and Mather (1953) showed that only a moderate advantage of the heterozygotes will prevent complete fixation. It is also a matter of observation that small populations, e.g., stocks of laboratory mice, do not in fact suffer much from the effects of slow inbreeding, although they might be expected to have accumulated such effects through time, when the number of breeding pairs is only 10 to 20 per generation. This indicates that natural selection is, in fact, at work.

What, then, are the effects of inbreeding, with respect to the measurable parameters of the population? Starting again with a single-locus model, we shall examine first the effect of inbreeding on the population mean. When the inbreeding coefficient is raised from the arbitrary zero level to a value F, the relative frequencies of the three genotypes are modified, according to well-known formulae, in the direction of increasing the homozygotes at the expense of the heterozygote. If the initial frequency of the A_1 allele is again p, and that of the A_2 allele q, the modified frequencies when the inbreeding coefficient stands at F are shown in Table 11.10. It should be noted that these are average frequencies which refer either to one particular locus in a population of many inbred lines or, alternatively, to the array of loci, similar in kind and magnitude of effect, within any one line.

Multiplying the frequency by the value and summating over the genotypes, we can derive the population mean (M_F) when the inbreeding coefficient is F and also compare it with the mean (M_0) that we previously derived for zero inbreeding. For one locus,

$$m_F = a\,(p - q) + 2dpq - 2Fpqd$$
$$= m_0 - 2Fpqd$$

Generalizing by summating over loci,

$$M_F = M_0 - 2F\,\Sigma pqd$$

This shows that the effect of inbreeding to coefficient F is to reduce the mean, in terms of our arbitrarily assigned values, by $2Fpqd$. This formula permits three conclusions:

1 The change in the mean is linearly related to F; this is a theoretical conclusion, and it would perhaps be true to say that experimental support for it is not overabundant.

Table 11.10

Genotype	Frequency	Value
A_1A_1	$p^2 + Fpq$	$+a$
A_1A_2	$2pq - 2Fpq$	d
A_2A_2	$q^2 + Fpq$	$-a$

2   If d, or rather Σd, is zero, the mean does not change on inbreeding. Thus, genes that act additively, although redistributed between genotypes, do not affect the population mean on inbreeding.

3   The minus sign in the formula indicates that the change in the mean is always in the direction of the recessive allele. To the extent that recessive genes tend to be deleterious, the effect of inbreeding on the population is also deleterious.

There is a plethora of literature on the formal study of mutant genes, underlining the generally deleterious effects of recessives. This accords well with experimental experience of inbreeding, the deleterious effect of which is often all to obvious.

The decline in the population mean on inbreeding is known as *inbreeding depression*, the knowledge of which preceded its understanding of many centuries. It is particularly obvious in the case of reproductive capacity and maternal performance, the most obvious components of natural fitness. It was noted earlier that the additive genetic variance of such characters is relatively small but that their dominance variance is correspondingly greater. It is the source of this variance, the dominance deviations, that is the main cause of inbreeding depression. Furthermore, it is not sufficient that these deviations exist. They must also affect the character predominantly in the one direction; i.e., the dominant alleles must have a tendency to increase the phenotypic measurement and the recessive alleles to decrease it, or vice versa. The character is then said to exhibit *directional dominance*. It is intuitively acceptable that directional dominance should be characteristic of the components of natural fitness; alleles have themselves evolved, and the favorable ones are supposed to have evolved a dominant expression.

The establishment of directional dominance is about the only strict genetic conclusion that can be derived from the study of the effects of inbreeding on the population mean. If the mean does not change, then either all the genes act additively or the dominance deviations, on average, cancel each other out, that is, $\Sigma d = 0$. A third reason could apply if the inbreeding is sufficiently slow, namely, selection favoring the heterozygotes. In this case, the inbreeding coefficient, as calculated, would not reflect accurately the stage of gene dispersion in the population.

Turning now to the effects of inbreeding on the genotypic variance and its components, there is surprisingly little that may be said with profit in this context. The reason for this is twofold. Firstly, the theory has not yet been very fully developed. Secondly, as we have seen in the context of experimental design, variance components are very difficult to estimate with any precision, so that the experimental evidence on the subject is not very revealing. But in a general way, the tendency is for the genotypic variance of a population to be repartitioned on inbreeding, until ultimately it vanishes within lines and it all appears as the component of variance between lines. If all the genotypic variance is additive, which can happen only when all the genes affecting a character act additively, then the expressions for the between- and within-line components of genotypic variance, for partial inbreeding, are as follows:

Between lines: $2FV_G$
Within lines: $(1 - F)V_G$
Total: $(1 + F)V_G$

Thus, when F reaches 1, the total genotypic variance is in fact twice what it was originally, on account of the increased number of homozygotes. But these expressions are true only when all the genotypic variance is additive. They do not hold for the additive effects of genes with dominance; they do not hold for the additive part of the genotypic variance if dominance variance also exists. The case of fully dominant genes, for instance, is quite different. Robertson (1952) showed that, in this case, the within-line variance rises until F is in the region of 0.5, as a result of the segregation of more homozygotes; it then declines. We need go no further to appreciate the futility of attempting to draw genetic conclusions from the study of variances during inbreeding, though empirical observations are always of interest if they are reasonably precise.

The subject of inbreeding should perhaps not be dismissed without a mention of its complement: *heterosis* or *hybrid vigor*. The two terms should be regarded as synonymous, for although some writers attach slightly different shades of meaning to them, there is no consistency among their practices. Heterosis, as a topic, rightly belongs in the provinces of plant and animal breeding, and it is difficult to see how its study, at this stage, can lead in any way to a deeper understanding of the genetics of a particular character. Suffice to say that heterosis among crosses can occur only for characters that display inbreeding depression, and its existence can therefore reflect no more than the presence of directional dominance.

It is sometimes mistakenly supposed that heterosis is somehow contingent upon the presence of overdominance, where the heterozygote exceeds in value either homozygote. This is not a prerequisite of heterosis; crosses between lines fixed for the recessive alleles at different loci are prone to exceed the level of either parental strain, given directional dominance. It is nevertheless true that overdominant loci, if they exist, may influence greatly the degree of directional dominance; a few loci overdominant in one direction may easily outweigh more loci that are dominant, or partially so, in the other. If and when it is present, overdominance may well have an overriding influence on the change in the mean during inbreeding and crossing. But the existence of overdominance in quantitative genetics is not easy to establish; it is difficult to distinguish from epistasis and impossible, in the short run, to distinguish from close repulsion linkage.

Heterosis, then, reflects directional dominance. It may occasionally be a means of providing research material if, for instance, uniform genotypes are required and inbred material proves to be too infertile or too susceptible to sources of environmental variation. This exploitation of heterosis, however, can be approached only empirically, where possible practical advantages can be envisaged.

RESEARCH WORK WITH INBRED MATERIAL
IN QUANTITATIVE GENETICS

The usefulness of isogenic material in many types of research needs no emphasis. In quantitative genetics, however, this usefulness is severely circumscribed by the limitation on the number of genotypes that are available. But even more important is the peculiar nature of inbred material and its derivatives.

The gametes of all the individuals of an inbred strain are all exact replicates of one another, except for the sexual dimorphism, which in this context is of negligible significance. Since no genes have been added to the population during inbreeding, it is possible that an exact replicate of this particular gamete could have been found in the base population of outbred individuals. In a sense, therefore, an inbred line can be considered as a representative of one gamete only from the base population, and an experiment involving, say, 10 lines is an experiment on a sample of 10 gametes out of a possible very large number, perhaps literally many millions. Although a high degree of precision could be built into the experiment, the information derived from it would be precise about the 10 "gamete equivalents," and it would be rash to generalize from such a narrow base. Any work on inbred lines refers, therefore, very strictly to those inbred lines only. Any conclusions from such work should not be deemed to apply to the species at large without supplementary evidence.

This conceptual restriction on the employment of inbred material is aggravated by its peculiar genetic composition. An inbred line cannot, by its very nature, contain any lethal genes; it is unlikely also to contain either semilethal or, as a result of sampling, rare recessive genes. Such genes will therefore not be found in a population of line crosses either. And yet lethal genes and rare recessives are an important feature of outbred populations, as they are of wild populations in the field. These considerations also detract from the usefulness of inbred material. To all this, we must add the disadvantage of peculiarities in gene frequencies. A cross between four inbred lines, for instance, means that no allele can have a frequency lower than 0.25 in the derived population. Yet, gene frequencies figure prominently in all our discussions of genetic parameters; changes in gene frequencies can radically alter all of them. The application of results from inbred lines and their derivatives to outbred populations should thus be exercised with extreme caution. An inbred line represents a unique and extraordinary situation in biology; line crosses are hardly less unique and extraordinary. An investigator, contemplating the employment of inbred lines, should therefore reflect seriously on the kind of information he hopes to obtain from them and ask whether it covers the range of variation in which he is basically interested.

The use of inbred strains has been mentioned twice in this chapter.

Firstly, it was said that differences between inbred strains kept under uniform conditions were evidence of genetic variation in a character. Secondly, and less certainly, it was suggested that isogenic material could be employed to assess the environmental component of variance. But beyond these two uses, the value of inbred strains in research work on quantitative characters must be questioned, unless interest rests on the strains themselves and on the peculiar combination of circumstances they represent.

SELECTION

We must now consider briefly the second agency whereby changes in gene frequencies may be brought about, namely, selection. One must distinguish, as in the case of inbreeding, between the usefulness of selection in the production of special strains and its usefulness as a research tool in quantitative genetics. In the former case, selection is a very valuable method; in the latter case, it has much less to contribute, for reasons that will be explained. Much of the theory of selection is devoted to the prediction of its results and to the evaluation of the efficacy of various methods of changing a character in a required direction. We shall consider here only a few of the salient features of the theory in order to illustrate the concepts involved.

Selection may be defined in terms of differential fertility, whereby individuals do not contribute equally to the next generation. Under experimental conditions, the population is usually truncated at some point on a phenotypic scale; the individuals on one side of the truncation point are allowed to breed, while those on the other side are not. The mean value of the selected individuals deviates from the population mean by a certain amount, S, termed the *selection differential*. This deviation is composed of genetic and nongenetic components, and the only part of it reflected in the mean performance of the progeny of selected animals is that due to the average effect of the genes. The relative importance of the average effect of the genes, compared with other sources of variation, was seen to be measured by the heritability. Therefore, the deviation of the progeny (R) from the population mean is given by

$$R = h^2 S$$

as derived previously. The deviation of the progeny is known as the *response to selection*.

This formula suggests another way of estimating the heritability. The selection differential and the response are both easily measured in practice, and from the relationship just given, the heritability can be calculated. To reduce sampling error, both S and R should be cumulated over several generations and plotted generation by generation. The slope of the least-squares regression line of R on S through these points then gives the heritability. The value so obtained should be termed the *realized heritability*, i.e., what is observed in practice.

One snag in this context is that, although the selection applied truncates the population sharply, the selected parents will still differ among themselves in their contribution to the next generation. To correct for this, the selection differential of each parent should be weighted by the number of progeny it contributes for measurements. It is the figures so weighted that should be cumulated to arrive at the cumulated selection differential mentioned above. With some species, such as Drosophila, which provide many offspring, it may be more convenient to enforce equal representation among the progeny by taking random samples of equal size from each mating. Sterile matings must, of course, be excluded in calculating the selection differentials.

The selection differentials are not exactly equal for males and females, even when pair matings are employed; the two should be averaged for each generation. As the females tend to limit the rate of reproduction, the selection differential can often be increased by mating one male to several females and thereby achieve high differentials on the male side. The limit in this direction is usually the need to avoid excessive inbreeding by the restriction so imposed on the population size. In the case of laboratory mice, it has been found in practice that, provided the number of parents does not fall below the equivalent of 10 pair matings and if provision is made for each fertile mating to be represented in the next generation, the populations as a rule do not accumulate the more obvious effects of inbreeding. But if individuals are selected irrespective of the family from which they derive, the number of parents should be doubled.

The value of the selection differential depends on the proportion of animals selected and also on the phenotypic standard deviation of the trait. It is convenient to measure the selection applied in terms of the *intensity of selection* (i), which is in fact the selection differential measured in standard terms:

$$i = \frac{S}{\sigma_P}$$

where σ_P is the phenotypic standard deviation. By equating S to $i\sigma_P$, the formula derived previously becomes

$$R = i\sigma_P h^2$$

This shows how the response to selection depends on three factors: the intensity of selection, the phenotypic standard deviation of the character, and the heritability. The intensity of selection, expressed in this form, depends entirely on the proportion of animals selected, though the relationship is not a linear one. There are available tables, based on the normal distribution, that give the value of i for a given proportion selected.

Responses to selection, as is well known, tend to be rather erratic. Although the response in the desired direction may be quite apparent when the progress over several generations is surveyed, fluctuations from generation to generation render short-term assessments unreliable. These fluctuations may arise either through accidents of sampling or as a result of environmental changes. The first cause is beyond our control, and its influence can be estimated only by running replicate selection lines concurrently. The second cause is all too frequently of unknown origin, but some of these environmental creases can be ironed out by the use of an unselected control population. The control and the selected lines should derive from the same base population and should be of identical structure, with the exception that in the control population the parents should be chosen and mated at random. The response to selection should then be measured as a deviation from the control line.

If separate and simultaneous selection is carried out for "high" and "low" expression of the character (two-way selection), one line acts as a partial control for the other. The effect of the selection can then be judged from the divergence between the two lines. This procedure, however, does not enable us to recognize a fairly frequent feature of selection programs, namely, the asymmetry of the response, which means that selection in one direction brings about a more rapid change in the mean than it does in the other. A behavioral example is discussed by Hirsch in Chapter 12. Without an unselected control line, the separate responses in the two directions cannot be adequately evaluated.

The pattern of the response to selection is itself of intrinsic interest. The subject is discussed in some detail by Falconer (1955). As mentioned, asymmetry of the response is common, and it may arise from a variety of causes, some of the more obvious ones being as follows:

1   Natural selection may oppose the artificial selection in one direction, while assisting it in the other.

2   Selection may favor heterozygotes in one direction, which of course segregate out homozygotes, thereby retarding the response.

3   The selection differentials attained in the two lines may not be equal, either as a result of the variances being different or because of differential fertility in the two lines.

4   If directional dominance affects the character, the line selected for the recessives will show a more rapid response.

5   Inbreeding depression may affect the mean of the character in the same direction in both lines, opposing one while reinforcing the other.

6   The characters may be of partially independent genetic origin. For instance, body weight may be increased by increasing the relative amount of adipose tissue, while selection in the opposite direction may soon reduce the fat to a minimum; progress in the low line would then become contingent upon a reduction in bodily dimensions.

Except in its pragmatic aspects, therefore, selection is not a potent method in genetic analysis. Short-term responses, if measured accurately, may act as a useful check on theory and on the accuracy of parameters estimated from some base population. In the absence of additive variance, the character will not respond to selection.

Continued selection invariably results, eventually, in the cessation of the response, and a limit is reached beyond which no further progress is obtained. This *limit to selection*, also sometimes referred to as a "plateau" or a "ceiling" or other odd words, should be approached asymptotically as the additive genetic variance becomes exhausted. Occasionally, a renewed response may appear in practice, through the introduction of new variance in the form of a mutation of a major gene or the formation of rare recombinants. But usually a selected line at the limit remains at a fairly constant average level despite perhaps sharp fluctuations from generation to generation. If the additive variance has in fact been exhausted, then selection can do nothing to shift the mean level of the line in either direction. In practice, however, this is not always the case, and selection is often required purely in order to maintain the level of the line. When the selection is relaxed, the average performance often tends to revert toward that of the base population. This may indicate either the opposing force of natural selection or that the artificial selection had favored heterozygotes. Under these conditions, selected lines at the limit often respond rather rapidly to reversed selection, again indicating that the additive variance had not, in fact, been exhausted. When one adds the possibility of some genetic independence between the "high" and "low" expression of the character, one can appreciate that the possible complexities of the situation again defy precise genetic interpretation.

Because of these difficulties, the nature of the limits to selection has not yet been fully explored. No doubt the exhaustion of the additive genetic variance is often an adequate explanation. However, the response sometimes ceases too abruptly for this to be plausible; frequently a line responds more or less linearly and then suddenly stops. This obviously does not accord with a model of asymptotic depletion of the variance. As mentioned, natural selection or selection for heterozygotes often prevents progress while a considerable amount of additive variance remains. Indeed, the likelihood that natural selection is at work is often all too conspicuous in the form of widespread sterility or reduced fertility in selected lines.

Despite what may appear to some as a plethora of literature on selection work, there is still insufficient information to indicate the expected magnitude of the response to selection and how long the response may be expected to continue. A theory of limits published by Robertson (1960) suggests that, in terms of our formulation, the expected limit of selection is a function only of the product Ni, where N is the effective size of the population and i is the intensity of selection. Robertson shows further that one-half of the total response should be attained in $1.4N$ generations for genes that act additively and that this may approach $2N$ generations for rare recessive genes.

Selection is usually considered in terms of selecting individuals on the basis of their own phenotypic merit and is most conveniently discussed in such terms. This is usually called *individual* or *mass* selection. However, selection need not and does not always take this form. If, for instance, an individual must be killed in order to assay some hormone, it cannot be used for further breeding; it cannot itself be selected, unless it has been possible to obtain and maintain offspring from all individuals liable to be selected, which would make excessive demands on facilities. Under these conditions, the stock must be propagated from the relatives, e.g., sibs, of "selected" individuals. In animal breeding, a phenotypic assessment of an individual is sometimes based on the performance of its progeny; e.g., bulls are selected on the milk yield of their daughters. Often information is available on the performance of relatives, and the question arises whether such information could and should be employed in assessing the genotypic merit of an individual. This is a big subject in its own right, though chiefly of relevance in the realms of animal breeding. In the laboratory, the question is usually simplified to the consideration of families of either full or half sibs. The question then is whether an individual should be selected entirely on its own merit, or whether it should be selected by its relative merit compared with other members of its family, or whether whole families should be selected on the basis of the family mean, without regard to individual deviations within the family. These three forms of selection are known as *individual* selection, *within-family* selection, and *between-family* selection, respectively. We shall not enter here into the algebra and statistics involved but merely indicate the main conclusions.

The general formula

$$R = i\sigma_P h^2$$

can be modified into

$$R_w = i\sigma_w h_w^2 \qquad \text{and} \qquad R_f = i\sigma_f h_f^2$$

where the subscripts w and f refer to within-family and between-family terms, respectively. Now, σ_w and σ_f can be expressed in terms of σ_P, and h_w^2 and h_f^2 in terms of h^2. By doing this, R_w and R_f can both be formulated in terms of $i\sigma_P h^2$ weighted by a term containing n, the number of individuals in the family, and t, the intraclass correlation. The relevant formulae are a bit complicated though not difficult to derive. When expressed in this way, R_w and R_f can then be compared with R, the response to individual selection, by substituting values for n and t. The intraclass correlation is the more important of the two, with the following general conclusion. Where t, and therefore the heritability, is low, the response from between-family selection exceeds the other two. Where t is high, and the heritability now may or may not be high, the maximum response is attained from within-family selection. But over a wide range of intermediate values of t, individual selection surpasses both of the more complicated methods. The critical values of t depend on n, the number in the family, and the reader is referred to Falconer (1960) for the exact formulae and their derivation.

A high intraclass correlation can result only from a proportionately large component of variance due to common environment, making full sibs similar. In these cases only should individuals be selected on the basis of their deviation from the family mean. The method has a disadvantage of utilizing only one-half of the additive genetic variance but it has one redeeming feature. As each family is represented in the succeeding generation, the effective population size is doubled, as explained by Falconer (1960), and the method is therefore economical in space and facilities. Between-family selection, on the other hand, is costly in terms of facilities, as many more families must be measured than are selected, the remainder being discarded. It is therefore fortunate that individual selection is frequently the one that gives the maximum response.

In concluding this section on selection, we should note that it is possible to combine information on an individual's own merit with that on its family to arrive at an index of the expected breeding value. The individual's phenotype (P), expressed as a deviation from the population mean, can be regarded as a sum of two parts; firstly, as a deviation (P_f) of the family mean from the population mean and, secondly, as a deviation (P_w) of the individual from its own family mean:

$$P = P_f + P_w$$

The appropriate weighting factors for the expected breeding value are as follows:

$$Exp\ BV = h_f{}^2P_f + h_w{}^2P_w$$

This then becomes the index of selection. It might be added that *combined selection*, as this method is called, is seldom worth the trouble. Its superiority over some other form of selection is never very great and is often quite trivial.

CORRELATED RESPONSES TO SELECTION

Selection for any character may result in a concomitant change in some other character or characters. Such a phenomenon is referred to as a *correlated response* to selection, and we must examine briefly its genetic causation.

As a model, we shall postulate that selection for some character X changes also the mean level of performance of some other character Y. The connecting bridge is the *genetic correlation* (r_A), which may be defined as the correlation between the breeding values for the two traits, each measured in every individual. Thus, if the breeding values for X and Y are measured in a number of individuals, each individual provides a pair of observations which can be correlated. Such a correlation can arise only if the two characters are affected by "common genes"; in the long run, this implies pleiotropy, while in the short run a correlation may easily arise as a result of linkage; i.e., for "common genes," read "common chromosomal segments." Correlations of the latter kind can be particularly prevalent in crosses between divergent strains, though they fade through time as equilibrium is established between the coupling and repulsion phases of linkage.

The principle of the formulation is that the phenotypic covariance between two characters, measured on the same individuals, can be partitioned into an additive genetic component and a (nonadditive +

environmental) component, in a way strictly analogous to that previously employed for variances. If the breeding values themselves were correlated, as suggested in the definition, the additive genetic covariance alone would be derived. In practice, however, this usually proves too cumbersome, and the less direct method of partitioning is employed, by relating observational components of covariance to causal components. The reader is referred to Falconer (1960) for details. The genetic correlation between X and Y $[r_{(A)XY}]$ is then obtained by relating the additive covariance $[cov_{(A)XY}]$ to the two additive variances, according to the usual formula for a correlation:

$$r_{(A)XY} = \frac{cov_{(A)XY}}{\sigma_{(A)X}\,\sigma_{(A)Y}}$$

The correlated change in character Y, when the selection is for character X, will depend upon the regression of the breeding value for Y on the breeding value for X. This regression is the ratio of the additive covariance to the additive variance of X.

$$b_{(A)YX} = \frac{cov_{(A)YX}}{\sigma^2_{(A)X}} = r_{(A)}\frac{\sigma_{(A)Y}}{\sigma_{(A)X}}$$

The direct response in X was given as

$$R_X = ih_X^2\sigma_{(P)X}$$

where $\sigma_{(P)X}$ is the phenotypic standard deviation of X. This can be rewritten in terms of the additive standard deviation:

$$R_X = ih_X\sigma_{(A)X}$$

where h_X is the square root of the heritability. Hence the correlated response in Y, (CR_Y), can be formulated as follows:

$$\begin{aligned}CR_Y &= b_{(A)YX}R_X \\ &= r_A\frac{\sigma_{(A)Y}}{\sigma_{(A)X}}ih_X\sigma_{(A)X} \\ &= ih_Xr_A\sigma_{(A)Y}\end{aligned}$$

Since

$$h_Y^2 = \frac{\sigma_{(A)Y}^2}{\sigma_{(P)Y}^2}$$
$$\sigma_{(A)Y} = h_Y\sigma_{(P)Y}$$

so that

$$CR_Y = ih_Xh_Yr_A\sigma_{(P)Y}$$

The only purpose of this brief formulation is to show the number of factors that may influence the correlated response in one character when selection proceeds for another. The presence of a correlated response demonstrates only that none of the five factors in the formula is zero. The likeliest one to be zero is, of course, the genetic correlation. But it is much more difficult to argue that the absence of a correlated response means that r_A is, in fact, zero, especially if either of the two heritabilities is low. And it can never be argued that, because a genetic correlation cannot be detected, the characters do not share some common genes. It is easily conceivable that one gene could increase one character and decrease the other, while another gene had the opposite effect, both genes thus masking each other. This underlines yet again the difficulty of drawing precise genetic conclusions from the responses to selection. The chief usefulness of correlated responses in genetic analysis is to act as a check on previous estimates of the heritabilities and the genetic correlation.

As a final remark in this section, it should be noted that, if the heritabilities are known, the formula given suggests a method of estimating the genetic correlation by measuring the correlated response. A better method, using selection data, is to select for each of the two characters separately and to observe the correlated response of the other character in each case. The formulae given can then be manipulated to show that

$$r_A{}^2 = \frac{CR_X}{R_X} \frac{CR_Y}{R_Y}$$

In accordance with our previous terminology, the estimate so obtained should perhaps be termed the "realized genetic correlation," to distinguish it from a priori estimates derived from the methods of partitioning the covariance.

CONCLUDING REMARKS

What I have attempted in this chapter is to present some basic concepts as they affect our thinking about the problem of quantitative variation in genetics and to indicate a few of the methods whereby the sources of such variation are explored. It has not been possible, in the space available, to develop some of these ideas very far; neither has it been possible to discuss at all the exploitation of the products of selection and inbreeding in genetic analysis. The latter objective would have been difficult in any case, as each situation demands a particular approach and a specific analysis, based on the investigator's understanding of the biology of the system with which he is working. Instead, I have concentrated on presenting the subject almost as an attitude of mind, at times, toward situations where genes are at work but where they cannot be identified individually. Very often, a lack of precision in the interpretation came to the surface. This is partly a reflection of the fact that I limited myself to certain objectives, but also it partly reflects the genetic system involved. For it is an accepted fact of life that genes hunt in packs. As such, the pack must be studied as a pack and this, by its nature, poses difficulties. While it is always of interest to learn about some of the quirks of isolated members of the pack, the action of these individuals, as individuals, is almost irrelevant for many purposes. They may have little bearing on the action of the pack as a whole. From this point of view, the quantitative aspects of genetic systems are fundamental and basic to many biological problems.

It will be obvious to most that the treatment in this chapter is far from being exhaustive; indeed, many important topics are not even mentioned. I should not like it to be thought either that the approach adopted is an exclusive one; I have employed the terminology, symbolism, and formulation with which I am most conversant. I should like to believe, however, that the concepts developed here are basic to any approach.

Much of the stimulation of quantitative genetics derives from theoretical studies, developed largely from the works of R. A. Fisher, J. B. S. Haldane, and Sewall Wright in the 1920s and 1930s. Some of these early theoretical papers can still be claimed to be the cornerstones of the subject. Primarily, however, quantitative genetics should be regarded as an empirical science, and it is from more experimental work, especially perhaps with characters as yet uninvestigated, that further impetus and progress should be expected.

CHAPTER TWELVE
BEHAVIOR-GENETIC ANALYSIS AT THE
CHROMOSOME LEVEL OF ORGANIZATION[1]
Jerry Hirsch

INTRODUCTION

It has become increasingly evident that polygenic inheritance plays an important role in behavioral variation. While we have begun to learn about some properties of the organization of polygenic systems, to date we have little, if any, detailed knowledge of the relations between the properties of such systems and behavioral variation.

In other chapters of Part III, methods are discussed for analyzing and describing the genetic correlates of trait variation in populations in terms of concepts that are essentially statistical, such as the additive, dominance, and interactive components of the genetic variance. In several "lower" organisms analysis can now be approached from another point of view, because techniques exist that permit more detailed study of genetic mechanisms. In the genetically well-studied species, *Drosophila melanogaster*, for example, the chromosome can be made the unit of analysis on the independent variable while the behavioral phenotype is observed as the dependent variable. In this way the genetic variance, which is associated with variation in the expression of a behavioral phenotype, can be partitioned into components assignable to specific chromosomes and to their interactions.

Genetic markers, chromosomal inversions, and balanced lethal systems provide powerful tools for genetic analysis. The phenotypic expression of certain easily discernible traits has been related to the presence of specific alleles of known genes, e.g., curly wings and stubble bristles in Drosophila, taillessness in mice, taste sensitivity to certain chemicals in man (McKusick, 1966), etc. When their chromosomal locus is also known, as was the case for the 479 Drosophila loci listed in the latest *Biology Data Book* (Altman and Dittmer, 1964), such genes can be used as convenient markers to label specific chromosomes. Then the presence or absence of the chromosome carrying a specific allele of the marker gene can be inferred from the expression of the phenotype with which it is associated.

[1] This research was supported by Grant NSF GB 487 from the National Science Foundation to the University of Illinois.

An inversion is a structural rearrangement of a chromosome in which the previous linear arrangement of its genes has been altered. When one or more blocks of genes on a chromosome have rotated by 180 degrees, that chromosome is said to contain an inversion. During meiosis when homologous chromosomes pair at synapsis, they frequently exchange corresponding segments. This exchange is called *crossing-over*. Because of crossing-over very few exact replicates of a chromosome (other than the male Y chromosome) will ordinarily be found over several generations in any population. Because of crossing-over and the independent assortment of chromosomes, both of which occur at meiosis, genotypes tend to be unique and non-replicable. Inversions, however, impede the pairing of homologous loci at synapsis and thus offer a means of circumventing the effects of crossing-over. When crossover products are eliminated through the use of inversions, the replication of a given chromosome in all the individuals of a population becomes possible.

There exist dominant genes with recessive lethal effect. When two, nonallelic, recessive lethal genes are used as dominant markers, one on each of the homologs in a chromosome pair containing an inversion, a balanced lethal system is created. A hypothetical example of such a system would be a chromosome pair having the genes Ab on one chromosome and the genes aB on its homolog, where both A and B are recessive lethal and dominant to their respective alleles a and b. Mating two such double heterozygotes would produce three possible zygotic genotypes: $AAbb$, $aaBB$, and $AaBb$, but only the last of these would be viable. Hence, a balanced lethal system together with inversions on a pair of homologs provides a mechanism for preserving the integrity of a chromosome pair in a population and maintaining it unchanged for generations.

GENETIC ARCHITECTURE
AND TRAIT DISTRIBUTION

The existence of these special genetic situations makes possible, by means of controlled matings, the synthesis of populations having almost any desired genetic composition. Heredity can be made the independent experimental variable and the chromosomes can become the units of analysis. In an exploratory study (Hirsch, 1959) a stock that carried inversion chromosomes maintained with balanced lethal systems was crossed to an unselected and free-mating laboratory stock. We produced three F_2 populations which differed in known ways with respect to both the degree and the kind of similarity in chromosome constitution existing among their members. One population was made isogenic (genetically uniform) for a first chromosome which was homozygous in all females, while the autosomes were left to random assortment. Another was made isogenic heterozygous for chromosome II, with chromosomes I and III left to random assortment. And a third population was made isogenic for both homozygous first-chromosome and heterozygous third-chromosome pairs, with chromo-

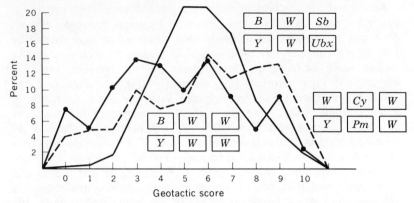

Figure 12.1   Distribution of geotactic scores in a 10-unit maze for males of three populations (described in text). W: chromosome varying at random; Y: the Y chromosome of males; other symbols represent marker genes and indicate chromosomes carried in identical form by all members of a population. (*Modified from Hirsch, 1959.*)

some II left to random assortment. Under these conditions two parameters of their geotactic behavioral distributions were controlled. The least dispersion occurred in the population isogenic for two of its three large chromosomes. The other two populations, isogenic with respect to a single chromosome pair, differed from one another in their central tendencies but not in their dispersions, which were twice the dispersion of the population isogenic for two chromosome pairs (see Figure 12.1). This study demonstrated that detailed manipulation of chromosomal genotype has measurable effects on the distributions of behavioral phenotypes.

CHROMOSOME ANALYSIS

Partial Assay

Subsequently, for each of three populations a partial assay was made of the role in geotaxis of the three major *D. melanogaster* chromosomes and their several interactions. The three populations consisted of one that was being selected for positive geotaxis, another being selected for negative geotaxis, and the unselected foundation population from which the two selected populations were derived. Each population was crossed to a multiple-inversion tester stock; then the resulting F_1 was crossed back to the population under analysis, in order to produce for behavioral comparison the eight backcross female genotypes shown in Figure 12.2. The assay depends on comparison of the behavioral effects of the chromosomes from the population under analysis with those of their homologs from the common tester stock. The homologous chromosomes from the different populations under analysis can then be compared with one another by means of this common standard of reference provided by the tester stock. Direct

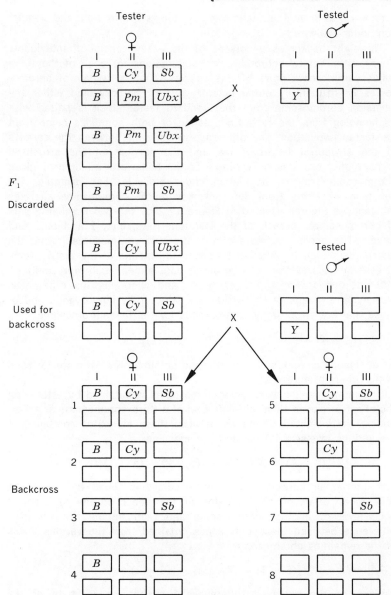

Figure 12.2   Mating plan for partial assay (see text for explanation).
(*Modified from Hirsch and Erlenmeyer-Kimling, 1962.*)

comparison of the chromosomes from different populations is difficult because ordinarily the chromosomes cannot be identified and followed in segregation. The chromosomes of the tester stock, however, contain inversions which reduce recombination in the F_1. They also contain the dominant morphological marker genes which permit us to follow through a backcross the segregation of the tester-stock inversion chromosomes from their unmarked homologs drawn from the population under analysis.

The eight backcross genotypes of the assay consist of individuals having either of two chromosome combinations for each of the three major chromosome pairs of this species. That is, a pair of chromosomes is structurally either heterozygous or homozygous; either the inversion chromosome from the standard tester stock is paired with its homolog from the tested population or both homologs come from the tested population. The difference in behavior between the carriers of the structural heterozygotes and the carriers of the structural homozygotes provides an estimate of the behavioral effect of a given chromosome from a population under analysis. For example, the genotypes resulting from the backcross in the chromosome analyses for geotaxis are represented in Figure 12.2 by the two numbered sets of chromosomes in each of the last four rows, e.g., 1 and 5, 2 and 6, etc. The pooled estimate of the effect of chromosome I from the tested population is obtained by taking the means of the distributions of geotactic scores from the backcross populations, each representing a different one of the eight genotypes, and averaging the differences between the means of those that are structurally homozygous and of those that are structurally heterozygous for the first chromosome.

$$I = \tfrac{1}{4}[(8 - 4) + (7 - 3) + (6 - 2) + (5 - 1)]$$

An analogous comparison is involved in obtaining the estimate for each of the other chromosomes.

The interaction between two chromosomes is obtained by estimating the effect of one of them at each level of a third chromosome and then finding the weighted difference between these estimates combined at the various levels of the second chromosome:

$$I - II = \tfrac{1}{4}\{[(8 - 4) + (7 - 3)] - [(6 - 2) + (5 - 1)]\}$$

The interaction between three chromosomes is found by taking the differences between the estimates of the effect of one of them at each level of another chromosome and then finding the weighted difference between these differences obtained from the various levels of the remaining chromosome:

$$I - II - III = \tfrac{1}{4}\{[(8 - 4) - (7 - 3)] - [(6 - 2) - (5 - 1)]\}$$

Each of the observed differences furnishes an estimate of the change in phenotypic score produced by the substitution of a chromosome from the population under analysis for its homolog from the tester population.

Ten replications of the assays were performed on the chromosomes from each of the three populations (Hirsch and Erlenmeyer-Kimling, 1962). Table 12.1 presents estimates of the effects of the three chromosomes and their four interactions obtained from these assays. For each population it also shows averaged estimates, together with their standard errors and the degrees of freedom from Student's *t* distribution for significant effects (that is, $p < .05$). Clearly, relative to their homologs in the tester stock, all the chromosomes have marked effects in some of the populations, and none of the interactions is important.

Table 12.1
Chromosome and interaction estimates

Population	Replications	I	II	III	I–II	I–III	II–III	I–II–III
Positive	1	1.50	2.27	0.12	−0.80	−0.29	0.18	−0.11
	2	0.53	1.82	−0.11	0.17	−1.42	0.80	−0.39
	3	1.58	2.49	0.13	0.35	−0.19	−0.29	−0.08
	4	1.27	2.23	−0.04	−0.71	0.31	−0.32	0.03
	5	0.95	1.62	−0.32	0.14	−0.23	−0.64	−0.59
	6	1.98	1.82	0.18	−0.37	0.15	−0.33	0.32
	7	1.49	1.19	0.52	−0.14	0.08	0.11	−0.27
	8	2.04	2.04	0.79	0.53	0.42	−0.30	0.29
	9	0.98	1.02	0.21	−0.30	−0.25	−0.23	−0.28
	10	0.94	1.46	−0.64	−0.19	−0.26	0.03	−0.01
	$\bar{X}$	1.39*	1.81*	0.12	−0.16	−0.10	−0.12	−0.14
	SE	0.13	0.14	0.12	0.13	0.18	0.11	0.08
Unselected	1	0.92	1.26	−1.28	0.03	0.13	−0.24	0.20
	2	−0.39	1.08	−0.66	−0.20	0.02	0.61	0.53
	3	0.77	1.74	−0.64	0.38	−1.08	0.60	−0.83
	4	1.18	1.51	0.17	−0.14	0.43	0.22	−0.63
	5	2.00	2.51	−1.33	0.43	−0.32	−1.06	−0.36
	6	1.08	1.76	−0.06	0.03	−0.49	0.36	0.44
	7	0.79	2.25	−0.18	−0.24	−0.02	−0.37	0.08
	8	1.22	1.40	0.35	0.18	−0.69	−0.24	0.72
	9	2.30	2.28	−0.19	−0.15	0.09	0.45	−0.62
	10	0.80	1.97	0.21	0.55	0.59	−0.37	−0.06
	$\bar{X}$	1.03*	1.74*	−0.29	0.05	−0.07	0.03	0.00
	SE	0.21	0.12	0.17	0.10	0.13	0.17	0.13
Negative	1	0.75	0.95	−0.77	−0.13	−0.52	−0.29	−0.28
	2	0.25	1.08	−1.17	0.03	0.31	0.07	−0.70
	3	0.23	0.64	−0.71	−0.18	0.52	−0.08	0.47
	4	0.65	0.76	−1.75	0.63	−0.13	−0.35	0.15
	5	0.28	0.13	−0.68	−0.94	0.58	0.01	0.12
	6	0.66	0.69	−1.41	0.00	−0.04	0.21	0.52
	7	1.01	1.00	−1.44	0.43	0.09	0.39	0.17
	8	−0.04	−1.32	−0.36	−0.58	0.69	−0.21	0.02
	9	1.11	0.14	−1.29	0.03	0.32	−0.54	−0.26
	10	−1.29	−0.39	−1.24	−0.42	−0.32	−0.23	0.01
	$\bar{X}$	0.47†	0.33	−1.08†	−0.12	0.14	0.06	0.06
	SE	0.17	0.20	0.16	0.11	0.13	0.11	0.13

NOTE: Individual and averaged estimates with standard errors for geotactic effects of chromosomes and their interactions from ten assay replications for three populations. Statistical constants and degrees of freedom refer to the behavioral samples (19 for the negative population and 18 each for the other two populations).
* $df = 17$.
† $df = 18$.

SOURCE: Modified from Hirsch and Erlenmeyer-Kimling, 1962.

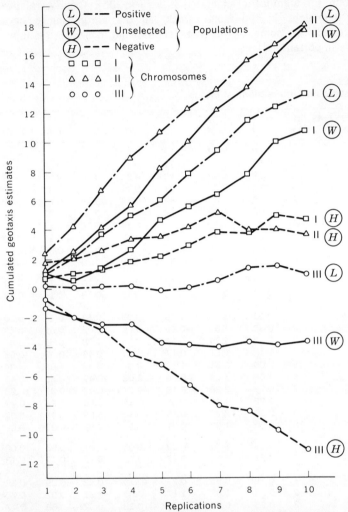

Figure 12.3 Cumulated geotaxis estimates for chromosomes by populations. (*Modified from Hirsch and Erlenmeyer-Kimling, 1962.*)

Figure 12.3 shows in cumulative curve form the behavior of the chromosome estimates over the 10 replications. In terms of the final height of the curves, the rank order of the chromosomes is II > I > III with the exception that the curve for chromosome II from the negative population is displaced below all three I curves; the rank order by populations for all three chromosomes is positive above unselected above negative. The figure shows that in the unselected population chromosomes I and II produce positive geotaxis and chromosome III is somewhat negative by comparison with the standard tester homologs. Selection for positive geotaxis has had little effect, if any, on chromosome II; might have increased the positive effect of chromosome I; and has changed chromosome III from negative to slightly positive. In general, selection for negative geotaxis seems to have had a greater effect on all three chromosomes than selection for positive geotaxis, a condition which should be directly related to the asymmetrical response to selection mentioned by R. C. Roberts (page 251). (In the selection experiment that produced these populations, the results of selection were markedly unequal in the two directions: Selection was far more effective for negative geotaxis than for positive geotaxis; see Erlenmeyer-Kimling et al., 1962.)

In the foregoing analysis not all the possible nonadditive relations can be measured. Therefore only certain limited inferences could be made about the nature of gene and chromosome effects. This is because the chromosomes bearing inversions and marker mutants appear in seven of the eight genotypes used for behavioral observation. Genes on the tested chromosomes, which are dominant or epistatic to genes on the tester chromosomes, must have the same phenotypic effects in the single kind of structural homozygote observed as they do in the structural heterozygotes and therefore remain undetected. Detection of such dominant and epistatic genes requires that the structural heterozygotes be compared with both kinds of structural homozygotes. The tester chromosomes have not been observed in homozygous combinations, however, because the most effective forms of the second and third chromosomes that were available for use as testers carried as marker genes the curly-wing and stubble-bristle mutant alleles, which are recessive lethal in the homozygous state.

Complete Assay

The breeding arrangement outlined next permits a more general evaluation of chromosomal interaction. The matings shown in Figure 12.4 illustrate how all possible combinations of chromosomes from pairs of contrasted populations can be obtained. This type of analysis has several advantages: Behavioral observations do not have to be made on individuals carrying the special chromosomes bearing inversions and marker mutants—unusual properties which might complicate the picture. The use of all possible combinations of chromosomes from two contrasted populations, in principle, permits the measurement of many kinds of interallelic, interlocular, and interchromosomal interaction.

If again we ignore the Y and the IV chromosomes, as in the previous analysis, and consider only the combinations of the three pairs of major chromosomes, then for each pair of chromosomes there are, in females, one heterozygous and two alternative homozygous combinations, i.e., $3 \times 3 \times 3 = 27$ possible genotypes. In males, since there are only two hemizygous alternatives for the first chromosome, there are 18 possible combinations. Following F. W. Robertson (1954) we shall now use a notation in which any genotype can be specified by three letters whose order corresponds to chromosome pairs I, II, and III. Thus chromosome pairs that are purely of negative geotactic or of wild-type origin are designated as HHH or WWW, respectively, and

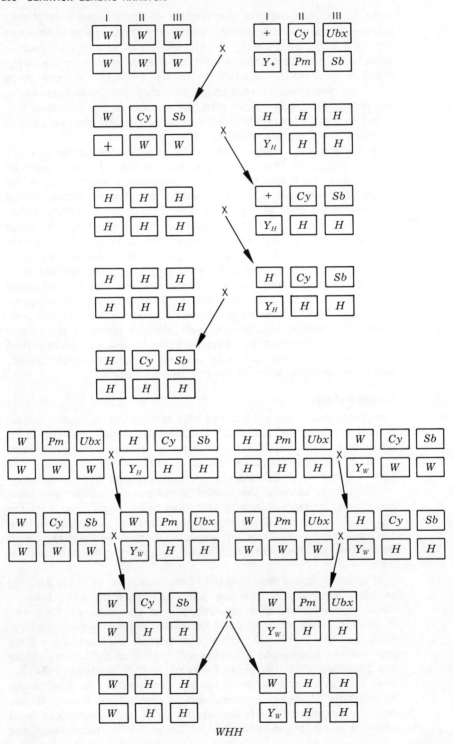

WHH

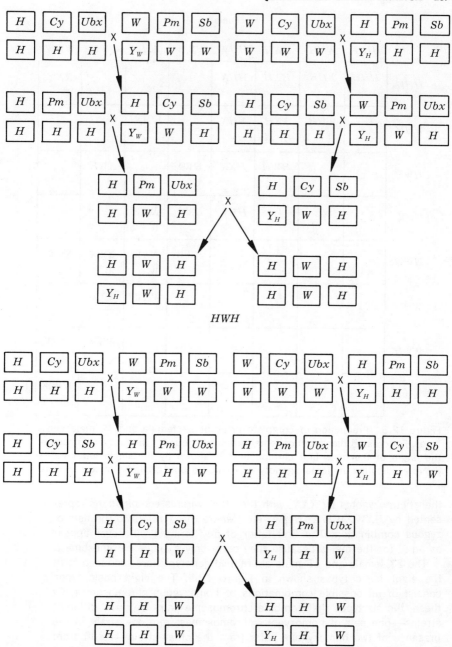

Figure 12.4   Mating plan to produce the eight basic types used in subsequent crosses shown in Figure 12.5. I, II, III refer to the first, second, and third chromosome pairs; *H*: chromosomes from negative-geotaxis population; +: wild-type chromosome but origin unknown; *W*: wild-type chromosome from unselected foundation population. The meaning of the other symbols is the same as or analogous to those in Figure 12.1 and in the text.

Females ♀

	HHH	WHH	HWH	HHW	HWW	WHW	WWH	WWW
HHH	HHH	XHH W	HXH	HHX				XXX W
WHH		WHH	XXH H	XHX H		WHX	WXH	
HWH			HWH	HXX	HWX		XWH W	
HHW				HHW	HXW	XHW W		
HWW					HWW	XXW W	XWX W	XWW W
WHW						WHW	WXX	WXW
WWH							WWH	WWX
WWW								WWW

Males →

Figure 12.5 Production of complete array of 27 female and 18 male genotypes. For crosses that produce females heterozygous for the first chromosome, the origin of the male first chromosome is shown by the single letter below the row indicating the female genotype (Y chromosome origin is not considered).

their heterozygotes by XXX, with the other various combinations represented by HXW, WWX, HHW, etc., where the X refers to the heterozygous combination. The first letter of any formula will never appear as an X for the males because they carry only a single chromosome I.

The 27 female and 18 male alternative genotypes are produced from the eight basic types shown in Figure 12.5. The eight basic types consist of all possible combinations of homozygous chromosomes. Of these, the six types that combine chromosome pairs from the different strains—one pair of homozygous chromosomes from one strain in the presence of two homozygous pairs from the other strain—result from the crosses shown in Figure 12.4, in which inversions are used to suppress recombination during the breeding generations required to bring together the desired combinations of homozygous chromosomes.

Data presently being analyzed from geotactic distribution of the 18 male and 27 female genotypes show that observations on a more complete spectrum of genotypes can, in fact, measure chromosome interactions. While the magnitude of the interactions measured so far is not large relative to the total genetic variance—9 percent in male data and 1 percent in female data (from the first two, of four, replications)—there is every reason to expect that higher values might yet be obtained as we improve our control over these difficult and somewhat laborious measurement procedures.

CONCLUSIONS

There has been a rather widespread belief that "it is impossible to show the presence of . . . groups of linked polygenes by normal Mendelian methods" (Sheppard, 1958, p. 109). Since the analyses just reviewed in this chapter have shown that measurements can be made on the individual chromosomes involved in behavioral variation that is already known to be of the polygenic kind (Erlenmeyer-Kimling et al., 1962; Ksander, 1966), it will be of interest to consider how far we might expect that analysis can eventually be carried.

Long ago Johannsen (1903) demonstrated the genetic discontinuities underlying a continuous phenotypic variable such as seed weight in bean plants. Through self-fertilization of individuals from a heterogenic population of variable seed weight, he established several classes of phenotypically true-breeding inbred lines. Later, Sax (1923) used the seed-coat color gene as a marker to classify chromosomes. In hybridization testcrosses carried through the F_3 generation, he was able to show apparent linkage between the major gene for seed color and polygenes associated with the bean-weight phenotype.

In the analysis of the polygenic correlates of bristle number in D. melanogaster, the elegant experiments of Breese and Mather (1957, 1960) go still further. They split a chromosome into a number of component pieces and showed that all of them contain some of the relevant polygenic factors.

More recently Thoday (1961) and his associates (Gibson and Thoday, 1963; Wolstenholme and Thoday, 1963) have refined the techniques of Breese and Mather to the point where they have apparently succeeded in pinpointing on the chromosome map individual second- and third-chromosome genes in the polygenic system related to bristle formation. Gibson and Thoday (1962) have even reported evidence for an apparent 20 map-unit position effect in this system. And, finally, Spickett (1963) has begun the description of ontogenetic effects on bristle pattern and number of two third-chromosome genes and of a single second-chromosome gene in the polygenic system.

There is at this time no obvious reason why analyses similar to those just reviewed for morphology cannot be performed for behavior as well. Furthermore, the accumulating evidence (Pontecorvo, 1958; Sonneborn, 1965; Tessman, 1965) suggests no apparent limit to the resolving power that may be attained as we refine our techniques for analyzing the relations between the genetic system and morphology, behavior, or any other identifiable feature of biological organization.

CHAPTER THIRTEEN
BEHAVIORAL HETEROSIS[1]
Jan H. Bruell

Mice seem to enjoy running in activity wheels. All mice run in wheels; we have yet to find a mouse that does not. But mice belonging to different inbred strains differ in their running performance (Bruell, 1964b). Some run up to 1 mile per hour; others run not more than 2,000 feet in the same time. What happens when mice from an active strain are mated to mice from a relatively inactive strain? Usually one finds that the F_1 hybrid offspring from such a cross outrun both parents. The hybrids display behavioral heterosis, popularly known as hybrid vigor.

Mice placed in a tunnel that exits into an illuminated open space have a choice: They can remain in the tunnel or enter the open field. Mice from some strains stay under cover for a long time; mice from other strains enter the lighted arena without delay. What happens when such strains are crossed? In most cases the hybrid offspring are slower than their impulsive parent but faster than their hesitating parent. In this case inheritance tends to be intermediate rather than heterotic (Bruell, 1965).

These phenomena of heterotic and intermediate inheritance of behavior, and their explanation in terms of genetic and evolutionary mechanisms, form the subject matter of this chapter. While our discussion will center around behavioral heterosis, or its absence, it is hoped that some of the ideas developed here will be of wider interest to students of behavior genetics.

ILLUSTRATIVE RESEARCH

The discovery of heterotic and intermediate inheritance of behavior was a by-product of behavior-genetic research designed to study other problems. So far no one has conducted experiments expressly devoted to an exploration of behavioral heterosis. Toward the end of this chapter, we shall offer some suggestions for the design of future investiga-

[1] Research supported in part by Grant G-14410 from the National Science Foundation, Grant HE-07216 from the National Heart Institute, and Grant HD-02589 from the National Institute of Child Health and Human Development, United States Public Health Service, to Western Reserve University.

tions of heterosis but, at present, we must draw on examples from available though not entirely satisfactory research.

Results of earlier genetic studies (Bruell, 1962, 1964a, 1964b) on the behavior of mice in activity wheels have been mentioned. However, in order to illustrate the design of so-called diallel studies, we shall describe here in some detail an unpublished diallel study of wheel running. Five inbred strains of mice (A/J, Balb/c, C3M, C57BL/10, DBA/1) and the 20 F_1 hybrid groups resulting from the complete intercrossing of these strains (see Broadhurst, Chapter 14; Broadhurst, 1960) were used. Each strain and each group of hybrids was repre-sented by 22 females and 22 males, so that a total of 1,100 mice were tested. The mean running score for each group (scores for females and males were pooled) is shown in diallel Table 13.1. The scores represent revolutions run per hour in an activity wheel having a diameter of 6 inches and thus a running surface of approximately 19 inches (cf. Bruell, 1962, Figure 4.6). The scores for the five inbred strains are shown on the left-to-right diagonal of the diallel table; scores for hybrids are entered above and below the diagonal. For example, the score for F_1 (A/J ♀ X Balb/c ♂) is entered in the right upper corner, and that of the reciprocal hybrid F_1 (Balb/c ♀ X A/J ♂) in the left lower corner of the table. Inspection of Table 13.1 reveals the strong tendency of F_1 hybrids to outdistance their inbred parents in activity wheels.

A graphic representation of these results is found in Figure 13.1. For the strains A/J and C3H and their hybrid offspring F_1 (A/J ♀ X C3H ♂) and F_1 (C3H ♀ X A/J ♂), the average of the parental scores, the so-called midparent, was ½ (1,717 + 1,981) = 1,849. The average of the hybrid scores was ½ (2,404 + 2,280) = 2,342. In Figure 13.1 this hybrid score is plotted against the midparental score; it is indicated by the first solid circle from the left. The two parental scores, 1,717 and 1,981, are indicated by the end points of the vertical bar under the circle marking the score of their hybrid offspring. We thus see at a glance that these hybrids scored outside the range of scores bracketed in by their parents' performances. We see similar results for all other hybrid groups except for F_1 (A/J X Balb/c) and their recip-rocals. In this case the scores shown were 1,717 and 2,358 for the parents, 2,037 for the midparent, and 2,353 for the hybrids.

Table 13.1
Spontaneous running in activity wheels (revolutions per hour) by five inbred strains of mice and their F_1 hybrid offspring

Strain of mother	Strain of sire				
	A/J	C3H	DBA/1	C57BL	Balb/c
A/J	1,717	2,404	2,417	2,394	2,360
C3H	2,280	1,981	2,352	2,638	2,775
DBA/1	2,273	2,415	2,072	2,461	2,744
C57BL/10	2,334	2,497	2,460	2,080	2,540
Balb/c	2,346	2,728	2,684	2,734	2,358

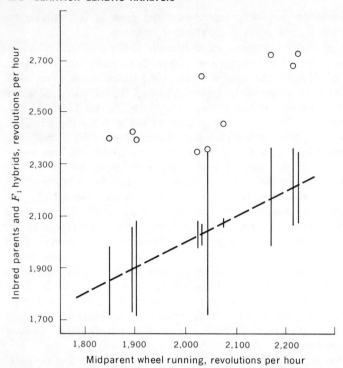

Figure 13.1 Wheel running by inbred and hybrid mice. Scores of hybrids are indicated by solid circles, and those of their inbred parents by the end points of the vertical bars; intersection of dashed line and vertical bar marks the midparental score. Families shown from left to right are A/J X C3H, A/J X DBA/1, A/J X C57BL/10, C3H X DBA/1, C3H X C57BL/10, A/J X Balb/c, DBA/1 X C57BL/10, C3H X Balb/c, DBA/1 X Balb/c, C57BL/10 X Balb/c.

Even without applying statistical tests of significance to the data of Table 13.1 and Figure 13.1 we can see that, at least in this group of inbred strains and hybrids, inheritance of wheel running was heterotic.

Inheritance of exploratory behavior of mice is also heterotic. This was found in an earlier study (Bruell, 1964a) and confirmed by the following unpublished diallel study. The same 1,100 inbred and hybrid mice that had been tested in wheels were used in this study. Each mouse was placed in a four-compartment exploration maze (cf. Bruell, 1962, Figure 4.8), and the number of times the mouse crossed the midline of a compartment was counted photoelectrically. The total number of midline crossings in a 10-minute test period constituted the exploration score of the mouse. The results of this diallel study are shown in Figure 13.2. Most groups of hybrids outperformed their inbred parents; inheritance of exploratory behavior was heterotic.

Quite similar results were obtained by Collins (1964) who conducted a 5×5 diallel study (total $N = 500$) of avoidance conditioning in

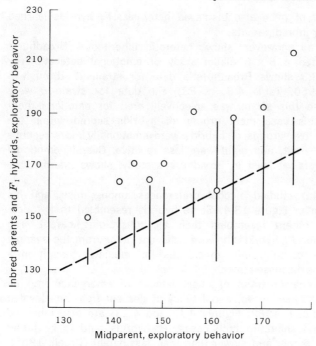

Figure 13.2   Exploratory behavior of inbred and hybrid mice. For explanation of symbols, see Figure 13.1.

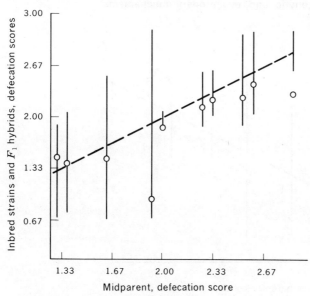

Figure 13.3   Emotional defecation by inbred and hybrid rats. For explanation of symbols, see Figure 13.1. (*Data adapted from Broadhurst, 1960.*)

mice. This trait of mice also displayed heterosis: F_1 hybrids learned faster than their inbred parents.

Not every trait, however, shows heterotic inheritance. Broadhurst (1960) conducted a 6 × 6 diallel study of emotional defecation by rats. Figure 13.3 shows Broadhurst's data for strains 1 through 5 (Broadhurst, 1960, Table 4.8, p. 83); the data for strain 6 were omitted because this strain was selectively bred for emotional non-reactivity. In this case most groups of hybrids scored within the parental range; two groups of hybrids were emotionally more reactive than their midparent, and eight were less reactive. Overall, emotional defecation by rats, at least in Broadhurst's study, shows intermediate inheritance.

Fuller (1964a) studied alcohol preferences among mice, and his data are shown in Figure 13.4. Some hybrids resembled their "alco-holic" parent; others resembled their alcohol-avoiding parent. The average score for F_1 hybrids, however, did not differ from the average score for their parents; inheritance was intermediate, though in a special way to be discussed later.

Two other behavior traits of mice, latency of emergence into an open field, mentioned above, and latency of descent from an elevated platform (cf. Bruell, 1962, Figures 4.11 and 4.13) are also inherited in intermediate fashion. In both cases, F_1 hybrids tend to be faster than their slow parent and slower than their fast parent (Bruell, 1965). To understand the results of these and similar nonbehavioral studies (e.g., Bruell, 1963, on inheritance of serum cholesterol level in mice; Bruell, 1964a, on inheritance of hematocrit percent in mice), we must now consider genetic and evolutionary mechanisms.

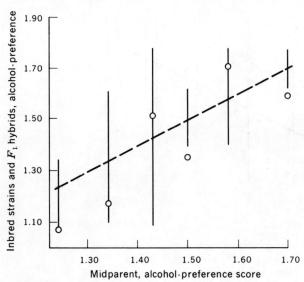

Figure 13.4   Alcohol preferences among inbred and hybrid mice. For explanation of symbols, see Figure 13.1. (*Data adapted from Fuller, 1964a.*)

GENETIC MECHANISMS

The theory of quantitative genetics is discussed in two other chapters of this volume by R. C. Roberts and P. L. Broadhurst. Therefore, this discussion can be limited to four special topics: We shall ask how, in theory, the genetic constitution of a laboratory population of inbred strains differs from the genetic constitution of the natural population from which the strains originated; we shall examine when, according to theory, the mean genotypic value of a quantitative trait in a population of inbred strains will differ from that of the same trait in the wild base population; we shall discuss the theory of inbreeding depression and heterosis; and we shall indicate how these theoretical considerations relate to the research presented above.

Let us compare first the genetic constitution of a population of inbred strains, S, with that of the original random-breeding population, R. To start with the simplest case, suppose we dealt with a monogenic trait, that is, a trait determined by genes at one locus $(l = 1)$, and suppose there existed only two allelic genes, A_1 and A_2, at that locus; we assume also that, in population R, allele A_1 occurred with frequency p, and A_2 occurred with frequency q, where $p + q = 1$. If all this were so, both males and females of population R would produce p A_1 gametes and q A_2 gametes, and random mating within the population would result in p^2 A_1A_1, $2pq$ A_1A_2, and q^2 A_2A_2 genotypes, as shown in Table 13.2 for the special case where $p = q = \frac{1}{2}$.

Let us now suppose that individuals drawn at random from population R were used in a program of inbreeding. After several generations of intensive inbreeding (Fuller and Thompson, 1960, p. 82, Figure 3.9) we would have two kinds of homozygous lines or strains: A_1A_1 strains and A_2A_2 strains. Since in population R, genes A_1 and A_2 occurred with frequency p and q, we would expect that our population of inbred strains S would consist of p A_1A_1 strains and q A_2A_2 strains.

When we compare population R (all of Table 13.2) with population S (left-to-right diagonal of Table 13.2) we see that the homozygous genotypes of S form a subset of the universal set of genotypes R.

Table 13.2 deals with a monogenic trait $(l = 1)$. Table 13.3 represents the case of a digenic trait, that is, a trait determined by genes at two loci $(l = 2)$. Again populations R and S are found in the same table, and again we see that S, on the left-to-right diagonal of the table, constitutes a subset of R.

Inspection of Table 13.2 indicates that population S contains p A_1 genes and q A_2 genes, and so does population R. Table 13.3 shows

Table 13.2
Random-breeding population R and population S ($l=1$; $p=q=\frac{1}{2}$)

Female gamete	Male gamete	
	A_1	A_2
A_1	A_1A_1	A_1A_2
A_2	A_1A_2	A_2A_2

that population S carries p A_1, q A_2, p B_1, and q B_2 genes, and so does population R. Thus, in theory, the genic constitution of population S does not differ from that of population R; but, most obviously, the genotypic makeup of the two populations does differ.

Populations of inbred strains are produced by inbreeding, that is, by systematic interference with random mating among animals drawn from a natural population R. What would happen if the restrictions on random mating were removed, that is, if inbred animals in population S were free to mate with any other animal in that population? Since, as we saw, the genic constitution of population S does not differ from that of R, such random mating could be represented in the same way as we represented random mating in population R in Tables 13.2 ($l = 1$) and 13.3 ($l = 2$). Looking at it this way, we see that one generation of random breeding within population S would reconstitute the original natural population R.

We now turn to the second question we proposed to discuss in this section: Why and when does M_S, the mean genotypic value of a trait in a population of inbred strains, differ from M_R, the mean genotypic value of the same trait in population R from which the inbred strains derived? To compute M_S and M_R we must adopt a system of denoting the genotypic values of homozygous and heterozygous gene pairs. Following R. C. Roberts (Chapter 11) and Falconer (1960, p. 113), we denote the genotypic values of homozygotes A_1A_1 and A_2A_2 by a_a and $-a_a$. In analogous fashion the genotypic values of B_1B_1 and B_2B_2, and C_1C_1 and C_2C_2 are designated by a_b and $-a_b$, and a_c and $-a_c$. These genotypic values of pairs of homozygotes, for example, A_1A_1 and A_2A_2, are measured on a scale that has its zero point midway between them; thus, values a_a and $-a_a$ are deviations from the mid-homozygote value (see Genotypic Value Scale, page 219; Falconer, 1960, Figure 7.1; Bruell, 1962, Figure 4.3).

The genotypic value of heterozygous gene pairs is indicated by the letter d with a subscript. For example, the genotypic values of A_1A_2, B_1B_2, and C_1C_2 are symbolized by d_a or $-d_a$, d_b or $-d_b$, and d_c or $-d_c$. The value of heterozygotes also is measured on a scale having its zero point midway between the corresponding homozygotes. If on an absolute scale the genotypic values of A_1A_1, A_1A_2, and A_2A_2 were 20, 18, and 10, respectively, then, on our scale, the corresponding values would be $a_a = 5$, $d_a = 3$, and $-a_a = -5$.

Table 13.3
Random-breeding population R
and population S ($l = 2$; $p = q = \frac{1}{2}$)

Female gamete	Male gamete			
	A_1B_1	A_1B_2	A_2B_1	A_2B_2
A_1B_1	A_1A_1 B_1B_1	A_1A_1 B_1B_2	A_1A_2 B_1B_1	A_1A_2 B_1B_2
A_1B_2	A_1A_1 B_2B_1	A_1A_1 B_2B_2	A_1A_2 B_2B_1	A_1A_2 B_2B_2
A_2B_1	A_2A_1 B_1B_1	A_2A_1 B_1B_2	A_2A_2 B_1B_1	A_2A_2 B_1B_2
A_2B_2	A_2A_1 B_2B_1	A_2A_1 B_2B_2	A_2A_2 B_2B_1	A_2A_2 B_2B_2

It is often convenient to express d_i, the value of heterozygous gene pairs, in terms of a_i, the value of the corresponding homozygotes. If this is done, several logical possibilities exist (Falconer, 1960, p. 113; Bruell, 1962, p. 50).

1 *Zero dominance*: The value of the heterozygote, for example, A_1A_2, corresponds to the average of the two homozygotes A_1A_1 and A_2A_2: $d_a = \frac{1}{2} (a_a - a_a) = 0$.

2 *Partial dominance of A_1 over A_2*: $0 < d_a < a_a$.

3 *Dominance of A_2 over A_1*: $d_a = a_a$.

4 *Overdominance*: $d_a > a_a$.

(Obviously, partial dominance of A_2 over A_1: $-a_a < d_a < 0$; dominance of A_2 over A_1: $d_a = -a_a$; and negative overdominance: $d_a < -a_a$, are also possible.)

Since d_i arises from the property of dominance among alleles at a locus (Falconer, 1960, p. 122), it is referred to as the "dominance deviation."

In the absence of dominance, when $d_i = 0$, we speak of additive gene action because, in this case, it is as if allelic genes had effects that added up when the genes combined to form homozygous or heterozygous gene pairs. To illustrate, if the effect of A_1 were $\frac{1}{2} a$ and that of A_2 were $-\frac{1}{2} a$, then, by addition, we could arrive at the genotypic values of A_1A_1: $\frac{1}{2} a + \frac{1}{2} a = a$; A_2A_2: $-\frac{1}{2} a - \frac{1}{2} a = -a$; and A_1A_2: $\frac{1}{2} a - \frac{1}{2} a = 0$.

Using this notation, let us first compute M_S and M_R for populations S and R of Table 13.2. Population S consists of genotypes A_1A_1 and A_2A_2, so that $M_S = \frac{1}{2} (a_a - a_a) = 0$. Population R consists of genotypes A_1A_1, $2 (A_1A_2)$, and A_2A_2, so that $M_R = \frac{1}{4} (a_a + 2d_a - a_a) = \frac{1}{2} d_a$.

$$M_S = 0 \qquad l = 1; p = q = \frac{1}{2} \tag{13.1}$$
$$M_R = \frac{1}{2} d_a \qquad l = 1; p = q = \frac{1}{2} \tag{13.2}$$

Analogous computations, applied to the populations shown in Table 13.3, lead to the following equations:

$$M_S = 0 \qquad l = 2; p = q = \frac{1}{2} \tag{13.3}$$
$$M_R = \frac{1}{2}(d_a + d_b) \qquad l = 2; p = q = \frac{1}{2} \tag{13.4}$$

We computed M_S and M_R for monogenic ($l = 1$) and digenic ($l = 2$) traits. Generalizing to polygenic traits, we see that

$$M_S = 0 \qquad p = q = \frac{1}{2} \tag{13.5}$$
$$M_R = \frac{1}{2}(d_a + d_b + \cdots + d_l) \qquad p = q = \frac{1}{2} \tag{13.6}$$
$$= \frac{1}{2}\Sigma d_i$$

Equations (13.5) and (13.6) provide a general answer to the question asked: M_R, the genotypic mean of a random-breeding population R, will differ from M_S, the genotypic mean of a population of inbred strains S, when Σd_i, the sum of dominance deviations, differs from zero. This, in turn, is possible only when most dominance deviations have had the same sign: Most of them must be either positive or negative; $|\Sigma d_{i(-)}|$ must not equal $|\Sigma d_{i(+)}|$. In the population as a whole, dominance must have direction. With "directional dominance" (Falconer, 1960, p. 213) M_R will differ from M_S.

M_R will *not* differ from M_S under two conditions: It will not differ when, in the population as a whole, dominance deviations have no direction; when positive and negative dominance deviations are equally frequent; when $|\Sigma d_{i(-)}|$ equals $|\Sigma d_{i(+)}|$. And M_R will *not* differ from M_S when, at all relevant loci, dominance is absent, that is, when $d_a = 0$, $d_b = 0$, $d_c = 0$, and so on. Stated differently, M_R will equal M_S when at all loci genic action is additive.

To summarize, three situations can be distinguished:

1   *Zero dominance*: $d_i = 0$ at all loci. In this case $M_R = M_S$.

2   *Balanced or random dominance*: $|\Sigma d_{i(-)}| = |\Sigma d_{i(+)}|$. Here too $M_R = M_S$.

3   *Directional dominance*: $|\Sigma d_{i(-)}| \neq |\Sigma d_{i(+)}|$. Only under these conditions does M_R differ from M_S.

These findings shed light on the problem of "inbreeding depression." When, through inbreeding, population R changes into population S, M_R changes into M_S. If, because of directional dominance, M_R differs from zero, the mean of the population subjected to inbreeding will change; it will approach zero, and one will observe inbreeding depression. If, on the other hand, directional dominance is absent in population R, inbreeding will not affect the mean of the population undergoing inbreeding.

Heterosis is inbreeding depression in reverse. We have seen that the genic constitution of population S does not differ from that of population R. If the members of population S are crossbred and population R is restored, M_S will change into M_R. However, the value of M_R will differ from that of M_S only when, in the population as a whole, directional dominance prevails. Only then will we observe heterosis, that is, a difference between M_S and M_R, the mean of the reconstituted population R.

These theoretical considerations have an obvious bearing upon the research presented in the first part of this chapter. They provide a rationale for the particular design of those studies, and they explain, to a degree, their findings.

The similarity between Table 13.1, a diallel table, and Tables 13.2 and 13.3 containing the genotypes of populations S and R (for $l = 1$ and $l = 2$) is evident. In a diallel study the experimenter completely intercrosses N inbred strains to reconstitute, in the laboratory, population R, the wild parent population of his strains. Obviously, in practice this goal is never achieved. First, it is improbable that the founders of the N inbred strains studied carried a fully representative sample of all trait-relevant genes of population R and that, during inbreeding, no genes were lost and no relative gene frequencies were altered. And

then, even with only two alleles per locus, the number of possible inbred strains increases rapidly as l, the number of loci influencing the
trait, increases. When $l = 2$, the number of possible inbred strains
$S = 2^l = 4$; with $l = 10$, the number of $S = 2^{10} = 1,024$. It is unlikely
that anyone will ever embark on a $1,024 \times 1,024$ diallel study.

The inbred strains of a diallel study are a sample of the universe of
possible inbred strains S. If the sample is representative, results of a
diallel study permit certain inferences about population S and population R. If we observed heterosis, we can hypothesize that in the wild
parent population $|\Sigma d_{i(-)}|$ differed from $|\Sigma d_{i(+)}|$; we can infer direc-
tional dominance. And if we observed intermediate inheritance, we can
assume that in the wild base population dominance was absent or
random.

Let us call the totality of genes affecting a given trait and carried by
a wild population the gene pool of that trait. The observation that
heterosis is trait-specific suggests that the characteristics of gene
pools differ from trait to trait. How can we account for these differences?
Genetic theory alone does not provide an answer. We must turn to evolutionary theory for an explanation.

EVOLUTIONARY MECHANISMS

This chapter deals with quantitative behavior traits, that is, traits that
permit one to place individuals on a continuum: Some individuals occupy a low, others an intermediate, and still others a high position on
the continuum. Presumably most quantitative behavior traits, like other
phenotypic characters of organisms, are subject to natural selection.
Selection may favor individuals scoring high on a trait continuum. Then
again, depending on the trait and the particular environmental conditions, it may favor low-scoring individuals, or just the average individual.

We must assume that selection for extreme phenotypes affects the
gene pool of a trait in other ways than selection for intermediate types,
and we must consider the probable differing characteristics of gene
pools that were shaped by one or the other kind of selection. However,
our discussion would not be complete if we neglected to consider also
the probable attributes of gene pools of "neutral" traits, that is, traits
that escaped selection because, under the prevailing environmental
conditions, they neither increased nor decreased the fitness of their
bearers.

In discussing the three types of gene pools we shall use terms not
introduced previously. We shall speak of trait-increasing "plus genes"
and trait-decreasing "minus genes"; it should be understood that, on an
absolute scale, both plus and minus genes may increase a response
tendency of their bearer. Only relative to each other is the gene with the
larger effect a plus gene, and that with the smaller effect a minus gene.
We shall also speak of dominant plus and minus genes. These terms
refer to genes that, in heterozygous gene pairs, shift the genotypic
value of the heterozygote above or below the mid-homozygote value.
Dominant genes, then, are those responsible for positive, $d_{i(+)}$, and

negative, $d_{i(-)}$, dominance deviations. Genes without dominance, $d_i = 0$, act additively.

We shall consider first gene pools of selectively neutral traits. Traits are response tendencies that manifest themselves under the proper environmental circumstances. For example, the tendency of some mice to drink and of others to avoid alcohol probably became manifest for the first time in the laboratories of psychologists. Alcohol does not flow in the natural environment of mice. Yet—and this must be noted and stressed—behavior-genetic analyses (Rodgers and McClearn, 1962; Fuller, 1964a) show that genes influencing alcohol consumption existed in mice, although, under natural conditions, they are neither adaptive nor maladaptive. Gene pools of traits arise through chance events of mutation; they are shaped by selection if, under fortuitous environmental "test conditions," their gene content becomes manifest. What, then, would be the content of "untested" gene pools, which in the past were not challenged by environmental forces? There seems to be only one logical answer to this question: Since genes arise by chance, untested and unchallenged pools presumably are filled with a random assortment of plus and minus genes, genes with additive action and genes with trait-increasing and trait-decreasing dominant action.

Next to be examined are the hypothetical consequences of selection for extreme phenotypes, that is, selection favoring individuals on one side of a trait continuum, either the high side or the low side, and discriminating against individuals at the other side. We consider genes with additive action first and imagine an allelic series consisting of alleles A_1, A_2, A_3, and A_4. In this series A_1 has the largest effect, A_2 a lesser effect, and so on. If selection were for extreme plus types, alleles A_4, A_3, and A_2 would be eliminated in that order, because, with additive gene action, even heterozygote A_1A_2 would be less extreme than homozygote A_2A_2. Allele A_1 would become fixed in the population; it would attain a frequency $p = 1$. "Additive variation" would have ceased to exist at locus A.

This obviously poses two problems. First, if we observed a population only after fixation of A_1 had occurred, we would in fact have no way of knowing that A_1 existed; variation is the mother of genetics. Second, and this point is more important, in reality we do observe additive variation, and if selection proceeded as just described, there should be none. How then is additive variation preserved? It is preserved, in part, through newly arising mutations but probably in a more significant way through the phenomenon of pleiotropy (Greek: $pleion$ = many, $trop\bar{e}$ = change). Most if not all genes affect more than one trait, and whether a gene is preserved or eliminated by selection depends on its overall adaptive value; it may be a minus gene in one of its phenotypic manifestations but a plus-plus gene in others, and so it persists. On the whole, however, selection for extreme types will eliminate most genes with additive action; it will keep additive variation at a low level. And since the severity of the selection pressures to which a trait is exposed is a direct function of its adaptive significance, traits most closely related to fitness will be left with the least amount of additive variation (Falconer, 1960, chap. 20).

How, in theory, would selection for extreme phenotypes affect dominant genes? The answer is: in essentially the same way as it affects additive genes, except that reduction of genetic variability would proceed at a slower rate (Falconer, 1960, chap. 2). If selection favored trait-increasing genes, dominant minus genes presumably would be eliminated most rapidly, but recessive minus genes would not escape selection. Dominant plus genes would be the only ones to persist. If A_1 were a dominant plus gene, and A_2 its recessive minus allele, A_2 would be exposed to selection each time it appeared in homozygous condition. The frequency of A_2 would thus tend to decline and lead to its eventual extinction, unless it were saved by an extraneous factor such as pleiotropy. Without extraneous "help," variation at a given locus would be preserved only in the case of heterozygote superiority, that is, overdominance $[d_{i(+)} > a_{i(+)}]$.

Finally, let us inquire into the hypothetical effects of selection for intermediate phenotypes. Presumably, in this case, dominant plus and minus genes would be eliminated first because of their centrifugal effects: They tend to shift genotypic values away from any central value. Extreme minus and extreme plus genes with additive effects would be eliminated for the same reason. But a residue of additive variation could persist at a locus: If the value of A_1A_2 were optimal, such heterozygote superiority would tend to protect both A_1 and A_2 from extinction. On the other hand, with selection for intermediates, gene pairs displaying true overdominance $[d_{i(+)} > a_{i(+)}]$ would be selected against because of their centrifugal effects.

Before summarizing our discussion, it should be stressed that, strictly speaking, the preceding applies only to local populations, that is, communities of potentially interbreeding individuals (Mayr, 1963, pp. 136ff.). Only such individuals share in a single gene pool, and to apply this term to the universe of genes found in a species does not make sense. The optimal position on a trait continuum depends on the environment to which a population is exposed, and so this local environment shapes the gene pool of interbreeding individuals. The characteristics of gene pools of populations separated in time and space will differ. This is a very important point, and we shall return to it later.

Keeping in mind that we speak of gene pools of local populations, we can sum up what was said so far: The gene pools of traits that have not been exposed to selection are likely to contain the whole gamut of possible gene types. Gene pools of traits selected for extremes contain genes contributing to additive variation. They also contain dominant genes, but only of one kind; depending on whether high-scoring or low-scoring phenotypes were favored by selection, the pool will contain only dominant plus or dominant minus genes. Finally, gene pools of traits with an intermediate optimum contain only genes with additive action; selection will have eliminated dominant plus and dominant minus genes.

To characterize the three types of gene pools, the concept of symmetry can be applied. Selection for extremes leads to asymmetrical gene pools; they contain only one type of dominant genes, plus dominants or minus dominants. Selection for intermediates leads to symmetrical gene pools; dominant genes are eliminated, and at loci where additive variation is preserved, each plus gene has its minus allele. Symmetry characterizes also the gene pools of neutral traits, but in this case the symmetry is one of randomness. Such pools contain dominant plus and dominant minus genes, while the symmetrical pools of intermediate traits contain no dominants.

We are now ready to integrate the experimental and theoretical material presented thus far. First we presented experimental data on heterotic inheritance of wheel running and exploratory behavior of mice, as well as data on intermediate inheritance of emotional defecation by rats (Broadhurst, 1960) and alcohol preferences among mice (Fuller, 1964a).

Next, we deduced from genetic theory that heterotic inheritance reflects an asymmetry in the gene pool underlying a trait $[|\Sigma d_{i(+)}| \neq |\Sigma d_{i(-)}|]$, while intermediate inheritance points to two kinds of symmetry in the gene pool: symmetry due to the absence of dominant genes, or symmetry due to balanced dominance $[|\Sigma d_{i(+)}| = |\Sigma d_{i(-)}|]$.

Finally we learned that each kind of gene pool points to a different history of selection: Selection for extreme phenotypes introduces asymmetry in a gene pool; it eliminates minus dominants from it and preserves plus dominants, or vice versa. Selection for intermediate phenotypes eliminates all dominant genes, plus dominants and minus dominants, from the pool. And absence of selection in the case of neutral traits, traits without adaptive value, preserves all types of genes in the pool, including plus and minus dominants, thus resulting in "symmetry of randomness."

Closing the circle of our argument, we can now hypothesize that activity level of mice—measured in activity wheels—and exploratory behavior of mice—measured in a novel environment, a maze—were subjected to selection favoring the most active and most curious animals. We can also hypothesize that emotionality of rats—gauged by emotional defecation in a strange environment—was subjected to selection for intermediate phenotypes: Neither the cowards nor the foolhardy survived. Finally, we can hypothesize that alcohol preferences of mice—measured in a situation where mice could choose between various concentrations—had not been exposed to selection in the evolutionary past of the species. Fuller (1964a) observed intermediate inheritance of alcohol consumption but a special kind of intermediate inheritance. There were indications of positive dominance in some strain crosses, and negative dominance in others, and overall intermediacy was attributable to a balance between the two. This is the picture one would expect in the case of traits that had not been exposed to selection; one would not be surprised if alcohol consumption by mice were such a trait.

If the reader is not convinced by the argument, he shares the misgivings of the writer. The argument may be logically correct, but it is based on insufficient data. If we want to know more about behavioral heterosis and behavioral inbreeding depression, we must conduct studies specifically designed to investigate these phenomena.

DESIGN OF FUTURE RESEARCH

None of the experiments summarized in this chapter was conceived or conducted to study behavioral heterosis or inbreeding depression. Thus all of them have shortcomings that can and should be avoided in future research. They all study only heterosis and *infer* inbreeding depression, and they all suffer from inadequate sampling of strains, populations, species, and behavior traits.

Most behavior-genetic studies suffer from an inadequate sampling of strains, and those reported here are no exception. We chose diallel studies for illustration because, in the case of such studies, the connection between genetic theory (cf. Tables 13.2 and 13.3) and experimental design (cf. Table 13.1) becomes particularly obvious. This should not obscure the fact that diallel studies, unless conducted on a much larger scale, are thoroughly inadequate for our purposes. A 5×5 diallel study samples 5 inbred and 10 hybrid genotypes, although inclusion of reciprocal hybrids increases the latter number spuriously to 20. But even if the trait studied were determined, in a decisive way, by only two alleles at each of five loci ($l = 5$), the number of possible inbred genotypes would be 32 (2^l), and the number of possible hybrids would be 496 $[\frac{1}{2} (2^{2l} - 2^l)]$. The complete intercrossing of N strains practiced in diallel studies, requiring as it does N^2 test groups, is wasteful. A more economical and more representative design would increase N, the number of inbred strains, and test only randomly chosen crosses between them. For example, at the price of a 5×5 diallel study, one could test 10, instead of 5, inbred strains and 15, instead of 10, F_1 hybrids. Studies moving in this direction but falling far short of the goal of representativeness have been reported elsewhere (Bruell, 1964a, 1964b; 1965).

Inbreeding depression is, in theory, the progenitor of heterosis. Whenever we observe heterosis we can assume that it was preceded by inbreeding depression. But this assumption—and in the case of behavior it is no more than that—should not be permitted to enter behavior-genetic literature without being subjected to an empirical test. Collins (1964) observed heterotic inheritance of avoidance conditioning in mice, and we described heterosis of wheel running and exploratory behavior. Hence we assume that learning, activity level, and curiosity of mice were depressed by inbreeding. To a degree this hypothesis is testable, and a study of R. C. Roberts (see Falconer, 1960, p. 255) could serve as an experimental paradigm. Roberts studied litter size in a random-bred population of mice and then observed changes in litter size during three generations of inbreeding and subsequent crossbreeding of 30 lines taken from that population. The expected depression of litter size on inbreeding and heterotic recovery on crossbreeding occurred. Behavior-genetic analyses modeled after Roberts' experiment would be economically feasible with many species. Without such studies many behavior-genetic interpretations would retain the flavor of the clinical postdiction which brought at least some branches of Freudian psychology into disrepute. But even if such studies were conducted they would not get to the core of the problem.

We pointed out that it does not make much sense to speak of the gene pool of a species. Thinking and experimentation must center around gene pools of local interbreeding populations for reasons discussed extensively by Mayr (1963, chap. 7). The inbred strains of most laboratory animals were not created with this in mind. Thus one cannot hope that strains chosen at random from those available through dealers would carry a representative sample of genes from any gene pool shaped by the forces of a common environment. The wild base populations from which the ancestors of our laboratory strains were drawn are shrouded by the mist of incomplete records and thus, in most cases, we do not know anything about the natural environmental conditions under which the founders of our strains evolved. This reduces inferences from breeding experiments about "the natural population" or "the natural environment" of our laboratory strains to the status of speculations. To illustrate the problem, what could one possibly hope to learn about the genetics of nest building and the evolutionary forces that contributed to it, if he experimented with strains that had among their ancestors an unknown mixture of individuals from hot and cold climates (cf. King, Maas, and Weisman, 1964)?

The remedy here seems to lie in a departure from current practice. Inbred strains have their important place in behavior-genetic research; they provide the best available estimates of environmental variance (R. C. Roberts, Chapter 11, page 225; Falconer, 1960, p. 130); because of their uniformity they are ideally suited for pilot studies designed to develop new tests and measuring techniques; and, as illustrated in this chapter, they may help one to develop tentative hypotheses about heterotic and intermediate inheritance of certain traits. But to study behavioral inbreeding depression and heterosis in a biologically meaningful way, it seems one should avoid inbred strains.

Work will have to start with proper samples of individuals drawn from local wild populations which evolved under distinct and well-known environmental conditions. Behavior traits of these individuals will have to be measured and changes in population means occurring during inbreeding and subsequent crossbreeding will have to be observed. This would add one dimension to Roberts' approach described above; while Roberts started with a random-bred population of mice, they were not wild mice in the true meaning of the word, and their natural environment remained unknown. But such knowledge appears to be essential. Obviously, to obtain insight into the modeling forces of the environment, it will not be enough to work with the descendants of one local population. Founder animals will have to be drawn from several distinctly different environments, and, while keeping animals of differing origin separate, parallel inbreeding and crossbreeding experiments will have to be conducted.

Research on behavioral inbreeding depression and heterosis certainly suffers from an inadequate sampling of species, genera, even phyla. In the past this may have resulted from the assumption that such research depended on the availability of inbred strains, but, as the preceding discussion indicates, such considerations must not stop us. On the contrary, progress now depends on work with wild animals, a wide variety of them, studied in as many biologically meaningful situations as possible.

Investigation of forms of behavior chosen on the basis of evolutionary criteria is probably the most critical requirement of future behavior-genetic analyses. In the past we often tended to study only those forms of behavior that were conveniently measured; our approach was test-oriented. This was all right at the start of our quest, but the time has now come to attend more to "the structure of a life as it is lived" (Allport, 1966). In the future, before embarking on genetic research, we should do our best to become well acquainted with the environment and the life as it is lived by the animal we study. Only such intimate knowledge will ensure that we select for detailed genetic analysis behavior whose adaptive significance, or lack of it, we understand. In this chapter we hypothesized, after the results of breeding tests were known to us, that certain forms of behavior had and others did not have adaptive significance. We also "postdicted" that in some cases selection had favored extreme phenotypes and that in others it had favored intermediates. In the future we should strive to state in advance of breeding tests what the tests will show, since healthy scientific inquiry depends on the corrective, restraining, and often sobering lessons of unconfirmed predictions. But to be useful, predictions must not be simply products of intuition; they must derive from a thorough knowledge of the animals with which we work.

In this final section we have looked ahead and offered suggestions for future research on inbreeding depression and heterosis. We suggested a wider sampling of existing inbred strains and their hybrids. This would increase our confidence in the generality of our findings. For example, we cannot be certain that Fuller's (1964a) results on intermediate inheritance of alcohol preferences among mice would hold if a larger sample of strains were tested. If some reader felt that Fuller's data (Figure 13.4) did not differ sufficiently from Broadhurst's data (Figure 13.3) to justify differing interpretations, it would be hard to argue the point.

We suggested that inbreeding be combined with crossbreeding experiments. We cannot very well continue to infer inbreeding depression from the occurrence of heterosis. For example, would wild mice actually learn faster than inbred mice, as Collin's (1964) study suggests, and would learning ability, or some other correlated ability, deteriorate as inbreeding progresses? Supposedly (Falconer, 1960, chap. 20) behavior traits of the highest adaptive value suffer most during inbreeding. We all have intuitive notions about what behavior is adaptive for an organism; inbreeding experiments could provide a proving ground for our intuition. The goal would be to sharpen our eye for characteristics of adaptive behavior.

Finally, we suggested the need for ecological and ethological studies as preparation for behavior-genetic research. To go beyond intuition and to understand the adaptive significance of a behavior, we must know the environmental conditions under which it occurs and how it varies as environmental conditions vary. A complete inventory, an ethogram, of the natural behavior of a species in its natural habitats should, in time, become a prerequisite of behavior-genetic investigations. How fruitful such an approach can be, with how many new insights it can provide us, was brilliantly illustrated by Scott's and Fuller's (1965) recent monograph *Genetics and the Social Behavior of the Dog*, a landmark of behavior-genetic literature.

All this will require a new breed of behavior geneticist: His training will have to be in zoology, ecology, ethology, quantitative genetics, and experimental psychology; and his outlook must be that of an evolutionary biologist. That much we hope to have shown. Although, ostensibly, we set out to discuss behavioral heterosis, between the lines we hope to have demonstrated that, in essence, our work—the work of behavior geneticists—is concerned with problems of evolutionary biology. The animals with which we work are products of evolution; they hold the key to the past. We must learn how to find it.

CHAPTER FOURTEEN
AN INTRODUCTION TO THE DIALLEL CROSS[1]
Peter L. Broadhurst

INTRODUCTION

The diallel cross is a way of using the classic method of genetics, crossbreeding of strains or varieties of a species, in such a manner as to maximize the information obtainable. Although the method has long been extant, the analysis has only recently been perfected within the general framework of biometrical genetics, with its techniques for the analysis of quantitative variation into its various components. It is the purpose of this chapter to introduce students of behavior to the method and the analysis and to indicate its possible utility in some problems that are encountered in this rapidly developing field. While the experimental procedures involved are straightforward, the analysis has its complexities which need to be considered carefully.

THE METHOD

The classic method of experimental genetics has been the hybridization of different strains or varieties of species. Abbé Mendel, just as the English horticulturist Thomas Knight did before him (Burt and Howard, 1956), crossed varieties of garden peas though, unlike his predecessor, he expressed his observations in numerical form and thereby demonstrated the essential truth of the hypothesis of particulate or, as we should say nowadays, "genic" inheritance. From our vantage point of

[1] Acknowledgments are due to Prof. J. L. Jinks for his criticism. The preparation of this chapter was supported in part by USPHS Research Grant MH-08712 from the National Institute of Mental Health, U.S. Public Health Service.

67 years after the rediscovery of Mendel's experiments and at a time when the importance of polygenic determination of behavior is at last coming to be appreciated, it is interesting to note that Mendel himself also anticipated the importance of compound characters and indeed suggested a mechanism, albeit imprecise, to explain the graded coloration of the crosses he made between white and colored varieties of peas.

The Mendelian method, as we know it, is based upon the crossing of individuals of known genotype which differ in phenotype. The products of such a cross (usually designated $P_1 \times P_2$) constitute the first filial generation (F_1); the offspring of F_1 parents, the second filial generation, or F_2. Crossing F_1 individuals with the parental strain is back-crossing; $F_1 \times P_1$ is designated B_1, and $F_1 \times P_2$ is designated B_2. Further mating possibilities are recognized as contributing still more information, but some of them are limited to self-fertilizing plants and are consequently not applicable with bisexual organisms. This is the standard genetic analysis as it has developed over the years. In its complete form it has rarely been applied to behavioral characteristics. Prior to 1959, Broadhurst and Jinks (1961, 1963) were able to find reported in the literature only four sets of this kind of data that were complete enough to make worthwhile reanalysis by the appropriate techniques of biometrical genetics. Three further sets which omitted either the backcrosses or the F_2 were found, and others have since come to notice (e.g., Foster, 1959; McGaugh, Westbrook, and Burt, 1961; McGaugh, Westbrook, and Thomson, 1962; Fuller, 1964b; Manosevitz, 1965) as well as some complete sets (e.g., McClearn, 1961; McClearn and Rodgers, 1961).

The results reported in our reanalyses were of sufficient interest to warrant a cautious optimism regarding the application of methods of quantitative analysis to behavioral data, especially since they were not collected from experiments designed for the type of analysis we applied. However, it is not proposed to discuss this type of design, with F_1, F_2, and/or backcrosses, further here; the interested reader is referred to the reviews cited for a fuller exposition of the biometrical analysis of the data derived from successive crossbred generations and to other sources for alternative approaches (for example, Bruell, 1962; Falconer, 1960; Scott and Fuller, 1965).

Instead, we shall note some of the limitations of this type of design in contrast with the diallel-cross method. In order to give point to such a discussion, it is necessary first to acquaint the reader with the fundamentals of the diallel cross. Perhaps the best way of doing so is to consider its origin. Schmidt (1919) probably originated the diallel cross, and he gave it the alternative name of the method of complete inter-crossings. In many ways this is a more informative term than diallel, since it specifies exactly what is done. Each strain is crossed with every other one in every possible combination: Given six strains of rats, A, B, C, D, E, and F, A would be crossed with B, with C, and so on through F; then B with C, with D, etc.; then C with D and so on; until the half of the diallel matrix shown in Table 14.1 above the leading diagonal, AA, BB, CC, etc., is completed. But each of the crosses listed above has its reciprocal; for example, if A (female) is mated with B (male), there is the other possibility of A (male) being mated with B (female). These matings are shown in the part of the matrix (Table 14.1) below the leading diagonal, and it is this feature that gives the name diallel to the table and the method. The word is derived from the Greek and means twice the complementary, referring to the presence of the two reciprocals. The entry of the parental genotypes along the diagonal completes the table, there being no reciprocals in these cases since both parents come from the same strain. Note that each row (or column, for that matter) contains only offspring from a single parental strain, crossed with each and every other strain in the diallel in turn, including itself. In the arrangement shown in Table 14.1, mothers generate rows, and fathers columns.

We now define the table formally. A diallel table is a square arrangement of the measures (means and variance) of the progeny of a diallel cross, that is, a set of all possible matings between several genotypes. It consists of n^2 measurements, where $n =$ number of genotypes, and comprises $n(n - 1)/2$ entries above, and $n(n - 1)/2$ below the diagonal which itself contains n measures.

It is important to notice that the progenies constituting the diallel cross are F_1s of all the possible crosses from the parental strains, indicated, for convenience, along the top and left-hand edges of Table 14.1. The parent strains are represented in the body of the table by the values on the leading diagonal. Thus the process of generating a diallel table is simplicity itself; F_1s are bred from every strain in turn with every other, and representative parental values are added. The latter need not be the values for the actual parents used in making the crosses but may be measures from representatives of the parental lines which are contemporary with the rest of the table, i.e., the F_1s.

Let us now contrast the diallel cross with the traditional analysis depending upon filial and backcross generations. The diallel table is a table of F_1 measures from which an amount of information about the genetic determination of the characters being measured in the strains used can be obtained by the analyses to be discussed later. The experimentation can stop at this point, though analyses using measures from

Table 14.1
Diagram of a diallel table, with an indication of the genetic constitution of the F_1 offspring of which it is comprised

		Strain of father					
		A	B	C	D	E	F
	A	AA	AB	AC	AD	AE	AF
	B	BA	BB	BC	BD	BE	BF
Strain of	C	CA	CB	CC	CD	CE	CF
mother	D	DA	DB	DC	DD	DE	DF
	E	EA	EB	EC	ED	EE	EF
	F	FA	FB	FC	FD	FE	FF

data from other generations have been developed and will be mentioned later. This clearly is a saving in time and trouble, especially since the increase in variance due to segregation often encountered in F_2 (see Chapter 20) and backcross generations makes it necessary to breed large numbers in these generations in order to achieve stable scores. Thus the same amount of effort as would be expended on the analysis of the genetic difference between two strains can yield comparable data on three more, at a conservative estimate, or a total of five strains.

Unlike the traditional method, the diallel cross is an extensive rather than an intensive method. While it may be necessary to investigate one or two among several F_1 crosses more closely by breeding further generations, at the present stage in the development of the field, when there is so much to be done with so many different species, survey methods of this type seem indicated. A further advantage, which thus far apparently remains unexploited, is the opportunity that the diallel method offers for genetic studies not believed possible before. Interspecies hybrids are often sterile, thus precluding any possibility of breeding the F_2 and further generations required for the standard genetic analysis. If several species or subspecies will mate and produce viable offspring for the F_1, a genetic analysis is now not only possible but can be as complete and as elegant as any using species or strains whose crosses are fully fertile.

ENVIRONMENTAL CONTROL

General

The usual precautions regarding the maintenance of a uniform environment, both in rearing all the subjects and testing them to obtain a measure or measures of the characteristics whose inheritance it is desired to investigate, apply with all the rigor possible. This point has been stressed so often, by, for example, Hall (1951), Scott and Fredericson (1951), and Broadhurst (1960), that it will not be elaborated again here beyond emphasizing the folly of attempting to investigate heredity without controlling environment. It should also be noted that, while it may appear desirable to obtain as many measures as possible on populations that have been bred, when several behavioral measures are taken there is always the possibility that order (or sequence) effects will influence the later measures. Furthermore, after the experimentation is completed, the carcasses of the subjects may also prove of use in the investigation of the inheritance of morphological and biochemical traits. Since any sort of crossbreeding experiment is inevitably laborious, the maximum of information that can be extracted from it should be carefully considered at the outset.

The diallel-cross method presents few special difficulties in connection with environmental controls. The time needed to complete a diallel cross clearly depends upon its size, i.e., the number of strains, n, involved. Since n increases arithmetically and n^2 geometrically, with $n = 2$ or 4, adding another strain nearly doubles the amount of work involved, though not perhaps the time needed to complete the testing.

In this respect it is unlike selection, which may involve long-range problems stemming from the need to maintain environmental constancy over many years. The hitherto standard analysis employing filial and backcross generations, which must necessarily be reared sequentially over a period of time, also entails a possible risk of influence due to environmental fluctuations. A similar risk may occur during the time needed to complete a diallel cross, but here the experimenter has the advantage of being able to balance them out to some extent by breeding crosses from various parts of the diallel table in a predetermined sequence, so that any change is not liable to affect solely the crosses involving one strain. That is, one can sample the time period involved in the whole experiment with a view to equalizing over time any genotype-environmental interaction which may occur as a result of systematic environmental fluctuations. A replication of the diallel table is desirable in many ways, and it has the further advantage that the balancing process referred to above can be carried out not only in relation to breeding different parts of the same table at the same time but also in relation to breeding the same crosses in the two replications at different times.

Maternal Effects

Another aspect of environmental control of special importance in mammalian inheritance is that relating to the possibility of maternal effects. These have also been stressed and the appropriate methods for testing them described elsewhere (Broadhurst, 1961). Basically, there are two possible times at which a maternal effect can be exerted. The first is during the prenatal period when the animal is in the mother's uterus and is physiologically dependent upon her, and the second is during the time before weaning when the animal is in intimate contact with the mother and still to some extent dependent upon her for milk. In addition, there is at this time the possibility of learning taking place, both from the mother and from any siblings present in a litter. The controls for the examination of these various possibilities follow.

Prenatal Maternal Effect

Firstly, let us consider the prenatal effect. The standard genetic technique is reciprocal crossing and the comparison of the two sorts of F_1 progeny, $A \times B$ versus $B \times A$. Since all F_1s bred from homozygous parents should be alike genetically (see below), any difference between them suggests an effect of the only factor in which they differ, the prenatal uterine environment. The diallel cross is ideally suited to assess this possibility, for, as we have seen, it involves breeding reciprocal crosses throughout. There is consequently a large amount of data available from which to estimate possible prenatal influence, and it can be assessed in the analysis of variance of the diallel table, to be discussed later. For the sake of completeness, two points about prenatal maternal effects should be noticed: If a substantial maternal effect of this nature is reliably detected, it may become necessary to exclude the possibility that it is due to other than prenatal uterine factors. There are several genetic possibilities, such as sex linkage and delayed inheritance which might give rise to what may properly be called maternal effects but which cannot be regarded as falling under the rubric of environmental effects in the way we have sought to define them. Fortunately, these effects are rather rare and are more likely to involve major gene effects not usually found in quantitative behavioral traits with their typically polygenic determination.

A second possible influence, which is not chromosomal, is extra-nuclear or cytoplasmic inheritance. This is a hereditary mechanism which has received little attention in relation to behavioral characteristics, but there are well-authenticated cases in the genetic literature of its occurrence in other phenotypes (Jinks, 1964). A probable mechanism by which this type of maternal inheritance could come about in mammals lies in the difference in the amount of cytoplasmic material contributed by sperm and ovum, the latter's contribution being many times greater. Should a maternal effect be established phenotypically, therefore, it becomes necessary to analyze it further, and in particular to investigate the possibility that it is due to a cytoplasmic involvement rather than to an intrauterine effect. The usual control for this is the technique of transplanting fertilized ova. The zygote is removed from the uterus or from its point of fertilization in the Fallopian tube and transplanted for implantation in the uterus of a female of the opposing strain, that is, the strain implicated in any effects previously detected in the F_1. For this implantation to be successful the receptor female must be in an appropriate state of physiological readiness to receive the donor egg; this implies that she herself must have been recently mated and probably with zygotes at the same or similar stage of development in her uterus. This may cause a difficulty for, if the two sorts of offspring developing simultaneously are not identifiable after they have been born, the point of the procedure has been lost. With mammals it is sometimes possible to rely on coat color markings, either already present in the strains being compared or deliberately introduced for the purpose of distinguishing the offspring in this way. The relevant techniques are discussed by McLaren and Michie (1956; 1959); they have not been applied in behavioral work for investigating cytoplasmic inheritance, though Hall (1947) has used them in a different connection. While there is no compelling reason to assume that cytoplasmic inheritance plays an especially prominent role in the inheritance of behavioral characteristics, it is well to be aware of the possibility that apparent maternal effects can be caused in this way.

Postnatal Maternal Effects

The control for the second type of directional effect mentioned, that liable to occur postnatally and before weaning, is achieved by fostering. Singletons, a predetermined fraction of each litter, or the whole litter are transferred at birth or soon after to a foster mother who then rears them until weaning. The requisite number of her own offspring are taken from her to make way for the fostered infants. A control for fostering itself is desirable, so that three types of postnatal environment can be envisaged: rearing by natural mothers, rearing by foster mothers of the same strain as the natural mother, and rearing by mothers of the opposing strain. Comparison of the characteristics of these various groups of offspring will then reveal the presence or absence of a post-natal maternal effect. Broadhurst (1961) gives references to some earlier uses of the techniques; others are to be found in the work of Ginsburg and Allee (1942), MacArthur (1949), and Christian and LeMunyan (1958). Fostering has been employed in behavioral work with increasing frequency in recent years; for examples, see Uyeno (1960); Keeley (1962); Hockman (1961); Thompson and Goldenberg (1962); Thompson, Watson, and Charlesworth (1962); Ader and Belfer (1962a, 1962b); Denenberg, Ottinger, and Stephens (1962); Denenberg and Whimbey (1963); Denenberg, Grota, and Zarrow (1963); Ottinger, Denenberg, and Stephens (1963); Thompson, Goldenberg, Watson, and Watson (1963); Ader and Conklin (1963); Ressler (1962, 1963, 1964); Beach and Wilson (1963); McQuiston (1963); Nichols (1964); Griesel (1964); Bignami (1965); Joffe (1965b); and Lagerspetz and Wuorinen (1965).

Another kind of postnatal effect which may occur in the preweaning stage with mammals is not a maternal effect at all but can conveniently be considered along with it.[2] As mentioned above, unless offspring are born and reared singly, or unless the litter size is artificially reduced to one, the possibility of an effect deriving from siblings must be considered. A control for the simple effect of litter size is to standardize by discarding offspring above a certain number. Broadhurst and Levine (1963) have suggested, on the basis of their own and previous work, that there may be an optimum litter size for laboratory rats. Controls for other sibling effects have not been employed hitherto, but it clearly would be possible to study the effect, say, of sibling activity of rats or mice by introducing into certain litters one or more animals of appropriate age from a strain known to be high in preweaning activity. However, this would introduce an almost intolerable complication into the analysis of the effects of preweaning environments, for the behavior of the alien offspring could induce changes in that of the mother of the sort found by Fredericson (1952) in aggressive behavior of mice, which could in turn alter her behavior to her own offspring, and so on. Ressler (1963) and Joffe (1965a) have recently demonstrated both the reality of these possibilities and the need for controlling them.

All these possibilities of control for postnatal maternal and other effects by fostering and standardizing litters or otherwise manipulating their environment have not been studied in the light of their application to the diallel cross. Clearly, it would, in principle, be possible to effect

[2] All *postweaning* effects can be subsumed under the general heading of environmental effects and treated as such by the usual methods of standardization of husbandry, diet, etc., which have not been stressed here, since they do not differ in their application to the diallel cross from that in connection with any other breeding method.

the necessary control, and Table 14.2 shows an example of a possible experimental fostering program for rats or mice, designed to give the three different postnatal environments prior to weaning, as defined above. It presupposes that each litter can be divided into half at birth, with one half remaining with the natural mother while the other is fostered as indicated. The procedure outlined in the table would obviously entail an extensive breeding program in order to ensure that the appropriate litters were available for fostering at birth or within a few days of each other; otherwise age differences would, at best, lead to complications in their subsequent handling, especially in the control of the postweaning environment, and at worst to a failure in the fostering procedure, with the foster mother rejecting the foster pups or her own, or both. The table, moreover, indicates the minimum case and is applicable only to diallel crosses in which n is an even number and which are replicated at least once. It also imposes a limitation on the procedure recommended above for balancing genotype-environmental interactions by breeding different parts of the replicated diallel tables at different times. The minimum unit within the diallel table that can be bred is now no longer the single litter but a replicated 2 × 2 diallel cross. This may be a drawback, but the advantages are clear.

Study of Table 14.2 shows that the suggested arrangement will allow the evaluation of (1) the effects of simple fostering on pure-bred offspring within each strain, e.g., by comparison of AA, reared by mother A, with AA', the second half of the same litter, reared by A'; (2) the effect of simple fostering on the reciprocal F_1 cross, e.g., by comparison of BA, reared by B, with the other half of BA, reared by B'; (3) the effect of cross-fostering between strains on pure-strain offspring, e.g., by comparison of AA, reared by A, and $A'A$, reared by B; and (4) the effect of cross-fostering between strains on the reciprocal F_1 cross, e.g., by comparison of AB, reared by A, with $A'B$, reared by B. The respective comparisons would be summated and evaluated statisti-

Table 14.2
Scheme to show fostering procedures in a replicated 2 × 2 diallel cross

		First replication				Second replication	
		Father				Father	
		A	B			A	B
	A	$AA/B'B$	AB/BA'		A'	$A'A/AA$	$A'B/AB$
Mother					Mother		
	B	$BA/A'B$	$BB/A'A$		B'	$B'A/BA$	$B'B/BB$

NOTES: 1. A and A' indicate different individuals of the *same* strain, as do B and B'.
2. The letters to the right of the solidus (/) indicate the genetic constitution (by letters as in Table 14.1) and parentage [by prime (')] of the offspring to be added to half of the litter shown in the same way on the left and to be retained by the natural mother and reared by her.
3. The mothers used to generate the offspring in any one horizontal line (array) of the diallel table need not be, as indicated here, the same individuals but merely of the same strain, and similarly with the fathers of both replications. But clearly mother(s) A cannot be the same as mother(s) A'.

cally in an appropriate analysis-of-variance design. For a 4 × 4 diallel cross, four such replicated subtables would be required, and for a 6 × 6 diallel, nine. Such a design would confound the sort of sibling effect noted above with any possible maternal effect, but since the two may interact in the way described it is not proposed to attempt at this time to seek a design that will separate them.

ANALYSIS

Scaling

The first problem that arises in any biometrical analysis of a quantitative measure relates to scaling. This is a familiar problem to psychologists, who must often transform their data from the original scale for various reasons, one frequently being the need to meet the assumptions of statistical methods, such as the analysis of variance, especially in relation to homogeneity of variance between various subgroups. But violation of the assumptions in this case has been shown not to have particularly serious consequences (Lindquist, 1953), and so if the appropriate scale transformation for eliminating undesirable features of a distribution cannot be found, the analysis can still proceed. In serious cases there is always the possibility of fractionating the data and performing separate analyses or having recourse to nonparametric methods.

The problem of scaling in biometrical genetics is a different one, and there do not appear to be any easy solutions to it. The assumptions relating to scale are fundamental to the biometrical model and cannot be dismissed lightly. This stems from an important difference between the procedures in biometrical analysis and, for example, the analysis of variance. In the latter the procedure, broadly speaking, is to partition observed variances into component parts and then to apply to them tests of significance to estimate whether certain components are larger than others. In biometrical analysis, on the other hand, the procedure is to use observed variances to calculate the value of various parameters based on theoretical models derived from Mendelian theory. Two most important assumptions are involved in these models: first, that no interaction between genotype and environment is present, and, second, that the gene effects are additive over the range of variation present. Let us consider these assumptions in turn in relation to the way they impinge on the diallel cross.

Genotype-Environment Interaction

The first scaling criterion is the usual one in genetic analysis; it requires that the variances of populations of nonsegregating generations, that is, the parental and F_1 generations (as opposed to the segregating generations, the F_2 and backcrosses), should not differ significantly from each other. Since the diallel cross is limited to the nonsegregating populations, we need not be concerned with this distinction. The logic of the requirement of equality of variance is widely appreciated but may perhaps be briefly restated here. It stems from the fact that the parental and F_1 generations are each genetically homogeneous: the former since they are chosen for crossing because they are pure, homozygous strains, and the F_1 because it is, in consequence of the homozygosity of the P_1 and P_2, uniformly heterozygous. The same reasoning underlies the analysis of reciprocal crosses, as was shown above. It is to be expected that three genetically homogeneous, but different, populations will have different mean values for any characteristic measured, but since they have all been reared in the same environment their variances should be equal, within the limits of sampling error. If they are not, this fact suggests that the different genotypes are responding to the (same) environment in different ways, that is, that the contribution of the environment to the phenotypic expression is not independent of the genotype, or, more succinctly, that a genotype-environment interaction is present.

Another way of looking at the same problem is to consider that, in the biometrical analysis, the value of the environmental component of variation, E, is defined as the variance of the nonsegregating populations, that is,

$$V_{P_1} = V_{P_2} = V_{F_1} = E_1$$

where V is the variance of the population indicated in the subscript and E_1 is the nonheritable variation of individuals (Broadhurst and Jinks, 1961; Mather, 1949). It is this equality that must be tested; if it is absent, rescaling efforts must be directed toward ensuring it.

In any diallel table there are $n\,P_1$ variances and $n(n-1)\,F_1$ variances, as shown above. A test of homogeneity of these n^2 variances can now be applied, using the standard methods, for example, Pearson and Hartley (1958), in which the ratio of the maximum to the minimum variances observed is tested for significance with $2n^2 - 1$ degrees of freedom (for a replicated $n \times n$ diallel cross). In an example of a replicated diallel cross of six strains of rats (Broadhurst, 1960), to which we shall refer again in what follows, one of the measures derived from the open-field test of emotionality (Hall, 1934), the ambulation score, showed satisfactory homogeneity of variance. On the other hand, the other score, the defecation measure of emotional elimination, did not do so. However, a simple square-root transformation resolved the difficulty and allowed the analysis to proceed.

It is not always to be expected that the outcome will be so fortunate, and perhaps no transformation will be adequate to meet the case. Then it may be necessary to consider other solutions, one of which is to reduce the size of the diallel table by discarding all the scores, that is, both horizontal and vertical arrays, for the strain in which the deviant scores occur. As we shall see, this solution has been advocated (Hayman, 1954a) for failure of the second scaling criterion which we must consider, and there seems to be no good reason why it should not be applicable to persistent difficulty in relation to the first scaling criterion. It may be a Draconian solution to discard a goodly portion of one's data, but if it is a choice between a satisfactory analysis of only part of the data or no analysis of all of it, then the course of action is obvious.

Genetic Parameters

After the adequacy of the scale has been investigated, insofar as it relates to genotype-environment interaction, the next task is to analyze the diallel table to assess the nature of the genetic system determining the phenotype measured. There are three stages in this process. The first employs an analysis of variance, the second the variance-covariance diagram, and the third consists of the partitioning of the observed phenotypic variation into its components. The first, though essential, is dealt with only superficially in this account because it is relatively straightforward for those possessing the usual statistical skills. The third, though important if the analysis is to be complete and the subtlest possible inferences are to be drawn from the data, is beyond the scope of this introductory discussion. Instead, the emphasis is placed on the second of these stages, the variance-covariance diagram as devised by Jinks, the immense utility of which in increasing understanding of the nature of underlying genetic mechanisms by means of simple graphical procedures is now widely recognized in the genetic literature. But first we must consider stage one.

Analysis of Variance of the Diallel Table

Wearden (1964) has reviewed the various forms of analysis available. Which one is used will depend on circumstances and the data available, but it is imperative that some variance analysis yielding tests of significance for genetic variation at least of the additive sort (D) be applied before proceeding further. Obviously if no demonstrable genetic variation is present, further analyses of the kind to be described below are hardly worth pursuing.

Of the analyses available, undoubtedly the most informative is that of Hayman (1954b), who constructed an analysis of variance for the (replicated) diallel table. Hayman gives a worked example of this analysis, and use has been made of it for behavior (Broadhurst, 1960). Computer programs for this and related analyses have also been written (Cooper, 1965). In addition to tests of the significance of a range of components of variation based on the biometrical model, this analysis of variance also allows a relatively precise test of the reciprocal differences, which, it will be remembered, enables the assessment of prenatal maternal effects. Thus the analysis is carried out on the mean values for the reciprocal crosses separately in the body of the replicated diallel tables, and the intertable difference forms an overall estimate of E, the environmental component of variation. This estimate of E is important in several ways: For example, it enables appropriate correction to be made to the location of the axes of the variance-covariance diagram discussed below. Broadhurst and Jinks (1966) give examples of the application of such a correction. Secondly, the estimate of the environmental contribution enables a calculation to be made of the heritability of the phenotypes under investigation. This employs the formula $D/(D + E)$ and hence is a measure of heritability in the narrow sense, though based on interfamily differences.

Thus one of the principal results of the analysis is an estimate of the environmental component of variation, and this E then enters both into the calculations of the third and final stage of a diallel analysis—estimates of additive and dominance variation which are more precise than can be derived from the graphical methods of the variance-covariance diagram—and into the estimation of the standard errors (Hayman, 1954a). It is obviously of great importance to determine the errors of estimation of the components we may partition from our analysis, but we shall not pursue this topic further here; rather, we shall turn to the graphical methods that constitute the second stage of the diallel analysis and that also have important implications for scaling.

Additivity and the Variance-Covariance Graph

The second criterion of scaling to be considered in connection with the analysis of quantitative inheritance relates to the assumptions upon which the model used for biometrical analysis is based. Perhaps the most important is that the gene effects are, on an average, simply additive (Mather, 1949). The interaction between nonallelic genes is primarily responsible for departures from this assumption, and so the failure of additivity is usually ascribed to nonallelic interaction. It should be stressed that *allelic* interaction between genes, that is, dominance effects. does not disturb the analysis, the model on which the biometrical analysis is based being devised in such a way as to allow dominance to emerge as an additive (which, of course, includes subtractive) effect (Fisher, 1918; Fisher, Immer, and Tedin, 1932; Mather, 1949). For the traditional analysis, Mather has devised a series of "scaling tests" which have been elaborated (Jinks and Jones, 1958) to provide components of variation from first-degree statistics. Examples of their application to behavioral data will be found in Broadhurst (1960) and Broadhurst and Jinks (1961). In addition, Cavalli-Sforza (1952) has provided a joint scaling test which combines the three of Mather's tests that are applicable to mammalian data.

However, the graphical analysis of the diallel table provides its own tests of additivity as follows: First, the table should be set up with the body of the table (or tables if replicated) formed by the mean scores for the various F_1 families, reciprocals being pooled and the resultant entries repeated both above and below the leading diagonal formed by the values for the parental strains. This procedure obliterates the distinction between rows (mother's offspring) and columns (father's offspring) previously made, but this will be considered again later. Then the variance (V_r) of each horizontal or vertical array in the table so formed is calculated. This scale provides the abscissa or x axis on the variance-covariance diagram shown in Figure 14.1; there will be n variances corresponding to the n arrays generated by the n strains in the diallel cross. From the same table the covariance (W_r) of each array with the leading diagonal is computed. The same paternal values on the leading diagonal are used for each covariance computation, only the values for the recurrent parent in the array change as each is taken in turn. Thus there are also n covariances, corresponding to each array and so to each parental strain. These are laid off on the ordinate of the

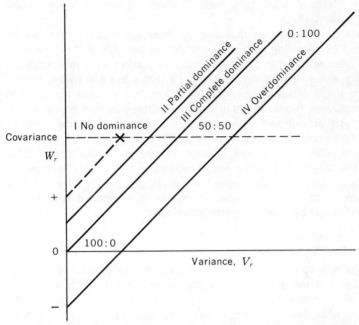

Figure 14.1   Variance-covariance graph. The figure shows a representation of the variance-covariance graph and indicates the various dominance relations that may be detected from it. Four possible regression lines of W_r and V_r are shown as cases I, II, III, and IV, and the proportion of dominant to recessive genes is illustrated on the line for complete dominance (case III).

variance-covariance graph, and the n points entered in the body of the graph.

In the genetic model upon which the diallel analysis is based, the mean of all offspring from one parent varies with the values for the parent itself, and its covariance can be indicated on the W_r axis by a value which is always twice that of V_r, within the limits of sampling error. Thus, if W_r gets bigger and if additivity holds, so must V_r, and proportionately so. Therefore, these (W_r, V_r) points should define a straight line, except in one special case which we shall discuss later; this straight line, moreover, should be of unit slope, $b = 1$, that is, make an angle of 45 degrees with the upright. If these conditions are not both fulfilled, then failure of additivity is indicated. There are statistical tests that can be applied to determine the extent of any suspected failure of additivity; they employ $W_r - V_r$ as the measure (Hayman, 1954a).

The first is essentially a test of the unity of slope of the regression line in the variance-covariance graph; the t-test devised for the purpose by Hayman will also be found in Broadhurst (1960). The second test can be used if the diallel has been replicated and consists simply of an analysis of variance of $W_r - V_r$, with arrays and blocks (replications) as main effects. A significant "array" effect indicates a failure of the additivity hypothesis.

Failure of additivity as detected by these tests can sometimes be countered by rescaling, but this may not be possible if the data have already been rescaled to meet the first scaling criterion of independence of genotype from environment. It may be that a scale adequate for the one purpose is unsatisfactory for the other. Two courses are open to the investigator faced with this situation: Either he can proceed with the analysis as outlined below, bearing in mind that it now probably relates to a complex genetic system (see Hayman, 1954a, for the various possibilities), or he can attempt to find a smaller diallel table within the larger and within which additivity holds. This is done by discarding arrays in turn. Sometimes a single cross, rather than a whole array, causes the trouble; it can be removed and the missing value estimated by a missing-plot technique.

Dominance and the
Variance-Covariance Graph

The second important contribution of the variance-covariance graph to the analysis of the diallel table is the further indication it gives of the nature of the dominance relationships among the strains in a diallel cross. In what has preceded, it has been assumed that the situation under investigation is one in which some degree of dominance is present, and its significance can be assessed from the analysis of variance. In the less usual case of a complete absence of dominance, the (W_r, V_r) points will not generate a straight line in the variance-covariance graph but will cluster round a single point in the diagram where $W_r/V_r = 2$ (case I in Figure 14.1). That is to say, within the limits of sampling error, these points are all estimates of a single point marked x. A line of unit slope drawn through this point will intersect the ordinate, the vertical W_r axis, at a point that will be equal to $\frac{1}{4}D$, D being the component of variation defining the additive genetic effect in the quantitative analysis. But in the cases where dominance occurs, the (W_r, V_r) entries generate a line, as we have seen. This is because V_r now contains a component that reflects the value of H, the dominance component of variation in the biometrical analysis, which consequently moves the points along the V_r axis. Since the proportionality of W_r to V_r must still obtain (unless there is a failure of additivity), the succession of points inevitably forms a straight line.

The position of the line on the graph and the position of the points along the line have important implications relating to dominance. Clearly the points can define the line in such a way that it intersects the W_r axis above the origin O (W_r's intersection with the abscissa, the V_r axis), or exactly at the origin, or below the origin (cases II, III and IV, respectively, in Figure 14.1). What do these differences signify? They all relate to the average degree of dominance present in the parental strains, and the greater the downward placement of the line, the greater the degree of dominance. Thus, an intersection above the origin indicates partial dominance (case II); intersection at the origin, complete dominance (case III); and intersection below the origin, overdominance (case IV). The average degree of dominance can be determined from the graph by reading the distance on the W_r axis between the intersection and the origin. This distance is equivalent to $\frac{1}{4}(D - H)$, H being the dominance component of variation. The point corresponding to $\frac{1}{4}D$ on the W_r axis, which is the intersection in the absence of dominance (see above), can also be calculated since it is one-fourth of the variance of the parental array on the leading diagonal of the diallel table, which we shall call V_1. The distance between these two points on the W_r axis is now equivalent to $\frac{1}{4}H$ and can be obtained by simple subtraction. Knowing $\frac{1}{4}H$ and $\frac{1}{4}D$, we can also calculate the ratio $(H/D)^{1/2}$, which therefore gives the average degree of dominance. In case II (incomplete dominance) it will be less than unity; in case IV (overdominance) it will be greater than unity, and in each case the figure will indicate the average degree of dominance or overdominance, respectively. In case III (complete dominance) it must equal 1.

Not only does the diagram provide information about the average degree of dominance in this way, but it also indicates the relative standing of the parental strains in the diallel cross in respect to the characteristic measured. Their position on the regression line—or their average position if the diallel is replicated—shows the strains' variation in the proportion of dominants to recessives in their genotypes. Again the progression downward implies greater dominance, so that the strains whose (W_r, V_r) points occur nearest the bottom of the diagram have a higher proportion of dominants to recessives, whereas those placed higher up have a lower proportion. It is possible to be more precise and to assign a numerical value to this location by drawing a line parallel to the V_r axis through the W_r axis at a point where $W_r = V_1/2$. The point of intersection with the regression line then defines the point at which dominants and recessives are in equal proportions (50:50); by laying off other appropriate divisions, the proportion present in any given strain can be graphically estimated in this way. This is true only in case III (complete dominance) where the lower end of the line, that is, where the proportion of dominants to recessives is 100:0, is defined by the origin, as shown in Figure 14.1. In cases II and IV it is necessary to construct a parabola from the equation $W_r^2 = V_r V_1$, and the intersections of the regression line with this parabola define its upper and lower ends and hence the points of 0:100 and 100:0, dominants: recessives, respectively. Broadhurst (1960) gives several examples of this calculation. Thus the dominance order for the strains in the diallel cross can be established; this can then be compared with their relative standing in the various characteristics with respect to which their F_1 progeny were assessed. Simple rank-order correlations are adequate for the purpose and will indicate, for example, the respective direction of phenotypic effect of dominant and recessive polygenes.

EVALUATION

The diallel cross is a powerful technique in the analysis of quantitative inheritance and can reveal much about the genetic control of behavior measured in a group of strains. So far it has rarely been applied to mammals (Craig and Chapman, 1953; Dickinson, 1954), most of the developmental work on the method having been done on plants. The first application to behavioral characteristics known is the 6 × 6 diallel cross in rats previously mentioned (Broadhurst, 1959, 1960), but others have since appeared (Collins, 1964; Parsons, 1964; Joffe, 1965b; Fulker, 1966).

In Broadhurst's experiment, the inheritance of rat emotionality, as measured by emotional elimination and exploratory activity in the open-field test, was investigated. The six strains—five pure-bred and one of considerable uniformity—were crossbred in a replicated diallel, the progeny reared, and the phenotype (behavior in this case) tested in the standard manner at the same age. In addition to the behavioral measures of defecation and ambulation (exploratory activity), measures of growth were obtained and have been analyzed separately (Jinks and Broadhurst, 1963). No insuperable scaling problems were encountered, and the analysis proceeded at least as well as in the cases of many of the published plant diallels. The principal findings were that both measures indicated a partial dominance, the defecation response (to fear) showing the more complete dominance. In neither case was there any definite evidence that dominant or recessive polygenes had a uniformly positive or negative effect on the phenotype measured, the indications being that they are probably equally distributed in this respect. For the ambulation measure it was possible to estimate the total number of dominant and recessive genes in *all* parents, and the outcome showed that dominants were more than twice as numerous as recessives.

The relative placement of the six strains on the variance-covariance graph in respect to their proportions of dominants to recessives varied for the two measures and enables certain speculations to be made regarding the possible evolutionary history of the behavior measured. Thus one strain which has a wild-type coat color showed a high proportion of dominants in ambulation, yet its score on this measure was typically intermediate. This suggests there may be a selective advantage for the middle as opposed to the extremes of the range of the behavior sampled, that is, that this may be an example of the selection for the optimum expression of a characteristic. A good case can be made for this view: Too little exploratory behavior in the wild, and the rodent would starve; too much, and it becomes an easy target for predators. Heritabilities turned out to be rather high: around 80 percent for ambulation and 60 percent for defecation. In general, the findings accord well with a selection experiment for emotional elimination currently in progress (Broadhurst, 1962).

The diallel analysis, however, has by no means reached its final form, and refinements and extensions can be expected as use becomes more widespread. Extensions to include measures from F_2 populations have already been reported by Hayman (1958) and by Jinks (1956) who also includes backcross data. Dickinson and Jinks (1956) have also developed an analysis which does not require that the parental strains be homozygous and which may therefore be of especial utility in animal work with species where pure strains do not yet exist. This, combined with the development of a method of analysis suitable for very small diallels employing individuals as parents, rather than strains, for example, 2 × 2 (Jinks and Broadhurst, 1965), may have applications in primatology where numbers are at a premium. The possibilities inherent in the analysis of incomplete diallel crosses are not yet fully employed; in particular, the analysis of single arrays, that is, for example, one male crossed with several females, may prove to have applications for the student of behavior.

The diallel cross also shows promise of having an important place in the analysis of stability of phenotypic expression. This phenomena relates to the fact that some quantitative characters are environmentally sensitive, to use Woodger's (1953) term, and show fluctuations with changing environment, whereas others are more stable. Thus, the flowering time of tobacco plants shows greater stability than does the plant height (Jinks and Mather, 1955). But this stability is itself under genetic control and hence susceptible to analysis by biometrical methods. Thus, in the example given, it was shown by means of a variance-covariance graph that stability of flowering time tended to be a dominant characteristic. There have already been suggestions in the behavioral literature of an awareness of this problem (Fuller and Thompson, 1960, pp. 91–93, 217; Broadhurst and Jinks, 1961, 1963) which has been viewed in the context of the analysis of the change in the genetic control of a behavioral measure with practice, but the biometrical analysis of stability as such is only now being attempted and its implications for the study of behavior evolution explored (Broadhurst and Jinks, 1966), though some body-weight and litter-size data have already been investigated in these respects (Jinks and Broadhurst, 1963). The systematic manipulation of environment during early life, combined with the diallel cross, may prove to be a powerful method for this study, and a basis for one possible approach to this type of analysis has already been laid (Allard, 1956).

The outlook, then, for this method is promising, but there are difficulties which must be faced. The assumptions that underlie the diallel analysis are complex, and while the necessary care with scaling, which has been stressed, will do much to meet them, it is not always clear whether or not the data and the experiment from which they were derived meet the assumptions. Indeed, it does not always appear to be known what are the consequences of failure to meet them. Perhaps the best policy for the psychologist and ethologist with problems for which this method appears to be specific is to proceed with caution and to evaluate the results with reservation. Corrections, modifications, and elaborations will doubtless follow the extension of the method to different behavioral characteristics and to different species.

SUMMARY

This chapter describes the diallel method of crossbreeding for genetic analysis and contrasts it with other methods available. The nature of the environmental controls necessary in behavior study, especially in relation to maternal effects in mammals, is examined and their application in the diallel cross explored. The importance of demonstrating the adequacy of the scale of phenotypic measurement used is stressed, and ways of doing this are outlined. The analysis of variance of the diallel table is touched upon and the variance-covariance graph introduced, both in connection with the problem of scaling and with the analysis of the variation into its various components. The chapter concludes with a brief evaluation of the diallel cross as a method of behavior-genetic analysis, together with some illustrations derived from previous use of it in a diallel cross of emotionality in rats.

PART IV CONCEPTUAL AND METHODOLOGICAL PROBLEMS

☐ It is well-nigh unthinkable that . . . laws and regularities . . . should happen to apply immediately to the behaviour of systems which do not exhibit the structure on which those laws and regularities are based. (Schrödinger, 1946, p. 3.)

☐ We can speak of the difference between the behavior of a trained and an untrained animal as learned, provided that their genotypes are similar, but the behavior of a single individual cannot be spoken of as learned. (Marler and Hamilton, 1966, p. 619.)

INTRODUCTION
Jerry Hirsch

Many students of behavior have had as their explicit goal the development of a science whose corpus would contain a body of general laws. In the first chapter of Part IV McClearn takes up the problem of general laws, not by means of an abstract discussion of the form that propositions in science should (=?) take, but rather by considering in sufficient detail the complexities and realities of the data of behavior. He discusses the uniformity of the material with which one can work and the consequences and uses of the genetic manipulations more commonly practiced. In the second chapter DeFries provides an overview and perspective on quantitative behavior-genetic analysis.

Next, it is fitting that discussions of behavior-genetic analysis and the information flow from genetics to the behavioral sciences should also include consideration of feedback to genetics; this is provided by the geneticist R. C. Roberts. Then, Thompson reviews tests and human behavior, multivariate analysis, and relations between test factors and genetic factors. Finally, there is an extended critique of the literature on behavioral differences between races by Spuhler and Lindzey.

CHAPTER FIFTEEN
GENES, GENERALITY, AND BEHAVIORAL RESEARCH
Gerald E. McClearn

INTRODUCTION

Perhaps the most compelling biological generalization is that living things vary. Much of the variability is partitioned among different species, genera, families, orders, classes, phyla, and kingdoms. Evolutionary theory, one of the most significant contributions to human thought, arose from the attempts of biological scientists to account for this distribution of the characteristics of living material.

It has become clear that phyletic differentiation has been, and continues to be, dependent upon intraspecific variability, for individual differences within groups form the raw material of evolution. Studies at the biochemical, cytological, and statistical levels have revealed the elegance of the genetic mechanisms which guarantee to a population heterogeneity of its members. In addition to these genetic mechanisms, the effects of differential environments, acting from conception to death, contribute to individuality. Individual differences are a basic fact of life, but their ubiquity has given an air of commonplace which in many contexts has obscured their fundamental import.

In Chapter 11, it was shown that the total phenotypic variance V_P of a population can be represented as the sum of genetic (V_G), environmental (V_E), and interaction (cov_{GE}) variance components. If one assumes, for simplification, that all genotypes are affected equally by given environmental factors and that the factors are randomly distributed over all genotypes, the situation may be represented as $V_P = V_G + V_E$.

The total measurable phenotypic variance V_P is a matter of central concern in behavioral investigations, determining as it does the magnitude of the standard error against which the effectiveness of the independent variable is assessed. In their concern to reduce this variance and increase the precision of their experimental procedures, psychologists have identified many of the environmental factors contributing to variability in various situations and have developed sophisticated techniques for coping with them. Perusal of almost any journal reporting behavioral research will reveal the concern with the control of such things as magnitude of reward, time of day of testing, intertrial interval, pretraining schedule, etc. Counterbalanced designs are employed, mazes are rotated, noisy relays are acoustically isolated, one-way screens are used, animals are tested in light-tight, airtight, sound-proofed chambers.

As the relevance and importance of these and many similar factors have been demonstrated, the criteria of adequate research have inevitably become stricter. Editorial referees examine submitted papers with an increasingly critical eye for lapses or inadequacies of environmental control. An entire new industry has arisen in the past few years to provide behavioral scientists with the instruments required for the ever more refined experimental procedures. It can be concluded, then, that the problems associated with V_E, although by no means eliminated, are fairly well in hand. The genetic variance V_G, on the other hand, has been treated with a curious neglect. Genetic variability, in the main, has either been regarded as an inevitable nuisance, about which nothing constructive could be done, or has been ignored entirely. In many standard psychology reference sources the text includes three or four pages in which the possibility of genetic involvement is considered in a rather embarrassed and futile manner. In many others the indexes do not even include the words "genetics," "heredity," or "inheritance." The primary source material is little better.

The relationship $V_P = V_G + V_E$ does not assert anything about the relative importance of V_G and V_E. Is the failure to consider the genetic nature of the experimental animals in behavioral research a serious omission, or does V_G make only a trivial contribution, the elimination of which would not be worth the effort and expense? The following section will present evidence that V_G, in many instances, is of critical importance and can be disregarded only at great risk.

Perhaps the most straightforward evidence on the role of heredity in a behavioral trait can be obtained from examination of differences between inbred strains. Such differences have been described with respect to a large variety of behavioral phenotypes. In some cases the differences refer to a parametric value. As an example, it is often desirable to characterize a species with respect to some trait. The research question might concern the alcohol preference of *Mus musculus*. The answer to the question is highly strain-specific. C57BL mice show preference for a 10 percent ethanol solution over water, whereas DBA/2 mice show a strong avoidance of the alcohol solution (McClearn and Rodgers, 1959). McGill (1962) found enormous differences in various aspects of the mating pattern of males of C57BL, DBA/2, and BALB/c strains of mice. Similarly, Sawrey and Long (1962) found large strain differences in rats in susceptibility to ulceration in a conflict situation. Other significant strain differences, in many cases so great as to show no overlapping of the distributions of the separate strains, have been described for activity (Bruell, 1962; McClearn, 1961; W. R. Thompson, 1953), learning (Carran et al., 1964; Collins, 1964; Lindzey and Winston, 1962), emotional elimination (Broadhurst and Levine, 1963), and so on.

Perhaps of more relevance to the present topic is evidence concerning strain differences in the effect upon behavior of an experimentally controlled environmental variable. Hughes and Zubek (1956) compared control groups with groups given glutamic acid on learning performance in a Hebb-Williams situation. Two strains of rats were used: In the bright strain there was no difference between groups; in the dull strain there was a significant difference in favor of the glutamic acid group. In a study on the effects of inhaled alcohol vapor on activity, comparisons were made between experimental and control groups of six strains in three different activity-testing situations. C57BL mice of the experimental group showed *less* activity than their controls, and the experimental animals of the C3H/2 group showed *more* activity than did the controls (McClearn, 1962; McClearn and Schlesinger, unpublished). Weir and DeFries (1964) studied the effects of trauma administered to pregnant female mice of the C57BL and BALB/c strains on the subsequent behavior of their offspring. In both strains, the effect of the experimental treatment was significant. However, in one strain the effect was a reduction in open-field exploratory behavior of the young, and in the other strain the effect was an increase in this behavior.

These examples should suffice to make the point that very different outcomes may be obtained for different genetic groups of experimental animals. The broad range of behavioral phenotypes for which such strain differences can be documented suggests the ubiquity of genotypic determination of behavioral properties.

GENES AND GENERALITY

In terms of statistical inference, a study is performed on a sample of individuals for the purpose of estimating parameters of the population which these sampled individuals represent. The examples cited in the preceding section make it clear that the range of permissible generalization may be very narrow indeed. The outcome obtained with one group of animals may be drastically different from that obtained with a different group. Failure to appreciate the implications of biological individuality has resulted in a state of affairs wherein many investigators expect that an obtained result has universal application—to all rats, or all monkeys, or even to all mammals. The explanation of discrepant results from other investigations is usually sought in terms of subtle differences in apparatus or technique, and the possibility is rarely considered that there exist different subgroups within a species to which different rules apply.

A great many of the major controversies of psychological theory may possibly be due to this orientation. The fantastic restraint of progress which this may have imposed in developing sound behavioral theory can be appreciated by considering that there may be response-learning rats, and there may be place-learning rats; there may be continuity learners and noncontinuity learners; there may be rats that can learn latently and those that cannot; some rats may learn under irrelevant drive states, others may not; indeed, there may be Hullian learners and

Tolmanian learners. Controversies which dominate animal research for long periods of time, which occupy the resources and energies of innumerable laboratories and researchers, may in many cases largely be tilting at windmills because of the genetic differences in the animal material employed by the various protagonists.

These genetic facts of life have more than the negative implication regarding the limitation on generalization, however. The purpose of the remainder of this chapter is to examine ways in which biological specificity of animal subjects may be turned to advantage, providing psychological researchers and theorists with new methodologies and new conceptual frameworks.

There are three basic mating procedures which are relevant to mammals: inbreeding, crossbreeding, and selective breeding, and, because the bulk of psychological research is performed on mammals, the discussion will be restricted largely to these procedures. Since practically all researches are ultimately evaluated in terms of means, variances, and covariances, or their nonparametric relatives, we shall be concerned with the effects of breeding procedures on these statistics. A refined treatment of most of the relevant principles has been provided in Chapter 11, and frequent references will be made to appropriate sections thereof. The present discussion will simply serve to summarize salient points which are of particular relevance to the present topic.

Consequences of Inbreeding

An instructive way to examine the consequences of inbreeding is to consider the changes that take place subsequent to the subdivision of a heterogeneous foundation population into brother-sister mating pairs. If, in each following generation, a single male and a single female are selected for mating from each family, the total population will be composed of a number of lines within each of which the theoretical inbreeding coefficient will approach unity.

As R. C. Roberts has explained in detail (Chapter 11), the mean of the total population of lines with respect to a trait that is influenced only by loci acting additively, or by loci whose dominance effects balance out in the plus and minus directions, would be unchanged as inbreeding progressed. For a trait for which there is directional dominance, however, there would be a shift of the total population mean in the direction of the recessive alleles. By virtue of the fact that inherited deleterious conditions are most often of a recessive nature, there would be a reduction in biological fitness of the overall population. Thus, body weight, life span, fertility, resistance to disease, and other attributes of fitness are generally lower in inbred strains. The different developing strains would not be equally affected by this inbreeding depression, however. Some would have accumulated fewer deleterious alleles in homozygous state than others. Furthermore, in practice, very few strains would survive the inbreeding process. In most, the inbreeding depression would likely become so extreme that the animals would fail to reproduce. The surviving strains would be a biased sample, representing the more fortunate; therefore, simply knowing the inbreeding coefficient of a strain does not permit an a priori estimate of its mean on any given characteristic. As Falconer (1957, p. 102) has put it, ". . . general statements about inbreeding depression are true only of the average level of a stock composed of many lines: the behavior of any one line on inbreeding is to a large extent unpredictable."

With regard to variability, inbreeding results in a redistribution of the genotypic variance, with the within-strain component approaching zero in the limit, and with all the genetic variance coming to reside in the between-strain component, that is, associated with mean differences among the strains.

Two principal questions arise concerning the extent to which these theoretical expectations concerning variability are realized in practice. The first regards the question of whether genetic uniformity can really be obtained by inbreeding; the second regards the relationship between reduced genotypic variability and phenotypic variability.

Realized Homozygosity of Inbred Strains   It has long been realized that the occurrence of spontaneous mutations would prevent inbred strains from ever achieving complete homozygosity. Lerner (1958) has argued for the existence of an additional mechanism, more direct and potent, which might prevent or delay the attainment of homozygosity. The superiority in fitness of heterozygotes which is often observed (see Lerner, 1954, 1958, for reviews) implies that, among the offspring of a sib mating, those individuals who possess a relatively large proportion of homozygous loci will be somewhat less likely to survive and reproduce than their more heterozygous sibs. It is usually the case in practice that a strain is maintained by setting up several sib-pair matings in each generation and retaining only the offspring of one, usually the most productive. It can readily be seen that the relatively more heterozygous animals would be selected in each generation, and the actual rate of increase in homozygosity during the development of an inbred strain could be appreciably less than that expected on the basis of computed probabilities.

However, heterozygote advantage in this situation should only retard the rate of inbreeding and should not prevent fixation. In any given line that has met the technical criterion of "inbred strain" the number of heterozygous loci will probably be a small fraction of the total genotype. Falconer (1960, p. 103) has stated:

☐ Under laboratory conditions the highly inbred strains of mice, after 100 or more generations of sib-mating, have a fitness not much less than half that of non-inbred strains. It is conceivable that they might have one locus permanently unfixed, but it is difficult to believe that they can have more. Complete lethality or sterility of both homozygotes at one locus means a 50 per cent loss of progeny; at two unlinked loci, a 75 per cent loss. A mouse strain with a mortality or sterility of 50 per cent can be kept going, but hardly one with 75 per cent.

On the other hand, the above considerations are based upon the assumption that only "Mendelian" phenomena occur in polygenic systems (Chapter 1). Some as yet undescribed dynamism perhaps exists which contributes to prolonged maintenance of heterozygous loci or blocks of loci (Lerner, 1958, p. 212).

The empirical data are not conclusive. Loeb et al. (1943) found evidence of heterozygosity of at least some loci in a rat strain which was the product of 102 generations of sib mating. Cock (1956) found similar evidence in inbred lines of chickens. Deol and collaborators (1960), however, searched for evidence of heterozygosity at loci affecting skeletal characteristics in inbred strains of mice and concluded that all strains studied were homozygous at all investigated loci. Wallace (1965) has reviewed evidence from a variety of sources, much of which indicates homozygosity of inbred strains, but some of which suggests some residual heterozygosity.

The issue is obviously not resolved. Sufficient theoretical and empirical reasons exist for circumspection in assuming absolute homozygosity in inbred material. It appears, however, that the assumption is useful as a working approximation. If it is wrong, the number of heterozygous loci will probably be relatively small and even then may have no appreciable effect on the particular trait being studied. Fervent hope, however, is not the researcher's only recourse in this situation. A strong test of the assumption that an inbred strain is homozygous *at all loci affecting a particular characteristic* is to breed selectively within the strain for high and low degrees of that characteristic. Absence of response to selection pressure is presumptive evidence of homozygosity. An example of this approach is provided by Kakihana (1965), who was able to show that unexpectedly large variation in alcohol preference in BALB/c mice was of environmental origin.

Genotypic Variance and Phenotypic Variance   The reduction of genotypic variance with inbreeding presumably should result in reduction of phenotypic variance. Inbred strains with approximately equal inbreeding coefficients are often found to have very different variances, however, and this suggests that different homozygous (or nearly so) genotypes differ in their susceptibilities to environmental forces. The requirements of bioassay research, where a small variance is very desirable, have led to empirical work on comparisons of inbred-strain variability with that of other genetic groups. In large measure, the concern is with the relative magnitudes of the variances of inbred strains and of F_1 hybrids derived from them. Further consideration of this issue will therefore be deferred until F_1s are specifically discussed below.

The reduction of genetic variance within inbred strains also has important consequences for covariances. In a genetically heterogeneous population, the phenotype of a given individual can be regarded as composed of a genotypic value and an environmental deviation. If some of the loci which influence phenotype A have a pleiotropic effect also on phenotype B, there will exist a genetic covariance between A and B in the population. In addition, there may exist environmental covariance.

Other loci and other environmental factors will affect only A or B and will not contribute to an association between the phenotypes.

By eliminating or drastically reducing genetic variance, the process of inbreeding eliminates or reduces the genetic covariance, and the only correlations that can be detected within a highly inbred strain will be induced by environment. For most purposes, therefore, an inbred strain is unsuitable for correlational analyses, since the correlations due to pleiotropic gene action, which surely must be regarded as of fundamental importance in an investigation seeking to define relationships among variables, cannot be assessed.

Several additional points may be made regarding the nature of differences among inbred strains.

Consider two unrelated highly inbred strains. Inevitably they will differ at a number of loci, but they will be alike at other loci. With regard to a particular phenotypic characteristic some loci may be regarded as relevant and others as irrelevant. The two strains may differ with respect to some of the relevant loci but be alike with respect to others. Thus, if we assume a simple case in which the loci A-a, B-b, C-c, and D-d are the only relevant ones, we could imagine two strains:

Strain 1: $AABBCCddeeFF\ GG\ HH\ ii \ldots$
Strain 2: $aabbCCDDeeFFgg\ HH\ II \ldots$

If capital letters are taken to indicate the $+$ allele (that is, the allele which makes for greater manifestation of the phenotype), it can be seen that neither of these hypothetical strains has a monopoly on $+$ alleles but that one has more than the other. We might reasonably expect, therefore, to find more extreme strains than the ones in hand.

On the phenotypic scale, strain 1 will have a higher mean than strain 2. It is important to keep in mind, however, that a given phenotypic value of a polygenic trait may be achieved in a variety of ways. If we assume, in our example, that each $+$ allele contributes the same amount of phenotypic expression (disregarding environment for the moment), then the mean of strain 1 could be obtained also by strains constituted $aa\ BB\ CC\ DD \ldots$, $AA\ bb\ CC\ DD \ldots$, or $AA\ BB\ cc\ DD \ldots$ In terms of physiological mechanism or in terms of behavioral subcomponents, these might represent substantially different situations. One maze-dull strain, for example, might be dull for essentially cerebral reasons, while another, with equivalent error scores, might be dull for emotional or motivational reasons.

It might often be sufficient for the immediate purpose to regard strain means as defining points on a univariate scale. From a heuristic point of view, however, it is essential to be aware that finer analysis would likely reveal that the points are distributed in multidimensional space.

The Uses of Inbred Strains

One of the prime advantages of employing an inbred strain is the relative reproducibility of the key feature of an experiment: the living subjects whose behavior is being examined. With some reservation, as we have seen, each individual member of an inbred strain may be regarded as a genetic replication of each other individual in the strain. Over a number of generations, a certain amount of genetic drift may be expected to occur as a consequence of selection, either natural or artificial, acting upon the residually heterozygous loci or upon new mutations. Inbred animals do not, therefore, provide the behavioral scientist with as absolutely invariant research material as, say, pure elements do for the physical scientist. However, the genetic changes occur gradually and slowly, and inbred animals provide standardized reference groups of incomparably greater genetic stability than the animals employed heretofore in most psychological research. The experimenter who employs inbred animals can be reasonably confident, therefore, that the fundamental biological nature of his subjects has remained constant throughout a series of experiments. This is a great advantage, indeed, over the situation in which studies are performed with animals from whichever supplier can make delivery at the appointed time.

The researcher who employs animals from his own colony is also working under a handicap if his colony is not inbred or, alternatively, deliberately and systematically outbred. All too often, the breeding program of a departmental colony involves such a small number of breeding pairs in each generation that the inbreeding coefficient will gradually rise to an appreciable value. Furthermore, favorite dams or sires, treasured for their fecundity or pleasant disposition, may be used over and over in the breeding schedule. The inbreeding and selection thus inadvertently employed will lead to a colony that is largely an unknown quantity and offers neither the advantages of the inbred animals nor the advantages (to be described later) of the deliberately heterogeneous groups but shares in considerable measure the disadvantages of each.

The principal shortcoming of the typical "private" or departmental colony lies in its lack of general availability. A fundamental requirement of scientific evidence is that it be reproducible; this implies that other investigators can replicate the experimental conditions. The evidence cited earlier on strain differences should make it clear that an attempted replication with animals from a different gene pool than those originally used is a risky, tenuous, and uncertain venture. Success, to be sure, is a valuable outcome under such circumstances, but failure to replicate results, when unspecified or nonstandard animals are employed, is conspicuously uninformative. One of the most valuable attributes of inbred strains, then, is their *standardness*. For example, many of the inbred strains of mice are available from large-scale production colonies, such as the Roscoe B. Jackson Memorial Laboratory, which can provide research animals to investigators throughout the world.

In many cases different sublines or substrains are maintained in various laboratories or colonies. The same kind of genetic drift that can occur over generations within a strain can also be expected between sublines. Here, again, the changes can be expected to be relatively slow and small, particularly if the original strain was highly inbred before the substrains became established. (See Jay, 1963, for a description of standard strains of mice, rats, and other mammals.)

The use of inbred strains also makes it possible for an investigator to make optimum use of information derived from other laboratories conducting research with the same strain or a closely related substrain. It has been reported, for example (T. B. Dunn, 1954), that retinal degeneration occurs within a substrain of C3H mice. This finding is of immediate interest to any behavioral researcher using C3H mice in studies of learning phenomena, activity level, social responses, and the like. Any behavioral uniqueness displayed by C3H mice might be explainable in terms of this visual deficiency. While it would be necessary to check specifically to determine if this anomaly occurs in the particular substrain of C3H being employed by another investigator, the clue has been made available, and it has specific reference to C3H mice. Without genetic specification of the research material, such information would have no particular referent, and the applicability to other researchers would be vague, obscure, and uncertain.

Over a period of time, and from various laboratories, a "strain picture" gradually emerges, displaying the interrelationships among a variety of behavioral and physical characteristics. It is probably inevitable that this picture will appear puzzling, at least in the early stages of investigation of a given trait, but if the situation resembles a disassembled jigsaw puzzle, the investigator at least has the comfort of knowing that the puzzle pieces all belong to the same puzzle and his results need not stand in the pathetic isolation typical of so many data from animal-behavior researches.

These strain profiles accumulate for a variety of different strains, and it is possible to characterize animals of a given strain with respect to a number of traits. This cumulative knowledge permits an investigator to select a strain to conform to the needs of his particular experiment. For example, if one were interested in testing the effects of some variable upon activity level, the A strain would be an unsuitable choice because its spontaneous activity level is so low as to leave little margin for further decrease. Likewise, in a project aiming to determine if disulfiram reduces alcohol intake of mice, the DBA/2 strains would be a poor choice because its base-line consumption level is practically nil.

An apparent disadvantage of an inbred strain is the lack of generality obtained from research upon it. This cannot be construed as an argument for, or a defense of, studies using nonspecified or nonstandard groups, however. There the generality is no greater, but many are deceived into thinking that it is. In fact, the narrowness of applicability of results from inbred animals can be regarded as a sterling virtue. A researcher familiar with various strains is never tempted to overgeneralize his findings. If data are available for only one particular strain, the issue is regarded as completely open with respect to any other strain.

One approach to the problem of increasing the generality of results of a research program is to add a strain dimension to studies routinely, investigating several strains simultaneously. Such a research strategy has several advantages: If the results are similar in all strains tested, the investigator has suffered no loss of information as a consequence of using several smaller groups rather than one large group. On the contrary, confidence in the generality of the effect investigated is greatly enhanced. If strain differences appear, the lack of generality has been immediately demonstrated and the futility of interlaboratory controversies based upon inconsistent results obtained with other groups of animals can be, in large part, avoided. If strain differences are pronounced, the investigator has, in effect, identified a powerful resource for an analysis of the phenomenon in which he is interested, for the investigation of the determinants of the strain differences can be expected to provide valuable information concerning the phenomenon itself.

Apart from the comparative studies of strains, an investigator can select a strain whose mean, variability, responsiveness to a particular independent variable, or some other characteristic makes it best suited for subsequent detailed examination.

Limitations of Inbred Strains

Although the multiple-strain approach can contribute immensely to the task of specifying the generality or the restriction of generality of behavioral-research results, there are certain inherent limitations. In the first place, as pointed out in Chapter 11, inbred animals are biologically unusual, in that their genotypes could not occur in nature. The high degree of homozygosity is a markedly artificial situation for a normally outbreeding species. The attendant deficiencies of "buffering" capacity, the reduction in fitness, the general effects of inbreeding depression, may produce behavioral phenomena which appear only rarely or not at all in the "normal," more heterozygous members of the species. Second, inbred strains are unique in that no lethal genes are included in their genotypes. In outbred stocks, lethal genes in heterozygotes might produce effects that could not occur in inbreds. Third, even the most ambitious multiple-strain program can assess only a tiny fraction of possible inbred genotypes which theoretically could be derived. A program of research on 10 strains is essentially a program of research on 10 genetically replicated individuals (or, more specifically, on 10 replicated duplicate gametes; see Chapter 11). Fourth, the inbred strains available cannot be regarded as random samples of all possible inbred strains derivable from the total gene pool of the species. Not only have inbred lines with low reproductive fitness been eliminated, but further restrictions on randomness have been imposed by selection for various characters. Furthermore, many of the standard strains have some common ancestry.

For certain purposes of genetic analysis these limitations of inbred strains may become of considerable importance. Although there is not, as yet, sufficient empirical evidence to make a positive statement, it would seem that, with respect to inbred animals as behavioral-assay material, the limitations are far outweighed by the many advantages. In spite of their unique genetic condition, they are living, behaving organisms, and general theoretical formulations of behavioral systems must account for their behavior.

Consequences and Uses of Crossbreeding

Detailed discussions have been provided by Bruell and Broadhurst (Chapters 13 and 14) of the methods of genetic analysis pertaining to generations derived from pairs of inbred strains. Here we need only recapitulate some essential features concerning means and variances.

The F_1 hybrid of a cross between two inbred strains will be heterozygous at each locus at which the parent strains differ, but, though heterozygous at many loci, all members of an F_1 will be identically heterozygous at those loci and homozygous in like state at all other loci. Thus there is no genetic variance in an F_1, provided that the assumption of complete homozygosity holds for the parent strains. In terms of number of $+$ alleles, the F_1 will be exactly intermediate to the parent strains and will therefore have a genotypic mean at the midparent value. The complications of nonadditive gene action are so ubiquitous, however, that this tells us little about the phenotypic mean to be expected of the F_1. Given an F_1 mean value, the expectations of F_2 and backcross means are somewhat more reliable, but, at best, crossbreeding of this kind can provide only a very approximate method of manipulating means. Once a particular outcome has been obtained in an F_1, however, it should be reproducible on subsequent occasions within the usual limitations of sampling error.

Thus F_1s share many of the advantages of inbred strains; genetically they are just as reproducible and just as uniform. In the absence of complicating interactions, one might therefore expect the variances of F_1s and of inbred strains to be the same. Such is not the case, however. For many morphological, physiological, and pharmacological traits, F_1 variance is substantially less than that of inbred strains. The evidence is not unambiguous, however (Chai, 1956; Brown, 1962). These findings are of great practical importance in research involving bioassay methods. In addition, they have given rise to lively theoretical discussions that involve concepts such as homeostasis, developmental buffering, and canalization (Biggers and Claringbold, 1954; Biggers et al., 1961; Chai, 1961; Lerner, 1954; McLaren and Michie, 1954). For behavioral traits, it has often been found that the F_1s are *more* variable than their parent strains (Caspari, 1958), and these results have generated further empirical and theoretical work (Fuller and Thompson, 1960; McClearn, 1965; Mordkoff and Fuller, 1959; Schlesinger and Mordkoff, 1963). Our present concern, however, is the practical one of deciding whether inbreds or F_1s provide the most uniform response in behavioral experimentation. At the present time no theoretical principle seems adequate to provide a general answer; the question must be answered empirically in each specific instance.

Intermating of F_1 animals produces an F_2 generation. Segregation occurs in the gamete formation of the F_1 parents, with the result that their offspring are genetically diverse. There will be no alleles in the F_2 that were not present in the F_1 and the parent inbred strains, and allelic frequency will remain unchanged if there are no complications of differential fitness among the F_2 individuals. If all loci act in a strictly additive manner, the phenotypic F_1 mean and F_2 mean will both coincide with the midparent value. With dominance, the F_2 mean is expected to deviate from the midparent value in the same direction but only half as far as the F_1 mean. This provides but a weak method of manipulating means, however, and the advantage of F_2 animals is the fact that their variance has a genetic as well as an environmental component. At the present time, the principal utility of F_2 animals or of subsequent generations in behavioral research is in tests of hypotheses concerning relationships among different traits. Research on inbred strains almost inevitably leads to hypotheses concerning correlations between characters, arising, for example, from observations that strain 1 is high on trait A and on trait B, whereas strain 2 is low on both traits. Such an association might be determined by pleiotropic gene action, with at least some of the loci that affect trait A also affecting trait B. This type of relationship is biologically fundamental, and its demonstration can be of great value in elucidating causal relationships. On the other hand, such an association could well be fortuitous. Given that two strains differ with respect to trait A, they each must have *some* value for trait B. A significant difference between the strains in trait B, in either direction, is likely to be suggestive of some hypotheses concerning common causal pathways for trait A and trait B. Yet there may be no genetic determinants in common between the two traits, with chance alone responsible for the apparent association.

An F_2 generation provides an opportunity to test meaningfully for a correlation between traits. If a large proportion of the loci influencing trait A also influence trait B, then a high correlation (either positive or negative) should be found in the F_2. If the association in the parent strains was due to fortuitous arrangement of independent loci for the separate traits, the correlation due to genetic communality should be zero. Linkage could, however, maintain the apparent association in the F_2 if it were sufficiently tight and if a relatively small number of loci were involved. This possibility can be assessed in subsequent randomly mated generations where a progressive decline in the value of the correlation would reveal the breaking up of linkage relationships and an approach to equilibrium.

Several examples of the use of genetically heterogeneous animals to assess hypothesized correlations are available. Stockard et al. (1941) found body type and temperament, thought to be causally related, to segregate independently in F_2 generations derived from different dog breeds. Rosenzweig et al. (1958) described some experiments show-ing that learning ability and activity of the enzyme acetylcholinesterase were negatively associated in two strains of rats. This relationship was later tested (Rosenzweig et al., 1960) in a genetically heterogeneous stock derived from an F_2 between the strains and maintained subsequently by random matings. The obtained correlations of various relevant indexes were either nonsignificant or were in the direction opposite to that predicted.

The utility of an F_2 population has been seen to depend upon its genetic heterogeneity. This heterogeneity is a relative matter, however, and groups of greater genetic variability can be constructed. A double cross (four-way cross) can be generated by mating dissimilar F_1s. For example, the F_1 of strains C57BL and A can be mated with the F_1 of BALB/c and DBA/2 strains. Even greater diversity can be obtained by mating the progeny of such a four-way cross with the progeny of another, dissimilar, four-way cross to provide eight-way-cross animals, and so on. Such populations could prove even more satisfactory than F_2s for testing hypotheses about correlations and should be the groups of choice for animal studies using multivariate analysis or any other procedure based upon covariances.

Populations derived from four-way- or eight-way-cross groups by random mating can be maintained for correlational purposes or as reference populations against which measures of central tendency and of variability of other experimental groups, such as inbred strains, can be assessed. Such populations will be subject to some genetic drift, but if a sufficiently large number of mating pairs is employed in each generation, a group of considerable stability should be maintainable for long periods of time. The genetic stability of such a group is different from that of inbreds or F_1s. In a heterogeneous stock of this kind, each individual would be a unique assemblage of alleles, and any sample chosen from such a population would be composed of individuals unlike each other and unlike any obtained in a previous or subsequent sample. The stability is in terms of gene frequencies, not in terms of homozygosity.

Another advantage of a heterogeneous population derived in a systematic manner, such as the four-way or eight-way procedure mentioned above, is that the procedure is repeatable. Given the "recipe" and the availability of the basic ingredients (the inbred strains), the population can be reconstituted, within the limits imposed by sampling considerations, at some other time and in some other laboratory.

Yet another use of heterogeneous populations is as foundation stocks for selective-breeding programs to be described next.

Consequences and Uses of Selective Breeding

Selective breeding, in contrast to inbreeding, is intentionally directional. Insofar as the variability in a population has an additive genetic component, the application of selection pressure will bring about changes in the mean level of the trait under consideration. The success of the selection program thus gives evidence of the existence of additive genetic variance in the base population, and, as R. C. Roberts shows in Chapter 11, various deductions concerning the genetic system may be made from the relationship of selection response to selection differential, from the symmetry of the response in upward and downward selected lines, and so on. For the present topic, however, the most important aspect of selective breeding is its capacity to generate groups of animals of differing behavioral characteristics.

Many successful selection studies have already been undertaken for behavioral traits. In a classic work, Tryon (1940) selected a maze-bright and maze-dull strain of rats. Heron (1935) was also successful in selecting for maze-learning ability in rats, and Bignami (1964) established strains of fast and slow avoidance learners. Rundquist (1933) obtained active and inactive strains of rats by selective breeding, and Hall (1938) successfully bred for "emotionality" as measured by the tendency to defecate in open-field situations. Broadhurst (1960) was able to repeat the successful breeding for rat emotionality. Lagarspetz (1961) bred for aggressiveness in mice. In the fruit fly, Drosophila, Erlenmeyer-Kimling, Hirsch, and Weiss (1962) selected for positive and negative geotaxis, and Hirsch and Boudreau (1958) selected for high and low degrees of phototaxis. Manning (1961) bred for mating speed in Drosophila.

These examples demonstrate the feasibility of selecting for a desired behavioral trait. Once the selected lines have become stabilized, they then constitute research material of extremely high potential for elucidating the behavioral trait under consideration. The studies of Rosenzweig et al. (1960) on neurochemistry and learning, for example, depended initially upon the availability of descendants of Tryon's maze-bright and maze-dull animals. Broadhurst (1960) has been able to test a number of behavioral theories concerning emotionality and motivation with his selected lines of rats.

With respect to the interpretation of studies with selected strains, two points should be made. First, the precaution noted with respect to the association of traits in inbred strains applies equally to selected strains. In the process of selection, as the selected lines diverge with respect to the selected character, all other traits must assume some value. Since many apparent associations of traits may be fortuitous, they should be tested by appropriate breeding experiments. Second, it must be remembered that there are many genotypic ways to achieve a particular phenotype. Two groups selected for a particular phenotype, even from the same foundation stock, might attain that phenotypic value by quite different combinations of loci and alleles. There is no reason, therefore, to expect that subsequent research using different selected lines will give the same results.

The success of selective breeding in general and the specific successes in behavioral studies encourage optimism that success may reasonably be expected in any attempt to select for any trait. The researches to date represent a trifling scratch on the surface of what can be accomplished. It is probably through the use of selective breeding, with its capacity to produce experimental animals to specification, that behavior genetics can make the greatest contribution to the methodology of the behavioral sciences in general.

CONCLUSION

The foregoing arguments should not be construed as implying that generalities applicable to more than one individual are not to be expected in behavioral research. They do imply that the breadth of the generalization is a matter for empirical investigation and not one for arbitrary a priori assumption. That is to say, the generality of the results obtained from a particular sample of organisms should not immediately be assumed to be broad. Confidence in the generality of results must be obtained by gradual accumulation of evidence over a range of genotypes. Uncritical generalizations across strains, species, genera, and even phyla are common in the behavioral literature. This appears to have been a general tendency of many researchers. By adopting a point of view which ascribes generality to a result only when it has been demonstrated, behavioral scientists should be able to forestall premature crystallization of opinion and simultaneously, by systematic exploration of animals of different genotypes, to open up fruitful avenues of research.

The utilization of genetics to provide control and manipulation of the animal material of research can provide an extraordinarily powerful analytical tool for uncovering behavioral laws. Genetic individuality is a fact of life. Ignoring it will not make it go away. If it cannot be licked, it can be joined with profit.

CHAPTER SIXTEEN
QUANTITATIVE GENETICS AND BEHAVIOR: OVERVIEW
AND PERSPECTIVE
John C. DeFries

It has long been clear that the question of whether a characteristic is hereditary or environmental is meaningless: "Every characteristic is both hereditary and environmental, since it is the end result of a long chain of interactions of the genes with each other, with the environment and with the intermediate products at each stage of development" (Lush, 1937, p. 77). Therefore, the existence of a characteristic implies a genetic basis; however, the mode of inheritance may be elucidated only when individual differences for the characteristic exist. Once such variation is found, various genetic analyses may be employed.

Most behavioral characteristics are biometrical in nature and are probably controlled by polygenic systems (Caspari, 1958, 1965); thus, the concepts of quantitative genetics are particularly applicable to the study of such characters. These concepts are now being applied to the study of behavior, partly because of the nature of the characteristics involved but perhaps also, in part, because of the publication of Falconer's highly readable text on the subject (Falconer, 1960).

Literature concerning the genetic analyses of quantitative behavioral traits will be selectively reviewed in this chapter. From such a review, it is hoped that worthwhile suggestions for future work in experimental behavior genetics will emerge.

INBRED-STRAIN COMPARISONS

Inbred strains provide a simple index of the genetic component. Partly because of their availability, much early work in mammalian behavior genetics employed strain comparisons. These studies, previously reviewed in some detail (Broadhurst, 1960; Fuller and Thompson, 1960; McClearn, 1965), have demonstrated a genetic component for a wide range of behaviors and have indicated that genetically different strains often display behavioral differences. In fact, as Hirsch (1963, p. 1439) has noted, "When different strains within a species are compared, it actually becomes a challenge *not* to find differences in one or more behaviors."

Strain comparisons, by themselves, yield little information regarding the mode of inheritance. However, when strain comparisons are followed by appropriate crosses, the mode of transmission may be determined. Gene differences at only one or two loci may be isolated. Or, as is more usually the case with behavioral characters, a polygenic system may be indicated. Such polygenic or "quantitative" traits usually exhibit a continuous variability, as contrasted with qualitative traits which may be readily assigned to one of a few distinguishable classes.

Although the value of using inbred strains has been questioned (Falconer, 1960, pp. 272–275), they may still serve a useful purpose in future behavior-genetics research. They will continue as a valuable and easy index of heritable differences and will also provide *prima facie* evidence for correlations of various behavioral, morphological, and physiological measures, although such correlations may prove to be fortuitous when analyzed in more detail. The apparent association between alcohol preference and alcohol dehydrogenase in inbred mice, recently described by McClearn (1965), is an excellent example. C57BL mice consumed substantially more ethanol when both a 10 percent ethyl alcohol solution and water were available than did mice of the DBA/2 strain, and they also had much higher levels of liver alcohol dehydrogenase activity. However, when an F_2 generation resulting from these two strains was tested, the correlation between alcohol preference and alcohol dehydrogenase was found to be essentially zero. Such fortuitous associations may frequently be found when only a few inbred strains are examined and, if a linkage disequilibrium exists, may even persist in the F_2 generation; however, this disequilibrium would be expected to decrease with additional generations of random mating.

Inbred strains are also useful for assaying for possible genotype-environment interactions, i.e., differential responses of different genotypes to environmental influences. For example, differential effects of prenatal maternal stress on offspring behavior in inbred strains of mice have been reported (Thompson and Olian, 1961; Weir and DeFries, 1964; DeFries, 1964). Such results illustrate the need of testing for generality before formulating general theories based solely upon environmental influences. Since the phenotype is a function of both the genotype and the environment, it would seem that both variables should be included in any general theory of behavior.

Inbred strains will continue to be useful for synthesizing random-mating populations of known gene frequency, which may be readily reconstituted at a later date. Much useful information may result when the techniques of quantitative genetics are applied to such populations, although problems of generality will still exist.

As is described in the chapters by Bruell and Broadhurst, various components of genetic variance may be estimated from appropriate crosses of inbred strains. Considerable caution should be exercised, however, when generalizing from the results of such studies. In order to ensure a genetic component, strains are often chosen which are known to display behavioral differences; thus, such studies may be expected to overestimate the genetic component, when compared with results from similar studies employing strains chosen at random or segregating populations.

HERITABILITY

It is appropriate to begin this section by quoting from a recent review by McClearn (1963, p. 234): "In terms of application of current genetic theory and procedure, behavioral genetics lags behind. For example, one of the central concepts of modern genetics is that of *heritability*. . . . Further development of behavioral genetics will require the precise estimation of the heritabilities of a broad range of behavior patterns."

Definition

Lush (1940) first defined the term "heritability" "as the fraction of the observed variance which was caused by differences in heredity." He continues, "This fraction is a statistic describing a particular population. It can be made larger or smaller if either the numerator or the other ingredients in the denominator can be altered. Thus it may vary from population to population for the same characteristic and may vary from one characteristic to another even in the same population." Thus it was explicitly stated in 1940 that heritability is a function of both the trait and the population in which it is measured, a point which has recently been emphasized by workers in behavior genetics (Hirsch, 1963; Hadler, 1964).

The concept of partitioning the total variance into genetic and environmental components is still older. For example, Wright (1920) analyzed the relative importance of heredity and environment in determining the piebald pattern in two stocks of guinea pigs. One stock was a random-mated control population, and the other was an inbred line derived from the same foundation population. From various correlations between relatives, it was deduced that about 42 percent of the variance in patterns in the control stock was determined by heredity, whereas only about 3 percent was determined by heredity in the inbred family. The environmental variance was about the same in the two stocks but accounted for about 97 percent of the variance in the inbred family and only about 58 percent of the variance in the random-bred stock.

This is an elegant demonstration of the fact that genetic variance is a function of the population in which the trait is measured. It is of historical interest to note that Wright introduced the symbol h^2 as the degree of determination by heredity in this 1920 paper.

In the first edition of Lush's classic text, *Animal Breeding Plans* (Lush, 1937), and in his 1940 paper, the distinction between the additive and nonadditive components of genetic variance was recognized. Lush (1949) later defined the ratio of the additive genetic variance to the phenotypic variance as "heritability in the narrow sense," and the ratio of the total genetic variance to the phenotypic variance as "heritability in the broad sense," where the total genetic variance includes the additive as well as the dominance and epistatic components. Others (see Chapter 11) prefer to use the terms "heritability" and "coefficient of genetic determination," respectively, for these ratios.

Fisher (1951) has criticized the concept of heritability on the grounds that the degree of managemental control and the accuracy of measurement are reflected in this ratio and refers to it "as one of those unfortunate short-cuts, which have often emerged in biometry for lack of a more thorough analysis of the data." However, Johansson (1961, pp. 9–10) has presented a rather convincing rebuttal.

Use

Heritability provides "a partial description of the causes of the variation" in a population and yields a "quantitative statement of the relative importance of heredity and environment" (Lush, 1943, p. 88). In addition to its descriptive role, heritability (in its narrow sense) is also predictive, since it is equivalent to the regression of the breeding value of an individual on its phenotypic value (see Chapter 11). Thus it is possible to estimate both the breeding value of an individual and the progress to be gained from various breeding systems, if the heritability of the trait is known. The predictive nature of this ratio was discussed by Lush in 1937 (Lush, 1937, p. 111).

The concept of heritability may have uses in behavior genetics, in addition to the classical ones previously described. Because of its predictive nature, it might be useful for genetic counseling, if the heritability of various quantitative traits of concern in the human population were known. It might also serve as an index of the susceptibility of a trait to improvement through environmental manipulation. For example, let us assume that two traits (X and Y) with similar phenotypic variances have heritabilities of 80 and 10 percent, respectively, when measured in the same population. This difference suggests that trait X is relatively less influenced by environmental fluctuations that impinge upon the population than trait Y. Therefore, if we begin to manipulate or "select the environment," trait Y may be expected to be more affected than trait X. If the intensity of this environmental selection were sufficiently high, considerable change might also be brought about in trait X. If accurate estimates of the components of the phenotypic variance were available with human traits, prediction equations could be constructed which would indicate the improvement that might be realized by controlling existing environmental variation. Such control would have to be maintained in order to sustain any improvement realized by this technique. These estimates might even suggest which characters are more amenable to improvement through new and different environmental regimes. Evidence that more heritable traits are less affected by extreme environmental modifications than less heritable traits has been reported (Lush, 1949, pp. 372–373; Johansson, 1961, p. 169).

As has been discussed by Falconer (1960, pp. 336–337) and Bruell (1964a), a knowledge of the size of the additive and nonadditive components of genetic variance may also be useful for inferring how closely a trait is related to fitness. Characters with a large additive component (high heritability) might be expected to have little adaptive value, whereas those with little additive genetic variance (low heritability) might be closely related to fitness, since the additive genetic variance associated with fitness traits would be expected to have been exploited by natural selection. In general, this seems to be the case; traits such as fertility have lower heritabilities than those apparently less closely related to fitness (Falconer, 1960, p. 337).

The degree of heterosis observed in hybridizing experiments with inbred strains of mice has been used by Bruell (1964a) as an index of the nonadditive genetic component. Results on two behavioral tests (exploration of a strange environment and wheel running) and two physiological measures (serum cholesterol and hematocrit percent) were compared. Heterosis was observed with the behavioral tests, whereas intermediate inheritance of the physiological measure was indicated, suggesting that the former are more closely associated with fitness than the latter.

Estimation

As discussed in Chapter 11, several techniques are available for the estimation of heritability. An excellent example of the use of correlations of individuals of known relationship is found in a paper by Willham et al. (1963), in which genetic variance in a measure of avoidance learning by swine is reported. Paternal half-sib correlations and full-sib correlations were estimated and indicated that approximately 50 percent of the variance among pigs within relatively homogeneous groups was genetically additive. The design and analysis employed in this study may serve as a model for future studies in which the heritability of some characteristic in a litter-bearing animal is to be estimated.

One assumption underlying the technique of estimating heritability from the correlation of relatives is that the environmental deviations of the relatives are uncorrelated. With laboratory animals, this can be achieved to some extent by randomization of the subjects within the range of environmental conditions usually encountered in the laboratory. Several useful dodges are available, however, when such correlations exist. For example, in the study mentioned above (Willham et al., 1963) several sires were each mated to several different females. From the resulting hierarchical classification, it was possible to estimate a component of variance attributable to sires (between sires) and a component resulting from differences between dams mated to the same sires (between dams within sires). A comparison of these components will indicate the importance of maternal effects which might result in an environmental correlation among offspring produced by the same female (see Falconer, 1960, pp. 172–176).

Data of parent-offspring comparisons may also contain environmental correlations, especially when data from a wide range of environmental conditions and time periods are analyzed, i.e., when parents and off-spring are both reared in the same herd, flock, or group, but when data from several such groups are included in the analysis. When several sires are each mated to several dams within the group, as is usually the case with domestic animals, this difficulty may be circum-vented to some extent by calculating parent-offspring correlations or regressions on an "intrasire basis," i.e., computing correlations or regressions within groups of offspring by the same sire (Lush, 1940). This restricts the analysis to that variance found within groups of females mated to the same sire and removes any environmental com-ponent between groups which might add to the covariance of parents and offspring. In order to reduce further these environmental causes of covariance, analyses on a within-sire, herd, breed, year-season, or other basis are now commonly employed in animal-breeding research.

Parent-offspring correlations for various human characters have been reported (Erlenmeyer-Kimling and Jarvik, 1963), but environmental correlations are almost certainly reflected in such data. However, as with data from domestic animals, this confounding of environmental and genetic differences may be circumvented to some extent by a more refined analysis. If, for example, data were available on the socioeco-nomic level and if environmental differences between socioeconomic groups were thought to be important, an analysis on a within-socioeco-nomic-group basis may yield a more precise estimate of heritability. Such an analysis would remove any environmental component between groups, which might have otherwise added to the covariance of the relatives, but would leave differences between the groups unanalyzed with regard to their genetic and environmental causes. Any other fac-tors, such as educational level, which might be considered important could also be included in such an analysis. Or, if these factors were sufficiently quantifiable, a multiple-regression analysis could be per-formed instead, provided that the variables were not hopelessly con-founded.

Although parent-offspring correlations are often reported, the regres-sion of offspring on parent may yield a more accurate estimate of herit-ability. For example, parent-offspring correlations will be decreased if the parents are a selected group. The regression of offspring on parent, however, will not be systematically biased by such selection (Lush, 1940).

The regression of offspring on parent is also preferable to the cor-relation for other reasons. The covariance of offspring and parental values contains one-half of the additive genetic variance (V_A), assuming random mating, regardless of whether single values or means are compared (Falconer, 1960, pp. 152–159). The variance of single offspring values and single parental values (if unselected) estimate the phenotypic variance (V_P), and, if mating is random, the variance of the midparental values will estimate $V_P/2$. However, the variance of the mean of N offspring will estimate $V_P[1 + (N - 1)t]/N$, where t is the phenotypic correlation of the offspring values which contribute to the mean. As expected, this value will equal V_P/N, if $t = 0$, i.e., if the offspring values are uncorrelated, and will equal V_P if $t = 1$.

From the above considerations, it follows that the regression of offspring on parent will estimate $\frac{1}{2} V_A/V_P = \frac{1}{2}h^2$, where h^2 is the heritability, when single parental values are compared either with single offspring values or with the mean of N offspring. The regression of offspring on the midparental value will directly estimate h^2, also regardless of the number of offspring included in each comparison.

Estimating heritability from parent-offspring correlations, however, is somewhat more complicated. When both single parental and single offspring values are compared, $h^2/2$ is again estimated. However, the correlation of single offspring values and midparental values is estimated to be $(\frac{1}{2})^{\frac{1}{2}} (h^2)$. The correlation of single parental values with the means of each of N offspring, i.e., where each offspring mean is based on the same number of observations, is estimated to be $(h^2/2)$ $\{N/[1 + (N - 1)t]\}^{\frac{1}{2}}$. The correlation of midparental values and the means of each of N offspring is estimated as $h^2\{\frac{1}{2}N/[1 + (N - 1)t]\}^{\frac{1}{2}}$. Therefore, h^2 can be estimated from such correlations. However, the phenotypic correlation of sibs must be known and, as assumed above, the same number of sibs must be included in each offspring mean.

If assortative mating has occurred, as is likely to some extent in human populations, the variance of the midparental values will be increased, as well as the covariance between full sibs. The regression of offspring on midparental value, however, is little affected and can still be used to estimate heritability (Falconer, 1960, p. 171). Various other factors which should be considered when estimating heritability from the resemblance between relatives, e.g., problems due to such factors as unequal variances in males and females or to maternal effects, have been discussed in some detail by Falconer (1960).

Heritability may also be estimated from the response to selection (Lush, 1940); this estimate has been called the "realized heritability." Guhl et al. (1960) and Craig et al. (1965) have utilized this technique to estimate the heritability of social aggressiveness in chickens. Manning (1961) has used a similar technique to estimate the heritability of mating speed in Drosophila. These studies will be discussed in more detail later.

Twin Studies

The value and limitations of twin data in the study of the inheritance of human mental characteristics have been examined in detail (McClearn, 1963). McClearn concluded that several possible biases may exist in such studies. In addition to the problems of diagnosis of zygosity, the validity of the assumption of equal environmental variance within sets of fraternal and identical twins is questionable.

The use of twin data for estimating the heritability of various traits in domestic animals has been discussed by Johansson (1961). When identical-twin data were first used to estimate heritability, it was assumed that the variance between pairs was entirely genetic, whereas variance within pairs was an estimate of the nongenetic variance. Therefore, the intraclass correlation of identical-twin pairs was considered as a direct estimate of heritability. However, Johansson concludes that neither of these assumptions is correct. The variance between pairs will include both the additive and the nonadditive components of genetic variance, as well as any variance due to genotype-environment interactions and to the effects of prenatal and postnatal environments common to members of the same pair. In addition, the variance within identical-twin pairs is probably an underestimate of the environmental variance in the population, since it contains none of the variance due to genotype-environment interactions and since members of the twin pair are more contemporary than pairs picked at random. Thus, the intraclass correlation of identical twins tends to yield an inflated estimate of heritability. Several such estimates from dairy cattle twin data clearly exceeded estimates obtained from analyses of field records.

As with inbred-strain comparisons, twin studies may still be of value in future behavioral analyses. They may be useful for assaying for a possible genetic component and, when coupled with other studies that yield more accurate estimates of the genetic and environmental variance, may be valuable material for analyses of the importance of early environmental factors.

GENETIC CORRELATIONS

Phenotypic correlations between *different* traits on the *same* individual may be due to genetic correlations (correlations of the breeding values) or environmental correlations. Genetic correlations may be due to pleiotropy (manifold effects of genes) or to a linkage disequilibrium, although the effects of the latter should be relatively temporary. The relative importance of these factors in determining the phenotypic correlations of various behavioral characteristics or of behavioral and physiological or morphological traits would seem to be of interest. Two such analyses involving behavioral traits have now been reported (Siegel, 1965; Hegmann and DeFries, in preparation).

The genetic correlation, like heritability, may be estimated from the resemblance of relatives or from the results of selection experiments. These techniques have been discussed by Falconer (1960, pp. 312–319).

SELECTION EXPERIMENTS

Mather (1941) has stated that the most appropriate way to approach the study of polygenic inheritance is through the analysis of the effects of selection. These effects are the result of both the nature of the selection applied and that of the variation available; thus, since the type of selection may be controlled by the investigator, inferences regarding the nature of this variation may be drawn from the results of selection experiments.

Although a carefully designed selection experiment will yield a precise estimate of various genetic parameters, many of the selection studies involving behavioral traits have merely demonstrated a response to selection. Such a response indicates a genetic component; however, such studies could have provided much additional information, e.g., estimates of realized heritabilities, genetic correlations, selection limits, and possibly even some estimate of the number of loci involved, if the techniques of quantitative genetics had been rigorously applied.

Maze Learning by Rats

Tolman (1924) reported the results of what was apparently the first selection experiment for maze learning by rats. The foundation population consisted of 82 white rats of heterogeneous ancestry. From this population, nine "supposedly bright" pairs and nine "supposedly dull" pairs were mated to produce offspring representing the first selected generation (S_1)[1] of the bright and dull strains, respectively. The criterion of selection for breeding was "a rough pooling of the results as to errors, time, and number of perfect runs" (Tolman, 1924, p. 4). A second selected generation was also produced "by further selective breeding within the bright litters and dull litters and in each case a brother was mated with a sister" (Tolman, 1924, p. 4).

In general, the results of the first generation of selection were remarkable. Because of the completeness of the data presented, it is possible to estimate both the selection differential (difference between the means of the bright and dull parents) and the response to selection (difference between the means of the resulting bright and dull S_1 offspring). The ratio of this response to the selection differential provides a rough estimate of the realized heritability regarding each of the three factors considered in the selection. These estimates are as follows: errors, 0.93; time, 0.57; and number of perfect runs, 0.61.

Differences in all measures in the S_2 generation, however, were less than in the S_1. As Tolman suggested, this may have been due to inbreeding or to some extraneous factor.

Although the selection experiment initiated by Tolman (1924) was terminated, the general problem was not dropped. A preliminary report of a subsequent selection experiment was published a few years later (Tryon, 1929). In order to improve the reliability of the measurement, a 17-unit T-maze which automatically delivered each subject into the entrance of the maze and collected it at the end was developed. Objectivity of scoring was ensured by a device that automatically recorded the path of the subject through the maze. A detailed description of this maze was published separately (Tolman et al., 1929).

A heterogeneous sample of rats was deliberately chosen to serve as the foundation population. Selection was based on the total number of

[1] The F_1, F_2, . . ., generation notation employed by Tolman (1924) now, by convention, refers to the number of successive generations resulting from a Mendelian cross rather than to the number of generations of selection.

entrances into blinds from days 2 to 19, following a preliminary run of 8 days to acquaint the subjects with the maze. Some intentional inbreeding was again practiced. Because of problems with fertility, the S_1 offspring of dull parents were pooled with a "median group" and the pooled mean was compared with that of the S_1 offspring from bright parents in order to assess the response to the first generation of selection. Presumably for this reason, the difference between these means was not large. Because greater selection pressure was applied in the next generation or possibly because information from relatives or "kin performance" was also considered in selection, a considerably greater divergence was obtained in the S_2 generation.

Tryon (1940) later presented a summary of the results of 18 generations of selection in this experiment. A fairly consistent divergence of the strains was noted through generation VII, at which time little overlap of the distributions of the strains occurred. However, only negligible additional response was noted thereafter.

Selection for maze-learning ability in rats has also been reported by Heron (1935, 1941). The criterion of selection was total errors in a 12-unit automatic maze; however, some selection for fertility and some deliberate inbreeding were also practiced. The results of the first four generations of selection were published in 1935, and a progressive divergence of the selected lines was observed. Heron (1941) later reported the data of generations V through XVI. A relatively consistent response to selection was obtained, and little overlap of the distributions of the strains was observed by generation XVI. However, some intergeneration variability was noted. Since the two strains tended to vary together, it was suggested that these fluctuations were probably due to uncontrolled environmental factors.

More recently, W. R. Thompson (1954) reported the results of six generations of selection for performance in the Hebb-Williams maze. Two-way selection resulted in a marked and consistent change in mean errors of both the bright and dull strains. It is of interest to note that deliberate inbreeding was again practiced. It was later discontinued due to infertility, but it was stated that "when numbers have been sufficiently built up again, inbreeding will be resumed" (Thompson, 1954, p. 218).

Spontaneous Activity of Rats

Rundquist (1933) conducted a two-way selection experiment for number of revolutions in a rotating drum. Selection during the first four generations was "purely phenotypic"; i.e., the selected lines were not closed. However, later selection was on a within-line basis. Systematic inbreeding was not practiced in this experiment. The means of both the active and inactive groups tended to increase through generation IV, although some divergence was obtained. The activity of the inactive strain decreased rather consistently after generation IV, although no consistent trend was noted with the active strain during this period. Nevertheless, by generation XII, little overlap of the distributions of the two strains was observed. It should be noted that the greatest response to selection occurred after the lines were closed, thus demonstrating the efficacy of selecting strictly on a within-strain basis.

The inactive strain died out after 25 generations of selection. The results of 10 generations of selection for a new inactive strain, as well as data from generations XXX to XXXV of the original active strain, have been subsequently reported by Brody (1950). From the results of this study, she concluded "that there is a major controlling gene pair which differentiates these active and inactive strains of rats but that this gene pair alone does not by any means account for all of the differences in activity" (Brody, 1950, p. 287). A similar conclusion was reached by Brody from the results of an earlier study (Brody, 1942), and it seems that the criticism by C. S. Hall (1951, pp. 323–324) of this conclusion would still apply. The results suggest that a multiple-factor hypothesis should be assumed until more definitive evidence for a major gene effect is obtained.

Emotionality in Rats

In his review article of studies in behavior genetics, Hall (1951) presented the results of 12 generations of selection for "emotionality" in rats. Individuals were placed for 2 minutes a day on each of 12 days in a large, brightly lighted circular open field. Urination and defecation scores were recorded, with total scores ranging from 0 to 12. A fairly consistent response during the first nine generations of selection was observed in the high-emotional line, but little additional change was realized thereafter. In contrast, the maximum response to selection was obtained in the low-emotional line in generation I, and the mean scores were rarely lower in later generations. However, both strains in the S_1 generation had lower scores than the parental population. Scores of both strains then increased in the S_2, decreased in the S_3, etc., although a progressive divergence of the strains occurred. Therefore, the apparently larger initial response in the nonemotional strain may, in fact, have been due to random environmental fluctuations which affected both strains in the same direction.

Broadhurst (1960) has reported the results of 10 generations of two-way selection for total number of fecal boluses dropped during four daily tests in an open field. However, some modification of the criterion of selection was required in the "nonreactive" strain during the later generations. Deliberate inbreeding was again practiced. There was considerable divergence in the defecation scores of the strains, although the response appeared to be greater in the nonreactive strain. Some divergence in ambulation scores of the two strains was also observed, although both strains tended to increase somewhat in this measure. Nevertheless, the possibility of a negative genetic correlation of bolus number and ambulation is suggested. Broadhurst states that this genetic correlation "could be investigated by the technique of selecting separately for the two traits on different populations and observing the correlated response in the other" (Broadhurst, 1960, pp. 54–55). However, it would appear that a genetic correlation could be estimated from the data of this experiment (see Falconer, 1960, pp. 318–319).

Saccharin Preference of Rats

Nachman (1959) reported the results of two generations of selection for saccharin preference among rats. In a dietary experiment, two groups of Sprague-Dawley albino rats were given daily tests of preference between a 0.25 percent sodium saccharin solution and tap water. No difference in saccharin preference between the two dietary groups was found, but considerable individual differences within the groups were noted. Therefore, extreme saccharin preferrers and extreme water preferrers were bred. Significant differences among the two resulting progeny groups were found, indicating a genetic component for saccharin preference.

Susceptibility to Audiogenic Seizures in Mice

Frings and Frings (1953) successfully produced by selective breeding four strains of mice which have predictable susceptibilities to audiogenic seizures. One strain had a high susceptibility (90 to 100 percent) to clonic-tonic seizures when tested daily from 15 to 50 days of age but rarely died in this type of seizure. A second strain had a high susceptibility to clonic seizures but had few clonic-tonic seizures. A third strain had a very low susceptibility to seizures during the testing period, whereas a fourth strain had seizures regularly from 17 to 27 days of age, but not thereafter. This successful production of strains with differing susceptibility and pattern of seizures demonstrates the magnitude of genetic variation that must have been present in the original albino random stock and the potential for developing strains with predictable behavioral differences.

Wildness and Tameness in Mice

An analysis of the inheritance of "wildness and tameness" in mice, as measured by the time required to run 22 feet in a runway, was reported by Dawson (1932). As a part of this analysis, the fastest animals in a wild strain and the slowest in a tame strain were each selected for four generations. Selection was found to increase the initial difference between the tame and wild stocks, primarily by reducing the running speed in the tame strain.

Aggressiveness in Chickens

Four generations of two-way selection for aggressiveness in chickens, based on the number of paired encounters won and also on social rank, have been reported by Guhl et al. (1960). Significant differences in percentage of encounters won and social rank were found, beginning with the S_2 generation. It is of interest to note that realized heritabilities were estimated in this study. These estimates for percent of flock dominated and percent of initial pair encounters won were 0.18 and 0.22, respectively. These values were calculated from unweighted means of individual-generation estimates, although cumulative estimates yielded similar values. Considerable intergeneration variability in these estimates was found, suggesting that caution should be exercised in generalizing from the results of only one generation, especially when the effective population size is small.

The results of five generations of selection for social dominance in initial pair contests in each of two breeds of chickens have been published (Craig et al., 1965). Large and apparently symmetrical responses were obtained in each breed. Estimates of realized heritability for dominance scores in the two breeds were as follows: White Leghorn, 0.16, and Rhode Island Red, 0.28. The results of this study suggested a polygenic mode of inheritance with intermediate gene frequencies in the foundation populations. The selected strains were also found to differ with regard to percentage of contests with aggressive behavior, percentage won, and physical severity of the contest.

Behavioral Traits of Drosophila

Some considerable methodological sophistication has recently been applied to the study of the inheritance of behavioral characters in Drosophila. Hirsch (1959) has developed an elegant multiple-unit classification maze for the mass screening of *Drosophila melanogaster* and has reported the results of a long-term, two-way selection experiment for geotactic score (Hirsch and Erlenmeyer-Kimling, 1961). Erlenmeyer-Kimling and Hirsch have also utilized the chromosome-assay method to study the role of the three major chromosomes of *Drosophila melanogaster* in geotaxis. An unselected and two selected populations were compared; the results indicated that "genes distributed over most of the genome . . . influence the response to gravity" (Erlenmeyer-Kimling and Hirsch, 1961, p. 1069).

Selection for phototaxis in *Drosophila melanogaster* has also been reported. Using another mass screening technique, Hirsch and Boudreau (1958) conducted a two-way selection experiment for phototactic response and observed a fairly consistent divergence between strains. From a comparison of the variances of the parental and selected lines, heritability was estimated to be approximately 0.6.

Hadler (1964) has reported the results of 30 generations of selection for phototaxis in *Drosophila melanogaster*. Two-way selection in each of two multiple Y-unit mazes resulted in a clear divergence of the selected strains. Heritability estimates were again computed from comparisons of variances in parental and selected lines and were similar in magnitude to that reported by Hirsch and Boudreau (1958).

The results of 25 generations of selection for mating speed in *Drosophila melanogaster* have been reported by Manning (1961). It is of interest to note that this selection experiment was replicated; i.e., two lines were simultaneously selected for slow mating speed and two were selected for fast mating speed. An unselected control was also maintained, but it was not measured during the first few generations of selection. The effects of selection were immediate and continued for about seven generations, with little consistent divergence in later generations. Realized heritability was estimated from the regression of the divergence of response on the cumulative selection differential through generation VII and a value of about .3 was obtained. Flies from generations XVI through XIX were also tested in a small "open arena" and scored for the number of centimeter squares entered. The two slow lines showed similar activity levels, as did the two fast lines, but the activity of the slow-mating lines was significantly higher than that of the fast-mating lines. Control flies, however, were not significantly different from the slow-mating lines in activity level. Differences among the lines in the "lag time," i.e., time from the introduction of a male into an observation cell containing a stock female to the beginning of courtship, and the "courtship intensity" were also noted.

Manning (1963) has also attempted to select for mating speed in Drosophila, based on the behavior of only one sex. Replicate lines in which males were selected for fast mating speed did not respond to 20 generations of selection. Also, no response was obtained with lines in which females were selected for slow mating speed. A response was realized in two lines in which the males were selected for slow mating speed, although differences in the magnitude and consistency of response in these two lines were noted. In agreement with the previous study, slow-mating flies were again found to have reduced sexual activity. However, slow-mating flies had a lower level of general activity in the later study, whereas slow mating speed was associated with increased activity in the previous study. This inconsistency was not considered to be surprising, since it was felt that the behavior under selection was highly complex and that slow mating speed could be achieved in several different ways.

Two attempts to select for high and low activity in Drosophila melanogaster have been reported by Ewing (1963). In the first experiment, the criterion of selection was the speed of traversing a series of six small tubes connected by glass funnels. Fifty flies of one sex were introduced into the first tube, and the first or last ten to emerge in the last tube were selected. Two high-active lines and two low-active lines were selected, and an unselected control was maintained. Offspring of the later generations were tested in three different apparatuses, two of which measured activity in groups of flies, whereas the third (the arena) measured individuals. The selected lines responded differently in the first two measures, but no significant difference was found with the third. It was concluded that the level of "reactivity" of the flies had been changed by selection but that "spontaneous activity" had not been affected.

In the second experiment, the criterion of selection was the speed with which *individual* flies traversed a series of chambers connected by funnels. Selection was again replicated, and a control population was also measured. An immediate response to selection was realized in the inactive strain, but little consistent response was noted in the later generations. Selection for increased activity, however, was not effective. All lines were later tested in the arena, but no significant difference between the selected and control lines was found. Additional experiments suggested that selection for inactivity had resulted in flies which were "claustrophobic," i.e., "which were unwilling to enter confined spaces, such as funnels," (Ewing, 1963, p. 377) due to their reaction to visual stimuli.

Implications for Future Selection Research

An evolution in the design of selection experiments is evident from this review. Most of the early studies employed two-way selection; however, the lines were not replicated, no unselected control was maintained, and deliberate inbreeding was often practiced. In contrast, the design and thorough analysis employed by Manning (1961) should serve as a model for future selection experiments.

The importance of maintaining replicated selection lines should be emphasized. Considerable intergeneration variability is encountered in selection experiments. When lines are replicated, the amount of variation between lines selected alike can also be measured. From limited evidence (DeFries and Touchberry, 1961; Falconer, 1960, pp. 208–212; Marien, 1958), it appears that this variability of response to selection may be large and is probably a function of the effective population size in each line. In addition, fortuitous correlations between the trait under selection and other characters may often occur when only one line is selected. However, if similar associations are noted in each of several replicates, the probability of this correlation being fortuitous is greatly reduced.

The importance of maintaining an unselected control population should be obvious. With such a population, the effects of intergeneration environmental fluctuations may be measured as well as any effects of inbreeding, provided that the effective population size in the control and selected lines is approximately equal. In addition, when a control line is measured in a two-way selection experiment, the response in each line may be measured by its deviation from the control; thus, the degree of asymmetry of response to selection may be ascertained. Once such an asymmetrical response is indicated, the interesting problem of causality may be studied, i.e., is it merely a reflection of scalar problems or is it due to some "physiological limit," etc.?

More long-term selection experiments for behavioral traits are needed. With such studies, "selection limits" should eventually be encountered. Reverse selection may then be attempted in order to determine whether any genetic variance for the trait is still present. Several possible explanations for the cause of selection limits have been offered (cf. Falconer, 1960, pp. 219–224); however, more work on this interesting and practical problem is required.

Estimates of both heritability and the genetic correlation may be obtained from the resemblance between relatives in unselected populations and also from the response to selection. With relatively little additional effort, such estimates may be obtained from the foundation population before selection and then compared with realized estimates based on the response to selection. Any serious discrepancies between these estimates might point to interesting problems for future research.

The problem of maternal effects in selection experiments has been discussed in some detail by Broadhurst (1960), who has criticized the earlier selection studies which did not control for maternal influences: "But Hall's study suffers from a fatal flaw, as, indeed, do most of the studies on mammals reviewed here, but one which is particularly relevant to the inheritance of emotional tendencies. It is the failure to control for the most important environmental factor of maternal influences" (Broadhurst, 1960, p. 17). As Falconer (1965) has indicated, maternal effects may cause some interesting problems in selection studies. However, it is questionable whether this problem is sufficiently serious to justify the rigorous control suggested by Broadhurst, i.e., the laborious technique of crossfostering. If a response to selection has been realized, it is evidence for a genetic effect, either maternal or fetal in origin. A simple reciprocal cross between lines selected in opposite directions will indicate the relative importance of the fetal and maternal genotypes in this response. Or, if a maternal effect is suspected or known when a selection experiment is begun, one may select on a within-litter basis in order to avoid any difficulties due to environmental differences between litters. Thus, it would seem that the technique of crossfostering is not a requisite of a well-designed mammalian selection experiment.

One disturbing practice has come to light from this review: that is the tendency in the early studies, as well as in a few more recent ones, to inbreed deliberately while selecting. Although the objective of these studies was the production of highly inbred lines with uniform behavioral differences, such inbreeding causes serious difficulties. First, inbreeding results in a decrease in the genetic variance within lines; thus, the potential response to selection is decreased by this practice. Second, a high rate of inbreeding will almost certainly result in some problems with fertility. In fact, because of the small number of parents that are usually selected each generation, it is difficult to avoid such problems, even when the rate of inbreeding is intentionally minimized.

Selection for a quantitative trait should eventually lead to homozygosity at loci influencing that characteristic, unless selection is for some nonadditive gene effects or is in opposition to natural selection. Such strains, if still heterozygous at some other loci, may not suffer the serious consequences of high inbreeding. Nevertheless, if the primary objective is the production of inbred strains with behavioral differences and if one is unwilling to wait for more extreme divergence, selection coupled with deliberate inbreeding may be appropriate. One justification given for the production of highly inbred selected strains is that a lack of homozygosity precludes "any possibility of making a genetic analysis" (W. R. Thompson, 1954, p. 216). However, much may be learned from the response to selection before homozygosity appears.

Strains resulting from selection experiments may be valuable material for future research. Examples of the use of such strains have been reviewed by McClearn (1963). This use alone may justify the time and labor required for a careful selection experiment. However, such selected lines should be periodically reexamined; thus, if the strains have drifted from their selected levels, it will be possible to avoid the assumption of a difference that does not exist.

CONCLUDING REMARKS

Studies in experimental behavior genetics began early in the history of quantitative genetics and, as evidenced by this volume, much work is currently in progress. Various quantitative behavioral characters have been subjected to genetic analyses; however, the more rigorous methodology of quantitative genetics has only recently been applied.

More sophisticated studies will be required for behavior-genetic analysis of human populations. As mentioned previously, estimates of the heritability of various human behaviors may be useful, but accurate estimates will be difficult to obtain. It was suggested that the regression of offspring on parental value, calculated on a within-group basis, may yield a more precise estimate of heritability when relatives share a common environment within a group and environmental differences exist between groups, e.g., socioeconomic groups.

If the relative magnitude of the genetic and environmental components of the phenotypic variance were known for some human behavior, e.g., performance on some mental test, one could predict the response that might be achieved if the environmental variation were controlled. Thus, individuals picked at random from the population and reared in exactly the same environment as an individual who scored 20 points above the population mean would be expected to have scores about $(e^2)(20)$ above average, where $e^2 = V_E/V_P$. Such predictions may be of doubtful value, because of the impossible requisite of complete environmental control. Nevertheless, estimates of these genetic and environmental components may suggest the potential for improvement through environmental manipulations. For example, such information might suggest which behaviors or components of behavior will be more amenable to improvement through remedial training.

All the behaviors examined in the studies reviewed above were quantitative in expression. Traits which do not manifest a continuous variation may, nevertheless, have a quantitative genetic basis. Schizophrenia is an example of an apparently discontinuous character. Although many studies on the genetics of schizophrenia have been conducted, the exact mode of inheritance remains unknown. Recessivity theories, dominance theories, and polygenic theories of inheritance have all been postulated (Fuller and Thompson, 1960). Some question exists as to whether schizophrenia is truly an "all-or-none" trait or whether there is a gradation of symptoms. However, even if the trait were truly discontinuous in expression, it might still have a quantitative genetic basis. Examples of "threshold characters," like some forms of disease resistance, which have discontinuous distributions but are multifactorial in inheritance, are well known (Falconer, 1960, pp. 301–311).

Haldane (1963) has suggested combining qualitative and quantitative genetic studies in human populations: "My plea is that in future work on any character—for example, on haptoglobin frequency—some attempt should be made to accumulate anthropometric data on the people concerned." He cautions that some spurious correlations will be found, due to the presence of genetic disequilibria, but "that criticism will not apply to differences between brothers or sisters. If, for example, a particular main haptoglobin gene were responsible for a change of 1 centimeter in height, on an average (and that is, of course, most unlikely), then we should find these differences on comparing members of the same family" (Haldane, 1963, p. 43). When and if such studies are attempted, it is hoped that behavioral data will be collected as well. Evidence for major gene effect on quantitative behavioral traits of laboratory animals has been reported (cf. DeFries, Hegmann, and Weir, 1966).

In conclusion, it is clear that new and refreshing approaches to quantitative behavior-genetic analysis are needed; however, techniques long available have not yet been fully exploited.

CHAPTER SEVENTEEN
IMPLICATIONS OF BEHAVIOR GENETICS FOR GENETICS
R. C. Roberts

It is becoming increasingly apparent that the word "genetics" covers an extensive area of inquiry drawing upon personnel, techniques, and ideas from a diverse range of disciplines. In this chapter, I have been asked to look over the wall into the field of behavior genetics and to speculate on the possible implications of its cultivation for other genetic work. As my own activities cover only a restricted band of the genetic spectrum, I shall perforce think chiefly in terms of implications for that band, namely, quantitative genetics. This perhaps is not entirely inappropriate, for it seems that a good deal of behavior is affected by many genes.

In a sense, the question resolves itself into the implication of a part for the whole. Behavior genetics falls logically and unambiguously within the domain of genetics, without a qualifying adjective. When we think of some of the traditional branches of genetics—biochemical genetics, cytogenetics, developmental genetics, physiological genetics, and quantitative genetics—it is immediately obvious that the whole gamut is involved in the genetics of behavior. As a cross classification of genetics, it is sometimes convenient to employ taxonomic criteria, as instanced by bacterial genetics or (occasionally) mouse genetics, which again can accommodate behavior in several classes. "Behavior" genetics provides an example of a third dimension of classification, namely, by phenotype or, rather, by a group of phenotypes that have an identifying common denominator, though not necessarily much else in common. In this way, the term "behavior genetics" has the same validity as blood-group genetics or color genetics. From this perspective, the question is whether the genetics of behavioral phenotypes, by virtue of their particular kind of biological organization, has any implication for other aspects of genetics.

The answer of course depends on the degree of uniqueness of the biological organization of behavioral characters. We should not be too hasty to retort, "Of course these characters are unique." In terms of the primary action of the genes, as revealed by the molecular biologists, some behavioral characters may not be particularly different in their biological architecture from characters like growth or fertility. However, the range of what are termed behavioral characters is enormous, and it seems plausible that a learning process, for instance, requires the animal to store information in a way that involves some quite peculiar aspect of its biology, albeit that its genes are fundamentally concerned in the process.

As I see it, the chief value of behavioral characters in genetic work is contingent upon the fact that they may, in some cases, have a biological basis unlike that of the characters which have been explored and exploited to date. Much of the experimental work on the quantitative genetics of animals has been performed on a restricted range of characters. Weight and body dimensions, various aspects of fertility, and the popular bristle characters of Drosophila pretty well exhaust the list, as far as laboratory animals are concerned. Domestic animals more or less cover a similar range. Exceptions to this list come to mind, probably because they are exceptions. In view of this, the number of characters offered by behavioral science, when considerable thought has already been devoted to the methods of measurement and what the measurements mean, offers interesting possibilities. In quantitative genetics, at least, there is a need for new characters, and it will be of considerable value to discover whether the current thinking on quantitative systems withstands the test of new evidence.

We can consider briefly two illustrations of the ways in which the genetics of behavior may bear on our understanding of quantitative genetics. The first concerns the response to artificial selection. In his book, Falconer (1960) examines a small sample of selection experiments which seem to indicate that the response to selection may be expected to continue for about 20 to 30 generations, a range that accords well with general experience. Since then, Manning (1961), selecting for mating speed in Drosophila, obtained all his response in seven generations. Erlenmeyer-Kimling, Hirsch, and Weiss (1962), on the other hand, selecting for geotaxis in Drosophila, found a response that seemed to continue for at least 45 generations. Since then, Hirsch (private communication) says that the response continued, in fact, for over a hundred generations. These two experiments with behavioral characters have both produced quite unexpected results, but in opposite ways, and obviously underline the value of behavior in supplying new material for genetic investigation.

The second instance where behavior has produced results of genetic interest concerns the relative variance of inbred strains and their crosses. As far back as 1946, Mather commented that F_1s may show less phenotypic variation than their parental inbred strains. A considerable volume of literature has since accumulated, suggesting the general application of this phenomenon for a wide variety of characters in several organisms. Lerner (1954) comprehensively reviewed the evidence up to that time. Indeed, a finding like Falconer and Bloom's (1961) that inbred and F_1 mice did not differ in their variances, with respect to induced pulmonary tumors, may still be regarded as being somewhat unusual, except, it seems, in a whole range of behavioral characters. Several examples, quoted by Fuller and Thompson in their book (1960), show the F_1 variance of behavioral characters to be equal to, or intermediate between, the two parental inbred strains. An example provided by McClearn (1961) showed that the variance of an F_1 in activity score conspicuously exceeded that of either parental strain, even after rescaling. Whereas "nonbehavior" geneticists would have found this surprising, McClearn found it necessary to note the fact only, without comment. In short, it seems to be widely accepted that inbred animals show a remarkably uniform behavior, compared with crosses.

The excess variability of inbreds over F_1s or, indeed, outbred animals for characters other than behavior is pertinent to the widely accepted view that the evolution of an organism's genotype has established particular norms of development, a topic reviewed by Waddington (1957) and to which he had previously given the term *canalization*. Inbreeding, involving a radical reorganization of the genotype, is believed to lead to a loss of buffering ability, with the consequent greater liability of development, reflected in the increased variation commonly found. Lerner (1954) extended these ideas and argued that natural selection, favoring intermediate values, leads to an obligate level of heterozygosity in a population. From the concept of *genetic homeostasis* which he developed, it is easy to argue back to the variability of inbreds. Why, then, should behavioral characters not conform to the general pattern? What is even more extraordinary is that the behavioral characters themselves cover a diverse range of biological function and yet, it seems, provide results largely consistent with one another. The solution to the quandary must be sought in evolutionary terms, but what it might ultimately be is as yet obscure.

The orthodox view, again, is that an organism must evolve a fair degree of phenotypic uniformity while it would be evolutionary suicide for it to exhaust its genetic variation. This is consistent with the evidence that wild populations, although seemingly well adapted to particular contemporary niches, nevertheless contain adequate evolutionary reserves in the form of latent genetic variation, revealed whenever such a population is subjected to genetic analysis. In other words, phenotypic uniformity is itself an adaptive trait. It might be superficially plausible to argue that for many behaviors, on the other hand, uniformity must give way to some diversity, if the species is to survive. It is not too hard to imagine, for instance, that excessive timidity would circumscribe an animal in its quest of food while a lack of timidity would expose it to undue predation. A certain amount of behavioral variation within the species might then be regarded as the passport to survival, and the loss of this variation on inbreeding, the loss of adaptiveness, would then become amenable to the usual interpretation. Yet, I find it hard to believe that behavior, in all its essential features, should not be subjected to the same biological laws as other characters and that the balance of timidity, for instance, should be found not between individuals but rather within individuals.

Another aspect of the evolution of behavior is of genetic interest. I have discussed elsewhere in this volume how the obvious components of natural fitness, like fertility, reveal but little additive genetic variance, which is presumed to have been largely exhausted by natural selection. As a result of this, such characters do not show much response to further artificial selection, although they show marked changes on inbreeding, which acts on the nonadditive part of their genetic variance. It seems reasonable to suppose that natural selection should have placed behavior in this category of characters, for an animal's behavior is fundamental to its survival and mating success. As an example, Wood-Gush (1954) found differences in the mating behavior of five Brown Leghorn cockerels, all from the one partly inbred strain, which markedly affected their mating success. An analogous situation in a wild population would clearly have repercussions on an animal's fitness. Yet, it seems that, almost invariably, the measurements of behavior employed in the laboratory indicate a considerable amount of additive genetic variance. They show good responses to selection. This raises two possibilities. Firstly, if natural selection with respect to behavioral characters favors an intermediate rather than an extreme level of expression, then a considerable amount of additive variance would remain, despite the characters being subjected to strong selection. Or, secondly, it may be that the laboratory measurements bear little relationship to the behavior of the organism in the wild. This of course does not reflect in any way on the operational validity of the laboratory tests as such, but it does raise far-reaching implications for the biological interpretation of a comprehensive series of measurements now labeled "behavior." As more detail accumulates on the genetic architecture of a number of behavioral characters, it should be possible to identify those closely related to the animal's fitness.

Quantitative genetics and evolutionary theory merge at many points, and the understanding of the one requires an appreciation of the other. Perhaps the greatest contribution of behavior genetics to genetics in general will stem from the elucidation of the evolutionary implications of behavior.

CHAPTER EIGHTEEN
SOME PROBLEMS IN THE GENETIC STUDY
OF PERSONALITY AND INTELLIGENCE
William R. Thompson

INTRODUCTION

It may be well to make explicit at the outset a general proposition to which we are committed in this chapter, namely, that behavior genetics is not merely a part of genetics but is a special discipline in its own right, with its own special set of problems. This idea, though seemingly rather trivial, is of great importance insofar as it underlies all our ensuing theoretical discussion. Some indication of how crucial it is may be afforded by contrasting it with the opposed notion held by some that behavior genetics is simply another aspect of genetics. From this viewpoint, the study of the inheritance of intelligence or learning ability, for example, would be simply analogous to the study of the inheritance of such characters as weight, litter size, or bristle length. In a very general sense, this may be true. But a critical difference lies in the fact that behavior is already studied intensively in its own right, whereas the traits mentioned, and others like them, are not.

Classical genetics has always had as its main goal the explanation of the nature of genes and the transmission of genetic information from one generation to the next. Traits or characters are thus used only as convenient and neutral indices that permit inferences to be made about heredity. Behavior genetics, on the other hand, has almost an opposite focus. Its primary end is the further understanding of behavior. Accordingly, genetics has an ancillary role. It simply supplies models and techniques that will make this task easier. This is not to say that behavior genetics does not have implications for formal genetics. Indeed, R. C. Roberts discusses some of them in Chapter 11. It does mean, however, that the behavior geneticist may do well to relegate certain kinds of problems to a primary, and others to a secondary, place. For example, the estimation of number of genes involved in the transmission of some psychological trait, though of interest, is not crucial information. Demonstration of linkage between two characters is again not of primary importance.

We regard as critical only those problems that can be expected to shed light on the general question of how behavior works. Since most of this chapter attempts to explicate the basic nature of such problems, it is unnecessary to present specific examples at this point. We wish merely to emphasize strongly that the use of genetics as a tool by which to explore behavior can prove at least as valuable to the behavioral scientist as has the use of physiology and biochemistry. We should never forget that variation in respect to almost any type of behavior is genetic as well as environmental in origin. Contrary to an implicit faith prevalent in many quarters in the psychological world, traits are not infinitely manipulatable environmentally. The genotype sets definite limits to the operations that can be performed on it. Furthermore, of greater importance is the fact that different genotypes set limits of different extent. That is to say, genotype and environment interact, and this interaction makes up a major source of variance of behavioral traits. As will become clearer in the course of the discussion, it is also probably the most important and interesting component of variance from the standpoint of psychology.

THE INHERITANCE OF PERSONALITY AND INTELLIGENCE

In order to tease out the major conceptual problems that face us in the field of behavior genetics, we shall first present a brief overview of presently available information about the inheritance of personality and intelligence.

From its incipience, in the latter part of the nineteenth century, up to the present, work in this area has been strongly empirical. Numerous studies have shown that a large number of measurable traits of normal and abnormal personality and of intelligence has genetic basis. However, hardly any of this work has involved testing either of genetic or psychological models. The main developments during this phase have represented primarily improvements in methodology. These have related to such dimensions as zygosity determination in twin studies, better control of environmental influences in family-resemblance experiments, the use of more reliable testing instruments, and more sophisticated statistical handling of the obtained data. An indication of these trends may be found in the review by Fuller and Thompson (1960). It might be added, emphatically, that methodology in the field is still far from perfect. Particularly in connection with psychopathology, for example, there are serious problems relating to diagnosis and control of environment that have hardly begun to be solved. Some recent work by Gottesman (1963), however, on the inheritance of some normal and abnormal personality traits gives us reason to be fairly optimistic about the future. The care this author has exercised in gathering and analyzing his data should set suitably high standards for future investigators.

Along with increasing sophistication in methodology has come a growing awareness that certain kinds of problems are central to the field and must be faced squarely if progress is to be made. These problems fall into two general classes. The first relates to the description of the phenotype. Before we are able to make statements about genetic sources of variance in intelligence, for example, we must know something about the composition and nature of this trait. More and more, in the history of the field, investigators have focused on this problem, as indicated most clearly by the ever-increasing use of factor-pure tests (see Fuller and Thompson, 1960; W. R. Thompson, 1966). The second class of problems relates to the tendency for behavior to fluctuate or change. In the field of psychometrics, fluctuation that commonly appears in respect to test scores has till quite recently been regarded more as a methodological nuisance than as a problem of interest in its own right. Clearly, unreliability in our descriptions of phenotype, whether these descriptions are in the form of tests or of clinical diagnoses, can only lead to worthless information. At the same time, tendency to change, or fluctuation tendency, is a primary property of behavior. From this point of view, it has more than nuisance value; in fact, as we shall shortly attempt to show, it involves problems that should be of great interest and concern to behavior geneticists.

Thus it is our contention that we can capitalize on problems that have usually been regarded simply as sources of difficulty in studying the inheritance of personality and intelligence. In fact, these problems relate to two properties of behavior so basic as to be of primary concern to every behavioral scientist. Let us look at them more closely.

One way or another, psychologists have always been concerned with two evident facts about behavior. First, it is *complex*, in the sense that it can be broken down or analyzed into components or subcomponents which may be dealt with separately or in a variety of combinations. This relates, in testing, to the validity problem. Second, behavior is *fluid*. That is, it is prone to change or fluctuate, either "spontaneously," for example, as it is governed by developmental factors or by biological clocks, or as the result of the imposition of environmental factors. This relates to the reliability problem. Different areas of psychology have given greater or less attention to one or another of these properties, and, as a result, the whole field has been split. Thus psychometrics has been largely concerned with the descriptive analysis and taxonomy of behavior. While psychometrics has not ignored the changes to which behavior is liable, it has, at least until recently, treated them as secondary methodological obstacles to be overcome before the main task can be satisfactorily accomplished. Because the psychometrician's model of behavior is complex but static, he employs mostly descriptive statistics.

The experimentalist, on the other hand, has focused on the deviations of fluctuations that occur in behavior as a systematic result either of time or of some variables he has manipulated. Learning is a typical case of such a behavior change. To study such changes, however, the experimentalist is usually content to ignore the complexity of behavior and to choose arbitrarily some simple and convenient unit of observation, for example, a bar press, a galvanic skin response, or an error score. The usual statistic is the inferential, and the prototypic situation traces to experimental physics.

The split in psychology is a real one and a serious one. It has been noted already by a number of authors (Cronbach, 1957; Ferguson, 1956) who have made pleas that some kind of reconciliation be made. However, there are rather few signs that this will be accomplished in the very near future. The difference in approach between the two fields has, in point of fact, a relatively long history. Testing which began with Galton and Binet was based on the notion that human beings had different abilities which were measurable and which should be revealed if talent was to be suitably utilized in society. Thus the basic orientation was humanistic and pragmatic. An additional assumption, at least implicit in the field, was that abilities and aptitudes were relatively fixed. Fluctuations were thus treated as error. It is somewhat paradoxical that psychometrics, which seems, at least in America, oriented strongly to an environmentalist position in respect to the etiology of traits and abilities, nonetheless has tended to treat them as if they were not readily amenable to change.

Modern experimental psychology and behavior theory have different origins entirely. They can be traced most directly to Pavlovian physiology whose basic notion was that organisms are made up of relatively few fixed or unconditional reflex units which can almost indefinitely be elaborated (or made conditional) by varying environmental circumstances. Thus interest focused on organismic change from one point in time to another. Present-day "behavior theory" traces directly to such an orientation.

It is clear that psychology, if it is to develop any unified thrust, must incorporate both these points of view. Thus we must be concerned with modes of describing behavior that aim at reducing its complexity, and equally we must pay attention to the fact that behavior fluctuates both systematically and randomly. As a part of psychology, behavior genetics has the same obligation. More than this, it is the writer's conviction that the most interesting problems for the behavior geneticist will emerge from research strategies that focus on these two basic properties of behavior. The ensuing discussion will attempt to develop this point of view.

THE DESCRIPTIVE TOOLS

Personality and intelligence are measured by means of tests, these being constructed so as to assess the particular domain under study. Since the early work of Binet, Ebbinghaus, J. McK. Cattell, and Galton, a great many different types of tests have been developed covering almost every imaginable form of behavior. Generally speaking, however, they may be classified into a number of different groups, depending on their structure and on the type of behavior they aim at measuring. Thus there are tests that measure achievement, that is, grasp of some material as a result of practice with it, for example, the well-known Stanford Achievement Test series. These may be contrasted with aptitude tests which are designed to determine basic potentials independent of special practice or opportunity. Another mode of classification divides tests into those measuring personality, intelligence, interests and attitudes, intellectual deterioration, perceptual skill, and other substantive domains of behavior. Such a division is usually somewhat arbitrary, being based on some a priori conception of the particular domain. Still another classificatory dimension is represented by the *format* of the test. Thus there are so-called objective tests, usually consisting of a number of items to which the subject must respond with one of a limited number of predetermined answers. The typical IQ test is of this type. Subjective, or more exactly projective, tests place no sharp limits on responses. Thus in the Rorschach test, the subject simply tells the tester what he sees in the ink blots presented to him. His responses are later scored by reference to a limited number of categories. The situational test represents yet another type. As the name suggests, such a test requires the subject to respond to a situation that is a replica or a part of some larger task on which success is being predicted. Performance in a simulated airplane cockpit, for example, was used widely in World War II as a predictor of ability to pilot a real plane. Many other examples could be cited from industry. Finally, tests may be classified as individual tests or group tests, depending on whether they are given to individuals separately or together.

It will be clear from a review elsewhere (Fuller and Thompson, 1960) that tests from almost all categories have been used in examining the heritabilities of psychological traits. It is equally clear from the data obtained that there is little reason to prefer any one type of test to any other. At least in terms of the genetic methods used, usually family correlation or twin methods, results have been comparable. Much the same has been true for those tests designed to assess personality and intelligence in animals. Thus we have as yet no real criteria for allowing us to predict that a particular measure will yield results that are especially fruitful from the standpoint of behavior genetics. It is with the development of such criteria that we must be concerned. This brings us directly to the general problem of trait complexity and the more general problem of test validity. These are discussed below.

THE COMPLEXITY OF PSYCHOLOGICAL TRAITS

As stated earlier, behavior is complex, in the sense that any given "piece" of it is usually capable of further division into subunits. Furthermore, such a division is usually arbitrary. There are seldom grounds for deciding that one unit is, in some sense, "better" than another. The techniques for classifying behavior traits into groups, that is, the factor-analytic techniques, are widely used and have obvious value in that they provide the psychologist with a more parsimonious description of behavior. This, at least, is a step in the right direction. As Thurstone, one of the fathers of factor analysis, has rightly pointed out (Thurstone, 1947), the law of parsimony is basic to all of science. The factor method represents one validating procedure.[1] To the degree that a test involves less complexity, that is, measures fewer unitary traits, it can be considered more valid. Nevertheless, this definition does not permit decisions about the relative merits of different units obtained by different procedures. This is another problem.

The second interpretation of the validity concept involves the congruence between tests designed to measure the same trait (congruent validity) or between a test and some performance that the test is intended to predict (predictive validity). For our purposes, these two notions may be treated as equivalent. Both refer to the basic problem of establishing what any given test is really measuring. Since the testing movement has had a rather practical orientation, it has dealt with validity mainly at a specific and local level, rather than as a problem of general scientific interest.

Essentially, validity relates to meaning, which, in turn, involves the relationship between a particular term or concept and a referent. Usually, the referents used to validate tests are specified in terms of other tests or of some performance that the tests have been designed to predict. The actual understanding of the trait or behavior being measured is therefore not greatly furthered by such a procedure. Knowing that a Stanford-Binet IQ score correlates with school marks tells us very little about the nature and operation of intelligence. In this sense, testing has been very much like Linnaean taxonomy, though more restricted in what it covers; that is, it has been overconcerned with purely empirical description and classification of individual traits.

In biology, the advent of Darwinian theory gave taxonomy and systematics a new significance. The simple description of characters gave way to an attempt to relate these characters to broader evolutionary developments. Thus a trait such as coloration, for example, came to be studied for its possible usefulness as a mechanism for survival, reproductive isolation, diversifying selection, and so forth. Such referents are at a higher or broader level of discourse than the terms whose meaning they explicate. The same cannot be said of such a relation as that between measured IQ and school performance. Although in psychology there is no theory of the same comprehensiveness as evolutionary theory in biology, there are perhaps a sufficient number of scientifically respectable frameworks to which traits and the tests that measure them can be related with profit. In the context of this chapter, the conceptual framework provided by the dimensions of heredity and environment is obviously the one on which we should focus our attention.

Having presented two major meanings of validity, we shall now discuss the relevance of both to behavior genetics. We shall consider first the problem of factorial validity and the suitability of behavioral factors as units for genetic study; secondly, we shall discuss the problem of arriving at procedures that can yield tests securely anchored to the

[1] For a full discussion of validity of tests, see Cronbach and Meehl (1955).

kinds of concepts that are of most interest to us. This is more concerned with predictive validity.

Factors as Natural Units

The general problem was first examined by the writer some years ago (Thompson, 1957a). The essential purpose of factor analysis is to reduce redundancy in a system of measures (Ferguson, 1956). The factors drawn out of a battery of tests permit the construction of a few new "purified" tests that measure the domain as well as the whole battery did originally. As we have already emphasized, this is a useful goal. However, there are many factor methods available and consequently many possible sets of factors for any given domain. If factors are to be regarded as "natural" or "basic" units, in any sense, there must also be available criteria for deciding which factors are more natural or basic than others. Factor analysis itself, since it is a purely mathematical or statistical procedure, offers no real criteria for such a choice. In the framework of the present discussion, we must ask whether, in fact, psychological factors (or any other kind, for that matter) should be regarded, in any sense, as more meaningful, from the standpoint of genetics, than complex tests. There is no doubt that the view that factors somehow are more "real" biologically than tests certainly has a good deal of intuitive appeal as a working axiom. But on closer analysis, it does not seem to make much sense, either on theoretical or empirical grounds. Let us now examine two approaches to the problem that have been taken by Royce and by Cattell, respectively.

Royce's model (1957), shown in Figure 18.1, supposes that complex general intelligence, for example, is carried by very many gene pairs and factors of intelligence by fewer gene pairs. This implies that progressive factorization would ultimately yield units that might be carried by single Mendelian genes. Royce says (1957, p. 369):

☐ How might we link the multiple-factor approach of the geneticist with the multiple-factor approach of the psychologist? I submit that the identity of terminology is more than accidental or analogous, and that a linkage is feasible, and even sensible. The identity of logic is certainly obvious— that there are a multiplicity of factors, both behavioral and genetic, which are determinants of variation.

The isomorphism with which Royce is dealing is between number of behavioral and number of genetic units. Thus it should follow that a behavioral factor should show a simpler mode of hereditary transmission (e.g., a single Mendelian gene) than that shown by a complex trait.

Actually, there seems to be no sound theoretical reason why units of behavior should correspond to genetic units. Indeed, the identity of terminology probably represents an analogy far more than it does a logical identity, contrary to Royce's suggestion. Certainly, both phenotypes and genotypes can be subdivided. But there is no reason to suppose that the resulting units in each case will somehow correspond, unless the procedures used for dividing are deliberately chosen to obtain such a correspondence.

Behavior Domain   Genetic Domain

Complex behaviors   Group factors   Genotype   Gene pairs

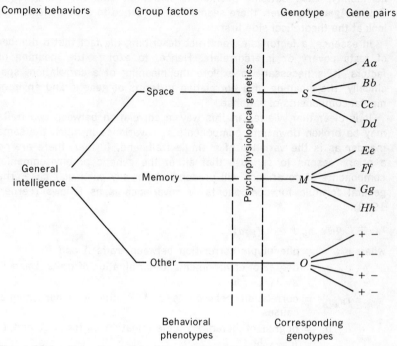

Behavioral   Corresponding
phenotypes   genotypes

Figure 18.1   Royce's model of the relationship between factors and genes.
(*From Royce, 1957.*)

Empirically, Royce's model has little support, so far as the writer knows. Some very complex traits, such as the hygienic behavior of bees (see Rothenbuhler in this volume) or human phenylketonuria, apparently are carried as entities by relatively simple genetic mechanisms. Other seemingly simple traits (e.g., activity level in rats and mice), on the other hand, often show quite complex transmission (see Fuller and Thompson, 1960).

Cattell's position on the relation between genetic and behavioral factors also supposes a kind of congruence between them but in respect to another dimension. Rather than dealing with the correspondence of units as Royce does, he suggest a relationship between factors and heritability, that is, relative degree of dependence on genotype as opposed to environment. Cattell has stated (1960, p. 370) in reference to the problem:

□ The position adopted in our own investigations [is] that the variance ratio of any factorially composite piece of behavior will be middling, but that more clearcut predominance of heredity or environment will be found for factor-pure source traits.

It is clear that this point of view makes no assumptions about the relative simplicity of *transmission* of factors as against traits. A trait could be carried by a great number of genes and show very high or low heritability. Thus Cattell's position is somewhat different from that of Royce. Again, however, there seems little evidence to support it. Let us look at the theoretical side first.

In essence, a factor is a construct describing the fact that a number of tests covary or intercorrelate. Hence, to explore the meaning of factors, it is necessary to explore the meaning of a correlation, specifically, the meaning of a correlation in terms of genetic and environmental components of variance.

It is clear that, viewed in this way, a correlation between two traits may be broken down into components of covariance in much the same manner as is the variance of a single trait, and, further, there are no a priori reasons to suppose that either the genetic or environmental components of covariance will predominate in any particular case. The general formula for components of covariance is as follows (Lerner, 1958):

$$r_{AB} = e_A e_B r_{E_A E_B} + h_A h_B r_{G_A G_B}$$

where r_{AB} = phenotypic correlation between traits A and B

e_A, e_B = degree of environmental determination of traits A and B, $e = \sqrt{1 - h^2}$

$r_{E_A E_B}$ = correlation between A and B due to environmental causes

h_A, h_B = degree of hereditary determination of traits A and B, $h = \sqrt{h^2}$

$r_{G_A G_B}$ = correlation between A and B due to genetic causes

The essential feature of the equation is the dependence of the genetic and environmental components on the heritabilities of the characters involved. The higher these heritabilities, the higher will be the genetic covariance component.

It is unnecessary to describe the precise operations by which the values in the equation are computed. They are readily available elsewhere (Lerner, 1958; Falconer, 1960). We wish merely to emphasize that any phenotypic correlation may be so broken down and that there are therefore no logical grounds to justify the idea that factors should show a different degree of heritability from tests. Empirically the results of partitioning correlations are sometimes surprising. Some examples from Falconer (1960) are shown in Table 18.1. It will be clear from

Table 18.1
Examples of phenotypic, genetic, and environmental correlations

Traits	Correlations		
	r_P	$x_{G_A G_B}$	$r_{E_A E_B}$
Milk yield X butterfat yield in cattle	.93	.85	.96
Body length X back-fat thickness in pigs	−.24	−.47	−.01
Fleece weight X body weight in sheep	.36	−.11	1.05
Body weight X egg weight in poultry	.16	.50	−.05

these data that the kinds of factors extracted from a matrix of environ-mental correlations may be very different from those taken from a matrix of genetic correlations. To the knowledge of the writer, such a com-parison has never been made for psychological traits. However, Smith, King, and Gilbert (1962) have attempted this, using a number of morphological parameters in bacon pigs. Since these workers used only a simple principal-axes method without rotations, their solutions are of less interest than they might otherwise be. Further analyses could very easily be done and should yield information of great importance.

Approaching the problem from a slightly different angle, we may com-pare two samples or pieces of behavior, A and B. Let us assume that A is composed of many parts or subunits, and B is a single unit. The variance of each of the subunits of A will depend on genetic and environ-mental factors plus interaction. The total variance for A, then, will be dependent on some kind of summation of the genetic and environmental variances of the subunits. Since these are by definition independent (if we consider only orthogonal factors), genetic components may be high in one subunit, low in the next, medium in the next, and so on. Hence, on a purely statistical basis, it seems most probable that the extreme values of the genetic and environmental variances will cancel and that those for the whole of behavior A will be in the middle range, that is, each around 50 percent. Certainly, the chances of, say, five subunits each showing high variance for the same component seem to be small. Consequently, Cattell's suggestion that for complex traits nature-nurture ratios will be "middling" may well be valid. But in the case of the unitary behavior B, there are no grounds for supposing that the situation will be much different, if we assume that genetic-environ-ment variance ratios are normally distributed for a population of such unitary traits.

Let us now turn to empirical data bearing on the position taken by Cattell. Some material is summarized in Table 18.2 taken from Thomp-son (1966). The figures shown indicate fairly clearly that heritability of factors tends to be of the same order of magnitude as the heritability of tests. It must be emphasized that there is nothing absolute about the estimates shown in the table. Thus, for some of the same factors, another worker (Vandenberg, 1962) has obtained quite different values.

Table 18.2
Comparison of heritability estimates for composite stand-ard tests, primary factors, and second-order factors

Composite tests		Primary factors		Second-order factors	
(Newman et al., 1937)		(Blewett, 1954)		(Blewett, 1954)	
Test	h^2	P.M.A. Factor	h^2	Factor	h^2
Stanford-Binet	.73	Verbal	.68	Composite 1	.339
Otis	.79	Space	.51	Composite 2	.594
Stanford Achievement	.61	Number	.07	Composite 3	.549
		Reason	.64		
		Fluency	.64		

SOURCE: From Thompson, 1966, Table 6.

The two factors showing the greatest discrepancy are number and reasoning, Blewett's (1954) values being .07 and .64, and Vandenberg's being .61 and .28, respectively. Accordingly, no definite conclusions, either positive or negative, can be educed from the available data in Table 18.2. Some limited support for the position of Cattell is afforded by an experiment of Eysenck and Press (1951). They found that a general neuroticism factor showed higher heritability (81 percent) than any of the single tests with loadings on it. However, work done by Cattell, Stice, and Kristy (1957) appears to contradict this finding. Their source trait U. I 23 (neural reserves versus neuroticism), which Cattell considered rather close to Eysenck's general neuroticism factor, showed strong environmental determination.

A recent attempt to deal with the same general problem has been made by McClearn and Meredith (1964). These investigators applied factor analysis (principal-axes solution) to an intercorrelation matrix derived from 25 variables relating to activity and emotionality. These were administered to a heterogeneous population of mice from a four-way strain cross. Five factors were identified: ambulatory activity, urination tendency, wall-seeking activity, defecation, and climbing. Unfortunately, the data gathered did not permit analysis of factors into genetic and environmental components. Thus the authors were unable to make any statement that might bear on the positions taken either by Cattell or Royce. Nonetheless, the research represents an important first step in the elucidation of the relation between factors and genes.

In summary, we may say that the empirical data available suggest that any a priori position on the relations between factors and genes is at least not feasible at present, and perhaps not even possible. Heritability values are not absolute but depend on many variables of a local character. As a consequence, genetic correlations will also be liable to differ strikingly in magnitude; hence the genetic and environmental components of factors will also vary.

We cannot assume arbitrarily that there are correspondences between factors and genes, either in terms of factors showing a simpler mode of inheritance or certain specified heritability values. However, it is certainly sensible to look for procedures by which such correspondences might be assured beforehand. The following section deals with this point.

Validation by Genetic Criteria

As we have already stated, factors of personality and intelligence are extracted according to mathematical criteria. Validation is thus internal rather than external. We may now ask whether there are available techniques designed to reduce the complexity of traits according to external criteria, specifically, genetic ones. We shall discuss three examples illustrative of such a procedure.

The first is a method developed by Eysenck, called criterion analysis. This is essentially a method of rotating factors in a battery of tests until they show maximal correlation with a criterion test that has been included in the battery. Factors are thus extracted that maximally reflect the criterion, whatever it may be. Eysenck (1951) has illustrated the method with a neuroticism factor which maximally reflects a criterion of presence or absence of psychiatric treatment. He has also suggested that the same method could be applied to a battery of tests using heritability as a criterion, for example, concordance in monozygotic twins. This has so far not been tried, though an application is presently being attempted by Eysenck and Broadhurst (Broadhurst, personal communication).

A second procedure is an iterative trial-and-error method by which different tests and test combinations are tried in turn until one is found that yields the most satisfactory genetic result. Ginsburg (personal communication) at the University of Chicago has been using this with dogs. Different measures of aggressive behavior, for example, are used on two breeds plus hybrids, and each of the measures is subjected to genetic analysis. The one giving the simplest and most consistent genetic result can then be chosen for further genetically oriented work, for example, studies involving selection, linkage, or physiological analysis.

The work of Chung and Morton (1959) and Morton and Chung (1959) represents a third approach to the problem. These investigators were concerned with the genetic transmission of muscular dystrophy, a disease that appeared to be inherited but whose mode of inheritance was still ambiguous. By discriminant functional analysis, Chung and Morton (1959) were able to separate the syndrome into three simpler entities, each involving a different genetic basis, as follows: (1) *facio-scapulohumeral form*, carried by a single dominant gene with a mutation rate of 5×10^{-7}; (2) *limb-girdle form*, carried by an autosomal recessive gene with a mutation rate of 3.1×10^{-6}; (3) *Duchenne form*, dependent on a single sex-linked recessive, with a mutation rate of 9×10^{-6}. The breakdown of an initially complex disease entity with an ambiguous hereditary mechanism into subunits each showing a simple and distinct type of transmission clearly represents a major achievement. It is to be regretted that this kind of analysis has not yet been applied to such behavioral syndromes as the major psychoses. Forty years of work on them has so far not even established to the satisfaction of many that they are definitely dependent on heredity; much less has it shown precisely what genetic mechanisms are involved (see Gregory, 1960).

The three methods discussed above all have in common the goal of establishing units defined in terms of genetic criteria. By "genetic criteria" we mean conformity to known models of genetic transmission, the simpler being preferred over the more complex. It should be noted, however, that such units need not correspond to units of selection. A certain trait may show a highly complex form of inheritance and yet be selected positively or negatively as a discrete unit. General intelligence or general physical appearance might be examples in human populations in which there is competition for mates. Such broad units of selection are of equal interest, though they would not necessarily correspond to units validated by conformity to models of genetic transmission.

THE FLUIDITY OF TRAITS

The second major property of behavior is its plasticity or fluidity. Psychological traits, more so than morphological or physiological characters, fluctuate over time. Thus repeated measures of the trait may yield quite different values. As we have already indicated, this may be treated as a methodological difficulty or as a substantive problem. Let us look at several different types of behavioral fluctuations that can occur and the bearing they have on the field of behavior genetics.

Random Fluctuations

We need not become involved in this context in the problem of whether so-called random fluctuations in behavior represent true randomness or simply lack of knowledge of the many variables that may be involved. For practical purposes, they may be simply treated as error. Since they cannot usually be held constant, magnitude of the score deviations they produce is calculated so that statistical control can be obtained. The result of this procedure provides increased accuracy of measurement. Thus, in testing, observed scores can be expressed as follows:

$$X = T + E$$

where X = observed score of an individual
T = true score
E = random error

Accordingly,

$$\sigma_x^2 = \sigma_t^2 + \sigma_e^2$$

The reliability coefficient r_{xx} can then be expressed in terms of these variances as follows:

$$r_{xx} = \frac{\sigma_t^2}{\sigma_t^2 + \sigma_e^2}$$

The terms on the right side of the equation have the same meaning as before. Thus r_{xx} is simply a ratio between the variance of true scores and the total variance due to both true and error scores. As σ_t^2 increases relative to σ_e^2, r_{xx} approaches a limit of 1.0. As we have already suggested, the purpose of estimating r_{xx} is to gain in accuracy of measurement. The methods available for obtaining such an estimate are of two major types, both of which involve the essential step of replication. In psychometrics, these are the split-half and the test-retest methods. They relate, broadly speaking, to the spatial (or nontemporal) and temporal aspects, respectively, of the trait under study.

The split-half technique in testing involves estimation of error by comparing two samples drawn, by hypothesis, from the same population and matched according to their statistical parameters, in particular, their means and variances. Thus, odd-numbered items may be compared with even-numbered items, those in the first half of the test with those in the second half, and so forth. An analog in genetic analysis of a morphological character is the repetition of measures on homologs. Many traits, for example, wing length, eye-facet number, bristle length, occur more than once in most animals. Thus reliability may be obtained by taking advantage of this fact.

Reliability by the test-retest method involves taking repeated measures of the same trait. It is assumed that the fluctuation of the trait over time within individuals will be random rather than systematic, that is, dependent on specific temporal and local conditions. It may often be difficult to grant this assumption in the case of behavior traits. The effects of the first measurement often carry over to the second measurement to produce a spuriously high reliability. Subjects may remember particular items on an IQ test, for example, and answer them the same way, whether correctly or incorrectly. For most physical or physiological traits, however, the repeated measurements are more likely to be independent.

The kind of gain to be expected from the use of repeated measurements is illustrated by data on bristle number in Drosophila shown in Table 18.3 from Falconer (1960). Taking two abdominal segments rather than one reduces the proportion of variance due to specific environmental factors by approximately 20 percent. In this case, the importance of genotype in the etiology of the trait rises correspondingly. Falconer (1960) has given an equational statement of this gain as follows:

$$\frac{V_{p(n)}}{V_p} = \frac{1 + r(n-1)}{r}$$

where $V_{p(n)}$ = total phenotypic variance of trait measured in a number of individuals n times

V_p = total phenotypic variance of trait measured once

n = number of measurements

r = correlation between measurements repeated in same individuals

As the writer has shown elsewhere (Thompson, 1966), the equation may be directly derived from the classical Spearman-Brown formula relating reliability to test length:

$$R = \frac{nr}{1 + (n-1)\,r}$$

where R = new reliability for test lengthened n times

n = factor by which test is lengthened

r = reliability of test, when $n = 1$

Table 18.3
Results of partitioning variations of bristle number in Drosophila, using one or two abdominal segments

Point of variance	Percent of variance	
	1 segment	2 segments
Total phenotypic	100	100
Additive genetic	34	52
Nonadditive genetic	6	9
Environmental, general	2	2
Environmental, special	58	35

SOURCE: Falconer, 1960.

The validity of Falconer's equation depends on the assumption that the variance of raw test scores is reduced by taking repeated measurements and that only the error portion of the total variance is involved in this reduction. It is also true, however, that a gain in accuracy can be achieved by increasing the true-score variance and holding error-score variance constant.

The above discussion does not really point up substantive problems as much as methodological ones. It indicates that specification of genetic and general environmental causation for a trait requires reliable measures of the trait. The work of Cattell et al. (1957) already cited has encountered this difficulty. The "factor-pure" tests these investigators used, perhaps because of their short length, had reliabilities ranging from .19 to .82 (for the general population) with 7 out of the 11 used having values under .50. Subsequent analysis was based mostly on corrected variances. According to the authors, the actual nature-nurture ratios are relatively little affected by such a procedure. However, there is some ambiguity involved.

We now turn to a consideration of those aspects of fluidity of traits that give rise to problems of more substantive interest. Specifically, these relate to systematic rather than random changes over time. We shall try to illustrate each of the theoretical possibilities with some empirical data.

Cyclical Fluctuations

A very large part of the behavior of organisms is subject to cyclic changes so that it may vary between high and low intensities. Such changes may be long-term or short-term, depending on the species and on the behavior. Breeding cycles, which we shall discuss in more detail below, furnish a good example of such variation. Migration and feeding and sleep are also typical cases of cyclic behavior. It is obvious that different periodicities for such basic drives must have adaptive value and hence, assuming their heritability, have considerable evolutionary importance. R. S. Miller (1955) has emphasized this point in his study of the activity rhythms of two sympatric species, the wood vole and the bank vole. The behavior of members of these two populations appears to be so cycled that they can occupy the same ecological niche with a minimum of competition and interspecific strife. Obviously many other examples could be cited.

Such cycles are not immune to environmental regulation, as Miller showed in the same study (1955). In fact, the biological clocks governing the behavior of some insect species may be set partly by environment (Pittendrigh, 1954). Much of the work done with animals, such as the lobster and the crab, in fact, appears to indicate the presence of two clocks governing cyclic behavior, one of these ("hard" clock) being relatively more insensitive than the other ("soft" clock) to environmental changes (Bunning, 1964; see Thorpe and Zangwill, 1961).

As we might expect, most human cyclic behavior has probably a much larger component of environmental variance, although there are some stable forms, such as the female reproductive cycle, and certain metabolic processes such as renal output (Lewis et al., 1956) which most likely are rather closely linked to genotype. Whether more complex forms of behavior, such as those falling under the nomenclature of personality and temperament, may also be closely linked to genotype is debatable. Some of the evidence, such as it is, has been reviewed by Thouless (1936), Fiske and Rice (1955), and Cattell et al. (1957). Until some conclusions can be established on a firm mensural basis, it will be futile to inquire into the heritability of temperament cycles or mood swings in human beings.

Among lower animals, we may refer to an experiment by Scott, Fuller, and King on the inheritance of breeding cycles in dogs, basenjis and cocker spaniels, and their F_1 hybrids (1959). These investigators found that basenjis tend to have an estrus period in September or early October, whereas cockers, like most domestic dogs, may have estrus in almost any month, with a slight excess occurring in the spring, and an approximate 6-month cycle. In respect to its timing, the basenji cycle is more like that of the dingo; it presumably coincides with what would be a period of increasing light in the Southern Hemisphere, from which both species originate. In this respect, both are like the wolf whose cycle is also timed for reproduction to occur in the spring or early summer (i.e., May or June in the Northern Hemisphere). Likewise, the basenji cycle is also like that of the wolf in its periodicity, being annual rather than irregular. Domestic species, presumably in the absence of strong selection pressures, show far more polymorphism with respect to reproductive cycles. The analysis of F_1, F_2, and backcross hybrids was based on rather small numbers but appeared to fit a model of recessive single-factor inheritance of the basenji condition, that is, autumnal estrus and annual breeding cycle. More work along these lines should prove very useful in aiding our understanding of the basic mechanisms involved in cyclical behavior.

Directional Changes

We consider now two types of directional changes, those involved in development and those involved in learning.

. The first has been discussed at some length in this volume by King. It seems clear that growth rate and time of emergence of different behaviors are under genetic control and represent adaptations to particular environmental situations. Thus, depending on the niche it occupies, it may be useful for an animal to develop or *differentiate* rapidly both on the input or sensory side and on the output or motor side as well. Examples of early development are seen in free-ranging species such as of the ungulate order. Many other animals, such as the great apes, man, many carnivores, and rodents, appear to develop rather slowly on both the sensory side and on the motor side. Learning may be thought of as the establishment of connections between stimuli and responses. Consequently, it seems likely that the kinds of learning of which a young organism is capable will vary considerably, depending on the developmental status or degree of differentiation of the stimulus and response systems at any particular age. This will also vary, depending on the genotype of the animal, according to the possibilities listed above.

In addition, whatever type of holding or memory mechanism is operating in any particular case, the amount and permanence of the environmentally produced change that can occur will also depend on genetic factors. In many species, the deviations that can be produced by environment are considerable and tend to override genetic differences. For example, in human infants, birth-weight differences are largely environmental, V_E being 82 percent as compared with 18 percent for V_G (Falconer, 1960). At adulthood, however, V_G appears to predominate. The same is true in many other species in which V_G rises from a low value at weaning to a much higher value later on. Similarly, with complex traits in human beings, there is a suggestion that the genetic component of variance, as reflected in familial correlation coefficients, goes up with age, as shown in Figure 18.2 (Jones, 1946). Since these data do not clearly separate genetic from environmental factors, they are not conclusive. But they suggest that the young organism may be rather "erratic" and that his early level of performance predicts his later level only poorly. Such a notion appears to be well borne out by many other studies summarized by Bloom (1964). Thus it seems entirely possible that hereditary potentialities may take time to become fixed, in the sense that phenotypic deviation between equivalent genotypes is at first great but much smaller later on.

We must also allow for the possibility that a disposition to vary, or to be "environment-sensitive," is itself genetically controlled. Evidence on this is likewise scanty, though the following animal experiments are suggestive. Hughes and Zubek (1956) and Cooper and Zubek (1958) studied the effects of early glutamic acid administration and enriched early experience, respectively, on the McGill bright and dull rat strains.

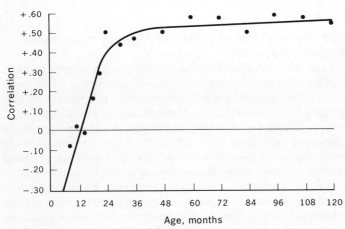

Figure 18.2   The correlation of children's intelligence with parental education computed at successive age levels. (*From Jones, 1946.*)

Their essential finding was that one of the strains (the "dulls") responded to early treatment, whereas the other (the "brights") did not. It is not known whether this lack of response on the part of the "brights" was due to their being already close to the physiological limit of rat intelligence, or whether it was due simply to their being less "environment-sensitive" than the dulls. An early treatment designed to depress maze scores rather than to elevate them would help to answer this interesting question.

An experiment by Olian and the writer (Thompson and Olian, 1961) bears somewhat on this question. Adrenalin was administered to pregnant female mice of three genotypes, high-, medium- and low-active. The effect of the pregnancy stress was to decrease the activity level of offspring from the high-active strain, increase activity of offspring from the low-active strain, and leave relatively unchanged the activity level of offspring of the medium-active strain. This result indicates that, with a general stress, deviations in behavior away from a physiological limit are most likely regardless of possible genetic differences in plasticity. A final experiment by Levine and Broadhurst (1963) obtained rather similar findings with respect to emotional reactivity. The response of a reactive strain to early manipulation was found to be different from that of the nonreactive strain for some adult test variables, though approximately the same for others.

A somewhat different possible basis for responsivity to environment—early or late—has to do with heterosis. It is by now well known that heterozygous genetic populations have a survival advantage over inbred populations. This gain is usually expressed in terms of greater resistance to stress, improved viability of offspring, larger size, better health, and other such characteristics (Lerner, 1954). However, it is not immediately obvious that an increased inter- and intra-individual variability should also be a property of heterozygous populations, either at the morphological or at the behavioral level. One can easily imagine cases where an increased or a reduced propensity to vary could have selective advantage. Thus, we can contrast species attraction in such birds as ducks, thrushes, and chickens with that in parasitic birds such as the European cuckoo and the American cowbird (Klopfer, 1962). In the former species, the young imprint strongly to the mother bird which, under normal circumstances, is the object to which they are most likely to be exposed during the critical period. This presumably ensures that later on they will be able to recognize and be attracted to their own kind. The mechanism is, however, somewhat "lazy," in the sense that the intrusion of some inappropriate stimulus during the critical period can disrupt the normal series of events. That is, the young will learn to follow the "wrong" objects, with whatever maladaptive consequences this may have. Such sensitivity could presumably result in an increase in behavioral variability. In this sense, these genotypes are environment-sensitive during early life. By contrast, the cuckoo and the cowbird, because of the parasitic habits of the mother, are exposed to alien species during early life. In spite of this, the offspring, after leaving the nest, still recognize and affiliate with others of their own species. In other words, the appropriate phenotypic behavior is fully fixed by geno-type and has, happily for these birds, little plasticity.

An experimental analog of such a species difference in environment sensitivity has been furnished by Winston (1964). This investigator showed that hybrid mice were less easily influenced by an early sound trauma than the parental strains. This "buffering" was presumably due to heterosis.

A number of direct empirical comparisons has been made of hetero-zygous with inbred populations with respect to behavioral variability, for example, by Bruell (1964a, 1965) among others, but they have yielded ambiguous results. Hence, Caspari's suggestion (1958) that the unex-pectedly high behavioral variance found by Tryon in his maze-bright, maze-dull hybrids is a heterotic effect may be true but cannot represent a general rule.

The second major type of directional change that is a property of most animal forms is learning. Like development, learning involves incre-mental changes in such behavioral indices as reduced errors or increased speed, these changes being common, as far as direction, to all members of the population involved.

We already know from a number of selection experiments (see Fuller and Thompson, 1960) that learning ability has a strong genetic com-ponent. But we know very little about the strength of expression of such a component at different points during the actual temporal acquisition sequence. It seems to be true that reliability increases with increasing number of trials. Hence it must be true that either the general environ-mental or genotypic components or both must increase, while the error variation arising from the special environmental conditions pertaining to each trial must remain constant or diminish. There are unfortunately

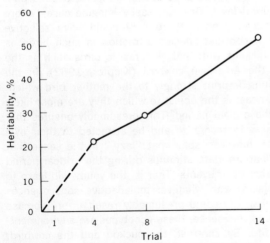

Figure 18.3 The increase in heritability of running time in mice in successive trials. (*From Vicari, 1929.*)

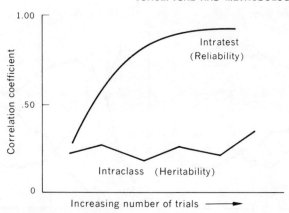

Figure 18.4 Heritability of behavior in several dog breeds tested over a number of trials. Note absence of change in heritability estimates. (*From Fuller and Thompson, 1960, p. 92.*)

rather few data available on this point. We may compare the results of two studies shown in Figures 18.3 and 18.4.

The Fuller and Thompson data indicate no relation between the temporal course of learning and genetic expressivity. The data of Vicari (1929), on the other hand, show an increase in heritability[2] with number of trials of exposure to the learning situation. Clearly, more experimentation on this interesting problem is needed. Again, we need not expect that such work will result in any universal or absolute conclusions. Some forms of genetically based adaptive behavior probably express themselves fully with a minimal exposure to environment, for example, imprinting. Other forms, however, may have a relatively delayed emergence time so that selection can produce a strong response only after the behavior has been shaped by environment over a relatively long time period.

Another slightly different approach to the problem of the genetic basis of learning involves analyzing the behavior into components whose relative importance varies from one time to another. Wherry's data (1941) can serve as an example. His results are summarized graphically in Figure 18.5. It is clear that different learning factors make different contributions to the total score, depending on the stage of the learning. Furthermore, there is a difference between "bright" and "dull" genotypes in the time course of the different factors. It would be useful to have heritability estimates over time for these factors separately. Unfortunately, they were not obtained by Wherry.

Since the converse of learning is forgetting, it is evident that a learning score is also to some extent a measure of speed of forgetting. At the same time, the two processes may have an independence such that any combination of rapid or slow learning and rapid or slow forgetting might

[2] Calculations made by Broadhurst and Jinks (1963).

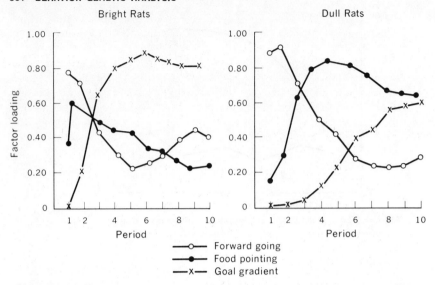

Figure 18.5   Changes in components of maze-learning ability during training shown by "bright" and "dull" rats. (*From Wherry, 1941, in Fuller and Thompson, 1960, p. 212.*)

have a special adaptive advantage and hence be selected in some niche. For example, for a species living in a terrain where the probability of occurrence of food in any particular place may change abruptly, a fast learning coupled with rapid forgetting may be adaptive. Furthermore, the other three combinations might also be useful in different ecological circumstances.

In human beings, the problem is more complex. However, it is probable that most intellectual and temperamental traits or factors are dispositions acquired by learning, that is, by exposure of a genotype to environment over a rather long time period, as suggested, in effect, by Ferguson (1956). Since our knowledge of how this occurs is at present rather slight, this is an area that needs research.

We emphasize again that the genetics of behavior change, whether random or systematic, is important in general and is especially crucial in the context of behavior genetics. Fluidity or plasticity is an essential property of behavior and can be studied as such. It should not be regarded simply a methodological nuisance.

SUMMARY AND CONCLUSIONS

In this chapter, we have attempted to point up two aspects of behavior which create methodological difficulties for genetic analysis but which also appear to represent conceptual problems of primary importance. They relate to two basic properties of behavior, namely, its complexity and its plasticity or fluidity. The so-called two disciplines of scientific psychology, delineated by Cronbach, psychometrics and experimental psychology, have tended to focus on one of these characteristics of behavior, somewhat to the exclusion of the other. It is suggested here that both can be fruitfully studied by the behavior geneticist.

One line of attack on behavior complexity has been by the technique of factor analysis. Several investigators have hypothesized a possible relationship between psychological factors and genotype. At present, however, such a notion finds little empirical support in the experimental literature and, furthermore, as presently articulated, is too primitive to make much theoretical sense. Several possibilities for further work on this question were suggested.

The second basic feature of behavior traits, namely, their tendency to change over time, was discussed under several headings: changes due to random factors, cyclical fluctuations, and several types of directional change. The last category, which includes alterations in some ability due to learning and memory and to development, is felt to be especially interesting for the behavior geneticist.

In general, a focus on these kinds of problems, the manner in which units of personality and intelligence reflect genetic units and the relation of changes in such traits to genotype, should do much to aid our understanding of the basic nature of behavior.

CHAPTER NINETEEN
RACIAL DIFFERENCES IN BEHAVIOR[1]
James N. Spuhler and Gardner Lindzey

This chapter begins with a brief discussion of the concept of race, an indication of what we mean by "race," and consideration of how racial differences may be detected. Following this we shall present a general discussion of empirical findings concerning the existence of racial differences in behavior. In conclusion, we shall comment briefly upon the utility of the concept of race for the behavioral scientist.

No reasonable discussion of race differences can commence without at least passing attention to present usage and the rather painful past of the concept of "race." The necessity of such an endeavor is underlined by the sizable number of contemporary scientists who are convinced that the concept of race has no legitimate place in the social or biological sciences. Scholars as diverse professionally as Klineberg (1954), Livingstone (1962), Montagu (1952), and Penrose (1951) argue that the concept only misleads and confuses and serves no legitimate scientific purpose. Although representatives of almost every social and biological science at one time or another have decried the use of this concept, it has displayed a tough viability and even now seems on the way to surviving the rash of attacks that followed Hitler's affinity for the term. One may note that the associative link between "race" and "Aryan" is even today strong enough in our society to account for a good deal of the discomfort that many scientists display in the presence of the term "race."

It was probably inevitable, once mankind had been divided into classes or categories, that some persons would think in terms of an order of quality or merit. It was equally ordained that dictators, demagogues, and elitists would employ such concepts for their own purposes and without regard for their technical meaning. This use of the race concept by "racists" scarcely seems a legitimate argument against the concept if it serves any useful purpose within an empirical discipline,

[1] Gardner Lindzey's contribution to the chapter was written largely while in residence at the Center for Advanced Study in the Behavioral Sciences. His work was facilitated by a grant from the Ford Foundation and Research Grant MH 11030 from the National Institute of Mental Health.

any more than the abuse of genetic concepts in the hands of these same individuals should lead to an abandonment or alteration of these concepts, or the misuse of probability concepts in the hands of an ardent advertising copywriter should influence statistical analysis in the social sciences. In considering the concept of race, we shall limit our attention to its empirical-theoretical significance within behavioral and biological science and largely overlook the social-political impact of the term. In another context the latter issue, of course, could be of primary importance. A number of informative essays dealing with the concept of race and its social and political implications are contained in the volume prepared by UNESCO (1961) and in the report of the recent UNESCO meeting on the biological aspects of race (1965).

THE CONCEPT OF RACE

As a technical term, the concept of "race" has been primarily the property of the anthropologist, although in recent years it has come to play an increasingly influential role in genetics. The original attempt of anthropologists to identify a specified number of races can be viewed as nothing more than an extension of Linnaean, zoological taxonomy. Given the physical diversity of mankind, it seemed only reasonable to subdivide the multitude into classes that were physically more homogeneous than the total population and that shared a common descent. According to this view, the category of human race is the next more specific category beneath the class *Homo sapiens*. Early concern with races was heavily influenced by the conception of "ideal" or "pure" races believed to represent the original ancestors of the many "mixed" classes of individuals which can be observed today. Such a viewpoint is no longer seriously entertained within biological or social science.

It was also logical that these early classificatory attempts should begin with those characteristics that appeared relatively invariant over time in the adult organism and were easy to measure objectively. In view of the link of physical anthropology to archaeology, it was not surprising that the search for a classificatory basis should focus upon attributes that could be measured, even though the subject was no longer living and might not have been living for many centuries. Thus, early efforts at classification emphasized physical measurements, particularly skeletal indices. A typical example is provided by the work by Dixon (1923) who used three skeletal indices (cephalic index, length-height index of the skull, nasal index) to classify races. Subsequent investigators were more concerned with "common descent" than Dixon, and many of them employed physical attributes that were not skeletal, such as lip thickness, skin color, eye color, and hair texture. Garn (1961) suggests, however, that significant links between the race concept and the theory of evolution have appeared only during the past 10 or 15 years.

The most important development in this area is the recent attempts to link racial classification with physical characteristics that are genetically well understood. The writings of Boyd (1950, 1960, 1963) and Mourant (1954) provide the best illustration of this approach, which has leaned heavily upon blood types as its classificatory base. Many arguments have been advanced in support of serological classification, but the decisive factor seems to be the additional power generated by understanding the genetic basis for the attribute. A somewhat enthusiastic statement of the case for a "genetic definition" of race is presented by Boyd (1953, pp. 495–496):

☐ A classification of men on the basis of gene frequencies has a number of advantages. (1) It is objective. Gene frequencies are determined by straight-forward counting or relatively simple computation from quantitative observations of clear-cut, all-or-none characters. The subjective element which complicates attempts to compare the skin colors of two peoples, for example, does not appear. (2) It is quantitative. The degree of similarity between two populations is not a matter of guess-work, but can be compared by calculating from the frequencies of the genes considered. (3) It makes it possible to predict the composition of a population resulting from mixture in any assigned proportions of two populations of known gene frequencies. (4) It encourages clearer thinking about human taxonomy and human evolution. Emotional bias is less likely to operate than in the case of physical appearances such as stature or skin color. There are no prejudices against genes. It permits a sharp separation of the effects of heredity and environment. In the case of a character like stature, it is difficult to say whether genes or food and climate have contributed more to making two populations alike. In the case of blood groups no such problem arises.

It should be noted that none of the classificatory systems that has any current popularity is strictly actuarial. Although they involve objective indices, these are intended for a trained observer who employs them with a weather eye for geographic location and such. Boyd's classification, for example, does not lean solely upon allelic frequencies but also seems to consider implicitly geographic location, lineage, and past classificatory practice. It appears that modern biometric and psychometric techniques for multivariate classification have not been employed extensively in this area, and the occasional applications have not received wide attention in secondary sources. Nonetheless, beginnings have been made (Spuhler, 1954), the most impressive being the recent studies of Cavalli-Sforza, Barrai, and Edwards (1964, and earlier papers).

Definitions of race fall into two broad categories, with those advanced by anthropologists usually emphasizing physical similarity and common descent (geographic factors), whereas the definitions proposed by geneticists typically focus upon differences in gene frequencies and breeding isolation. The continuity between the concept of race as applied to man and the concepts of breed or strain as applied to lower animals has often been identified.

An interesting sidelight to problems of definition of race concerns the tendency of many anthropologists of a decade or so ago to assert that the attributes to be used in classifying races must be nonadaptive, that is, make no contribution to the process of natural selection. This emphasis was defended on the grounds that only characters unrelated to natural selection would possess reasonable stability over time. It seems likely that these observers must have been influenced also by an implicit concern lest racial distinction be used as a basis for racial prejudice or racial superiority. The infirmity of their reasoning was pointed out by geneticists, particularly Dobzhansky (1950, 1962), and this position has been largely abandoned (cf. Boyd, 1953, 1964).

Often the genetic biography of a species is not uniform over its full distribution in space at some particular time. Given the raw materials of inherited variability, the genetic structure of a species depends largely on local selection intensities, on the one hand, and gene flow between different local areas, on the other. If there is much gene flow, local races cannot develop; if there is less, clines may be formed; if still less, local races may differentiate. When the genetic biography is not uniform, it is useful to recognize, and sometimes to name, the races or subspecies within the species.

The biological and anthropological notion of race (= subspecies) which we use is based on stipulative rather than lexical or ostentative definition. In the first instance, "racial differences" are those genetic differences which define "races." It does not follow that the geographic distribution of other genetic traits will be concordant with that of the selected set of traits used to define the racial groupings. It is a separate problem to find out if other genetic differences between populations are "racial." But as Simpson (1961, p. 175), in commenting on misapprehensions about subspecies, wrote:

□ . . . subspecies do not express the geographic variation of the characters of a species and are only partially descriptive of that variation. They are formal taxonomic population units, usually arbitrary, and cannot express or fully describe the variation in those populations any more than classification in general can express or fully describe phylogeny. They are not, for all that, any less useful in discussing variation.

Most species of animals with a wide distribution over a variable geographic habitat exhibit more or less distinct geographic races. The existence of such racial differentiation is a token of the biological adaptation of local populations to their local environment. Mayr (1965) reports: "Such races are sufficiently distinct in about one-third of the better known species of animals to have been designated as different subspecies. Some widespread species of birds and mammals may have as many as twenty or thirty or, in exceptional cases, more than fifty rather well-defined races."

The fundamental processes underlying the anthropological concept of race are bisexual reproduction and adaptation through natural selection. A given individual is connected by gametes to representatives of a set of persons which we stipulate as a subspecies, that is, what anthropologists usually call a race. Unlike the individual, this set of individuals (a partially isolated local breeding population or race) is self-perpetuating. This set of persons, largely of common descent, has potentially unlimited longevity; and, unlike the species, this racial population is open to gene flow from the outside.

In the human and other bisexual species, different individuals are heterozygous for different genes even though the gene pool of the breeding population is at equilibrium with the evolutionary forces causing change in gene frequencies. With the trivial exception of monozygous multiple births, all human individuals are unique in genotype. Under random mating with dominance and a few alleles at each locus, much of the variation is hidden in heterozygotes. This large store of hidden variation may be associated with a short-run disadvantage to the local population because of reduced fitness of homozygous deleterious recessives in the prevailing environment while at the same time providing a long-run advantage in the possession of genetic flexibility in meeting environmental change. Since the change in gene frequency over space is gradual, any attempt to separate people sharply into races having exact boundaries and different gene frequencies is arbitrary. The number of races recognized is thus a matter of convenience. It follows that a race does not have an average genotype and therefore it does not have an average phenotype. Thus it is misleading to try to picture a typical or average member of a race. Rather, a race should be defined in terms of the relative frequencies of some of the alleles contained in its gene pool.

Having glanced at the history of the race concept and its typical definition, we now ask: What are the races of man? It is characteristic of questions concerning human taxonomy that they lack clear or definitive answers. Consistently, the human races are sometimes identified as three in number (almost always Negroid, Mongoloid, Caucasoid), sometimes as six (for example, Negroid, Mongoloid, Caucasoid, Australoid, American Indian, and Polynesian), and in one authoritative volume as thirty (Coon, Garn, and Birdsell, 1950). At the present time there is no objective basis for agreeing upon how many racial categories should be employed, although it seems clear that few investigators will wish to be burdened with 30 different classes. Even Garn and Coon (1955) in a more recent publication have suggested that contemporary evidence supports a classification involving no more than 10 major races, and still more recently Garn (1961) suggested distinguishing between geographic races, local races, and microraces. He has also pointed out (1960) that differences in the number of races proposed by investigators are largely to be understood not in terms of disagreement or conflicting claims but rather in terms of different levels or principles of classification that have determined the various systems. His geographic races are nine in number and include Amerindian, Polynesian, Micronesian, Melanesian-Papuan, Australian, Asiatic, Indian, European, and African. Undoubtedly the most interesting classification in the present context is that developed by Boyd (1950, 1953) on the basis of approximate gene frequencies, primarily for blood types. The distinction he offers is between Early European (Basque), European (Caucasian), African (Negroid), Asiatic (Mongoloid), American, and Australian. In more recent publications Boyd (1960, 1964) has elaborated this scheme to include 13 races as follows: Early European, Lapp, Northwest European, Eastern and Central European, Mediterranean, African, Asian, Indo-Dravidian, American Indian, Indonesian, Melanesian, Polynesian, and Australian.

It should be noted that the setting apart of the Basques as members of an "Early European" race is of doubtful validity. The Basques who have been blood-typed, being contemporary, are not "earlier" than other living peoples of Europe. They speak a relict language, but the oldest evidence for Iberian language is not as old as that for other languages still spoken in Europe. They have a high frequency of the Rh blood type, but if one uses as many as 12 gene frequencies of the red blood cellular antigens to measure biological affinity, the Basques are not as distinctive from other European populations as are, for example, the Irish and the Sicilians.

THE DETECTION OF RACIAL DIFFERENCES IN BEHAVIOR

Before proceeding with our discussion of behavioral differences between races, it is necessary to consider the general relation between genetic variation and behavior. The genetic component of behavioral traits may be considered under two headings: (1) discrete, discontinuous, or qualitative variation controlled by major genes, or oligogenes, and (2) continuous, quantitative variation controlled by minor genes, or polygenes.

1   Discrete phenotypic variation with a large genetic component usually depends on the action of major genes at one or a few chromosomal loci. Most human populations are polymorphic for nearly 50 known sets of major genes; that is, two or more phenotypes controlled by these major genes occur in the population with appreciable frequency. By the criterion of numbers, these are "normal" genes. Phenotypes with reduced fitness are controlled by a second variety of major genes; these genes are rare and thus "abnormal" in most populations.

Major genes may be common in a breeding population for at least two different reasons: (a) The evolutionary forces are such that their frequencies are stable at an intermediate level, and (b) the evolutionary forces are such that a favorable gene is spreading in the population, replacing its less favorable allele, and thus resulting in a transient polymorphism. In later sections we shall discuss in some detail the racial distribution of four sets of major genes with distinctive behavioral consequences: albinism totalis, phenylketonuria, phenylthiocarbamide taste reaction, and red-green color blindness.

2   Continuous or quantitative phenotypic variation with a large genetic component depends on the action of minor genes at several or many loci. The contribution of individual minor genes to the phenotype may be additive, or there may be interaction within and between loci. Although some polygenes are known to be associated with all-or-none traits by means of a threshold, usually polygenic characters are continuous or semicontinuous from one extreme to the other so that individuals differ in degree but not in kind of their polygenic attributes. This is the most frequent and the most important type of hereditary variation. Most variations of body size and shape, of viability and fertility, and of behavior are polygenic. The environmental component is often larger in

polygenic than in oligogenic traits. The distinction between the action of oligogenes and polygenes is not absolute: Phenotypes whose variation is controlled largely by major genes may be modified by minor genes at other loci, and major genes may contribute to polygenic variation. Although individual polygenes have been identified in Drosophila by Thoday (1961) and his associates, techniques are not now available for their identification in man. In a later section, we shall return to the discussion of polygenes concerned with performance on intelligence tests.

By "race differences" we mean differences in gene frequencies between two or more breeding populations. The study of race differences therefore implies that we can identify phenotypes with genotypes and with genes in individuals and that we can assign these individuals to breeding populations regarded as members of races. It also implies that we can recognize that genes are the "same" or "not the same" in two or more populations. There are at least three different meanings of the word "same" in this context: (1) We may refer to one and the same particular gene on separate occasions; (2) we may refer to different genes which have the same function, and (3) we may refer to two genes which are identical by descent. The existence of recurrent forward and back mutation makes it impossible to be certain that genes identical in function are identical by descent or that genes descended from a common ancestral gene are identical in function. Our main concern in this chapter is with genes that are identical in function.

In practice, one may have varying degrees of confidence in the evidence that genes have the same function in different populations. At a low level of confidence, we may assert identity of genes in function because the modes of inheritance for similar phenotypes are the same in the two populations, e.g., single alleles with dominance. The argument for identity is a little stronger if the genes are shown to be located in the same pair of chromosomes in the two populations, e.g., X-linked single alleles with dominance. The evidence for identity in function is quite strong if a trait inherited as an autosomal recessive in each of two populations occurs in the offspring of a racial cross involving parents heterozygous for the recessive gene, or if all offspring of parents with the recessive phenotype in such a racial cross are themselves recessive.

The strongest argument at present for functional identity of genes in diverse breeding populations is to show that the genes in question control identical sequences of amino acid residues in the gene product. We know, for example, that the genes for hemoglobins A, C, and S are functionally identical in certain populations at the level of the exact sequence of amino acid residues in the globin proteins. We also know that phenotypic similarities even at the molecular level may be misleading in regard to identity of genes. Samples of hemoglobins D, G, and O each showed identical patterns of electrophoretic mobility between different populations and thus were assumed to be controlled by genes identical in function. Later studies of the three hemoglobin varieties at the amino acid level showed each to contain two or more distinct varieties and thus demonstrated the genes are not identical in function. In the future, we may know the structure of genes at a more fundamental level, the sequence of base pairs in the desoxyribonucleic acid molecules of heredity. Genes identical in the sequence of the DNA base pairs would be identical in function (subject, of course, to the action of modifier genes at other loci) but not necessarily identical by descent. For a recent discussion of the problems of gene taxonomy at the molecular level, see Zuckerkandl (1963).

As will be shown in later sections, the evidence for between-population identity in the function of the genes discussed here is limited to the results of formal genetics and thus is not as direct as the evidence based on biochemical genetics.

We may distinguish three possible ways in which two populations may differ in gene frequencies at each chromosomal locus:

1   None of the alleles is common to the two populations. This is a difference in kind.

2   One or more alleles present in one population may be absent in the other, while at least one allele is common to both populations. This is a restricted difference in kind.

3   All the alleles are common to the two populations, although in different numbers. This is a difference in degree.

Prior to the twentieth century, many anthropologists and biologists assumed the differences betwen races were of kind, as in the first case above. For example, in 1871 Quetelet, the founder of anthropometry, assumed the genetic material of (pure) races was homogeneous within races and diverse in kind between races; all variability within races was taken to be of environmental origin. Today, differences of kind are not known to hold for any human race. Further, on theoretical grounds, such differences would not be expected between subgroups of any sexually reproductive species.

For a small number of loci restricted differences are known to hold between some pairs of human races. For instance, the gene associated with blood group A of the ABO series is present in all known European and African populations, as is the gene for blood group B, but gene A is absent in some, and gene B is absent in many American Indian populations. Some of the gene loci to be discussed below fall in this category.

Variations in degree rather than in kind are the most common type of difference observed between local populations and geographic races of man both for normal, major genes and for nearly all the identified deleterious genes.

Let us illustrate some of the above reasoning with the well-known character of albinism. Individuals homozygous for the autosomal recessive gene *albinism totalis* (complete generalized albinism) are deficient in an enzyme necessary for the rapid conversion of tyrosine to melanin pigment. In such homozygotes there is a generalized absence of the mesodermal melanin and a deficiency of the ectodermal melanin; their skin is pale, they cannot tan, their irides are usually pink, and they are hypersensitive to strong light, nystagmus is present, and strabismus frequently occurs. For some environments and some items of behavior, there is nothing in the behavior of an albino to set him apart from those who have the functioning enzyme. But albinos act differently when exposed to strong sunlight. As Dobzhansky (1950, p. 146) said: "The behavior of an albino on a sunny day is obviously influenced by the gene for albinism. The fact that no peculiarities may be noticeable in the behavior of the same albino on a cloudy day does not change the fact that the gene for albinism modifies the behavior of its carriers. Examples of this kind can be multiplied at will."

A number of references in the literature suggest an association between albinism and mental deficiency (cf. Waardenburg et al., 1961). In some cultures albinos experience their lack of melanin as a social handicap and develop feelings of inferiority, but the association with mental defect is by no means general. Among Europeans (Waardenburg et al., 1961), Negroes (Beckham, 1946), and American Indians (Keeler, 1953; Stewart and Keeler, 1965), albinos show about the same range of variations in mental ability as the general population.

As would be expected from the number of gene- and enzyme-controlled metabolic steps between tyrosine and melanin (Fitzpatrick, 1960) a number of different genes may result in partial or total albinism. However, there is presumptive evidence that a homologous, recessive, autosomal gene is responsible for complete generalized albinism in the Caucasoid, Mongoloid, and Negroid races. The best evidence comes from Negro-white and Chinese-Malayan crosses (Davenport and Davenport, 1910; Gates, 1946).

The data in Table 19.1 indicate there are significant differences in gene frequencies between human breeding populations from Africa, America, Asia, and Europe. Woolf (1965) found 17 out of 23 Indian

Table 19.1
Gene frequencies for albinism totalis

Population	Gene frequencies, %
Apache Indians	0.00
N. E. Switzerland	0.45
Baschi, Congo	0.50
Scotland	0.68
North Italy	0.71
Sicily	0.82
United States	1.00
Norway	1.02
Baluta, Congo	1.58
Navaho Indians	1.63
Nigerians	1.87
Warego, Congo	5.00
Zuni Indians	7.07
Hopi Indians	7.07
Caribe Cuna Indians	8.36
Jemez Indians	8.45

SOURCE: Data from Waardenburg, Franceschetti, and Klein, 1961, and Woolf, 1965.

populations of the American Southwest had no albinos. When the mutant gene for albinism is present, its frequency is observed to vary more than eighteenfold, from 0.45 to 8.45 percent. Reed (1965) has suggested an explanation for the high frequency of the gene for albinism in small inbred populations. He postulates a reproductive advantage of hetero-zygotes for albinism due to a heterotic effect of genes closely linked with the albinism locus. The heterotic effect would disappear when the breeding isolate is broken or would decrease slowly as crossing-over decreases the gene differences adjacent to the albinism locus.

It seems evident that albinism is regularly associated with distinctive behavior in certain environments; and there is no doubt that there are racial differences in the frequency of the gene for albinism. Thus, we must conclude that racial differences exist for *some* genes associated with behavioral differences.

BEHAVIORAL DIFFERENCES
BETWEEN RACES

Very few of the investigations concerned with racial differences in behavior have employed explicitly the classifications discussed earlier. Most of them have been concerned with specific comparisons involving two races, and, not surprisingly, the bulk of such studies has centered about comparison of the European and African races.

It may be appropriate at this point to emphasize that whatever may be said in this chapter concerning the possibility or probability of racial differences in behavior does not imply a corresponding likelihood of there being generalized racial inferiorities or superiorities that would lead to a meaningful hierarchy of races. Indeed, many of the reasons that might be used to defend the possibility of such differences could be used with equal cogency to argue against such generalized superiorities or inferiorities. Most important of all, we should like to state unequiv-ocally the lack of any meaningful association between the existence, or lack of existence, of racial differences in behavior and political-legal de-cisions in regard to civil liberties, equal opportunities, or personal free-doms. The latter issues are rooted in moral, ethical, evaluative con-siderations that can never be derived from scientific fact and should not be confused with empirical questions. To blend the two issues is to risk the likelihood that both will suffer. The quality of research may suffer because certain findings are likely to assume an odious and ethically objectionable quality that makes it difficult for most investi-gators to work in the area or to report their findings bluntly. On the other hand, what may be a straightforward moral or ethical issue can become hopelessly confused if an attempt is made to demonstrate that it is somehow derivable from a set of scientific findings.

In simplest terms, no sentient observer could ever have argued that all men are created equal, or, as many still absurdly assert, all men are created *potentially* equal. All but the bigot know that, whether we are talking about members of the European or African races, there are im-portant differences between individuals that have a great deal to do with their potential capacity to perform many interesting and significant acts. For example, are there really people who believe that, given the opportunity and appropriate training, Albert Einstein could have proved a successful fullback on a professional football team; or conversely that all successful pro fullbacks with proper stimulation and opportunity might have revolutionized modern physics. The obvious point is that no one expects that these two very different individuals, with their very different genotypes, would have the same potential no matter how identical their environments. What is important is not that society and their fellow men treat them in the same manner and expect them to perform in the same fashion, but rather that each be permitted to realize his potential capacity without regard for characteristics that are nonrelevant to the performance in question.

In brief, while we endorse in the strongest possible terms the importance of equal rights for all men, it is our personal conviction that none of the evidence in regard to differences in behavior between races has the slightest implication for this principle, which is rooted in moral and ethical considerations and not in individual, racial, or species differences. In addition, we have implied that existing evidence and theory within biology and behavioral sciences suggest that it is most unlikely that whatever differences in behavior (and capacity) exist between races would support the conception of a generalized superiority or inferiority of one race over another.

It is important to distinguish between two types of statements concerning differences between races. One may simply assert that races differ in certain attributes without specifying whether the differences are due to learned behavior or to genetic factors: "There are differences between races A and B in intellectual behavior." The other type of statement asserts differences that are due to genetic factors: "There are racial (genetically determined) differences between A and B in intellectual behavior." The first kind of statement is true of many or most behavioral traits (including particularly items that show cultural variation). The second kind of statement is known to be true for only a very limited number of behavioral traits. Our principal interest here is in genetically determined racial differences in behavior but we shall discuss a number of findings where the relative importance of genetic and environmental determinants is not clearly established.

Sensory-Motor Processes

Although the earliest psychological comparisons of races dealt with simple sensory processes and modes of response, there has been relatively little systematic work in this area until very recently. Indeed the decades immediately before and after 1900 probably saw more pertinent investigation of this variety than we have seen in the ensuing 50 years.

One of the earliest experimental comparisons of races was conducted by the pioneer American clinical psychologist, Lightner Witmer (Bache, 1895). He compared Caucasians, American Indians, and a group of mixed African-Caucasian descent in reaction time to visual, auditory, and tactile stimuli. The American Indian subjects had the lowest average latency, followed by the African-Caucasian group, with the Caucasian subjects the slowest to react.

The well-known Cambridge Anthropological Expedition to the Torres Straits included among its staff the psychologist-anthropologist Rivers and the psychologists Myers and McDougall. These investigators attempted to complete a "census" of the sensory capacities of their primitive subjects and to compare their performance with that of European subjects. Considering the relatively primitive state of experimental psychology of this era, they conducted what is, in many respects, an exemplary investigation. Rivers (1903), who directed the psychological studies, focused his own efforts upon the study of vision. He found that the nonliterate subjects were generally superior in visual acuity to typical European norms but he expressed strong doubts concerning the dependability of this racial difference. He also reported a much lower incidence of color blindness among his Murray Island subjects than that customarily encountered in European groups.

Myers (1903a) studied auditory acuity and tone discrimination, finding that the natives of the Murray Islands tended to be generally inferior to European subjects, whereas there appeared to be no appreciable difference between the two groups in the upper limit of their hearing. A similar study concerned with olfactory discrimination (Myers, 1903b) and taste discrimination (Myers, 1903c) led to the conclusion that "we may say of these Murray men that their sense of touch is twice as delicate as that of Englishmen, while their susceptibility to pain is hardly half as great" (Myers, 1903c, p. 195). McDougall (1903b) also found that the Melanesians were much more sensitive in discriminating weights and at the same time were much more susceptible to the size-weight illusion than their European counterparts. Professor Myers (1903d), in a study of reaction times, found that the European subjects were faster in responding to visual stimuli, whereas there was no group difference in reaction time to auditory stimuli.

A study closely parallel to the Torres Straits investigation was carried out by Rivers (1905) with the Todas of southern India. Once again an attempt was made to assess the sensory capacities of the nonliterate subjects, and the findings were compared with findings for Melanesian and Caucasian subjects. In comparison with Caucasians, the Todas were superior in visual acuity and tactile discrimination, whereas they were inferior in auditory discrimination. There was also evidence for a higher pain threshold and inferior olfactory discrimination for the nonliterate subjects, but Rivers felt that the influence of nonsensory factors here was too great to assign much weight to these findings. He also found a much higher incidence of color blindness (12.8 percent among male subjects) than was characteristic of European subjects. Perhaps the best known of his findings was the significantly lower magnitude of effect of the Müller-Lyer illusion among the Todas when compared with either Melanesian or Caucasian subjects.

The St. Louis World Fair in 1904 provided an unusual opportunity for the psychological comparison of races because of the large number of performers recruited from nonliterate societies around the world. The results of the psychological study of approximately 400 subjects of diverse racial background are discussed in general terms by Woodworth (1910), and detailed findings in the area of audition are reported by Bruner (1908). The most dependable generalizations refer to Caucasian, American Indian, and Malayan Filipino subjects who were the most numerous of the groups studied. Bruner's results suggested the relative auditory superiority of his Caucasian subjects and thus reversed the belief, common at that time, that primitive man was superior to civilized man in simple sensory performance. Bruner (1908, pp. 111–112) concluded:

☐ The one fact standing out most prominently as a result of these measurements is the clearly evident superiority of Whites over all other races, both in the keenness and in the range of the hearing sense. The evidence is so clear and striking as to silence effectually the contention that the hearing function, inasmuch as it is of relatively less utility in the pursuits attending modern social conditions than those surrounding the life of the savage has deteriorated and is degenerating.

Woodworth reported clear-cut findings indicating the superiority of Caucasian subjects in comparison with the other racial groups in color discrimination and somewhat more ambiguous findings suggesting a lower pain threshold in Caucasian subjects. In addition, the American Indian and Malayan subjects appeared superior to the Caucasian subjects in visual acuity. The author points explicitly to the many cultural factors that might have influenced performance on all of these measures and the consequent difficulty in being sure that only racial factors are responsible for the difference in performance. Somewhat surprisingly, in view of the data he summarizes, Woodworth (1910, p. 177) concludes: "On the whole, the keenness of the senses seems to be about on a par in the various races of mankind."

Although there are significant empirical flaws associated with most or all of the studies we have reviewed, the fact remains that these data, in general, suggest the possible existence of interesting and appreciable racial differences in behavior. While one might argue that the shortcomings in design and method vitiate the findings, it is difficult to see how these studies could be used as the basis for claiming an *absence* of race differences. And yet this is actually what has occurred. More often than not these studies are cited, empirical flaws and all, as providing definitive evidence for the absence of any significant race differences in behavior. For example, Anastasi (1958), in her exemplary text, describes the St. Louis Fair studies and concludes (p. 578): "On such controlled tests of sensory acuity, the primitive groups did no better than the white norms. Subsequent investigations on many different groups have corroborated these findings." This statement accurately mirrors the convictions of most contemporary social scientists, but it does not mesh smoothly with some of the results we have just examined. Nor does the implication that this is an area that has been subjected to extensive and definitive investigation seem warranted. Much of the discussion of these studies, both by the investigators and subsequent reporters, has been marred by focusing upon the question of whether there are dramatic sensory superiorities displayed by the nonliterate subjects. Given this orientation, striking race differences, even those indicating a central tendency advantage for the "primitive" subjects, are treated as negative findings because of the overlapping distributions observed for the various groups studied.

The studies we have noted represent a significant and challenging beginning to an area of investigation that could have contributed importantly to the emerging field of social psychology and, at the same time, emphasized the continuity of this speciality with general psychology. It is difficult to know whether the subsequent loss of interest in comparing races by means of simple tests or measures was primarily a consequence of the emphasis upon negative findings by some of these early investigators, or whether it was more importantly related to the general success which greeted the study of the individual differences by means of complex achievement variables (Binet) rather than simple processes (Cattell). In any event, the great majority of investigations during the past four or five decades has searched for race differences primarily through the use of instruments designed to assess molar attributes such as intelligence or personality traits.

Although there were several rather isolated British studies of racial differences in "phenomenal regression" during the 1930s (Beveridge, 1935, 1939; Thouless, 1933), only in the past decade has there appeared any substantial evidence of interest in racial and cross-cultural differences in perception. Allport and Pettigrew (1957) compared European subjects, acculturated Zulu, and nonacculturated Zulu subjects in the incidence of perception of the trapezoidal illusion. They found that under optimal conditions there appeared to be little difference in the performance of the three groups, but under marginal or nonconducive conditions the unacculturated subjects reported the illusion less often than the other groups of subjects.

As part of a similar but much more extensive and systematic program of research Campbell and Segall (Campbell, 1964; Segall, Campbell, and Herskovits, 1963, 1966) have reported a study of racial and cultural differences in the incidence of various perceptual illusions (Müller-Lyer illusion, horizontal-vertical illusion, Sander parallelogram). This investigation is in many ways a model of how to go about making comparisons across cultures or races, particularly because of the painstaking efforts made by the investigators to distinguish between variation in the effectiveness of communication and actual differences in the psychological process under study. The research involved the collaboration of a large number of anthropologists and psychologists and spanned a substantial number of nonliterate tribes as well as American subjects. The investigators reported marked group differences, as illustrated in Figure 19.1, where we find, for example, that the incidence of the Müller-Lyer illusion is almost four times as frequent among American subjects as among Bushmen subjects.

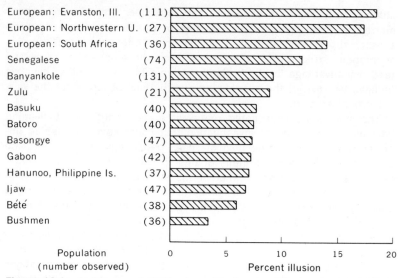

Figure 19.1   Incidence of Müller-Lyer illusion in different racial groups. (*Adapted from Campbell, 1964.*)

Bonte (1962) compared European, Bashi African, and Mbuti Pygmy subjects in susceptibility to the Müller-Lyer illusion, using the same procedure employed by Rivers (wooden apparatus manipulated by the subject), and failed to find any differences between the three groups. However, when a separate sample of European and Bashi subjects were compared, using the pictorial method devised by Campbell and his collaborators, he found that the African subjects were significantly less susceptible to the illusion. This same finding of an increased susceptibility to perceptual illusions on the part of European subjects in comparison with African subjects is also reported by P. Morgan (1959). Thus, four relatively modern studies provide a general confirmation of the early report of Rivers that European subjects were more susceptible to certain perceptual illusions than various nonliterate groups.

It should be noted that none of the American investigators adopts a nativistic frame of reference in accounting for the racial differences in the frequency with which these illusions are reported. They consider the most reasonable explanation of the differences to be in terms of variation in the environment or experience of the different races. For example, Campbell (1964) advances the "carpentered-world" hypothesis to account for observed differences. He reasons that the world of the western European subject is much more heavily saturated with straight lines and right angles as a result of man-made and carpentered objects and consequently acute and obtuse angles tend to be perceived as right angles extended in space. Such a tendency would generally predispose the subject to experience the illusions Campbell studied. This interpretation does not fit neatly with all the data for, among other considerations, there is the fact that young children are more susceptible to the illusion than older subjects in our culture; one might reason that, if experience in the "carpentered" world produced the illusion, the illusion should become more common with greater exposure to this environment. Thus, even with such a relatively simple process as a perceptual illusion, it appears difficult to identify the relative contribution of environmental and biological determinants, although not nearly so difficult as in the case of intellectual and personality variables.

Several recent studies indicate *auditory differences* between races. Hinchcliffe (1964) found Jamaican women to have poorer hearing levels than a comparable group of Scottish women, although there were no significant differences between similar groups of male subjects. The investigator suggests that the difference may be attributed to differential incidence of sensorineural deafness. Tanner and Rivette (1964) report an interesting observation of "tone deafness," or the inability to discriminate auditory stimuli on the basis of frequency, in a small number of Indian subjects. Tentatively, they suggest the deficiency may be attributed to cultural differences, particularly in language, between these subjects and the usual European subjects. A series of studies comparing the Mabaan tribe of southeast Sudan with subjects from Wisconsin, New York, Dusseldorf, and Cairo reveals that with aging the Sudanese maintain considerably higher levels of hearing than the other groups (Jansen et al., 1964). Although the investigators do not discount the role of genetic factors, they consider the relatively noise-free environment of the Sudanese to be the most likely determinant of this difference.

An interesting and important exception to the general tendency to study complex processes is represented by work with the genetically determined inability to taste phenylthiocarbamide and related compounds. This research has seldom been considered relevant to the topic of race differences in behavior, in spite of the fact that it undoubtedly represents the best-established behavioral correlate of race yet observed. Presumably, this slight on the part of persons interested in race differences derived partly from a preference for the "null hypothesis" in this area and partly from their focus upon complex achievement variables. Surprisingly, the only treatment of these data as relevant to race differences in behavior appeared in the psychological literature (Cohen and Ogden, 1949), with the authors emphasizing the variability of findings and their failure to differentiate racial groups. This conclusion is distinctly at variance with the position of observers such as Boyd (1950), Valls (1958), and Stern (1960), all of whom report marked differences between races in PTC tasting. Boyd (1953) actually lists this gene as one of eight that differentiate his genetically determined races.

Taste blindness for creatine was discovered in 1926 by Lasselle and Williams; its mode of inheritance is not known. Fox (1931) found four-tenths of an American sample could not taste the synthetic compound p-ethoxy-phenylthiocarbamide (PTC)—also called phenylthiourea—while the other subjects regarded it as very bitter even in low concentrations. When taste thresholds are measured over a wide range of concentrations, a sharply bimodal distribution appears, with an antimode which allows classification of individuals into two phenotypes, *tasters* and *nontasters*, with few borderline cases (Setterfield et al., 1936; Harris and Kalmus, 1949). A bimodal distribution like that for PTC was found for 16 out of 17 other substances all containing the group

$$= N - C -$$
$$\overset{||}{}$$
$$S$$

but this correlated dimorphism is not observed for a variety of other compounds, some containing sulfur in other groupings (Harris and Kalmus, 1949). The details of the molecular and cellular basis for the PTC taste reaction are unknown; it is not due to a lack of all bitter receptors, for nontasters for PTC are sensitive to other bitter substances, e.g., quinine (Kalmus, 1959).

The statement commonly found in the literature that the ability to taste PTC is due to a single, autosomal dominant gene requires some qualification. There is no doubt, however, that most of the variation in PTC taste reaction is genetically determined.

Studies on 800 American families by L. H. Snyder (1932), 124 Negro families by Lee (1934), 2,447 Chinese and Taiwanese families by Rikimaru (1937), 126 Bengalese families by Das (1958), and 60 Norwegian families by Merton (1958) show no significant differences between the observed and expected numbers of nontasters from matings where one or both parents are tasters (cf. Spuhler, 1951). There are marked departures, however, between the observed and expected numbers of the two phenotypes from parents who are both nontasters. In such families, according to the hypothesis, all children should be nontasters. Snyder (1932) found 5 exceptions among 223 children from 86 matings, and Rikimaru (1937) reported 53 exceptions (29.8 percent) among 178 offspring from 47 families. These results may be due to the fact that (1) some of the individuals are wrongly classified as to paternity or taste reaction, (2) the taster gene lacks full penetrance in some individuals, or (3) the genetic hypothesis is wrong. Similar conclusions are supported by twin studies; Ardashnikov et al. (1936) determined that 2.2 percent of 137 pairs of monozygous twins were discordant for PTC taste reaction, and Rife (1938) observed that 3.6 percent of 194 pairs of monozygous twins were discordant for this behavior trait.

A considerable part, perhaps a majority, of the discrepancy between theory and result in the family material may be due to the method of testing used in the earlier studies, that is, in the use of PTC crystals or paper impregnated with PTC rather than serial dilutions of PTC in a water solution. Harris and Kalmus (1949) and Kalmus (1958) have shown that an objective classification of the PTC taster genotypes is possible by means of a sorting technique employing individual taste thresholds.

We may summarize the available genetic information on taste reaction by stating that nontasting for PTC is an "almost recessive auto-

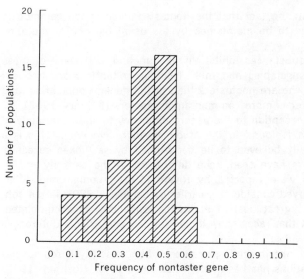

Figure 19.2   Distribution of phenylthiocarbamide non-taster gene frequencies in samples from 50 populations tested by means of the sorting or equivalent technique. (*From Saldanha and Nacrur, 1963.*)

somal trait." The greater part, but not all, of the variation in PTC taste reaction is genetic. Family studies give presumptive evidence that the mode of inheritance for the PTC taste reaction is the same in the three major races of mankind (cf. references cited above and Pons, 1960).

The frequencies of nontasters for PTC show striking race differences. Schwidetzky (1962) tabulated the results of the PTC taste reaction (including all methods of testing) in nearly 150 population samples. She found the frequencies of the nontaster phenotype varied from 0 to 56.9 percent. Gene-frequency estimates based on the Harris and Kalmus sorting technique or its equivalent are summarized for 49 population samples by Saldanha and Nacrur (1963). They found the frequency of the recessive, nontaster gene ranges from 0.111 to 0.658; Figure 19.2 is a histogram showing the number of populations by PTC nontaster gene frequency in the samples assembled by Saldanha and Nacrur.

Table 19.2 shows the distribution of gene frequencies within and between the major races of mankind as identified by Garn (1961). The numbers listed may be regarded as representative of high and low gene frequencies for the nontaster allele within each race but they do not necessarily represent the minimum and maximum values reported in the literature for the various races.

The forces that determine changes in gene frequency of the PTC taster alleles between human breeding populations are not known. It is highly probable, however, that natural selection leading to a balanced polymorphism at different levels according to local environmental conditions is an important factor in the determination of gene frequencies.

This follows from the fact that the nontaster allele is too common in many populations to be maintained by the usual balance of mutation and selection.

Several SCN substances inhibit thyroid function, and some of these are present in vegetables and milk. Nodular goiter is more frequent among patients who are nontasters than in the general population, and toxic diffuse goiter is more common among tasters (Kalmus, 1959).

An additional exception to the general tendency to shun the study of simple processes is found in the study of *color perception*. This is a character generally believed to be determined by sex-linked recessive genes and, as we have seen, race differences in the capacity to discriminate colors were reported by Rivers (1901); comparable differences were observed considerably earlier by Schöler (1880). In the subsequent years, a great deal of additional evidence has accumulated, leaving no doubt that races show important and appreciable differences in this attribute.

The sex-linked inheritance of partial color blindness (dyschromaposia) in human populations has been recognized since 1777. Rushton (1962) has shown that normal color vision in man depends on the presence of three light-sensitive pigments in the cones of the retina. The blue-absorbing pigment cyanolabe occurs in the blue cones; chlorolabe, the visual pigment with an absorption maximum in the green, occurs in the green cones; and the red-sensitive cones contain both chlorolabe and the red-absorbing erythrolabe.

The two main groups of sex-linked partial color blindness are called *protans* (red blindness) and *deutans* (green blindness); the members of

Table 19.2
Distribution of high and low gene frequencies for PTC nontasters within and between Garn's (1961) major races

Race and population	Author	Gene frequency
African, Batutsi and Bahutu	Hiernaux, 1954	0.114
African, Bantu, Kenya	Lee, 1934	0.285
Amerindian, Caraja, Brazil	Junqueira et al., 1957	0.000
Amerindian, Eskimo, Labrador	Sewall, 1939	0.639
Asian, Chinese, Singapore	Lugg and White, 1955	0.142
Asian, Malayan	Thambipillai, 1955	0.400
Australian, natives, southern	Simmons et al., 1954	0.702
Australian, natives, central	Simmons et al., 1957	0.707
European, Lapps, Finland	Allison and Nevanlinna, 1952	0.253
European, Danish	Mohr, 1951	0.584
Indian, Hindu, Riang	Kumar and Sastry, 1961	0.403
Indian, Mala-Vedan	Büchi, 1958	0.732
Melanesian, New Hebrides	Simmons et al., 1956	0.303
Melanesian, Pygmies, New Guinea	Simmons et al., 1956	0.713
Micronesian, Turkese	Simmons et al., 1953	0.428
Micronesian, Kapinga	Simmons et al., 1953	0.529
Polynesian, Easter Islanders	Simmons et al., 1957	0.281
Polynesian, Cook Islanders	Simmons et al., 1955	0.404

SOURCE: Data from Schwidetzky, 1962, and Saldanha and Nacrur, 1963.

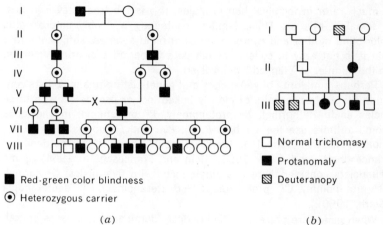

Red-green color blindness

Heterozygous carrier

Normal trichomasy

Protanomaly

Deuteranopy

(a)   (b)

Figure 19.3 (a) Inheritance of red-green color blindness in eight generations. X = marriage between a red-green-color-blind man and a
carrier woman with apparent inheritance from father to son. (*From
Horner, 1876.*) (b) Crossing-over between deuteranopia and protanomaly. (*From Vanderdonck and Verriest, 1960.*)

a third group *tritans* (blue blindness) are more rare than the other two
groups and probably show an autosomal rather than a sex-linked mode
of inheritance. Two subtypes are distinguished in each of the common
groups. Within the protans, those with *protanomaly* have anomalous
trichromatic vision with a deficiency of perception at the red end of the
spectrum; those with *protanopia* have dichromatic vision and lack the
red-sensitive cones containing both chlorolabe and erythrolabe. Within
the deutans, those with *deuteranomaly* have trichromatic vision with a
deficiency in green perception (this is the mildest and the most frequent
of the four types); those with *deuteranopia* have dichromatic vision and
lack the green cones containing chlorolabe. Protans confuse red with
green and cannot see a red figure on a black background; any color with
a red mixture is perceived as though red were absent so that purple
appears blue and orange yellow. Deutans see green objects as gray;
they distinguish only the yellows and blues in the spectrum.

There is general agreement that the four subtypes of red-green color
blindness are controlled by genes located in the long arm of the X
chromosome. Males have only one of the set of alleles which occupy
each X-linked locus and females have two; males manifest the mutant
gene for color blindness if it is present, while heterozygous females may
carry a recessive mutant gene for color blindness without its being manifest in their phenotype. Color-blind males receive their mutant gene from
their mothers and pass it to their daughters. Color-blind females must
receive a mutant gene from each parent. Figure 19.3a illustrates
the inheritance of red-green color blindness in eight generations of a
Swiss family.

It is controversial whether just one chromosomal locus or two closely
linked loci are involved in X-linked partial color blindness. The condition

is much rarer in females than in males; if q is the frequency in males, in females it is q^2. Thus, families containing a doubly heterozygous color-blind mother and two or more color-blind sons are sufficiently rare that it is not easy to collect enough data to permit a clear-cut decision on the number of X-linked loci involved.

Perhaps a majority of geneticists and geneticist ophthalmologists conclude there are two pairs of closely linked loci or pseudoloci with three alleles each: P^+ normal, P^L protanomaly, P^P protanopia, in one locus (some authors use the symbols P, P_1, P_2), and D^+ normal, D^L deuteranomaly, D^P deuteranopia in the other locus (D, D_1, D_2) (Waardenburg, Franceschetti, and Klein, 1963; Klein and Franceschetti, 1964). Some geneticists consider there is a single locus with five alleles: C^+ normal, C^{PL} protanomaly, C^{PP} protanopia, C^{DL} deuteranomaly, C^{DP} deuteranopia (Stern, 1960).

When genes are present in double dose, dominance decreases in both groups in the order normal, anomaly, anopia. Females heterozygous for mutant protan and deutan genes have normal color vision, the genes being complementary between groups. This controversy regarding the number of loci in X-linked partial color blindness is analogous to the three- (c,d,e) versus one- (r) locus controversy regarding the rhesus (Rh) blood groups.

Decisive evidence against the one-locus hypothesis could be obtained through the study of the sons of many mothers heterozygous for those combinations of mutant genes that result (according to the two-locus hypothesis) in normal color vision. The occurrence of sons with normal color vision in such families would establish the existence of two recombinable loci controlling the four types of red-green color blindness. In the one such family (see Figure 19.3b) thus far reported, the protan mother has two deutan, one protan, and two normal sons (Vanderdonck and Verriest, 1960). There are other possible explanations of unique cases, and at present there is insufficient evidence to decide between the two hypotheses. The biochemical genetics of color vision is very poorly understood (Kalmus, 1959). Actually, the problem regarding number of loci is not of great importance for present purposes. Everyone agrees the four most common types of red-green color blindness are controlled by genes with fairly high penetrance (Knox, 1958) and fairly constant expression. If the one-locus hypothesis is correct, the phenotypic frequencies have a one-to-one correspondence to the gene frequencies; if the two-loci hypothesis is correct, the correspondence is not one-to-one, but it is approximately that. Rather than equate phenotypic frequencies in hemizygous males to gene frequencies, we would equate phenotypic frequencies in males to four different mutant "chromosomal types" in the hemizygous, male population.

The distribution of X-linked partial color blindness is now known for more than 100 different populations. Unfortunately, only a few population surveys (all, aside from one from Japan, are from European peoples) allow estimation of the frequencies of the four subtypes; these are summarized in Table 19.3. In this set of populations, the range in frequency of partial color blindness in males goes from 3.9 to 9.0 percent.

Table 19.3
Frequencies of the four subtypes of red-green color blindness

Population	Author	Number	Grand total	Protan group			Deutan group		
				Prot-anomaly	Prot-anopia	Sub-total	Deuter-anomaly	Deuter-anopia	Sub-total
Japan	Sato, 1935	249,014	3.90	0.50	0.65	1.15	1.72	1.02	2.75
France	Hebert, 1957		6.58	0.58	0.58	1.16	4.64	0.77	5.41
Germany	Heinsius, 1941	4,406	6.64	0.97	0.59	1.56	4.02	1.06	5.08
United States	Schmidt, 1955		6.91	1.13	0.57	1.70	3.96	1.25	5.21
United States	Newhall, 1958	323	7.26	0.63	0.32	0.95	4.42	1.89	6.31
Germany	Schmidt, 1936	6,863	7.76	0.68	1.10	1.78	4.01	1.97	5.98
Switzerland	von Planta, 1928	2,000	7.95	0.60	1.60	2.20	4.25	1.50	5.75
Norway	Waaler, 1927	9,049	8.01	1.04	0.88	1.92	5.06	1.03	6.09
Switzerland	Wieland, 1933	1,036	8.20	1.16	0.96	2.12	5.12	0.96	6.08
Belgium	Francois, 1957	1,243	8.29	1.05	0.96	2.01	4.91	1.37	6.28
England	Nelson, 1938	1,338	8.82	1.27	1.27	2.54	5.08	1.20	6.28
Switzerland	Bally, 1954	1,000	9.00	1.10	1.10	2.20	4.70	2.10	6.80

SOURCE: Data from Sato, 1935, and Waardenburg, Franceschetti, and Klein, 1963.

Of the mutants, the deutan alleles (or chromosomes) are the most frequent in all populations, the frequency of deuteranomaly varying from 1.7 to 5.1 percent, and that of deuteranopia from 0.8 to 2.1 percent. The two deutan subtypes together account for about three-fourths of the mutant genes at this locus (or these loci) in both the European and Japanese populations.

The frequency of C^{PL} is equal to or larger than that of the other protan allele, C^{PP}, in 9 of the 12 populations, C^{PL} varying from 0.6 to 1.3, and C^{PP} from 0.3 to 1.10. Post (1962) in a summary of the distribution of the protan and deutan alleles in 32 populations found the frequency of $C^{PL} + C^{PP}$ varies from 0 to 3.3 percent and that of $C^{DL} + C^{DP}$ from 0 to 9.7 percent.

The considerable variation in the population frequencies of X-linked color blindness among one or more representatives of the nine major races recognized by Garn (1961) is shown by Table 19.4. The extreme value for the Kotas probably is an example of the marked fluctuations of the frequencies of major genes in very small breeding populations. A range in gene frequencies for X-linked partial color blindness from essentially zero to 15.0 percent is well established. Samples of 111 are sufficient to establish differences in proportions between two populations with frequencies of 1.0 and 10 percent, respectively, at the 5 percent level of significance.

The distribution of gene frequencies for protans + deutans in 114 populations summarized by Post (1962) plus a few additional samples from other sources is presented in Figure 19.4. The distribution is bimodal with modes at 2 to 3 and at 7 to 8 percent.

It is interesting to speculate on reasons for changes in color-vision-gene frequencies between different human populations. It is doubtful

Table 19.4
Frequency of color blindness (protan and deutan groups) in males

Population	Race (Garn, 1961)	Author	No.	Frequency, %
Fiji Islanders	Melanesian	Geddes, 1946	200	0.0
Brazilian Indians	Amerindian	Mattos, 1958	230	0.0
Bagandas	African	Simon, 1951	537	1.9
Navaho Indians	Amerindian	Spuhler, 1951	163	2.4
Australian natives	Australian	Mann and Turner, 1956	378	3.3
Marshall Islanders	Micronesian	Mann and Turner, 1956	268	4.1
Turks, Istanbul	European	Garth, 1936	473	5.3
Chinese, Peking	Asian	Chang, 1932	1,164	6.9
Tonga Islanders	Polynesian	Beaglehole, 1939	67	7.5
Belgians	European	Francois, 1956	1,243	8.6
Russians	European	Flekkel, 1955	1,343	9.3
V.N.B. Brahmins, Bombay	Indian	Sanghvi, 1949	100	10.0
Americans	European	Shoemaker, 1943	803	11.4
Todas, India	Indian	Rivers, 1905	320	12.8
Dutch, Brazil	European	Saldanha, 1960	97	15.5
Kotas, India	Indian	Sarkar, 1958	28	61.0

SOURCE: Data from Waardenburg, Franceschetti, and Klein, 1963, and Post, 1962.

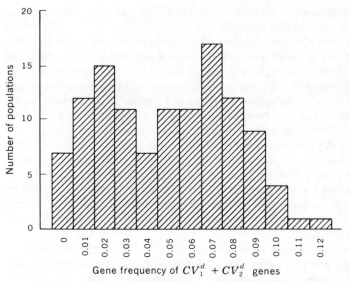

Figure 19.4   Distribution of protan and deutan color-blind genes in samples from 114 populations. (*From Post, 1962, with additions.*)

that variation in mutation rates acting alone can account for the observed variation. If all the selection is assumed to occur in males, the gene frequency at equilibrium, $\hat{q}$, is a function of the mutation rate per gene per generation, u, and the selection coefficient against the mutant gene, s (where the fitness of the mutant hemizygote is $W = 1 - s$), is as follows (Li, 1955, p. 287):

$$\hat{q} = 3u/s$$

If $u = 10^{-5}$, then $s = 0.003$ for $\hat{q} = 0.01$, $s = 0.0006$ for $\hat{q} = 0.05$, and $s = 0.0001$ for $\hat{q} = 0.1$. Thus, for this proposed balance between mutation and selection, it takes 6 times the selection required to keep the gene frequency at 0.1 to reduce it to 0.5, and 30 times to reduce it to 0.01. These considerations led Post (1962) and Neel and Post (1963) to speculate that natural selection under the rigorous conditions experienced by hunting and gathering peoples (say $s = 0.003$ or larger) holds their mutant-gene frequencies low, whereas the less rigorous environment in agricultural and civilized populations permits a relaxation of selection (say to $s = 0.0001$) with a consequent elevation in gene frequency as the new equilibrium point is approached. This example suggests that the selective disadvantage of red-green color blindness in terms of differential survival and fertility varies today in the human species from something like 1 to 1,000 to a little less than 1 in 10,000.

It seems evident that the variable performance of individuals taking the standard tests for color perception constitutes variation in behavior. Consequently, data we have just reviewed on the distribution of the X-linked forms of color blindness among the different human populations clearly establishes the existence of racial, i.e., genetic, differences in behavior for the human species.

If we except PTC tasting, color vision, and certain perceptual illusions, there appears to be little compelling evidence at present either for racial differences or racial equality in simple sensory or motor processes. This dearth of pertinent investigation exists in spite of promising beginnings and occasional exhortations by respected figures. For example, in the mid-1930s, Florence Goodenough (1936) recommended a renewed interest in the study of sensory differences between races, and at the same time she questioned the potential contribution of studies concerned with intelligence and personality. Her reasonable analysis and plea seem to have had very little effect, for in subsequent years the bulk of investigation focused upon just those variables in which she saw so little promise. It is only with the recent and promising work of Segall, Campbell, and Herskovits that there seems to be some basis for optimism on this score.

On the basis of the evidence discussed in this section we conclude that there are particular forms of behavior that exhibit genetic differences between both local and major races. At the same time, it appears that these forms of behavior have only limited functional significance within human societies.

Intelligence

There is probably no area of psychological investigation that has been accompanied by greater passion and more strenuous activity and yet has led to less in the way of definitive findings than the study of intellectual differences between races. Decades of active research have resulted in nothing more than a tentative preference for the null hypothesis.

Perhaps the best place to begin the present discussion is to turn to the end point in the deliberations of some representative experts in this field. First, we have the statement on the race concept prepared by the 1951 UNESCO Committee (1952, p. 13):

☐ When intelligence tests, even non-verbal, are made on a group of non-literate people, their scores are usually lower than those of more civilized people. It has been recorded that different groups of the same race occupying similarly high levels of civilization may yield considerable differences in intelligence tests. When, however, the two groups have been brought up from childhood in similar environments, the differences are usually very slight. Moreover, there is good evidence that, given similar opportunities, the average performance (that is to say, the performance of the individual who is representative because he is surpassed by as many as he surpasses), and the variation round it, do not differ appreciably from one race to another. . . . It is possible, though not proved, that some types of innate capacity for intellectual and emotional responses are commoner in one human group than in another, but it is certain that, within a single group, innate capacities vary as much as, if not more than, they do between different groups. . . . The normal individual, irrespective of race, is essentially educable. It follows that his intellectual and moral life is largely conditioned by his training and by his physical and social environment.

Otto Klineberg, a social psychologist who has contributed actively to this area of research for many years, arrives at the following generalizations (1954, pp. 320–321):

☐ As far as intelligence tests are concerned, it was formerly believed by many psychologists that racial differences had been demonstrated, but the present consensus is to the effect that so many environmental factors enter into the comparisons that no conclusion as to innate ability is justifiable. The superiority of northern over southern Negroes argues in favor of the environmental determination of the test scores, since there is no definite evidence for the selective migration of a superior group. The discovery of individual Negro children with intelligence quotients at the extreme upper end of the distribution, the excellent showing made by American Indian children adopted into superior White homes, as well as the marked improvement following a rise in economic level and educational opportunities, also testify to the absence of innate ethnic differences in intelligence.

To these conclusions, which are obviously slanted in the direction of expecting no significant differences between racial groups, could be added others that emphasize present ignorance coupled with the rational likelihood that such differences eventually may be observed. Still others survey the existing literature and conclude that there already exists satisfactory evidence to demonstrate racial differences in intelligence. For example, Shuey (1966, p. 520), following a survey of 240 individual investigations, suggests:

☐ The remarkable consistency in test results, whether they pertain to school or preschool children . . . to high school or college students, to enlisted men or officers in training in the Armed Forces . . . to gifted or mentally deficient, to delinquent or criminal; the fact that differences between colored and white are present not only in the rural and urban South, but in the Border and Northern states . . . the fact that relatively small average differences were found between the IQ's of Northern-born and Southern-born Negro children in Northern cities . . . the evidence that the mean overlap is between 7 and 13 per cent; the evidence that the tested differences appear to be greater for logical analysis, abstract reasoning, and perceptual-motor tasks than for practical and concrete problems . . . the fact that differences were reported in practically all of the studies in which the cultural environment of the whites appeared to be similar in richness and complexity to that of the Negroes . . . all taken together, inevitably point to the presence of native differences between Negroes and whites as determined by intelligence tests.

The empirical evidence that has led to these inconclusive conclusions has been summarized in a number of places (for example, Anastasi, 1958; Cryns, 1962; Dreger and Miller, 1960; Garth, 1931; Klineberg, 1935; Pettigrew, 1964; Shuey, 1958; Tyler, 1956). It seems clear, however, that with all this investigation the position of most modern observers is at least as much influenced by prior belief as by present findings. What we can say with confidence is that racial groups differ in intelligence as measured by existing instruments. The extent to which these differences are to be attributed to biological factors (race) rather than to experiential (particularly cultural) factors remains largely unknown.

The variety of procedural problems that face the investigator in this area has been discussed by Anastasi (1958), Dreger and Miller (1960), Thompson (1957b), and Tyler (1956), among others, and the weight of these difficulties is not encouraging in regard to the likelihood of quick production of revealing data. The major difficulty facing the investigator is the extent to which cultural (particularly language) differences are associated with race differences; thus a comparison of different races has almost always involved a comparison of groups exposed to different cultures and different socialization experiences. Also, it has proved singularly difficult to select groups for study that reasonably can be assumed to be equally representative of their respective races.

It would be both repetitious and tedious to give a detailed survey of the hundreds of studies that have been conducted in the attempt to compare the intelligence of different races. However, it may be worthwhile to comment upon a few points of general agreement concerning these studies, to discuss some of the central issues, and to cite some illustrative data.

As noted previously, the bulk of research has dealt with the comparison of European and African subjects in North America. One important conclusion is that at least some observers, including both those on the left (Klineberg, 1963) and those on the right (Shuey, 1966) agree that a broad view of the existing research in this area suggests that an average IQ of roughly 85 or 86 is an appropriate index for the American Negro while existing evidence points toward 100 as representative of the average American white subject. Moreover, representatives of both groups would agree that such a difference transcends any questions of statistical significance and, more importantly, it represents an appreciable and significant potential handicap for the Negro. Obviously the difficulties between egalitarian and elitist conventionally have entered in the process of attempting to interpret these differences.

As the above conclusions imply, intelligence differences (as measured by conventional tests) between Negro and white in our society are dependably present. At the same time a variety of environmental parameters significantly influence these racial differences. Those who are motivated to argue against racial differences in intelligence, as measured by American tests of intelligence and defined by American psychologists, point to the obvious impact of cultural or environmental determinants and suggest that whatever differences are observed between races can be accounted for by means of the unequal operation of these determinants upon the two races in question. Those who believe in the existence of biologically determined differences in intelligence between races take comfort in the fact that alteration or control of such factors as socioeconomic status and educational level almost never eliminates the difference between white and Negro subjects.

It may be illuminating to examine as illustrative the results of Charles' (1936) investigation of differences in intelligence quotients (IQ) in American Negroes and whites. The Kuhlman-Anderson Intelligence

Tests were given to 172 Negro and 172 white boys aged 12 to 16 years selected in about equal numbers for each yearly age group from schools in different parts of St. Louis, Missouri. A summary of the results follows:

IQ Scores

Race	Mean	Standard deviation	Range Observed	Range 6 sigma	Coefficient of variation
White	98.31	12.25	135 − 60 = 75	135.06 − 61.56 = 75.50	12.46
Negro	88.60	11.00	114 − 55 = 59	121.60 − 55.60 = 66.00	12.42
Difference	9.71	1.25	16	9.50	0.04

The observed difference in the mean IQ is 9.71, and the difference between means is significant beyond the 0.001 level ($t = 7.706$ with 342 degrees of freedom).

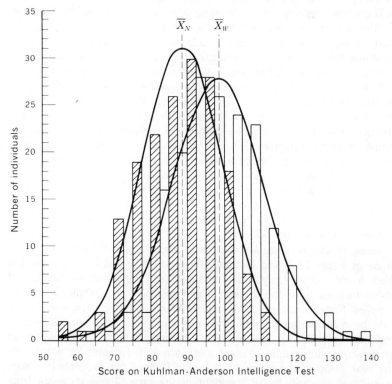

Figure 19.5 Distribution of Kuhlman-Anderson Intelligence Test scores of 172 Negro and 172 white boys of ages 12 to 16 years in schools of St. Louis, Missouri. Frequencies for Negro scores are shown by bars to the left (shaded) and those for whites to the right (unshaded) for each interval of 5 points. The mean for Negroes is marked X_N and that for whites X_W. The ordinates of the normal curves were obtained from the estimated variances. (*From Charles, 1936.*)

Before considering the possible genetic or racial significance of these results, let us look at the distribution of the IQ scores of the two groups in the histograms given in Figure 19.5.

Those who stress the differences between whites and Negroes in IQ scores tend to pay particular attention to differences between the means, or the degree of "overlap," that is, the percentage of Negroes with scores above the white mean. A fuller interpretation indicates the following:

1   21.75 percent of Negroes are superior in IQ to the white mean IQ.

2   3.48 percent of Negroes have an IQ below 70.

3   1.16 percent of whites have an IQ below 70.

4   77.64 percent of whites are superior in IQ to the Negro mean IQ.

We should also note that the difference of 10 IQ points is due chiefly to the differences for boys 15 and 16 years of age (10 and 17.5, respectively) as the difference is only 6.5, 3.5, and 4.3 for those 12, 13, and 14 years, respectively (Charles, 1936, p. 505).

These observations are fully consistent with acceptance of the hypothesis that the mean observed IQ is about 10 points lower for Negroes than for whites. However, these results, and all comparable results now available, neither prove nor disprove that there is a racial, that is, a genetic or inherited, difference in IQ between Negroes and whites.

It would be correct to speak of a quantitative genetic difference between two human populations measured in the same environment or to speak of a quantitative environmental difference between genetically identical populations measured in different environments. However, the members of no two races live in identical environments, and the members of no two races living in different environments are identical in genotype. This is the reason population geneticists have long said it is impossible to give a correct, *general* answer to the question of whether heredity or environment is more important in determining the variation in a quantitative trait, such as intelligence, in human populations (cf. Hogben, 1939, pp. 95–97).

Some of the reasons for the above statement will be clear after a study of Table 19.5 and Figures 19.6 and 19.7. Table 19.5 gives the median and mean Army Alpha Intelligence Test scores of Negro and white draftees from the 23 states and the District of Columbia which supplied Negro draftees during World War I. The medians and means have been estimated from the data given in Tables 205 (pp. 689–690) and 266 (p. 730) in Yerkes, 1921. Column 8 summarizes estimated expenditures for public primary and secondary schools in dollars for the year 1900 in the various states and District per estimated number of children, ages 5 to 18, resident in the area. These data are from Statistical Abstract of the United States, Table 134 (pp. 437–439), 1902. Draftees of ages 22 to 35 in 1917 would be of school ages 5 to 18 in 1900.

There is a difficulty in handling the results from the District of Columbia. The maximum scores for whites (median = 78.8 and mean =

85.6) are from the District and are based on a sample of only 77 draftees. The D.C. median is 11.6 alpha units and the mean 12.6 units above the next highest scores. The overall means of the medians and their standard deviations are $\bar{x}_1 = 50.79$ and $\sigma_1 = 11.03$, if the D.C. data are included, and $\bar{x}_2 = 49.13$ and $\sigma_2 = 9.45$, if the D.C. results are excluded. (Here and in the summary of major results given below the subscript 1 indicates the statistic is based on data including the District of Columbia and subscript 2 indicates the statistic is based on data excluding the District of Columbia.) The D.C. median is 29.57, or 3.13 σ_2 above $\bar{x}_2$, and 28.33, or 2.57 σ_1 above $\bar{x}_1$. The overall mean of the means excluding the District is 56.00 with $\sigma_2 = 8.50$. The D.C. mean score is 29.60, or 3.48 σ_2 above $\bar{x}_2$. The sizes of three other samples included in Table 19.5 are smaller than the D.C. sample for whites (i.e., 55, 57, and 67) although the means and medians based on these three smaller samples are well within the range of the means and medians based on the larger samples, the most extreme case being

Table 19.5
Median and mean Negro and white Army Alpha Intelligence Test scores and school expenditures per child aged 5 to 18 years by state

State (1)	White			Negro			School expend- itures (8)
	N (2)	Me- dian (3)	Mean (4)	N (5)	Me- dian (6)	Mean (7)	
Alabama	779	41.3	49.4	271	19.9	27.0	1.51
Arkansas	710	35.6	43.3	193	16.1	22.6	3.09
Florida	55	53.8	59.8	499	9.2	15.3	4.68
Georgia	762	39.3	48.3	416	10.0	17.2	2.68
Illinois	2,145	61.6	66.7	804	42.2	47.9	13.46
Indiana	1,171	56.0	62.2	269	41.5	47.6	11.75
Kansas	861	62.7	67.0	87	34.7	40.6	10.58
Kentucky	837	41.5	48.6	191	23.9	32.4	4.57
Louisiana	702	41.1	49.0	538	13.4	20.8	2.52
Maryland	616	55.3	60.2	148	22.7	30.7	8.44
Mississippi	759	37.6	43.7	773	10.2	16.8	2.63
Missouri	1,329	56.5	61.9	196	28.3	34.2	8.54
New Jersey	937	45.3	52.9	748	33.0	38.9	14.04
New York	3,300	58.4	63.7	1,188	38.6	45.3	19.22
North Carolina	702	38.2	45.9	211	16.3	22.1	1.51
Ohio	2,318	67.2	73.0	163	45.4	53.4	12.13
Oklahoma	865	43.0	50.6	98	31.4	35.9	5.50
Pennsylvania	3,280	62.0	67.1	790	34.8	40.5	12.85
South Carolina	581	45.1	51.1	334	14.2	19.2	1.93
Tennessee	710	44.0	52.0	504	29.7	35.9	2.71
Texas	1,426	43.5	50.2	854	12.2	18.2	4.38
Virginia	506	56.3	60.5	57	45.6	52.0	3.39
West Virginia	423	54.9	60.8	67	26.8	28.5	6.79
Subtotal (23 states)	25,774	49.53	56.00	9,399	26.09	32.30	6.91
District of Columbia	77	78.8	85.6	30	31.2	34.3	17.78
Total	25,851	50.75	57.23	9,429	26.43	32.39	7.36

SOURCE: Data from Yerkes, 1921, and *Statistical Abstract of the United States,* 1902.

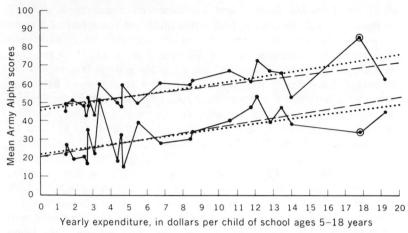

Yearly expenditure, in dollars per child of school ages 5–18 years

Figure 19.6   Regression of mean Army Alpha scores on yearly expenditures in dollars per child of ages 5 to 18 years resident in 23 states and the District of Columbia. The points for each pair of scores and expenditures are listed (from left to right) in the order designated (from top to bottom) in Figure 19.7; the points for Negroes are below and those for whites above; the points for the District of Columbia are encircled. The dotted regression lines were estimated including, and the dashed lines excluding, data from the District of Columbia. (*Alpha scores based on data from* Yerkes, 1921; *school data from* Statistical Abstract of the United States, 1902.)

the Negro scores from Virginia where the state mean score is 1.69 σ_2 above the overall mean. As might be expected, the men drafted from the District of Columbia were not typical of those drafted from the 23 states. The difficulty of the extreme D.C. median and mean scores is met by doing all the calculations with and without inclusion of the D.C. results.

The minimum median (35.6) and mean (43.3) scores for whites are from Arkansas and are based on a sample of 710 draftees; those of Negroes (9.2 and 15.3) are from Florida and are based on a sample of 499 men. The overall range$_1$ in median scores is $78.8 - 9.2 = 69.6$, and the range$_1$ in mean scores is $85.6 - 15.3 = 70.3$ alpha units. The range$_{1,2}$ of expenditures for schools is $19.2 to $1.5 per child of ages 5 to 18 per year. The difference$_1$ between the overall median scores of Negroes and whites is 23.44, and that for overall mean scores is 23.70 alpha units.

Figure 19.6 shows the relationships of the intelligence test scores of the two races and expenditures for schools by state and District. The vertical axis represents mean scores as listed in Table 19.5. The horizontal axis gives the estimated expenditures for primary and secondary schools in dollars per child of ages 5 to 18 per year in each of the states and the District. The results for whites are given in the top lines and those for Negroes in the bottom lines. Regression lines for the two racial groups and the two sets of data, with and without District of Columbia, were fitted by least squares.

In the summary of the major results of the Army Alpha Test scores given below, subscript n refers to Negro and subscript w to white statistical values:

Means:

$\bar{x}_{w1} = 50.75$
$\bar{x}_{w2} = 49.53$
$\bar{x}_{n1} = 26.43$
$\bar{x}_{n2} = 26.09$

Regression of white mean Alpha score, Y_w, on expenditures, X:

$Y_{w1} = 46.13 + 1.51X$
$Y_{w2} = 47.30 + 1.26X$
$\quad \sigma_y = 5.84 \qquad \sigma_b = 0.25 \qquad t = 5.06 \qquad P < 0.001$

Regression of Negro mean Alpha score, Y_n, on expenditures, X:

$Y_{n1} = 15.44 + 1.48X$
$Y_{n2} = 20.73 + 1.68X$
$\quad \sigma_y = 8.85 \qquad \sigma_b = 0.38 \qquad t = 4.45 \qquad P < 0.001$

Interclass correlations within states (all significantly different from zero at the 0.001 level):

	r_1	r_2
Negro alpha with white alpha	$+ 0.61 \pm 0.14$	$+ 0.73 \pm 0.10$
Negro alpha with \$/child/year	$+ 0.67 \pm 0.12$	$+ 0.70 \pm 0.11$
White alpha with \$/child/year	$+ 0.79 \pm 0.08$	$+ 0.74 \pm 0.10$

It is clear that mean Alpha scores are related significantly to expenditures for schools in both the Negro and white populations. On an average, the mean Alpha scores of the white draftees in the 24 areas were elevated 1.51 alpha units and that of the Negro draftees by 1.48 alpha units for each dollar of yearly expenditures on schools.

Figures 19.5 and 19.6 may be used to illustrate the problems in interpreting the relative importance of racial and environmental factors in determining the median Alpha scores. For any one of the states the vertical distance between the points for Negro and white Alpha scores corresponds to what commonly is considered a difference due to race or heredity. The differences in median scores within races but between states correspond to what commonly is considered a difference due to environment. These interpretations are wrong. We cannot assume the environmental factors are identical for Negroes and whites in any of the states, and we cannot assume either racial group in one state is identical genetically with the corresponding group in other states.

Figure 19.7 shows there is no one general answer to the question of how much of the variation in Alpha scores is due to heredity and how much to environment. The answers are quite different, depending on which state we consider. The total length of the columns corresponds to the difference between the overall maximum and minimum in median

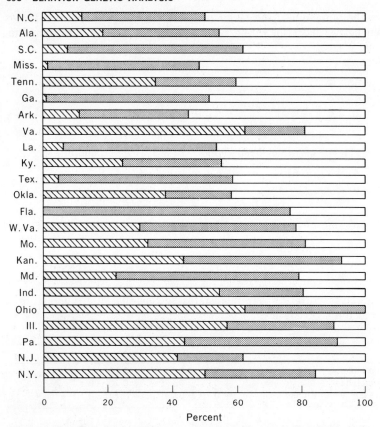

Figure 19.7   Percentage differences between Negro median Army Alpha scores in each of 23 states and lowest overall median score (cross hatching at left of each bar), between Negro and white median scores (shaded area in the middle), and between highest overall score and median white scores in the several states (at right of each bar). The states are listed from top to bottom in increasing rank of school expenditures per child of ages 5 to 18 during 1900. The length of the middle section of each bar (shaded area) is often taken to estimate the part of the variation due to genetic factors, and the lengths of the end sections the part due to environmental factors. (*Sources of data as in Figure 19.6.*)

Alpha scores. The length of the shaded areas corresponds to the differences between Negro and white scores in each state expressed as a percentage of the total length of the column. These differences are commonly considered to represent hereditary factors. The percentages "due to race" vary from 20.00 percent in the case of Virginia to 83.36 percent in the case of Florida. The length of unshaded areas corresponds to what commonly is considered differences "due to environment." For the Negroes these percentages vary up to 68.04 percent in the case of Virginia when the median score for Florida is set at zero percent.

For the whites these percentages vary up to 50.65 percent in the case of Arkansas, when the median score of Kansas is taken as the standard of comparison. The data from 10 of the 23 states support the "conclusion" that racial factors are most important; the data from 13 of the states support the opposite "conclusion" that environmental factors are most important. We must conclude that the question of the relative importance of nature and nurture as regards these data on Army Alpha Intelligence Test scores is without definite meaning.

If we make all possible groupings and comparisons, these paired observations from 23 state samples can be made to provide exactly 44,152,005,855,224,745 different estimates of the proportion of the variation in Alpha scores due to genetic factors. And none of these would provide a general answer to the question of the relative importance of nature and nurture for intelligence. There is no general answer. Both genotype and environment must be specified in every individual case.

We have stipulated that by "racial" differences we mean "genotypical" or "inherited" differences. Intelligence test scores and other phenotypes are not inherited as such. Only gametes containing genes are inherited, and genes have the capacity in certain environments to interact with the environment to produce phenotypes. Typically a large number of genes at different loci contribute to the variation of metrical characters such as IQ values.

Measurements and observations of individuals provide phenotypic values; from these we obtain means, variances, and covariances as well as other statistics. Before we can talk about racial differences in IQ scores, we need to find out what part of the variation in IQ scores is inherited, a difficult task. The problem is to divide the phenotypic values observed for IQ scores into a genetic (G) and a nongenetic (E) component where, at this stage, E is all the nongenetic components of P:

$$P = G + E$$

In short, we consider that G determines a phenotypic value, and E causes a deviation from it.

The methods for estimating the genetic and environmental components are discussed in Chapter 11; in the immediately following paragraphs we use the symbols and certain of the results of quantitative genetics summarized in that chapter.

For more than one locus there may be an interaction or epistatic deviation, I_{12}. Let A_1 be the additive value for locus 1 and A_2 the additive value for locus 2:

$$G = A_1 + A_2 + I_{12}$$

where I_{12} is the deviation from the additive combination of A_1 and A_2. (Thus the meaning of "additive action" for one locus is that there is no dominance, and for two or more loci that there is no epistasis.)

The genetics of a quantitative character such as the intelligence quotient is most conveniently studied in terms of its variance. We want to partition the phenotypic variance into its genetic and nongenetic components. The following symbols will be used:

Symbol	Variance component	Value whose variance is measured
V_P	Phenotypic	Phenotypic value
V_G	Genotypic	Genotypic value
V_A	Additive	Breeding value
V_D	Dominance	Dominance deviation
V_I	Interaction	Epistatic deviation
V_E	Environmental	Environmental deviation

With certain qualifications to be made below, the total or phenotypic variance is the sum of the components:

$$V_P = V_G + V_E$$
$$= V_A + V_D + V_I + V_E$$

For some purposes it is useful to divide the environmental variance into components.

In experimental genetics it is always easier to make sure members of a given group of organisms are identical (or nearly enough so) in genotype than it is to be sure that they are subjected to identical environments. A genetically uniform or isogenic group of bisexual organisms may be produced by making an F_1 cross between highly inbred lines derived from the same stock. The V_G in such a group may be taken as zero, and thus if we know the phenotypic variance V_P in the original randomly mated stock, we can find V_E from the equation $V_E = V_P$; V_G is the difference in variance between the heterogenic and isogenic groups, the ratio V_G/V_P is a measure of the variation of the character in the heterogenic group attributable to genetic differences between individuals, and $1 - (V_G/V_P)$ is a measure of the variance attributable to nongenetic differences.

Monozygous twins are two individuals identical in genotype. But since different pairs of twins differ in genotype and environment, and the environment of twins may be quite different from the environment of the singletons which make up the majority of the general population (Price, 1950), the use of twin data in the partitioning of variance is not fully trustworthy. We shall return to twins later.

Even in the best practicable experimental designs made to partition phenotypic variance into its genetic and environmental components, there are three circumstances that introduce uncontrolled error into the estimates:

1 Environmental variance may, and certainly in human populations does, vary between genotypes. Thus, the environmental variance of isogenic groups may differ from that of heterogenic groups, leading to an underestimation of the variance component due to genetic differences.

2 There may be a correlation between genotypic values and environmental deviations. Certainly for human populations, those who stress the

importance of the genetic component of variation also freely admit that, supposedly, better genotypes are often selectively exposed to better environments. If this is the case, the phenotypic variance is more than the sum of the genetic and environmental variance by twice the covariance of genotypes and environments: $V_P = V_G + V_E + 2cov_{GE}$. Neglect of the covariance term will lead to an overestimation of the genetic component.

3   There may be, and almost certainly is in human populations, an interaction of genotype and environment. The environmental response as regards a specific environmental difference may differ markedly for different genotypes. Haldane (1946) has shown that m different genotypes for a quantitative character such as intelligence quotients exposed to n different environments may result in (mn) $!/m!n!$ different responses. For the simplest case of 2 genotypes and 2 environments, 6 different responses are possible; for 3 genotypes and 3 environments, the number of possible responses is 10,080; and for 10 genotypes and 10 environments, the number of possible responses is 7.09×10^{144}. The six possible rankings of two genotypes exposed to two environments are given by the cells of 2×2 tables where the rows are the genotypes and the columns are the environments:

$$\begin{bmatrix} 1 & 2 \\ 3 & 4 \end{bmatrix} \quad \begin{bmatrix} 1 & 3 \\ 2 & 4 \end{bmatrix} \quad \begin{bmatrix} 1 & 4 \\ 2 & 3 \end{bmatrix} \quad \begin{bmatrix} 1 & 2 \\ 4 & 3 \end{bmatrix} \quad \begin{bmatrix} 1 & 3 \\ 4 & 2 \end{bmatrix} \quad \begin{bmatrix} 1 & 4 \\ 3 & 2 \end{bmatrix}$$

The abstract idea that phenotypic variance of quantitative traits, such as intelligence quotients, in all natural populations has components attributable both to genetic and nongenetic factors is fundamental for an understanding of the dynamics of polygenic traits. However, since for human beings there are available neither isogenic lines nor even approximately uniform environments, the idea does not allow us to make definite statements about the relative importance of genetic factors in determining the variation of the intelligence quotient.

The particulate theory of inheritance allows us to make definite predictions regarding resemblance of various degrees between biological relatives, on the supposition of definite modes of inheritance. In order to reveal the genetic causes of the resemblance between biological relatives, it is necessary to partition the genetic component of variance into its additive, dominance, and epistatic components.

Starting with the components of phenotypic values

$$P = G + E$$
$$= A + D + I + E$$

it has been shown in Chapter 11 that

$$V_P = V_A + V_D + V_I + V_E$$
$$V_G = V_A + V_D + V_I$$

The additive variance V_A is the chief cause of resemblance between relatives and therefore a fundamental determinant of the observable genetic properties of a breeding population. The most useful division is to separate the additive genetic variance from the nonadditive genetic and environmental variance. This division may be expressed as heritability, h^2, or the ratio of additive genetic to total phenotypic variance:

$$h^2 = V_A / V_P$$

Heritability may be estimated from a knowledge of the resemblance between relatives, given a particulate mode of inheritance. It should be stressed that h^2 is not a property of individual traits; rather it is a property of a particular gene pool and a particular environment. Any change in V_A, V_D, V_I, or V_E will change h^2.

From Chapter 11 we see that the phenotypic variance may be partitioned into *causal* components of variance denoted by the symbol V. The phenotypic variance may also be partitioned into *observational* components corresponding to the grouping of individuals into families and denoted by the symbol σ^2.

Resemblance may be expressed either as intraclass correlation, t, or as a regression of relative x on relative y, b_{xy}:

$$t = \sigma_B{}^2 / \sigma_B{}^2 + \sigma_W{}^2$$

where $\sigma_B{}^2$ is the between-group, and $\sigma_W{}^2$ the within-group component,

$$b_{xy} = \sigma_{xy} / \sigma_y{}^2$$

where σ_{xy} is the covariance of x and y.

The covariances of offspring and parent, cov_{OP}, of full sibs, cov_{FS}, and of half sibs, cov_{HS}, are related to variance components as follows:

$$cov_{OP} = \tfrac{1}{2} V_A$$
$$cov_{FS} = \tfrac{1}{2} V_A + \tfrac{1}{4} V_D$$
$$cov_{HS} = \tfrac{1}{4} V_A$$

These relationships may be used to estimate the genetic components of variance. For example, if there is no variance due to common environment, $V_G = V_A + V_D$, and

$$V_D = cov_{FS} - 2(cov_{HS})$$
$$= \tfrac{1}{2} V_A + \tfrac{1}{4} V_D - 2(\tfrac{1}{4} V_A)$$

Thus, by subtraction we have an estimate of the additive variance, $V_G - V_D = V_A$, which leads directly to an estimate of the heritability, $h^2 = V_A / V_P$. (If $V_{EC} \neq 0$, the procedure gives the upper limit for V_D.)

In experimental and applied genetics the best estimate of an individual's additive genetic value (A) is the product of his phenotypic value (P) and the heritability (h^2):

$$A_{(\text{expected})} = h^2 P$$

where both A and P are measured as deviations from the population mean. Thus, h^2 expresses the reliability of a phenotypic value (e.g., IQ score) as a guide to the additive genetic value.

We have not found published estimates on heritability of intelligence-test values based on family material where similar methods of testing and of analysis were used for two or more racial groups. The most economical and convenient method of estimating h^2 requires observations of half sibs; these are rare outside Hollywood and other small breeding populations.

Lacking family data, we now consider the large amount of data available on the "heritability" h_T^2 of intelligence in twins, using the estimate

$$h_T^2 = \sigma_D^2 - \sigma_M^2 / \sigma_D^2$$

where σ_D^2 and σ_M^2 are the within-pair variance for dizygous and monozygous twins, respectively. This statistic is not the equivalent of h^2 as defined above, inasmuch as h_T^2 is an estimate, not of the extent to which a trait is genetically determined, but of the proportion of variation in a metric character that is genetically determined.

The variance components between and within the two kinds of twins are (Falconer, 1960):

	Between pairs	Within pairs
Monozygous	$V_A + V_D + V_{Ec}$	V_{Ew}
Dizygous	$\frac{1}{2}V_A + \frac{1}{4}V_D + V_{Ec}$	$\frac{1}{2}V_A + \frac{3}{4}V_D + V_{Ew}$
Difference	$\frac{1}{2}V_A + \frac{3}{4}V_D$	$\frac{1}{2}V_A + \frac{3}{4}V_D$

Thus, we see

$$h_T^2 = \frac{(\frac{1}{2}V_A + \frac{3}{4}V_D + V_{Ew}) - V_{Ew}}{\frac{1}{2}V_A + \frac{3}{4}V_D + V_{Ew}}$$

while

$$h^2 = \frac{V_A}{V_A + V_D + V_{Ew} + V_{Ec}}$$

Erlenmeyer-Kimling and Jarvik (1963) reviewed 52 studies giving data on the correlation of relatives for tested intellectual abilities. Over two-thirds of the correlation coefficients were obtained for IQ tests; the others were based on other intelligence tests, for example, the Primary Mental Abilities Test of Thurstone. The medians of the correlation coefficients for 10 categories of relatives are given in Table 19.6. The

Table 19.6
Correlations for intellectual ability observed and expected among relatives on the basis of Mendelian inheritance

Correlation between	Number of studies	Median correlation, r Observed	Expected
Unrelated persons, reared apart	4	−.01	0
Unrelated persons, reared together	5	+.23	0
Foster parent–child	3	+.20	0
Parent-child	12	+.50	+ .50
Siblings, reared apart	2	+.42	+ .50
Siblings, reared together	35	+.49	+ .50
Dizygous twins, opposite sex	9	+.53	+ .50
Dizygous twins, same sex	11	+.53	+ .50
Monozygous twins, reared apart	4	+.75	+1.00
Monozygous twins, reared together	14	+.87	+1.00

SOURCE: Observed data from Erlenmeyer-Kimling and Jarvik, 1963.

smallest sample size in the pooled data was 125 for sibs reared apart, and the comparisons for unrelated persons living apart and the twins are based on samples over 1,000 pairs. The correlations between unrelated persons reared together ($r = +.23$) and those between monozygous twins reared apart ($r = +.75$) provide coefficients of determination of $r^2 = .0529$ for the unrelated and $(1 - r)^2 = .0625$ for the twins, suggesting that up to one-sixteenth of the variability in measured intellectual ability may be due to nongenetic factors.

The fairly good fit between observed values and those expected on the basis of Mendelian inheritance for unrelated persons reared apart, for parents and children, and for monozygous twins reared together, suggests genetic factors are of considerable importance in the determination of tested intellectual ability.

These pooled data are not suitable for making an estimate of h^2 (for example, we have no way of measuring the component of variance due to common environment within families). Table 19.7 gives estimates of h_T^2 from the Michigan Twin Study. These results indicate significant h_T^2 values for some components of general mental ability (e.g., numerical and verbal skill) but low values for other important components (e.g., reasoning and memory). In fact, except in the most approximate way, we do not know what proportion of the supercharacter general intellectual ability in any natural, human population is under genetic control.

Undoubtedly all behavior geneticists would conclude that both genetic and environmental components contribute to variation over the normal range of measured general intelligence. Probably all would conclude that the genetic component of intelligence is polygenic (Fuller and Thompson, 1960). The days when a C. B. Davenport could assert seriously the monofactorial inheritance of a violent temper or a wandering habit are long gone (Dunn, 1962).

Morton (1963) has estimated that genes at 71 loci are concerned with low-grade mental defect in human populations. His argument is too complex to present here, but it involves assumptions which if wrong would increase the estimated number of loci. Thus we may assume that at least 71 loci are occupied by alleles controlling normal mental ability.

Table 19.7
Heritability (h_T^2) of Thurstone's Primary Mental Abilities Test scores in Michigan twins

Name of test	Number of twin pairs		h_T^2
	Dizygous	Monozygous	
Number	37	45	.61*
Verbal	37	45	.62*
Spatial	46	45	.59†
Word fluency	35	44	.61†
Reasoning	37	45	.28ns
Memory	34	45	.20ns

* Significant at the .01 level.
† Significant at the .05 level.
ns, not significant at the .05 level.
SOURCE: Vandenberg, 1962.

We have seen that the best experimental procedures for determining the fraction of variability due to genetic factors in polygenic traits are not available for human populations; the general populations of American Negroes and of American whites have not been reared in identical environments. The next best estimates would come from the application of the methods of quantitative genetics to the analysis of measured intelligence in families, including comparisons of parents and offspring, full sibs, and half sibs. These methods would provide estimates of heritability (h^2) but such estimates could not be extended to other populations which might differ in genotypes or environments. Even so, such estimates are not available. The next best thing is to use estimates of heritability ($h_T{}^2$) based on the twin method. These estimates are not fully trustworthy for extension to the general population simply because twins are not representative of the general population. The available twin studies give estimates of the proportion of the variation in tested intelligence attributable to heredity varying from 60 percent (Woodworth, 1941), through 65 to 80 percent (Newman et al., 1937), 66 percent (Burks, 1928), 78 percent (Leahy, 1935), to 77 to 88 percent (Burt, 1958). Finally, we have seen that estimates based on a comparison between races within states of the World War I Army Alpha scores vary from 18.4 to 76.9 percent.

We have presented the argument on how to detect racial differences in polygenic traits (where they exist) in considerable detail in order to explain why we have reached a negative conclusion on the problem of racial differences in intellectual ability. Given the genetic theory outlined and the available observational data for American Negroes and whites, we must conclude that we do not know whether these two races are genetically different in general intelligence as measured by the Army Alpha or other tests of intellectual functioning. The evidence for other possible racial comparisons is weaker or nonexistent.

There is much evidence that measured intelligence in both American Negroes and whites increases markedly up to a comparatively high level as more money is spent on public education or, in broader terms, as the educational setting and social environment are improved. The recent survey by Bloom (1964) shows the preschool environment to be very important for attained intellectual ability in later years. The general conclusion supported by this evidence is supplemented by the data from longitudinal studies of the same individuals in differing environments, from identical twins reared apart, and from unrelated people reared together. These observations suggest that environmental factors may change individual IQ scores by as much as 20 points, an amount greater than the usual difference in mean IQ between samples of American Negroes and whites. These findings lead many biological and social scientists to assert there are no racial (genetic, inherited) differences in general intellectual ability. They lead us to conclude that we do not *know* whether there are significant differences between races in the kinds and frequencies of polygenes controlling general intellectual ability.

A further word might be said concerning culture-free tests of intelligence, since they have appeared to many to represent the simplest approach to a meaningful assessment of racial differences in intelligence. Virtually all sophisticated observers (for example, Anastasi, 1958; Dreger and Miller, 1960; Goodenough and Harris, 1950; Thompson, 1957b) agree that such tests are fictional creations with no correspondence to reality. It is impossible even to conceive of a test of intelligence that is *free* of culture. More reasonable is the attempt to construct a test that is culture-common, that relies upon cultural elements that are largely shared with the population to be studied. This is what the creators of the Binet and Wechsler tests have attempted, with partial success, within our own culture. The enormous diversity of modern cultures makes it difficult to conceive of such a test that could go very far in an extensive cross-cultural or cross-racial investigation. However, a test might rely upon such simple cultural elements that the subject could briefly be brought up to date or "socialized" within the "test culture" through an instruction phase, even if the test activity was not a part of his habitual experience. This is largely what Rulon (1953) attempted in his symbol identification test, although, as Thompson (1957b) points out, it is unlikely that there would not be real and important cultural advantages in connection with such an instrument.

If one is willing to consider infrahuman evidence mildly relevant to this general question, it is clear that here there is evidence for "racial" differences in intelligence or capacity to learn. This is true both for strains (races) selectively bred for maze learning (Heron, 1941; Tryon, 1940) as well as for inbred strains and genetically distinct hybrid offspring that have not been selectively bred for this character (Lindzey and Winston, 1962; McClearn, 1958; Vicari, 1929; Winston, 1964). Although these findings provide unmistakable evidence for strain (genetically determined) differences in various learning capacities, they also provide limited evidence suggesting a relatively high degree of "buffering" of this attribute. For example, separate investigations carried out in the laboratories of McClearn and Lindzey reveal only one strain that is significantly different from all other strains in appetitive learning, while for many other behavioral attributes there are a very large number of strain differences. In addition, the single strain that is reported to be inferior by both McClearn and Lindzey has been studied in the two laboratories by means of sublines that have been maintained in breeding isolation for decades, so that there has been ample time for genetic drift and mutation to produce significant genetic differences. In spite of this, the learning performance of the two populations has remained remarkably similar, suggesting again that this character does not seem readily influenced by genetic change.

In summary, it may be repeated that there is no firm evidence to support the existence of racial differences in intelligence that are independent of environmental variation. Whether one chooses to consider the likelihood of demonstrating such differences probable or improbable depends upon prior convictions, awareness of the technical problems just discussed, and certain issues that we shall consider in a subsequent section.

The general picture is somewhat different, however, if we turn from general intelligence to the restricted range of intellectual functioning involved in mental retardation. *Phenylketonuria*, which, typically, involves severe mental retardation, is perhaps the best known case in man where a major gene is known to affect behavior. Phenylalanine is an essential amino acid in man and many mammals. Its only known function is as a constituent of proteins. The minimum daily requirement is about 1.1 grams per day, whereas nearly all proteins contain about 6 percent phenylalanine. Since more is taken in the food than is needed for protein synthesis, the excess is oxidized to tyrosine. The conversion of phenylalanine to tyrosine is catalyzed by the enzyme phenylalanine hydroxylase. Individuals who are homozygous for the phenylketonuria gene are unable to oxidize phenylalanine to tyrosine because of the primary absence of phenylalanine hydroxylase enzyme activity. Individuals with at least one of the normal alleles of the phenylketonuria gene oxidize phenylalanine to tyrosine. Tyrosine in turn may be used to make melanin pigments, thyroxine, and other compounds, or may be broken down, in a number of steps, to carbon dioxide and water, which are excreted. The homozygotes for the phenylketonuria gene reabsorb the excess phenylalanine in the renal tubules, and the blood level of phenylalanine may rise as high as 50 mg/100 ml, which is some 50 times higher than the normal level.

The pathological effects of phenylketonuria are due to secondary metabolic blocks which result from inhibition of enzymes, including amino acid decarboxylases, by one or more of the compounds that form with increased levels of phenylalanine. One of the metabolites interferes with the production of 5-hydroxyindole-acetic acid. The amino acid decarboxylases give rise to primary amines which appear to be important to normal neurological function. Interference with decarboxylase activity might be responsible for the mental defect that is almost universally associated with phenylketonuria (Sutton, 1961).

Available information does not permit accurate estimates of frequencies of the gene for phenylketonuria because the statistics refer to institutionalized defectives in different countries. Such data are difficult to extend to the general population because of different practices in admission of mental defectives for institutional care and because of variations in the efficiency of diagnosis.

Carter and Woolf (1961) ". . . suggest that the gene for phenylketonuria is most common in Norway, common in Sweden and Denmark, the United Kingdom, and North America, present in central and southern Europe, and rare in countries with populations which are not of European extraction." The gene frequency in Ireland is about 0.014, which is nearly four times as high as the gene frequency for southeast England, about 0.004. Although phenylketonuria has not been reported for native Africans nor for Asians except Japanese, a survey by Tanaka et al. (1961) suggests the gene frequency in Japan is about 0.0002, or about one-eightieth of the estimated frequency for Ireland.

Jervis (1937) reported the racial origin of 50 patients with phenylketonuria in New York. He found that, compared with a control series of 100 patients with other types of mental deficiency, 12 (24 percent) of the phenylketonurics were Irish in comparison with 9 percent in the control series. In contrast there were no phenylketonuria cases of Jewish or Negro parentage whereas these made up 20 and 12 percent, respectively, of the control series. It is of interest to note that the observations of Hara and Yahiro (1958) on a hybrid Japanese-American child with phenylketonuria give presumptive evidence that the same genetic locus is involved in the mutant forms in the two races.

One may conclude that there is strong presumptive evidence supporting the likelihood of racial differences in the gene for phenylketonuria, although definitive data are clearly lacking. Such gene-frequency differences obviously would have implications for only a tiny proportion of any human race, and the consequences of the genetic differences are further lessened by the relatively successful programs of treatment that have been introduced in recent years.

Personality

The study of personality differences between races has led to an even more disappointing outcome than the study of intellectual differences. Nor is this surprising. In comparison with studies of intellectual functioning, the amount of personality research that has been conducted is appreciably lower, and there is generally much less confidence in the instruments employed. The trait domain of personality is far from being satisfactorily mapped, so that we have not even achieved agreement upon the important variables to be examined in such inquiries. The demonstrated usefulness of tests of personality is much below that of comparable measures of intellectual functioning, and typically these instruments are much more dependent upon local culture and language. Partial surveys of the relevant investigations have been presented by Dreger and Miller (1960), Garth (1931), and Klineberg (1935).

It is easy to find reports in the literature of race differences in personality traits, attitudes, and values as measured by such conventional instruments as the Cornell Selectee Index, California Test of Personality, Pressey XO Test, Strong Vocational Interest Blank, and the Study of Values (for example, Anderson, 1947; Felton, 1949; Milam and Sumner, 1954; Pressey and Pressey, 1933). The procedural or design problems in these studies are so apparent, however, that no critical observer would consider attempting to link obtained differences to biological factors, rather than to experiential or cultural factors. The extent to which the personality questionnaire or inventory is embedded in a single language would alone disqualify these instruments from most serious studies of race differences.

In view of the culture specificity of personality questionnaires or inventories, it is not surprising that recent years have seen a heavy emphasis upon the use of projective techniques in cross-cultural (often cross-racial) research. Employing projective techniques to identify personality differences between cultures or races is by no means without hazards. Most users of these tests would agree that the responses they evoke are influenced by a wide range of factors, which include temporary emotional states, stimulus factors, response sets, ability and performance factors, definition of the testing situation, situational factors, and the relation between examiner and subject. Thus, while these tests may be somewhat easier to apply in different cultures than personality inventories, there is still a host of problems confronting the person who might wish to link personality to variation in race. These very difficulties led Lindzey (1961), after a relatively exhaustive survey of cross-cultural studies using these instruments, to conclude that, while there was no doubt that individuals from different cultures (and races) responded differently to most projective techniques, there was little basis for believing that these differences in response reflected personality differences rather than variation in language, test-taking attitudes, rapport, or other factors not related to personality. The state of findings involving rating scales, situational tests, and special devices is in no respect superior to that which one encounters in connection with questionnaires and projective tests.

Easily the most ambitious investigation that has been carried out in this area is Karon's (1958) study involving the Picture-Arrangement Test (Tomkins and Miner, 1957). The study employed a very large number of subjects and utilized modern sampling and polling techniques. Thus, it offers a larger and more representative sample and more objective scoring than is the case for any comparable study. Unfortunately, for present purposes, the study focused upon differences between northern and southern Negro subjects, and it was only in connection with a pilot study that incidental comparisons of Negro and white subjects were performed. Karon reports differences between both southern and northern Negroes and between Negro and white subjects in personality dimensions, particularly those associated with the domain of aggression. He attributes these group differences to environmental variation.

We may conclude, then, that there is little other than opinion, reasoning from results with other species, naturalistic observation, or findings in regard to other human attributes, to support the contention that here are racial differences in personality attributable to genetic factors.

THE FUTURE OF THE RACE CONCEPT

There seems little doubt, in spite of disclaimers by many distinguished anthropologists, that the concept of race will continue to prove useful to the physical anthropologists and indeed to all investigators of the human organism who are interested in evolution. In fact, as information concerning human genetics increases, there is every reason to expect that classification by race may become more precise and the links to other domains of biological science more numerous and significant.

Whether the concept of race will prove useful to the student of *behavior* is an altogether different question. A satisfactory answer to this query demands information that is not at present available. If there are *no* significant behavior differences between races that can be attributed principally to biological factors, as many scientists have asserted and the current climate of public opinion urges, it is difficult to see how such a classification could be of much interest to the psychologist, sociologist, or cultural anthropologist. If the undeniable genetic differences between races are accompanied by significant differences in behavior, then there is every reason to expect that the concept may eventually assume significance, possibly considerable significance, for the student of human behavior. Under these circumstances the behavioral scientist might come to employ race as a parameter attached to most empirical relationships in much the same manner that variables such as sex and age are now employed.

At present the educated public and many trained scientists are so intensely concerned lest scientific findings be used to support ideas of racial superiority or race prejudice that it seems unlikely much energy will be expended in the kinds of research that might document the behavioral importance of the race concept. If it were possible to overcome the enormous procedural difficulties to which we have alluded and explore systematically this unpopular area, it is at least possible that enough differences would be discovered, of the sort exemplified by color vision and the PTC "taster" findings, so that race would become an important parameter for the behavior theorist or investigator. Those meager findings already available seem sufficiently robust to disconfirm statements such as the following, which is taken from an authoritative UNESCO report:

□ . . . *races* or *ethnic groups* differ in their psychological inheritance. For that there is no evidence. On the contrary, every racial group contains individuals who are well endowed, others who are inferior, and still others in between. As far as we can judge, the range of capacities and the frequency of occurrence of various levels of inherited ability are about the same in all racial groups.

The scientist knows of no relation between race and psychology. (Klineberg, 1961, p. 452)

If we are willing to concede that sensory processes are a legitimate part of the corpus of psychology, it is clear that we *do* know of associations between "race and psychology."

For a variety of reasons the dramatic strain differences in behavior commonly observed in laboratory animals are unlikely to be approached by human race differences. Nonetheless, there seem to us no convincing reasons for expecting a complete absence of significant differences in behavior between races. This conclusion is by no means evident, however, for we find distinguished scientists occupying directly opposed positions on this issue. For example, Mather, in criticizing the UNESCO statement on race, asserts (*The Race Concept*, UNESCO, 1951, p. 25):

□ *I, of course, entirely agree in condemning Nazi race theory, but I do not think that the case against it is strengthened by playing down the possibility of statistical differences, in, for example, the mental capacities of different human groups. They may not be there, though this would surprise me, but the fact that we have at present no evidence does not mean that they are not there.*

R. A. Fisher, who probably has had as much impact upon the inference process within social and biological science as any man, advocates an even stronger position, suggesting that "human groups differ profoundly 'in their innate capacity for intellectual and emotional development' " (*The Race Concept*, UNESCO, 1951, p. 61).

The cytogeneticist Darlington (*The Race Concept*, UNESCO, 1951, p. 26) suggests:

☐ By trying to prove that races do not differ in these respects (mental characteristics) we do no service to mankind. We conceal the greatest problem which confronts mankind . . . namely how to use the diverse . . . gifts, talents, capacities of each race for the benefit of all races. For if we were all innately the same how should it profit us to work together? And what an empty world it would be.

All this points to the importance of investigating systematically that which largely has been ignored, debated against, or studied through inappropriate means. In contrast to this position, we find a number of experts who minimize the likelihood of important racial differences in behavior in spite of the significant gene differences between races. This viewpoint has been expounded most effectively by Dobzhansky (1957), Dobzhansky and Montagu (1947), David and Snyder (1951), and Fuller and Thompson (1960). It suggests that man has evolved through natural selection in such a way as to increase trainability, plasticity, and his capacity to learn. Moreover, the argument goes, in all human races such characters have had a selective advantage so that there is no reason to think that selection would have led to significant differences between races in the relevant gene frequencies. David and Snyder (1951, pp. 71–75) summarize and endorse this position:

☐ . . . it is difficult to conceive of any human social organization in which plasticity of response, as reflected by ability to profit from experience (that is, by intelligence) and by emotional and temperamental resilience, would not be at a premium and therefore favored by natural selection. It therefore seems to us highly improbable that any significant genetic differentiation in respect to particular response patterns, personality types, temperaments, or intellectual capacities among different populations or races has occurred in the history of human evolution . . . the whole of human *social* evolution has occupied only a microscopic fraction of the geologic time scale, but it is hardly probable, either on theoretic grounds or on the basis of inferences from human history and archeology, that the biologic basis of human abilities or behavioral potentialities, whatever these terms may mean, has appreciably changed during this period.

A similar position is advocated by Fuller and Thompson (1960, pp. 323–324):

☐ Do the well-established differences in gene frequencies imply psychological differences as well? Strains of animals show behavioral differences correlated with their diversity of genotypes and it can be argued that the same must be true of human races. Such a view need not imply racial superiority, merely racial differences. The evidence to prove this point one way or the other does not exist, nor is it likely to be obtained in the near future. Nevertheless, there are reasons to discount the likelihood of such differences being very important. The most diverse human cultures have common features related to the perpetuation of the species. It is difficult to conceive of a society in which intelligence, cooperation, and physical vigor would not have positive selective value. Hence it is likely that natural selection tends to oppose the establishment of major heritable behavior differences between races.

Having argued against race differences in behavior, the authors go on to suggest that the same argument does not apply to social-class differences. Consequently, they believe it is more reasonable to expect genetically determined differences in behavior between classes than between races. This latter point is in opposition to the conclusions of David and Snyder (1951) but is congruent with a viewpoint advanced by Tryon (1957) and Stern (1960).

In spite of the weight of authority behind the above statements, it seems to us there are significant flaws in the reasoning. First, Dobzhansky has emphasized the importance of natural selection in man in relation to culture transmission and yet he fails to underline the enormous discrepancies between races in the efficiency with which culture is transmitted (for example, the difference between literate and nonliterate societies). Some of these differences are closely associated with race differences, have existed for many thousands of years, and presumably have been accompanied by very different selection pressures in regard to characters potentially relevant to culture transmission, such as "intelligence."

Second, most of the above reasoning appears to assume that intelligence is a unitary variable concerned solely with plasticity or modifiability of behavior. Yet most specialists today would argue for the existence of components or factors of intelligence (for example, verbal comprehension, reasoning, perceptual speed, associative memory) that are not highly intercorrelated and certainly are not related by any simple function to plasticity. Presumably these various components of intelligence may have had quite different selective advantages in particular ecological settings and cultures. We know, for example, that numerical reasoning or quantitative capacity is not strongly associated with all other types of intellectual capacity, and we also know that races have differed enormously in their utilization of, and contributions to, the number system. Is it not likely that there would be a selection advantage for genes relevant to this variable (if such exist) in one race and no such advantage in other races? Indeed, even within a single culture, existing evidence suggests that different factors of intelligence may be subject to different degrees of genetic control. Blewett (1954) working with identical and fraternal twins in England found coefficients of heritability ranging from .07 to .68. As already discussed, Vandenberg (1962) has reported comparable findings based upon an American sample. The newest study in this area (R. C. Nichols, 1965) employed the National Merit Scholarship Qualifying Test and found heritability coefficients for the five subtests ranging from .27 to .80. Such findings certainly argue against treating intelligence as a unitary variable that is subjected to uniform selection pressure in all cultures.

Third, the argument advanced by Fuller and Thompson that different cultures have common attributes related to survival scarcely seems to differentiate the human race from an animal strain or species, where there are also common demands related to survival and yet pronounced "race" differences in behavior. It seems reasonable to expect that attributes such as physical vigor are much more regularly related to survival across lower animal strains than across human races and societies. One may go even further and suggest, contrary to the assertions of these authors, that what we know of different races, their associated cultures, and their physical settings suggests that physical vigor, cooperation, and (components of) intelligence may have had quite different selective significance in different races. Social structure, specifically kinship system and marriage rules, could have enormous effects upon selection within different societies and possibly races. Moreover, cultural (and racial) differences might also lead to very different patterns of assortative mating which, if accompanied by differential selection, again could produce systematic differences in the rate or type of change in the gene pool. A careful analysis of these variables in relation to natural selection would be essential in order to be sure that there were uniform opportunities for natural selection to influence allelic frequencies in different races.

Eventually this issue must be resolved on the basis of empirical evidence, albeit evidence that is singularly difficult to obtain. In the meantime, however, it seems to us surprising that one would accept present findings in regard to the existence of genetic, anatomical, physiological, and epidemiological differences between races (Garn, 1960, 1961; Boyd, 1950) and still expect to find *no* meaningful differences in behavior between races.

In addition to advocating a more serious inquiry into the possible existence of psychological race differences, we might further observe that the proper study of racial differences may necessarily involve a return to the kinds of simple processes and modes of response that characterized the study of individual differences prior to the Binet revolution. Not only is it likely that important associations with underlying genetic factors will be visible easily only in connection with simple behavior processes, but also the possibility of at least partially disentangling the effect of culture from the effect of race is greatest with instruments or techniques that minimize the importance of language and other complex processes that are heavily influenced by the socialization process. This should not be understood to imply that there are no exceedingly difficult empirical problems associated with studying simpler processes in different races. The problems remain severe and demand careful and sophisticated investigation if there is to be any likelihood of unambiguous findings. Nonetheless, in comparison with intellectual and personality variables, sensory-motor processes present difficulties of a much lower magnitude.

CONCLUSIONS

In conclusion, we may say that the concept of race has little importance for the student of human behavior at present. There are areas of human behavior where it is as misleading to refer to "the human being" as it is in comparative animal psychology to refer to "the rat," "the mouse," or "the monkey"; these include the areas of the sensory processes or of the enzyme deficiencies where major genes have high penetrance and nearly constant expression in most environments. In these domains, major genes with clear-cut behavioral consequences may control in most environments a large proportion of the observed variation within and between races. These are the areas in behavioral science where the concept of race—or its biological equivalent—is needed to give a satisfactory account of the observed variation. But the behavioral consequences of such known major genes are not very important for human behavior considered broadly. For the areas of human behavior that are vital in everyday life, for the varieties of behavior that allow individuals to participate satisfactorily in their society, there is no comparable evidence for genetically determined racial differences. Indeed there is at least the possibility that selection acting over the past 2 or more million years has made genes adaptive for symbolic behavior, for behavior associated with language, and consequently has made it very unlikely that such racial differences exist.

The concept of race is likely to remain of small general importance for behavioral science until anthropologists and other students of human biology replace the typological and taxonomic notion of race with a dynamic notion based on the genetic theory of evolution. The possibility of future change in the status of the concept is dependent upon increased activity in an area of research that is procedurally difficult, politically dangerous, and personally repugnant to most psychologists, sociologists, and anthropologists.

EPILOG

CHAPTER TWENTY
BEHAVIOR-GENETIC ANALYSIS[1]
Jerry Hirsch

☐ . . . among all the numerous experiments made, not one has been carried out to such an extent and in such a way as to make it possible to determine the number of different forms under which the offspring of hybrids appear, or to arrange these forms with certainty according to their separate generations, or definitely to ascertain their statistical relations (Mendel, 1866, p. 8).

Where are we? This volume represents but a stage in the growth (or perhaps convergence) of two fields, behavior study and the study of heredity. Prior to the work it considers, both had been developing independently (almost in reproductive isolation). To what extent or in what ways may we now maintain that there has been, will be, or must be a synthesis of concepts and techniques from the two fields into the kind of joint effort I call behavior-genetic analysis? One affirmative answer presents itself as soon as we realize there are problems that can now be analyzed and understood better from the new synthetic point of view than was possible previously.

THE TRYON EFFECT

As the DeFries survey and the McClearn discussion show, selective breeding and inbred strain comparison have hitherto been the methods of choice in laboratory behavior-genetic analysis. Since almost all selection and strain comparison studies have failed to yield comprehensible results in the later steps of their analyses, it will be instructive to reexamine them from a point of view that may appear new to many people today, though in fact it derives directly from Mendel's 1866 paper. As Sir Gavin de Beer has sadly observed (1964, p. 192): "Little apology is felt to be needed for going back to a source of information so elementary, because it is to be feared that readers of Mendel's original paper are not now numerous." Ever since Mendel, a standard pro-

[1] The work on this chapter was done with the support of the National Institute of Mental Health Training Grant 1 TO1 MH 10715-01 BLS to the University of Illinois for Research Training in the Biological Sciences.

cedure has been to hybridize strains showing extreme expression of some trait and to look for the segregation of genetic factors in the F_2 generation.

Because chromosomes were unknown to Mendel, much of his discussion is formulated in terms of (what are now called phenotypic) characters. In the following quotations from the translation (Bennett, 1965, pp. 27 and 44) of Mendel's classic paper I have substituted contemporary cytogenetic terminology in order to transpose his argument from the level of the phenotype to the level of the genotype. For each substitution, Mendel's original language is given first in italics and in parentheses, and is followed immediately by my substitution, in brackets.

☐ If n represents the number of (*differentiating characters*) [chromosome pairs] in the two original stocks, 3^n gives the number of (*terms of the combination series*) [genotypic combinations], 4^n the number of (*individuals which belong to the series*) [cells in the genotype matrix], and 2^n the number of unions which (*remain constant*) [produce complete homozygotes].

☐ . . . the (*series*) [matrix] in each separate experiment must contain very many forms, since the number of (*terms*) [genotypes] . . . increases with the number of the (*differentiating characters*) [chromosome pairs] as the powers of three. With a relatively small number of experimental plants the result therefore could only be approximately right, and in single cases might fluctuate considerably. If, for instance, the two original stocks differ in seven (*characters*) [chromosomes], and 100-200 plants were raised from the seeds of their (*hybrids*) [F_1s] to determine the grade of relationship of the offspring, we can easily see how uncertain the decision must have become since for seven (*differentiating characters*) [chromosome pairs] the combination (*series*) [matrix] contains 16,384 individuals under 2187 (*various forms*) [different genotypes]; now one and then another relationship could assert its predominance, just according as chance presented this or that form to the observer in a majority of cases.

One of the best known and most widely referenced experiments in the behavioral science literature is the Tryon selective-breeding study of maze learning in rats, begun in 1925 and continued until 1940. Many aspects of Tryon's simple experiment have been replicated in the same and different behaviors in the same and different species. One of his replicable results has stood as an enigma in the field for over 27 years. It has even been embalmed in the literature like a classic and labeled the "Tryon effect" by Scott and Fuller: "The F_1's and F_2's show the same amount of variance and [they] overlap both parent strains" (1965, p. 264).

Tryon bred rats selectively on the basis of their error scores in learning a multiple T-maze. Three times between the eleventh and twenty-second generations of selection the reproductively isolated "bright" and "dull" strains were testcrossed to produce F_1 and F_2 progeny. According to (naïve) Mendelian theory, because of segregation in the F_2, its variance should be detectably larger than that of the F_1. All three times this expected result failed to appear. This failure to obtain an increase in variance from the F_1 to the F_2 generation has been called the "Tryon effect." It has happened often, though not invariably, in behavior-genetic studies.

Tryon (1940, p. 116) sought possible explanations for his findings, and, long before the advent of the computer, he developed ingenious Monte Carlo methods for generating models with dice. Quite correctly, he believed that he was dealing with a polygenic situation, then called multiple-factor determination. In 1951 Hall, one of Tryon's first students, attributed the Tryon effect to lack of sufficient inbreeding in the selected lines: "The ambiguity that results from crossing strains that are not homozygous . . ." (1951, p. 322); of course he was correct. But that is only one aspect, and not the most important. In 1958 Caspari (p. 119) suggested that behavior may show properties that are different from morphology. Possibly behavioral heterosis is expressed in increased variability while morphological heterosis shows greater uniformity. Quite wisely, he qualified his remarks by indicating the need for more evidence. In 1960, in the text that established the field, Fuller and Thompson made use of both the Hall and Caspari interpretations. In 1962, I too described the results of a selection and hybridization study (of phototaxis in Drosophila) in the following terms (Hirsch, 1962, p. 16):

☐ Our first attempt at analysis for phototaxis seemed successful. The F_1 hybrid was no more variable than the selected strains. . . . Furthermore, the F_2 was more variable than the F_1. . . . We immediately attempted to replicate this important result. Our replication foundered on the same shoals as much of the previous mammalian work. The F_1 hybrid was as variable as the F_2.

In what is becoming an extensive literature on behavior genetics there are no explanations of why these experiments were doomed to failure from the outset. Typological thinking had blinded us all to the nature of the biological situation in the Tryon paradigm. We were thinking of *the* rat and hoping to map its chromosomes in the same reductionistic way that brains were supposed to have been mapped against behavior in physiological studies.

If we stop thinking about the archetypal rat, however, and focus our attention on a species population of unique organisms having a karyotype of 21 pairs of chromosomes, the flaw in the reasoning about these studies becomes embarrassingly obvious. If, for simplicity, we treat the chromosomes as major indivisible genes, a little Mendelian algebra will show the nature of the experiment that breeds F_1 and F_2 generations by hybridizing two strains having different sets of chromosomes. (To simplify matters further, we can make the *un*realistic assumption that the strains being crossed are perfectly homozygous for different forms of each chromosome, and we shall ignore the difference between sex chromosomes and autosomes.)

The number of different kinds of gametes that can be produced is a number equal to 2 (for the two chromosomes in each pair) raised to a power equal to the number of chromosome pairs in the set. Since the rat has a 21-chromosome-pair karyotype, it can produce $2^{21}(=2,097,152)$, or over 2 million, kinds of gametes. For the rat the matrix of genotypes analogous to Bruell's Tables 13.2 and 13.3 (page 273) would have to contain $4^{21}(=4,386,046,511,104)$, or over 4 trillion, cells. Such a matrix would contain $3^{21}(=4,782,969)$, or about $4\frac{3}{4}$ million, different genotypes. Of these, 2^{21} or slightly over 2 million are homozygotes and appear only once, in the main diagonal of the 4-trillion-cell matrix, if and only if the theoretical distribution of proportions is realized exactly. Otherwise, unless the matrix is replicated, many will not appear at all. Any experiment intended to sample the spectrum of possible genotypes must be planned so that there is a statistically sufficient number of replications of the appropriate genotype matrix. In other words, these experiments never had the slightest chance of making the measurements for which they were intended—a limitation that applies equally to the four-way and eight-way crosses discussed on page 319, as analogous calculations would show!

Thus the Tryon effect can be attributed to the design of experiments having insufficient magnitude—a problem clearly understood by Mendel over a century ago, but not often since then. We realize today that it amounts to ignoring species karyotype in breeding for recombinant genotypes. Therefore, many conceivable experiments are a practical impossibility—a conclusion reached by Bruell (on page 279) when he considers an analogous problem in the context of the diallel analysis.

HERITABILITY, BEHAVIORISM, NORM OF REACTION, AND BEHAVIORAL "LAWS"

Perhaps nothing has vitiated more work than the misunderstanding of the heredity-environment pseudoquestion. It usually involves nontechnical use of the heritability concept. Since it has been asserted as official dogma of the American Psychological Association that John B. Watson's work "has been one of the vital determinants of the form and substance of modern psychology" (*Science*, 129:198, 1959), it is essential that we understand the fallacious reasoning on which our behavioristic heritage (excessively anti-intellectual) has been based.

Johannsen's demonstration—that variation (1) *within* pure lines is attributable to environmental rather than to genetic influences and (2) *between* pure lines shows independence of environment and can be attributed to genotypic differences—became a milestone of biosocial science, being the first clear separation of nonheritable from heritable variation. Watson reported Johannsen's experiment correctly and then announced to his gullible psychological progeny (Watson, 1914, p. 161): "The results of this experimental work . . . prove conclusively that the vast majority of the variations of organisms are not inherited"—a fantastic *non sequitur*. The very observations that, to biology, meant non-inheritance of environmental influences (Lamarck's theory still had adherents) and that variation could be analyzed intelligibly into genetic and environmental components provided psychology, through Watson's *mis*interpretation, with the procrustean frame that was to trap us in typological thinking for another half century.

In more recent times Fuller and Thompson (1960) have most wisely pointed out that "heritability is a property of populations and not of traits." (Conceived of in any way other than as an observable attribute of the individual members of specific populations, a trait becomes a platonistic reification. It is precisely for this reason that the heritability concept can apply only to a population—that which may have more or less trait-relevant genetic variation.) Yet, somewhat later, even Scott and Fuller (1965) seem to have slipped back into repeated attempts to estimate "the effect of heredity upon behavior" (see Hirsch, 1965, p. 819).

Ironically, now that the pernicious influence of American behaviorism has been controverted, the likelihood of indiscriminate use of the concepts of heredity presents an even greater danger than when their use was to be expected mainly in a more specialized literature. For example, in *Science* within a 3-year period we have witnessed a neo-Watsonian "restatement of radical behaviorism" (*Science*, 140:95, 1963) followed by a post-eleventh-hour disavowal of Watsonian extremism expressed in hopelessly vague language: "not denying that a substantial part of behavior is inherited," and a specialist in learning wanting "to emphasize what could be done in spite of genetic limitations," and "any genetic trait," etc.; then added to all that was a statement of hubris: "no reputable [sic!] student of animal behavior has ever. . . ." (*Science*, 153:1205, 1966). Such behavioristic confusion, unfortunately quite representative, contrasts painfully with a clarity to be found in the treatment of that same topic by two nongeneticists in the finest animal-behavior text of the postwar period (the latter is admittedly, perhaps, a unique example, but therefore all the more important): "When a geneticist speaks of an inherited trait he refers not to a characteristic of one individual but to the difference between two individuals or groups of individuals, or populations. . . . At no point is the inference drawn that a particular trait in a given individual animal is inherited; rather a certain difference between the traits of the two individuals is shown to be inherited." In the same discussion these authors make the observation, most insightful and important theoretically, that an analogous argument applies to what has too often been vaguely called acquired or learned traits: "We can speak of the difference between the behavior of a trained and an untrained animal as learned, provided that their genotypes are similar, but the behavior of a single individual cannot be spoken of as learned" (Marler and Hamilton, 1966, p. 619).

In the past, when behaviorists wished to pay token respect to heredity but to justify concentrating all their efforts on studying the "more important" environmental conditions, some such statement would be made as: Heredity sets the limits but environment determines the extent of development within those limits. Paradoxically that statement is at once true and misleading. Its truth lies in its expression of the norm-of-reaction concept: The phenotypic development of each genotype *is* determined by its ontogenetic environment. [The norm of reaction was beautifully illustrated by Stockard's experiment (1907) showing that, when magnesium salts are added to water, eggs of the "normally" two-eyed minnow *Fundulus* develop into an "abnormal" cyclopean form.] The misleading aspects of the statement are due to typological thinking. Because there is no place for individual differences in the typological frame (uniformity is axiomatic), a true statement has been misconstrued as justifying the impossible, that is, the study of environmental influences *per se*. What I call impossible (theoretically) might have been practically feasible (loosely speaking) if the variation pattern for responding to the limitless set of conceivable environmental conditions were exactly the same for all possible genotypes. Since genotypic diversity and genotype-environment interaction are apparently ubiquitous, attempts to study the laws of environmental influence have been grasping at shadows.

The 50-year fiasco that was behaviorism, what the Brelands (1961) correctly called "a clear and utter failure of conditioning theory," resulted from a blind fixation on the impossible task of trying to generalize about "laws" of environmental influence. More than a decade ago it had been pointed out, "It is patent . . . that environmental influence must be an influence on something and therefore the laws of such influence must differ as the object influenced differs" (Hirsch and Tryon, 1956, p. 403). When taken seriously, those much bruited "laws" must be descriptions of a limitless set of genotypic norms of reaction. Since there is no a priori reason to expect any two genotypes to have the same norm of reaction, whether or not any, in fact, do have must always remain an empirical question.

Heritability measurement itself is usually approached from the variance-component point of view, as illustrated by several chapters in this volume. The danger of confusion is reasonably small so long as it is discussed in biometrically sophisticated circles. More general use of the heritability concept, however, runs the serious risk of misinterpretation, because of the typological thinking prevalent in many fields that have still to assimilate the population concept now diffusing from evolutionary genetics throughout the biosocial sciences; e.g., see Mather's (1967) account of how the Nobel laureate in physics and professor of engineering science "William B. Shockley . . . falls into one of the classical genetical errors of confusing the apportionment among contributory agencies of the causation of a character itself with the apportionment of causation of the variation it is observed to show" (p. 126).

If heritability is approached with a proper appreciation of the norm of reaction, however, some of the pitfalls may be avoided. The norm-of-reaction concept applies to genetic systems at all levels of organization, from the alleles of a single locus and the total genotype of an individual to the population gene pool. The effects of any genetic system depend upon the context in which it is acting: the genetic background or genotype in which individual genes occur, the population of which an individual genotype is a part, and the environment in which genes, individuals, and populations exist.

When these considerations are taken into account, we see why it is impossible to generalize about the contribution to a phenotype of either heredity or environment. The interpretation usually made is that the average expression of a trait in a specific population can be changed by selection pressure when heritability is high or by environmental selection when heritability is low. This means that high heritability indicates the possibility, through selective breeding, of accumulating alleles favoring a desired expression of a trait, whereas low heritability indicates the possibility that exposing all individuals to the same selected environmental conditions will promote their uniformity with respect to some desired expression of that trait.

Accepting the foregoing as a reasonably correct version of what is often *said* does not complete the story, however. Nature proves to be far more complex than such a relatively simple characterization implies. It is important to understand how a measured heritability is obtained in order to understand what it means. It is obtained from the measurement of the expression of some trait by a certain set of genotypes in a certain set of environments. Statistical analysis (based on very explicit additivity assumptions) of such measurements then yields an estimate of the percentage of trait variance that is inferred to be related to the additive contemporary genetic variance. Naturally such measurement requires a perfectly balanced experimental design: all genotypes (or their trait-revelant components) measured against all environments (or their trait-relevant components). Few, if any, behavioral studies have been so thorough, certainly not any human studies. [The heritabilities estimated in our simple Drosophila phototaxis experiment (Hirsch and Boudreau, 1958) and in Dobzhansky and Spassky's (1967) were based on measurements made under a *single* set of environmental conditions.]

Only when we consider the number of possible genotypes and the number of potential environments that may influence trait expression do we begin to realize how narrowly limited is the range of applicability for any obtained heritability measure. J. B. S. Haldane has shown that for m genotypes and n environments "we have $(mn)!/m!n!$ possible types of interaction; for example, if $m = n = 10$, there are 7.09×10^{144} types. Even for the simplest case but one, of two genotypes in three environments or three genotypes in two environments, there are sixty types of interactions" (Haldane, 1946, p. 202). Furthermore, in the recurrent reports about the achievement of strength through adversity by "exceptional" cases, for example, there is a strong suggestion (though obviously not proof) that it is futile to attempt to characterize an environment as generally favorable or unfavorable. Since each genotype has its unique norm of reaction, there is no a priori reason to expect any environmental condition to be universally beneficial or harmful. Some average measure of an environmental influence, therefore, is applicable only to those genotypes affected by it in the same way or, at least, in the same general direction. This means that the characterization of a genotype-environment interaction can only be *ad hoc*.

It should also be pointed out that one can *not* infer from a high heritability that environmental selection is hopeless. Tryon's rats showed a very high heritability, but this did not mean that their performance was "genetically determined." They showed a genetic difference in their responsiveness to the training conditions that Tryon had administered to them. Later McGaugh, Jennings, and Thompson (1962) demonstrated that, for two strains differing in the way Tryon's did, a change in training procedures can eliminate the difference by raising the level of performance of the low strain to that of the high strain (also, see Caspari's discussion of "the genetic control of a character," page 5).

For the foregoing reasons, I cannot seriously consider heritability to be "one of the central concepts of modern genetics." Heritability estimation of course can sometimes serve as an early step in the description of genotype-phenotype relations. Perhaps it is most useful for comparing several polygenic traits in a given population having a known breeding regimen under well-specified conditions.

HOMOZYGOUS PURITY

Ever since Johannsen's work many people have considered the use of homozygous pure lines to be an experimental ideal. Sometimes the failure to use homozygous lines has been criticized as leading to uninterpretable results, e.g., Hall's critique of the Tryon effect cited earlier; also in this context see McClearn's discussion in Chapter 15 of the advantages of control and replicability offered by inbred strains and R. C. Roberts' warning in Chapter 11 that they represent merely a single gamete.

In 1921 Wright reported calculations for the approach to homozygosity under various inbreeding regimens *in the case of two alleles at a single locus.* Wright's calculations show that 95 percent expected homozygosity is reached before the eleventh generation of sib mating. For reasons to be discussed, these calculations have been widely interpreted as applying to the entire genome.

That this is no longer the safest interpretation can be seen in several ways. If we consider first the matrix of genotypes referred to previously in discussing the Tryon effect, there are 4^{20} cells in the matrix for an animal like the mouse having 20 chromosome pairs. Of these, all are heterozygous for from 1 to 20 chromosomes, except the 2^{20} different types represented in the main diagonal of the appropriate genotype matrix. The likelihood of a randomly chosen mating pair having the same homozygous genotype is $\frac{1}{2}^{60}$ (their sex-chromosome difference is ignored in this argument). The probability of a given individual being homozygous is $2^{20}/4^{20} = \frac{1}{2}^{20}$; the probability that a second individual will have the same genotype as the first one chosen is $\frac{1}{4}^{20} = \frac{1}{2}^{40}$; and the probability of the simultaneous occurrence of these two independent events is the product of their separate probabilities: $\frac{1}{2}^{20} \times \frac{1}{2}^{40} = \frac{1}{2}^{60}$. Moreover, when an inbred line is begun with noninbred individuals, it is reasonable to assume that every chromosome pair in the karyotypes of the two individuals mated is heterozygous for at least some genes and therefore that the individuals are not isogenic with respect to any pair of chromosomes.

If, for simplicity, we focus on each chromosome as an indivisible supergene, the chromosome constitution of each mating pair will be of the nature $(A_1A_2B_1B_2C_1C_2 \ldots S_1S_2T_1T_2) \times (A_3A_4B_3B_4C_3C_4 \ldots S_3S_4T_3T_4)$, where a letter stands for a chromosome, different letters represent heterologous chromosomes, and the different numerical subscripts attached to the same letter indicate the nonidentity (heterozygosity) of homologs. Clearly, a mating of this type produces only heterozygous offspring. For each chromosome the results of such a mating will be $A_1A_2 \times A_3A_4 = A_1A_3$ or A_1A_4 or A_2A_3 or A_2A_4. When these 4 genotypes are then intermated, they produce 10 genotypes in the next generation, 4 homozygotes and 6 heterozygotes: A_1A_1, A_2A_2, A_3A_3, A_4A_4, A_1A_2, A_1A_3, A_1A_4, A_2A_3, A_2A_4, and A_3A_4. For 20 chromosomes each having 4 such alternative homologs, the genotype matrix has 10^{20} cells. Of these, 4^{20} cells (the main diagonal) contain pure homozygotes, each one of which is a different combination of 20 homozygous pairs of chromosomes. Thus it is clear that the likelihood of two completely identical homozygous genotypes arising simultaneously and combining in a sib mating early in an inbreeding program is infinitesimally small.

Li has observed that the work on the percentage of homozygosity attained (or, what is equivalent, the residual heterozygosity) after n generations of inbreeding "can be interpreted in three ways. The simplest interpretation is with respect to the proportion of Aa in the general population for one pair of genes. The second is with respect to heterozygous pairs (of all loci) in a population. But for most practical breeding purposes the third interpretation is the most important—it measures the degree of genetic uniformity of individuals subjected to selfing or crossing among themselves" (Li, 1955, p. 106; also see p. 118). In other words, we can focus our attention on a single locus in a large population and ask what percentage of the individuals in that population will carry homozygous alleles at one particular locus after n generations of a given inbreeding regimen or we can focus attention on a single inbreeding line and ask what percentage of all its loci carry homozygous alleles after a similar n generations of a given inbreeding regimen.

At the present time there is another question to which an answer is needed. Many laboratories today carry inbred stocks. Therefore, there are, throughout the world, populations of inbred lines from several species. Because the genetic correlates of so much behavior are proving to be polygenic, it is now very important to know the probability of complete genome homozygosity after n generations of a given inbreeding regimen for any randomly selected line, i.e., the proportion of all lines that can be expected to have attained complete homozygosity in n generations. That is, instead of considering either one gene across a population of lines or the population of genes comprising the genome of one line (two mathematically equivalent problems), we must now consider a population of genes (some estimate of the total number of genes in a species genome) across a population of inbred lines.

The following expressions derived by Prof. Rafael Hanson of Long Beach State College, California, permit an approximate calculation of the approach to homozygosity for l loci over n generations; the first equation assumes each gene in the genome to be heterozygous at the start for two alleles, for example, $A_1A_2B_1B_2 \ldots N_1N_2$, and the second equation assumes heterozygosity at the start in every mating pair for four alleles of each gene, for example, $(A_1A_2B_1B_2 \ldots N_1N_2) \times (A_3A_4B_3B_4 \ldots N_3N_4)$. For the two-allele case, $g_n \approx [1 - 0.7236(0.8090)^n]^l$ and for the four-allele case, $g_n \approx [1 - 1.1708 (0.8090)^n]^l$. These approximations are satisfactory for $n \geq 5$ and $g_n = 0$ for $n = 0$.

After $n = 20$ generations of sib mating, therefore, the percentage of homozygosity expected for $l = 1$, 4, 20, and 39 loci, respectively, is $g_n \approx 98.96$, 95.91, 81.13, and 66.51 for two alleles, and $g_n \approx 98.31$, 93.42, 71.12, and 51.45 for four alleles. These cases might be considered to represent, respectively, one gene and the chromosome complements of the fruit fly, the mouse, and the dog. An analogous calculation shows that for the mouse over 65 generations of sib mating are required to attain 99 percent homozygosity, on the assumption that there might be 10,000 independent genes having only two alleles each.

WHAT'S IN A NAME?

At various times different names have been proposed for the kind of research and thinking we have been discussing: psychogenetics, genetics of behavior, behavior genetics, behavioral genetics, and behavior-genetic analysis.

Adoption of the term psychogenetics would be confusing for several reasons but mainly because it has long had almost the opposite meaning to that its proponents now wish it to convey. Psychogenetics means the study of psychogenesis, which, in turn, has the following meanings: (1) the genesis or origin of the soul or mind or of a mental function or trait, (2) development from psychic as distinguished from somatic origins (i.e., psychogenic), and (3) development from mental factors operating through the central nervous system. These are the meanings that tradition and usage by quite distinguished authors have assigned to the family of terms derived from the combination of psycho and genesis. There is almost perfect agreement on this matter among an array of authorities: *Webster's Third International Dictionary*, *Comprehensive Dictionary of Psychological Terms* by English and English, Piéron's *Vocabulaire de la Psychologie*, Lalande's *Vocabulaire Technique et Critique de la Philosophie*, the 1966 *Encyclopaedia Britannica*, and *The Oxford English Dictionary*. The last gives an ample set of quotations and references.

It is in the typological frame with its reductionist ideal that the temptation to speak about the genetics *of* behavior is perhaps strongest. Paradoxically many people, as well as most genetics textbooks, assimilate genetics itself to the typological frame of reference: There are typologically conceived traits and typologically conceived genes, and wherever a correlation between the two can be established one has a reductionistic causal explanation. As Caspari explains, the genetic correlate(s) "of a character in a population depend(s) on the genetic structure of the population rather than on the genetic nature of the character itself" etc. (page 5). To be concrete: It is one thing to find a gene like the one associated with the phenylketonuria condition in man, but it would be folly to claim that the "bad" allele of this gene produces mental deficiency and its "good" allele produces mental sufficiency. Mental deficiency can occur in many ways—without the presence of the phenylketonuria condition. Furthermore, the "right" genotype is no guarantee of either the presence or the absence of phenylketonuria (see Connell et al., 1962). The expression of any gene depends upon the prevailing genetic background, the prevailing environmental conditions, and their many possible interactions.

I began my own experimental work by asking the question: How far can we carry genetic analysis *of* behavior. We chose the fruit fly in the hope that we could take advantage of the great body of knowledge and techniques accumulated in the study of an animal that had been so well analyzed genetically.

While many of us have tried to study the genetics *of* behavior, as a result of much discussion and experimentation over more than a decade we have come to realize that it is impossible to study the genetics *of* behavior. We can study the behavior of *an* organism, the genetics of *a* population, and individual differences in the expression of some behavior by the members of *that* population. For this reason it is less confusing to speak of behavior-genetic analyses, understanding by that expression simultaneously the analysis of well-defined behaviors into their sensory and response components, the reliable and valid measurement of individual differences in the behaviors and in their component responses, and *then* subsequent breeding analysis or, for man, pedigree analysis by the methods of genetics over a specified set of generations in the history of a given population under known ecological conditions. We know full well that both the behavioral and the genetic properties can and will vary over time, over ecological conditions, and among populations. Furthermore, there will be no simple isomorphism between the two.

THE STUDY OF MAN[2]

Next, we shall consider the study of man from the standpoint of behavior-genetic analysis. By and large, human behavior genetics has long consisted of a series of rounds in the heredity-environment controversy. A spectrum of behavior traits has provided the battlegrounds for numerous repetitions of this sterile controversy.

In their summary of behavior genetics through 1960, Fuller and Thompson (1960, p. 95) observed:

☐ The distinction between human and animal behavior genetics is more than a matter of the species studied or the techniques which are feasible in the two fields. The primary objectives of the workers in the two areas

[2] The material in this section is drawn from a paper presented first to the conference on the Behavioral Consequences of Genetic Differences in Man of the Wenner-Gren Foundation for Anthropological Research, Burg-Wartenstein, Austria, Sept. 16–28, 1964 (Spuhler, 1967) and developed further for an invited address to the American Psychological Association (Hirsch, 1967).

are different. Animal experimenters use genetics as a device to study the nature of the variables which determine behavior. In getting such information, traits and subjects are selected for study because of experimental convenience, not because wheel running, maze learning, and audiogenic seizures in rodents are socially or economically important.

In contrast, workers in human behavior genetics have concentrated on problems of social significance and accumulated a great body of observations on intelligence (particularly mental defect), psychoses, and other psychiatric problems. The desire to put the newly discovered science of genetics at the service of human welfare led some early twentieth century scientists to make excessive claims for the importance of heredity in the origin of social maladjustment. The anti-heredity movement was equally one-sided in the opposite direction. Though the battle between hereditarians and environmentalists no longer rages conspicuously, the concern with applied problems still persists among most human geneticists who deal with psychological characters.

It is only when we understand both genetics and the complexity of the genotype-environment system of interactions which is responsible for everything that can be observed at the level of the phenotype—and the subject matter of the behavioral sciences resides almost exclusively at the level of the phenotype—that the problems of genetic analysis and behavioral analysis can be separated rationally. Because of the mosaic nature of the genome, because of mutation, segregation, independent assortment, recombination, and the consequent family transmission pattern for hereditary endowment, today we understand both the ubiquity of individual differences and the importance of their study. Before we understood Mendelian genetics, the usual reason for studying individual differences was a hope of relating those differences to variations in environmental conditions. Because so many genes have pleiotropic effects, because the genetic correlates of so many phenotypic traits are polygenic, because phenocopies do occur, because of the complexity of norms of reaction, etc., it is now important to separate the problem of inferring the nature of the details of specific genetic systems in man from the study of human phenotypic variation.

It is now realized that the description and phenotypic analysis of human variation are legitimate—and important—scientific tasks in their own right (see Brozek, 1966). The question of the relationships that prevail for particular phenotypes between components of their genotype and components of their environment is a separate question and a most difficult and complex one. For many reasons it is now clear that only a biosocial science, that is built upon a solid foundation of the most thorough understanding possible of genetics and population structure, can be completely free of any preoccupation with the heredity-environment pseudoquestion.

I shall now argue for the desirability of a pure-science approach to behavior-genetic analysis in man as well as in animals. Near the end of their book, Fuller and Thompson (1960, p. 318) observed: "Possibly the most significant contribution of behavior genetics is its documentation of the fact that two individuals of superficially similar phenotypes may be quite different genotypically and respond in completely different fashion when treated alike." While Fuller and Thompson's observation may be self-evident to evolutionary biologists who fully appreciate the population concept, it remains incomprehensible to many behavioral scientists still trapped in the typological mode of thought. It certainly suggests that behavior measurement offers an approach with extremely high resolving power for the analysis of human diversity.

A pure-science approach to behavior-genetic analysis in man has as its objective the discovery and understanding of natural units. In animal ethology, there is a face validity to according natural-unit status to such behaviors as nesting, courtship, and predation—activities that have obviously been molded by the prolonged interaction of the species genome and the forces of natural selection. Man too is an animal whose characteristics have evolved through natural selection. But, as we now study his behavior in civilization, there is no comparable face validity permitting us to apply the label "natural" to most of the units we observe.

In contemporary behavioral science, far more attention has been paid to man's social roles than to his biological properties. In industrial psychology, for example, tests are devised to select individuals who will most skillfully perform those tasks for which they are needed by industry. Because of the speed of cultural evolution, man cannot possibly have been subjected to intense natural selection for his technological skills, though man must employ the capacities he has evolved in the exercise of skills. The great challenge now before the behavioral sciences lies in the behavior-genetic analysis of man's biological properties and the elucidation of their modus operandi in a sociotechnological context.

A review of the world's literature on the relationship between heredity and tested intelligence by Erlenmeyer-Kimling and Jarvik (Table 19.6, page 403) has revealed the remarkable fact that, for the distributions of correlational measures collected over the past half century, a most consistent pattern exists. The value of the intelligence correlation between relatives increases as the degree of biosocial relationship becomes closer. This supports, though certainly does not prove, the suggestion that behavior may provide one of our most sensitive measures of the human diversity we now know exists.

Clearly, data like those in the review merely demonstrate the heritability of a trait. That tested intelligence or most other human characteristics will show some measurable heritability, however, is knowledge that should no longer evoke surprise. Since heritabilities are population-, situation-, and generation-specific, studies that merely estimate their magnitude contribute knowledge of little general significance at this time.

The great challenge referred to a moment ago lies in the identifications and behavior-genetic analysis of the phenotypic dimensions of human variation. Clearly, those dimensions will not be identified through the exclusive study of complex behaviors, because we would expect them to be relatively simple, numerous, and largely uncorrelated. Estimates place the number of human genes at "probably . . . not less than 100,000" (McKusick, 1966, p. ix). Over several generations and across a world population of more than 3 billion individuals, by the nature of the genome, very many of these genes must be assorting independently. It is therefore very unlikely that we shall learn much about their behavioral correlates by omnibus tests of intelligence or personality. The trouble with broad-spectrum tests is that they measure too much. While they may be useful instruments of classification to serve the practical needs of society, because of their omnibus nature and their focus on social categories, they cannot be very precise measures of biological differences, which, because of the mosaic nature of the genome, should prove to be relatively fine-grained.

Intelligence testing classifies individuals on the basis of their test performance: (1) Idiots score below 20, (2) imbeciles 20 to 49, (3) morons 50 to 69, (4) borderline deficients 70 to 79, (5) dullards 80 to 89, (6) normals 90 to 140, and (7) geniuses over 140. Since there are practically an unlimited number of ways of obtaining any score, lumping together all individuals who fall in the same category on cultural tests undoubtedly obscures many biological differences. When heritabilities are calculated, heredity and environment are interpreted as "accounting for" the estimated proportions of the variation over the range of test scores.

Forty-two years ago H. J. Müller, the second Nobel laureate in genetics, studied "mental traits and heredity" in a pair of female monozygotic twins reared apart. His comments on those observations are still relevant (Müller, 1925, pp. 532–533):

□ The responses of the twins to all these tests—except the intelligence tests—are so decisively different almost throughout, that this one case is enough to show that the scores obtained in such tests indicate little or nothing of the genetic basis of the psychic make-up . . . it is necessary to institute an intensive search for ways of identifying more truly genetic psychic characters . . . despite the diverse reactions to almost all the non-intellectual tests used . . . there are really many other mental characteristics in which the twins would agree closely could we but find appropriate means of measuring them. Thus . . . the twins both seem possessed of similar energy and even tensions, in their daily activity, with a tendency to "overdo" to the point of breakdown; both have similar mental alertness and interest in the practical problems about them, but not in remote or more purely intellectual abstractions and puzzles; both are personally very agreeable (as indicated by their popularity); both display similar attitudes throughout in taking the tests, even to such detail as lack of squeamishness in blackening the fingers for the fingerprints, and in being pricked for the blood tests—but turning away before the needle struck. The tastes of both in books and people appear very similar. It would seem, then, that the operations of the human mind have many aspects not yet reached by psychological testing, and that some of these are more closely dependent upon the genetic composition than those now being studied.

Thus, there is reason to question the validity of prevailing testing procedures for the measurement of biosocially significant properties both when heritability is present and when it is absent.

Another factor that worked against discovering the socially relevant biological dimensions of behavioral variation has been the preoccupation with typological reifications such as learning, perception, and motivation; even Frank Beach (1965) has deplored the faculty-psychology organization of our textbooks. The classical approach has been to measure the performance of groups of Ss on specified tasks, average their scores, and infer the nature of some "process" from properties of an average curve. The Ss are obtained through schools, hospitals, industry, military installations, etc. Rarely are observations made on their kin. Consequently, two conditions essential for uncovering the culturally significant biological dimensions of human variation are hardly ever satisfied in most behavioral science research: careful analysis of human differences and tracing through kinships whatever segregation such differences show.

Both theory and observation point to the need for a radically different approach to behavior study. Theoretically, it is implied by our modern picture both of the mosaic organization of the individual genome and of the heterogenic nature of human populations. The complex of characteristics that comprises the total phenotype of each unique member of a population is the developmental result of thousands of genes, most of which, owing to crossing-over, undoubtedly assort independently.

Empirically, individual differences have been measured in the phenotypic expression of many traits and, for some, observations have been made on the similarities and differences both within and between families. On seven variables related to autonomic nervous function (vasomotor persistence, salivary output, heart period, standing palmar conductance, volar skin conductance, respiration period, and pulse pressure) measured in children from three relationship categories (monozygotic twins, siblings, and unrelated individuals), Jost and Sontag (1944) found that score similarity increased with genetic similarity. Fivefold within-family threshold differences have been found (von Skramlik, 1943) with taste stimuli that showed identical thresholds in monozygotic twins (Rümler, 1943). With somewhat more complex measurements, the fine structure of auditory curves was found to exhibit high intrapair concordance among monozygotic twins, intrafamilial similarities, and significant differences among unrelated subjects (von Békésy and Rosenblith, 1951).

Individual differences in memory span have been measured among adult Caucasian men (Wechsler, 1952) as well as among Chinese boys and girls (Cheng, 1936). When attention span (immediately after oral presentation of a series of digits, demonstration by written reproduction of the number of digits remembered, and introspective report that this performance did not involve mental grouping of digits) and memory span (again, oral presentation and written reproduction, but with mental grouping permitted) were studied in the same individuals, both measures yielded well-dispersed arrays having a rather low correlation and very little overlap. The importance of the mental-grouping operation is shown by the almost uniformly higher scores for memory span than for attention span (Oberly, 1928). With practice, some intelligent subjects can improve their memory span by learning to group items. Practice does not, however, seem to affect their attention span, which apparently sets a limit to the number of items they can combine into a group (Martin and Fernberger, 1929).

It is not at all uncommon to find that only some individuals in a diagnostic category receive extreme scores on specific trait measures. In 1914, Binet and Simon noticed the incapacity of some mental defectives to discriminate points on the skin, and they incorporated a test of this in their early intelligence measures (O'Connor and Hermelin, 1963). Birch and Mathews (1951) found poor auditory discrimination for tones above 6 kilocycles in many, but not all, mental defectives. O'Connor (1957) found the incidence of color blindness in imbecile males to be higher than that in males of the general population. Berkson et al. (1961) found that the blocking of the EEG alpha rhythm, following the presentation of a bright light, lasted longer in some mental defectives than in normals. Siegel et al. (1964) found that, following ethanol loading, some alcoholics showed significantly different plasma amino acid patterns from normals.

Many interesting family correlations can also be obtained. The relative paucity of available data can more likely be attributed to lack of interest in family studies on the part of behavioral scientists than to their being unobtainable. Lennox et al. (1940) found that, although only 10 percent of normal subjects showed occasional EEG abnormalities, some 60 percent of the relatives of known epileptics had abnormal rhythms. Lidz et al. (1958) reported marked distortions in communicating among many of the nonhospitalized parents of schizophrenic patients. McConaghy (1959), using an objective test for irrelevant thinking, found that at least one of the nonhospitalized parents of each of 10 schizophrenic in-patients showed significant thought disorders. Of the 20 "normal" parents, 12, or 60 percent, scored within the range of responding characteristic of their offspring, whereas of 65 normal controls, only 6 individuals, or 9 percent, scored in the range characteristic of the schizophrenics.

The most radical change in behavior study now called for involves a shift of focus away from insubstantial abstractions such as learning, perception, and motivation to concern with consanguinity relations among the subjects observed and to the study of the simplest possible units of intellectual functioning among individuals of known ancestry.

THE "RACE PROBLEM"

In their extended discussion of the "race problem" that has so long caused so much trouble for so many people, Spuhler and Lindzey emphasize what a great mistake it is to try to use race to study behavior. The confusion over the "race problem," like that connected with the heredity-environment and the validity-of-introspection pseudoquestions, is due to several common fallacies: (1) a uniformity assumption that recognizes no individual differences within a racial group, (2) a belief in the universal applicability of the analysis-of-variance model, (3) a failure to understand one important genetic reason for many empirical correlations among traits, and (4) a platonistic world view engendering the typological reification of concepts (behavioral or other).

First of all, from the platonistic "general laws" of theoretical psychology to the "psychic unity of mankind" of cultural and evolutionary anthropology, the uniformity assumption has survived as a pre-Mendelian, pre-Darwinian typological vestige of a belief in the fixity of species. Prior to Mendel there was no intelligible way of accounting for variation (individual differences)—a problem that gave Darwin no end of trouble. When the Mendelian mechanism and Darwinian evolution became part of our intellectual arsenal, however, the uniqueness of individuals and the evolutionary divergence of the populations they comprise became immediately comprehensible.

Secondly, in their striving for objectivity and scientific respectability, the biosocial sciences embraced a procrustean statistical methodology which became the *lingua franca* for the description, evaluation, and discussion of their observations. Unfortunately, it smuggled in a host of oversimplifying assumptions, which prolonged their commitment to the already discredited uniformity postulate. The overpopular analysis-of-variance model, which has been so indiscriminately employed, assumes normality of form and homogeneity of variance in the distributions to be compared (Eisenhart, 1947). Empirically, the invalidity of both assumptions has now been demonstrated many times (see Fisher, Immer, and Tedin, 1932; Hirsch, 1961b; King, 1963; Yamaguchi et al., 1948). Because of those two assumptions, the central tendency of a trait distribution becomes the "typical" value for each population, all variation around it must be "error," and the permissible comparisons among populations are limited to mean values. Such an approach ignores both the ubiquitous individual differences, which are not error, and the nature of empirical distributions, so many of which show neither normality nor homogeneity of variance.

It is one thing to take advantage of the central-limit theorem for statistical inference and estimation (i.e., the *sampling* distributions for many statistics tend to be normal even when the statistics are computed from samples drawn on populations having a wide range of nonnormal distributions) but it is quite another matter to infer that, for a given trait, the populations being sampled have the particular form of distribution called normal and are to be represented by the familiar mathematical expression for the normal curve.

Thirdly, although we may pay lip service to the truism that correlation does not mean causation, most of the time we have nothing but correlations with which to work. Our attempts at reductionism, e.g., brain and chromosome mapping, and our search for laws (platonisms?) have usually been based on an unquestioning (and possibly unrealized) acceptance of (1) the counterfactual uniformity postulate—what holds for one individual, the "representative" organism, holds for all—and (2) the belief that typologically conceived behavioral entities, e.g., courtship, intelligence, etc., can be analyzed into (invariably?) correlated components that are causally interrelated.

In his 1909 lecture to psychologists Jacques Loeb (1964) insisted

"We must . . . accept the consequences of Mendel's theory of heredity, according to which the animal is to be looked upon as an aggregate of independent hereditary qualities." Unfortunately those consequences have neither been accepted nor understood. As indicated elsewhere (Hirsch, 1963, 1967), they mean that, of the characteristics we identify and study, many will show correlations but few will be correlated because they share a common biochemical mechanism. The reason for this becomes evident when we consider the joint distribution of the alleles of two (or more) independent genes. With random mating, even though both loci attain their respective Hardy-Weinberg equilibria at once (in the first generation), their joint distribution does *not*. It approaches equilibrium asymptotically over several generations (see Li, 1955, pp. 86ff). This means that traits related to the action of independent genes will show fortuitous correlation at least for several generations (with random mating) and perhaps indefinitely (without random mating).

The concept of random mating is a very useful assumption in some mathematical models. However, it should be noted that it means that every conceivable kind of heterosexual union occurs and produces progeny with equal likelihood; that this implies incest both between members of the same generation and across generations; that inbreeding is as equally probable as outbreeding; that matings between distant individuals occur as often as between close individuals, no matter how distance may be defined—geographically, socially, economically, in terms of education, etc. Clearly, this assumption, which is so convenient for mathematical models, represents a condition there is no reason to believe ever has prevailed or is ever likely to prevail, especially in human populations. Since random mating is the exception rather than the rule, significant but biologically *un*important correlations between functionally independent traits may be maintained indefinitely. Many of the trait correlations that distinguish racial, ethnic, and national groups can be of this fortuitous nature, maintained by reproductive isolation and nonrandom idiosyncratic systems of matings.

A final fallacy involves the typological reification of behaviors. Labeling faculties (see Beach, 1965, on texts again), such as intelligence, motivation, anxiety, and the id, may occasionally afford some descriptive convenience, but it becomes positively misleading when it encourages the belief that the same "thing" can always be observed in different species or in different cultures. Merely because we can administer the same (operationally defined) test procedure to a pigeon and a rat or to a Harvard student and a Kalahari Bushman does not mean that we are measuring the same behavior—a modern version of William James' "psychologist's fallacy."

As was explicitly shown several years ago (Hirsch, 1963, p. 1441), without these fallacious assumptions there is no "race problem." Races are populations that differ in gene frequencies. On the average, members of one racial population share more ancestors in common with each other than they share with the members of another racial population. Since behavior is a property of organisms, and organisms show great variety, there are definite theoretical limitations to the use of organisms for the study of some typologically conceived reification called behavior. Behavior, however, is proving to be one of the most sensitive measures of individual and population diversity.

Members of the same species share the same genes. Even though reproductive isolation between populations may be incomplete, the relative frequencies of different alleles of genes in their gene pools are almost certain to differ; mutations and recombinations may be expected to occur at different places, at different times, and with differing frequencies. Furthermore, selection pressures also vary (Post, 1962). Because of the individual differences within populations, because different populations have different gene pools, and because the genotypic correlates of most phenotypic traits are mutually independent (because of crossing-over or chromosomal heterology), there is no reason to expect two populations ever to be the same. Therefore, in analyzing distributions of observations (behavioral and other) from different populations, we must learn to ask, not whether they are different, but, rather, in what ways they differ. Comparisons must be made with respect to trait distributions, and distributions may differ in any or all parameters— central tendency, dispersion, skewness, kurtosis, etc.

Furthermore, since careful analysis of empirical distributions has already revealed the existence of differences with respect to any one or any combination of parameters, no longer can any single parameter be considered exclusively important. That is why we now know enough to expect varying combinations of similarities and differences in the several parameters of distributions when we compare populations with respect to one or more traits or when we compare traits in one or more populations.

A most unfortunate speech in 1967 by A. R. Jensen (who came under Shockley's influence while at Stanford during 1966–67) illustrates the dangers of inappropriate use of both the concept of heritability and that of race by the biometrically unsophisticated. Following Shockley, Jensen suggests that, since intelligence is largely genetically determined (sic!), in order properly to carry out Project Head Start for enrichment of the lives of the culturally disadvantaged, we need to know (= the government should spend large amounts of money to learn) the proportion of difference (of mean difference, naturally!) in intelligence between Negroes and whites that is genetically determined, i.e., its heritability. As though intelligence were some "thing" which different (pure?) races shared in unequal proportions. As though "it" had a (determinate) heritability. As though such "knowledge" would enable a teacher to do a better job in training either a Negro or white or any other "type" of child!

THE FUTURE

In conclusion I should like to point out certain trends that are now developing. As the social, ethnic, and economic barriers to education are lowered throughout the world and as the quality of education approaches a more uniformly high level of effectiveness, heredity may be expected to make an ever larger contribution to individual differences in intellectual functioning and consequently to success in our increasingly complex civilization. Universally compulsory education, improved methods of ability assessment and career counseling, and prolongation of the years of schooling further into the reproductive period of life should serve to increase the degree of positive assortative mating in our population. From a geneticist's point of view, the attempt to create the great society might prove to be the greatest selective-breeding experiment ever undertaken.

Some might fear that these trends can only serve further to stratify society into a rigid caste system and that this time the barriers will be more enduring, because they will be built on a firmer foundation. On the other hand, at least two conditions should prevent that result:

1   Undoubtedly a significant contribution is made to intellectual functioning by the unique organization of each individual's total genotype and by its idiosyncratic environmental encounters. Furthermore, mutation, recombination, and meiotic assortment plus the inability to transmit more than a small part of individual experience as cultural heritage guarantee new variation every generation, to produce the filial regression observed by Galton and to contribute to the social mobility discussed by Burt (1961).

2   The ever-increasing complexity of the social, political, and technological differentiation of society creates many new niches (and abolishes some old ones) to be filled by each generation's freshly generated heterogeneity.

BIBLIOGRAPHY

Abood, L. G., and Gerard, R. W. Phosphorylation defect in the brains of mice susceptible to audiogenic seizure. In H. Waelsch (ed.), *Biochemistry of the Developing Nervous System*, pp. 467–472. New York: Academic Press Inc., 1955.

Adam, A. Bau und Mechanismus des Receptaculum seminis bei den Bienen, Wespen und Ameisen. *Zool. Jbr. Anat.*, 35:1–74, 1913.

Adam, Brother. In search of the best strains of bees. *Bee World*, 32:49–52, 57–62, 1951.

―――. Bee breeding. *Bee World*, 35:4–13, 21–29, 44–49, 1954a.

―――. In search of the best strains of bee: Second journey. *Bee World*, 35:193–203, 233–244, 1954b.

―――. In search of the best strains of bee: Concluding journeys. *Bee World*, 45:70–83, 104–118, 1964.

Ader, R., and Belfer, M. L. Emotional behavior in the rat as a function of maternal emotionality. *Psychol. Rep.*, 10:349–350, 1962a.

―――― and ―――. Prenatal maternal anxiety and offspring emotionality in the rat. *Psychol. Rep.*, 10:711–718, 1962b.

―――― and Conklin, P. M. Handling of pregnant rats: effects of emotionality on their offspring. *Science*, 142:411–412, 1963.

Alber, M., Jordan, R., Ruttner, F., and Ruttner, H. Von der Paarung der Honigbiene. *Z. Bienenforsch.*, 3:1–28, 1955.

Alexander, R. D. The evolution of mating behavior in arthropods. In K. C. Highnam (ed.), *Insect Reproduction*, pp. 78–94. London: Royal Entomological Society Symposium No. 2, 1964.

Allard, R. W. The analysis of genetic-environmental interactions by means of diallel crosses. *Genetics*, 41:305–318, 1956.

Allen, R. J. The detection and diagnosis of phenylketonuria. *Amer. J. Publ. Health*, 50:1662–1666, 1960.

Allport, G. W. Traits revisited. *Amer. Psychol.*, 21:1–10, 1966.

―――― and Pettigrew, T. F. Cultural influence on the perception of movement: the trapezoidal illusion among Zulus. *J. Abnorm. Soc. Psychol.*, 55: 104–113, 1957.

Altman, P. L., and Dittmer, D. S. (eds.). *Biology Data Book 1964*. Washington, D.C.: Federation of American Society of Experimental Biologists.

Anastasi, A. *Differential Psychology*. New York: The Macmillan Company, 1958.

Anderson, E. The hypothalamus and adrenocorticotrophic hormone release. In H. D. Moon (ed.), *The Adrenal Cortex*, pp. 157–175. New York: Paul B. Hoeber, Inc., medical book department of Harper & Row, Publishers, Incorporated, 1961.

Anderson, R. H. The laying worker in the cape honeybee, *Apis mellifera capensis. J. Apicult. Res.*, 2:85–92, 1963.

Anderson, W. E. The personality characteristics of 153 Negro pupils, Dunbar High School, Okmulgee, Oklahoma. *J. Negro Educ.*, 16:44–48, 1947.

Andrew, R. J. The displays given by passerines in courtship and reproductive fighting: a review. *Ibis*, 103a(3):315–348, 549–579, 1961.

Ardashnikov, S. N., Lechtenstein, E. A., Martynova, R. P., Soboleva, G. V., and Postnikova, E. N. The diagnosis of zygosity in twins. *J. Hered.*, 27:465–468, 1936.

Auerbach, V. A., Waisman, H. A., and Wycokoff, C. B., Jr. Phenylketonuria in the rat associated with decreased temporal discrimination learning. *Nature*, 182:871–872, 1958.

Ayala, F. Y. Factors which control the population size and growth in *Drosophila serrata*. *Genetics*, 52:426 (abstr.), 1965.

Bache, R. M. Reaction time with reference to race. *Psychol. Rev.*, 2:475–486, 1895.

Baerends, G. P. Ethological studies of insect behavior. *Ann. Rev. Entomol.*, 4:207–234, 1959.

Bajusz, E. Neuroendocrine relationships. *Progr. Neurol. Psychiat.*, 15:233–251, 1960.

————. Neuroendocrine relationships. *Progr. Neurol. Psychiat.*, 16:233–253, 1961.

————. Neuroendocrine relationships. *Progr. Neurol. Psychiat.*, 17:222–259, 1962.

————. Neuroendocrine relationships. *Progr. Neurol. Psychiat.*, 18:306–347, 1963.

Baldwin, J. M. A new factor in evolution. *Amer. Nat.*, 30:441–451, 536–553, 1896.

Baltzer, F. Einige Beobachtungen über Sicheltänze bei Bienenvölkern verschiedener Herkunft. *Arch. Julius Klaus-Stift.*, 27:197–206, 1952.

Barker, J. S. F. Sexual isolation between *Drosophila melanogaster* and *Drosophila simulans*. *Amer. Nat.*, 96:105–115, 1962.

Barnett, S. A. *A Study in Behaviour*. London: Methuen & Co., Ltd., 1963.

Barondes, S. H. Relationship of biological regulatory mechanisms to learning and memory. *Nature*, 205:18–21, 1965.

Barth, W. (ed.). *Starting Right with Bees*, 11th ed. Medina, Ohio: A. I. Root Co., 1956.

Bastian, J. R. Primate signaling systems and human languages. In I. DeVore (ed.), *Primate Behavior: Field Studies of Monkeys and Apes*. New York: Holt, Rinehart and Winston, Inc., 1965.

Bastock, M. A gene mutation which changes a behavior pattern. *Evolution*, 10:421–439, 1956.

———— and Manning, A. The courtship of *Drosophila melanogaster*. *Behavior*, 8:85–111, 1955.

Bateman, K. G. The genetic assimilation of the dumpy phenocopy. *J. Genet.*, 56:341–351, 1959.

Baxter, C. F. Personal communication and in preparation.

Baylor, E. R., and Shaw, E. Refractive error and vision in fishes. *Science*, 136:157–158, 1962.

Beach, F. A. Review of S. C. Ratner and M. R. Denny, *Comparative Psychology: Research in Animal Behavior. Contemp. Psychol.*, 10:345–346, 1965.

———— and Wilson, J. R. Effects of prolactin, progesterone and estrogen on reactions of nonpregnant rats to foster young. *Psychol. Rep.*, 13:231–239, 1963.

Beadle, G. W., and Tatum, E. L. Experimental control of developmental reaction. *Amer. Nat.*, 75:107–116, 1941.

Beckham, A. S. Albinism in Negro children. *J. Genet. Psychol.*, 69:199–215, 1946.

Beerman, W. Ein Babiani-Ring als Locus eines Speicheldrüsenmutation. *Chromosoma*, 12:1–25, 1961.

Begg, M., and Packman, E. Antennae and mating behavior in *Drosophila melanogaster*. *Nature*, 168:953, 1951.

Bennett, J. H. (ed.). English translation of Gregor Mendel, *Experiments in Plant Hybridisation*, with commentary and assessment by R. A. Fisher. Edinburgh and London: Oliver & Boyd Ltd., 1965.

Bergman, P., Sjogren, B., and Hakansson, B. Hypertensive form of congenital adrenocortical hyperplasia. *Acta Endocrinol.*, 40:555–564, 1962.

Berkson, G., Hermelin, B., and O'Connor, N. Physiological responses of normals and institutionalized mental defectives to repeated stimuli. *J. Ment. Def. Res.*, 5:30–39, 1961.

Berlin, R. Addison's disease: familial incidence and occurrence in association with pernicious anemia. *Acta Med. Scandinav.*, 144:1–6, 1952.

Berliner, D. L., and Dougherty, T. F. Hepatic and extrahepatic regulation of corticosteroids. *Pharmacol. Rev.*, 13:329–359, 1961.

Best, J. B., and Rubinstein, I. Environmental familiarity and feeding in a planarian. *Science*, 135:916–918, 1962.

Beveridge, W. M. Racial differences in phenomenal regression. *Brit. J. Psychol.*, 26:59–62, 1935.

————. Some racial differences in perception. *Brit. J. Psychol.*, 30:57–64, 1939.

Bickel, H., Gerrard, J., and Hickmans, E. M. Influence of phenylalanine intake on phenylketonuria. *Lancet*, 2:812–813, 1953.

Biggers, J. D., and Claringbold, P. J. Why use inbred lines? *Nature*, 174: 596–597, 1954.

————, McLaren, A., and Michie, O. Choice of animals for bioassay. *Nature*, 190:891–892, 1961.

Bignami, G. Selection for fast and slow avoidance conditioning in the rat. *Bull. Brit. Psychol. Soc.*, 17:5A (abstr.), 1964.

————. Selection for high rates and low rates of conditioning in the rat. *Anim. Behav.*, 13:221–277, 1965.

Birch, J. W., and Mathews, J. The hearing of mental defectives: its measurement and characteristics. *Amer. J. Ment. Def.*, 55:384–393, 1951.

Birke, G., et al. Familial congenital hyperplasia of the adrenal cortex. *Acta Endocrinol.*, 29:55–69, 1958.

Blair, W. F. A study of the prairie deer-mouse populations in Southern Michigan. *Amer. Mid. Nat.*, 24:273–305, 1940.

————. Population dynamics of rodents and other small mammals. *Adv. Genet.*, 5:1–41, 1953.

————. Mating call in the speciation of American amphibians. *Amer. Nat.*, 42:27–51, 1958.

Blest, A. D. The evolution of protective displays in the Saturnoidea and Sphingidae (Lepidoptera). *Behaviour*, 11:257–309. 1957.

———. The concept of ritualization. In W. H. Thorpe and O. L. Zangwill (eds.), *Current Problems in Animal Behaviour*, pp. 102–124. London: Cambridge University Press, 1961.

Blewett, D. B. An experimental study of the inheritance of intelligence. *J. Ment. Sci.*, 100:922–933, 1954.

Bliss, E. L. (ed.). *Roots of Behavior: Genetics, Instinct, and Socialization in Animal Behavior.* New York: Paul B. Hoeber, Inc., medical book department of Harper & Row, Publishers, Incorporated, 1962.

———, Migeon, C. J., Branch, C. H. H., and Samuels, L. T. Reaction of the adrenal cortex to emotional stress. *Psychosom. Med.*, 18:56–76, 1956.

Bloom, B. S. *Stability and Change in Human Characteristics.* New York: John Wiley & Sons, Inc., 1964.

Boch, R. Rassenmässige Unterschiede bei den Tänzen der Honigbiene (*Apis mellifica* L.). *Z. Vergl. Physiol.*, 40:289–320, 1957.

——— and Shearer, D. A. Identification of geraniol as the active component in the Nassanoff pheromone of the honey bee. *Nature*, 194:704–706, 1962.

——— and ———. Production of geraniol by honey bees of various ages. *J. Insect Physiol.*, 9:431–434, 1963.

——— and ———. Identification of nerolic and geranic acids in the Nassanoff pheromone of the honey bee. *Nature*, 202:320–321, 1964.

———, ———, and Stone, B. C. Identification of isoamyl acetate as an active component in the sting pheromone of the honey bee. *Nature*, 195:1018–1020, 1962.

Bondy, P., Cohn, G., Herrman, W., and Crispell, K. The possible relationship of etiocholanolore to periodic fever. *Yale J. Biol. Med.*, 30:395–405, 1958.

Bongiovanni, A. M. Unusual steroid pattern in congenital adrenal hyperplasia: deficiency of 3β-hydroxydehyrogenase. *J. Clin. Endocrinol. Metab.*, 27:860–862, 1961.

——— and Eberlein, W. R. Clinical and metabolic variations in the adrenogenital syndrome. *Pediatrics*, 16:628, 1955.

Bonte, M. The reaction of two African societies to the Müller-Lyer illusion. *J. Soc. Psychol.*, 58:265–268, 1962.

Borchers, H. A. Effects of certain factors on the nest-cleaning behavior of honey bees (*Apis mellifera*). Master's thesis. Iowa State University, 1964.

Borror, D. J., and DeLong, D. M. *An Introduction to the Study of Insects*, rev. ed. New York: Holt, Rinehart and Winston, Inc., 1964.

Bösiger, E. Sur la role de la sélection sexuelle dans l'evolution. *Experientia*, 16:270–273, 1960.

Boyd, W. C. *Genetics and the Races of Man.* Boston: Little, Brown and Company, 1950.

———. The contributions of genetics to anthropology. In A. Kroeber (ed.), *Anthropology Today*, pp. 488–506. Chicago: University of Chicago Press, 1953.

———. Genetics and the races of man. In S. M. Garn (ed.), *Readings on Race*, pp. 17–27. Springfield, Ill.: Charles C Thomas, Publisher, 1960.

————. Four achievements of the genetical method in physical anthropology. *Amer. Anthrop.*, 65:243–252, 1963.

————. Modern ideas on race, in the light of our knowledge of blood groups and other characters with known mode of inheritance. In C. A. Leone (ed.), *Taxonomic Biochemistry and Serology*, pp. 119–169. New York: The Ronald Press Company, 1964.

Bradshaw, A. D. Evolutionary significance of phenotypic plasticity in plants. *Adv. Genet.*, 13:115–155, 1965.

Brant, D. H., and Kavanau, J. L. Exploration and movement patterns of the canyon mouse *Peromyscus crinitus* in an extensive laboratory enclosure. *Ecology*, 46:452–461, 1965.

Breese, E. L., and Mather, K. The organization of polygenic activity within a chromosome in *Drosophila*. I. Hair characters. *Heredity*, 11:373–395, 1957.

———— and ————. The organization of polygenic activity within a chromosome in *Drosophila*. II. Viability. *Heredity*, 14:375–399, 1960.

Breland, K., and Breland, M. The misbehavior of organisms. *Amer. Psychol.*, 16:681–684, 1961.

Brian, M. V. Caste determination in social insects. *Ann. Rev. Entomol.*, 2:107–120, 1957.

Broadhurst, P. L. Application of biometrical genetics to behaviour in rats. *Nature*, 184:1517–1518, 1959.

————. Experiments in psychogenetics: applications of biometrical genetics to the inheritance of behaviour. In H. J. Eysenck (ed.), *Experiments in Personality*, vol. I, *Psychogenetics and Psychopharmacology*, pp. 1–102. London: Routledge & Kegan Paul, Ltd., 1960.

————. Analysis of maternal effects in the inheritance of behaviour. *Anim. Behav.*, 9:129–141, 1961.

————. A note on further progress in a psychogenetic selection experiment. *Psychol. Rep.*, 10:65–66, 1962.

———— and Jinks, J. L. Biometrical genetics and behavior: reanalysis of published data. *Psychol. Bull.*, 58:337–362, 1961.

———— and ————. The inheritance of mammalian behavior re-examined. *J. Hered.*, 54:170–176, 1963.

———— and ————. Stability and change in the inheritance of behaviour in rats: a further analysis of statistics from a diallel cross. In preparation, 1966.

———— and Levine, S. Litter size, emotionality and avoidance learning. *Psychol. Rep.*, 12:41–42, 1963.

———— and ————. Behavioural consistency in strains of rats selectively bred for emotional elimination. *Brit. J. Psychol.*, 54:121–125, 1963.

Brøchner-Mortensen, K. Familial occurrence of Addison's disease. *Acta Med. Scandinav.*, 156:205–209, 1956.

Brodish, A. Diffuse hypothalamic system for the regulation of ACTH secretion. *Endocrinology*, 73:727–735, 1963.

Brody, E. G. The genetic basis of spontaneous activity in the albino rat. *Comp. Psychol. Monogr.*, 17 (5); 1942.

————. A note on the genetic basis of spontaneous activity in the albino rat. *J. Comp. Physiol. Psychol.*, 43:281–288, 1950.

Brooks, R. V. Disorders of biosynthesis in man. *Brit. Med. Bull.*, 18:148–153, 1962.

Brown, A. M. Strain variation in mice. *J. Pharm. Pharmacol.*, 14:397–428, 1962.

Brown, R. G. B. Courtship behaviour in the *Drosophila obscura* group. Part II. Comparative studies. *Behaviour*, 25:281–323, 1965.

Brozek, J. (ed.). The biology of human variation. *Ann. N.Y. Acad. Sci.*, 134:497–1066, 1966.

Bruell, J. H. Dominance and segregation in the inheritance of quantitative behavior of mice. In E. L. Bliss (ed.), *Roots of Behavior*, pp. 44–67. New York: Paul B. Hoeber, Inc., medical book department of Harper & Row, Publishers, Incorporated, 1962.

————. Additive inheritance of serum cholesterol level in mice. *Science*, 142:1664–1666, 1963.

————. Inheritance of behavioral and physiological characters of mice and the problem of heterosis. *Amer. Zool.*, 4:125–138, 1964a.

————. Heterotic inheritance of wheelrunning in mice. *J. Comp. Physiol. Psychol.*, 58:159–163, 1964b.

————. Mode of inheritance of response time in mice. *J. Comp. Physiol. Psychol.*, 60:147–148, 1965.

Bruner, F. G. The hearing of primitive people. *Arch. Psychol.*, 17(11):1–113, 1908.

Bryan, A. L. The essential basis for human culture. *Current Anthropol.*, 4:297–306, 1963.

Buettner-Janusch, J. (ed.). *Evolutionary and Genetic Biology of Primates, A Treatise in Two Volumes*, vols. 1 and 2. New York: Academic Press Inc., 1963–1964.

Bullock, T. H., and Horridge, G. A. *Structure and Function in the Nervous Systems of Invertebrates*, vols. 1 and 2. San Francisco: W. H. Freeman and Company, 1965.

Bunning, E. *The Physiological Clock*. New York: Academic Press Inc., 1964.

Burks, B. S. The relative influence of nature and nurture upon mental development. *Natl. Soc. for Study of Educ. Yearbook*, 27:219–316, 1928.

Burt, C. The inheritance of mental ability. *Amer. Psychol.*, 13:1–15, 1958.

————. Intelligence and social mobility. *Brit. J. Statist. Psychol.*, 14:3–24, 1961.

———— and Howard, M. The multifactorial theory of inheritance and its application to intelligence. *Brit. J. Statist. Psychol.*, 9:95–131, 1956.

Bush, I. E. The separation, identification and determination of adrenocortical steroids. In F. Clark and J. K. Grant (eds.), *The Biosynthesis and Secretion of Adrenocortical Steroids*, pp. 1–23. London: Cambridge University Press, 1960.

————. Chemical and biological factors in the activity of adrenocortical steroids. *Pharmacol. Rev.*, 14:317–445, 1962.

Butler, C. G. *The Honeybee*. London: Oxford University Press, 1949.

————. *The World of the Honeybee*. London: Collins Press, 1954.

————. The source of the substance produced by a queen honeybee (*Apis mellifera* L.) which inhibits development of the ovaries of the workers of her colony. *Proc. Royal Entomol. Soc., London*, (A)34:137–138, 1959.

————. The scent of queen honeybees (*Apis mellifera* L.) that causes partial inhibition of queen rearing. *J. Insect Physiol.*, 7:258–264, 1961a.

————. The efficiency of a honeybee community. *Endeavour*, 20:5–10, 1961b.

————, Callow, R. K., and Johnston, N. C. Extraction and purification of "queen substance" from queen bees. *Nature*, 184:1871, 1959.

———— and Fairey, E. M. The role of the queen in preventing oögenesis in worker honeybees. *J. Apicult. Res.*, 2:14–18, 1963.

———— and Free, J. B. The behaviour of worker honeybees at the hive entrance. *Behaviour*, 4:263–292, 1952.

Butler, R. A. Discrimination learning by rhesus monkeys to visual-exploration motivation. *J. Comp. Physiol. Psychol.*, 46:95–98, 1953.

Cairns, J. (ed.). The genetic code. *Cold Spring Harbor Sympos. Quant. Biol.*, 31, 1966.

Calhoun, J. B. A method for self-control of population growth among mammals living in the wild. *Science*, 109:333–335, 1949.

Callow, R. K., Chapman, J. R., and Paton, P. N. Pheromones of the honeybee: chemical studies of the mandibular gland secretion of the queen. *J. Apicult. Res.*, 3:77–89, 1964.

Campbell, D. T. Distinguishing differences of perception from failures of communication in cross-cultural studies. In F. S. C. Northrop and H. H. Livingston (eds.), *Cross-cultural Understanding: Epistemology in Anthropology*, pp. 308–336, New York: Harper & Row, Publishers, Incorporated, 1964.

Cannon, J. F. Diabetes insipidus. Clinical and experimental studies with consideration of genetic relationships. *Arch. Intern. Med.*, 96:215–272, 1955.

Carman, C. T., and Brashear, R. E. Phenochromocytoma as an inherited abnormality. Report of the tenth afflicted kindred and review of the literature. *New Engl. J. Med.*, 263:419–423, 1960.

Carpenter, C. R. A field study of the behavior and social relations of howling monkeys. *Comp. Psychol. Monogr.*, 10(48), 1934.

————. Territoriality: a review of concepts and problems. In A. Roe and G. G. Simpson (eds.), *Behavior and Evolution*, pp. 224–250. New Haven, Conn.: Yale University Press, 1958.

Carran, A. B., Yeudall, L. T., and Royce, J. R. Voltage level and skin resistance in avoidance conditioning of inbred strains of mice. *J. Comp. Physiol. Psychol.*, 58:427–430, 1964.

Carter, C. O., and Woolf, L. I. The birthplaces of parents and grandparents of a series of patients with phenylketonuria in South-East England. *Ann. Hum. Genet.*, 25:57–64, 1961.

Carthy, J. D. *The Behaviour of Arthropods*. San Francisco: W. H. Freeman and Company, 1965.

Caspari, E. On the selective value of the alleles Rt and rt in *Ephestia kühniella. Amer. Nat.*, 84:367–380, 1950.

————. Pleiotropic gene action. *Evolution*, 6:1–18, 1952.

————. Genetic basis of behavior. In A. Roe and G. G. Simpson (eds.), *Behavior and Evolution*, pp. 103–127. New Haven, Conn.: Yale University Press, 1958.

————. Mechanisms of the genetic control of animal behavior. *Amer. Psychol.*, 15:500 (title only), 1960.

————. Implication of genetics for psychology. Review of J. L. Fuller and W. R. Thompson, *Behavior Genetics*. New York: John Wiley & Sons, Inc., 1960. *Contemp. Psychol.*, 7:337–339, 1961a.

————. Some genetic implications of human evolution. In S. L. Washburn (ed.), *Social Life of Early Man*, pp. 267–277. Chicago: Aldine Publishing Company (Wenner-Gren Foundation for Anthropological Research. Viking Fund Publications in Anthropology, no. 31), 1961b.

————. Relevance of behavior genetics to psychiatric problems. Presentation to Department of Psychiatry, Stanford University, 1962.

———— (ed.). Symposium on Principles and Methods of Phylogeny. *Amer. Nat.*, 97:261–352, 1963a.

————. Selective forces in the evolution of man. *Amer. Nat.*, 97:5–14, 1963b.

————. Genes and the study of behavior. *Amer. Zool.*, 3:97–100, 1963c.

————. Perspectives in behavior genetics. In *Genetics Today*, Proceedings of Eleventh International Congress on Genetics, The Hague, vol. 3, pp. 817–821. New York: Pergamon Press, 1965.

Cattell, R. B. The multiple abstract variance analysis equations and solutions: For nature-nurture research on continuous variables. *Psychol. Rev.*, 67:353–372, 1960.

————, Stice, G. F., and Kristy, N. F. A first approximation to nature-nurture ratios for eleven primary personality factors in objective tests. *J. Abnorm. Soc. Psychol.*, 54:143–159, 1957.

Cavalli-Sforza, L. L. An analysis of linkage in quantitative inheritance. In E. C. R. Reeve and C. H. Waddington (eds.), *Quantitative Inheritance*, pp. 135–144. London: H. M. Stationery Office, 1952.

————, Barrai, I., and Edwards, A. W. F. Analysis of human evolution under random genetic drift. *Cold Spring Harbor Sympos. Quant. Biol.*, 29: 9–20, 1964.

Chai, C. K. Comparison of two inbred strains of mice and their F_1 hybrid in response to androgen. *Anat. Rec.*, 126:269–282, 1956.

————. Choice of animals for bio-assay. *Nature*, 190:893–894, 1961.

Charles, C. M. A comparison of the intelligence quotients of incarcerated delinquent White and American Negro boys and a group of St. Louis public school boys. *J. Appl. Psychol.*, 20:499–510, 1936.

Chase, H. B., and Chase, E. B. Studies on an anophthalmic strain of mice. I. Embryology of the eye region. *J. Morphol.*, 68:279–301, 1941.

Cheng, P. L. (A preliminary study of range of perception.) *J. Testing* (in Chinese), II, no. 2, 31 pp., 1935, from the *Psychol. Abstr.*, no. 5217, 1936.

Childs, B., Grumback, M. M., and Van Wyk, J. J. Virilizing adrenal hyperplasia: a genetic and hormonal study. *J. Clin. Invest.*, 35:213–222, 1956.

————, Sidbury, J. B., and Migeon, C. J. Glucuronic acid conjugation by patients with familial nonhemolytic jaundice and their relatives. *Pediatrics*, 23:903–913, 1959.

———— and Young, W. J. Genetic variations in man. *Amer. J. Med.*, 34:663–673, 1963.

Chitty, D. Population processes in the vole and their relevance to general theory. *Canad. J. Zool.*, 38:99–113, 1960.

Christian, J. J. Effect of population size on the adrenal glands and reproductive organs of male mice in populations of fixed size. *Amer. J. Physiol.*, 182:292–300, 1955.

———. The roles of endocrine and behavioral factors in the growth of mammalian populations. In A. Gorbman (ed.), *Comparative Endocrinology.* New York: John Wiley & Sons, Inc., 1959.

——— and Davis, D. E. Endocrines, behavior and populations. *Science*, 146:1550–1560, 1964.

——— and LeMunyan, C. D. Adverse effects of crowding on reproduction and lactation of mice and two generations of their progeny. *Endocrinology*, 63:517–529, 1958.

Chung, C. S., and Morton, N. E. Discrimination of genetic entities in muscular dystrophy. *Amer. J. Hum. Genet.*, 11:339–359, 1959.

Clark, J. D. The prehistoric origins of African culture. *J. African Hist.*, 5:161–183, 1964.

Clark, P. J. Grouping in spatial distributions. *Science*, 123:373–374, 1956.

Clark, W. E. Le Gros. Letters to the editor. *Discovery*, 25:49, 1964.

Cleveland, W. W., Nikezic, M., and Migeon, C. J. Response to an 11-hydroxylase inhibitor (SU-4885) in males with adrenal hyperplasia and in their parents. *J. Clin. Endocrinol. Metab.*, 22:281–286, 1962.

Clever, U. Genaktivitäten in den Riesenchromosomen von Chironorms teutens und ihre Beziehungen zur Entwicklung. *Chromosoma*, 12:607–675, 1961.

Coates, S. Results of treatment in phenylketonuria. *Brit. Med. J.*, 1:767–771, 1961.

Cock, A. G. Segregation of hypostatic colour genes within inbred lines of chicken. *Poult. Sci.*, 35:504–515, 1956.

———. Genetic aspects of metrical growth and form in animals. *Quart. Rev. Biol.*, 41:131–190, 1966.

Cohen, J., and Ogden, D. P. Taste blindness to phenyl-thiocarbamide and related compounds. *Psychol. Bull.*, 46:490–498, 1949.

Cole, L. C. Biological clock in the unicorn. *Science*, 125:874–876, 1957.

Collins, R. L. Inheritance of avoidance conditioning in mice: a diallel study. *Science*, 143:1188–1190, 1964.

Connell, G. E., Moore, B. P. L., Partington, M. W., and Walker, N. F. Phenylketonuria in one of twins. *Univ. Toronto Med. J.*, 39:257–263, 1962.

Cooch, F. G., and Beardmore, J. A. Assortative mating and reciprocal difference in the blue-snow goose complex. *Nature*, 183:1833–1844, 1959.

Coon, C. S., Garn, S. M., and Birdsell, J. B. *Races: A Study of the Problems of Race Formation in Man.* Springfield, Ill.: Charles C Thomas, Publisher, 1950.

Cooper, B. E. *Statistical Fortran Programs.* Document ACL/R 2. London: H. M. Stationery Office, 1965.

Cooper, R. M., and Zubek, J. P. Effects of enriched and restricted early environment on the learning ability of bright and dull rats. *Canad. J. Psychol.*, 12:159–164, 1958.

Corkins, C. L., and Gilbert, C. H. A comparative test of the Caucasian with the Italian race of honeybees. *Univ. Wyoming Agric. Exp. Sta. Bull.* 186, 1932.

Cowen, J. S. A hippocampal anomaly associated with hereditary susceptibility to sound induced seizures. Ph.D. thesis. University of Chicago, 1965.

Craig, J. V., and Chapman, A. B. Experimental test of predictions of inbred line performance in crosses. *J. Anim. Soc.*, 12:124–139, 1953.

————, Ortman, L. L., and Guhl, A. M. Genetic selection for social dominance ability in chickens. *Anim. Behav.*, 13:114–131, 1965.

Crane, J. The comparative biology of Salticid spiders at Rancho Grande, Venezuela. IV. An analysis of display. *Zoologica*, 34:159–214, 1949.

————. A comparative study of innate defensive behavior in Trinidad Mantids (Orthoptera, Mantoidea). *Zoologica*, 37:259–293, 1952.

————. Imaginal behavior in butterflies of the family Heliconiidae: changing social patterns and irrelevant actions. *Zoologica*, 42:135–145, 1957a.

————. Basic patterns of display in fiddler crabs (Ocypodidae, genus *Uca*). *Zoologica*, 42:69-82, 1957b.

Crome, L., Tymms, V., and Woolf, L. I. A chemical investigation of the defects of myelination in phenylketonuria. *J. Neurol. Neurosurg. Psychiat.*, 25:143–148, 1962.

Cronbach, L. J. The two disciplines of scientific psychology. *Amer. Psychol.*, 12:671–684, 1957.

———— and Meehl, P. E. Construct validity in psychological tests. *Psychol. Bull.*, 52:281–302, 1955.

Crossley, S. An experimental study of sexual isolation within a species of *Drosophila*. Ph.D. thesis. Oxford University, 1963.

Cryns, A. G. J. African intelligence: A critical survey of cross-cultural intelligence research in Africa south of the Sahara. *J. Soc. Psychol.*, 57:283–301, 1962.

Dancis, J., and Balis, M. E. A possible mechanism for disturbance in tyrosine metabolism in phenylpyruvic oligophrenia. *Pediatrics*, 15:63–66, 1955.

Darling, F. F. *Bird Flocks and the Breeding Cycle: A Contribution to the Study of Avian Sociality.* London: Cambridge University Press, 1938.

Das, S. R. Inheritance of the P.T.C. taste character in man: an analysis of 126 Rarhi Brahmin families of West Bengal. *Ann. Hum. Genet.*, 22:200–212, 1958.

Daumer, K. Reizmetrische Untersuchungen des Farbensekens der Bienen. *Z. Vergl. Physiol.*, 38:413–478, 1956.

Davenport, G. C., and Davenport, C. B. Inheritance of albinism. *Amer. Nat.*, 44:705–731, 1910.

David, J., Morrill, R., Fawcett, J., Upton, V., Bondy, P. K., and Spiro, H. M. Apprehension and elevated serum cortisol levels. *J. Psychosom. Res.*, 6:83–86, 1962.

David, P. R., and Snyder, L. H. Genetic variability and human behavior. In J. H. Rohrer and M. Sherif (eds.), *Social Psychology at the Cross Roads*, pp. 53–82. New York: Harper & Row, Publishers, Incorporated, 1951.

Davidson, J. M., and Feldman, S. Adrenocorticotropin secretion inhibited by implantation of hydrocortisone in the hypothalamus. *Science*, 137:125–126, 1962.

——— and ———. Cerebral involvement in the hibition of ACTH secretion by hydrocortisone. *Endocrinology*, 72:936–946, 1963.

Dawson, W. M. Inheritance of wildness and tameness in mice. *Genetics*, 17:296–326, 1932.

DeAlvarez, R. R., and Smith, E. K. Congenital adrenal hyperplasia: endocrine patterns in women and effects of cortisone treatment. *Obstet. Gynecol.*, 9:426, 1957.

DeBeer, G. Mendel, Darwin, and Fisher (1865–1965). *Notes & Records Roy. Soc. (London)*, 19:192–226, 1964.

DeFries, J. C. Prenatal maternal stress in mice: differential effects on behavior. *J. Hered.*, 55:289–295, 1964.

———, Hegmann, J. P., and Weir, M. W. Open-field behavior in mice: evidence for a major gene effect mediated by the visual system. *Science*, 154:1577–1579, 1966.

——— and Touchberry, R. W. The variability of response to selection. I. Interline and intraline variability in a population of *Drosophila affinis* selected for body weight. *Genetics*, 46:1519–1530, 1961.

Denenberg, V. H., Grota, L. J., and Zarrow, M. X. Maternal behaviour in the rat: analysis of cross-fostering. *J. Reprod. Fertil.*, 5:133–141, 1963.

———, Hudgens, G. A., and Zarrow, M. X. Mice reared with rats: modification of behavior by early experience with another species. *Science*, 143:280–381, 1964.

———, Ottinger, D. R., and Stephens, M. W. Effect of maternal factors upon growth and behavior of the rat. *Child Develpm.*, 33:65–71, 1962.

——— and Whimbey, A. E. Behavior of adult rats is modified by the experiences their mothers had as infants. *Science*, 142:1192–1193, 1963.

Deol, M. S., Grüneberg, H., Searle, A. G., and Truslove, G. M. How pure are our inbred strains of mice? *Genet. Res.*, 1:50–58, 1960.

Dethier, V. G. Communication by insects: physiology of dancing. *Science*, 125:331–336, 1957.

———. *The Physiology of Insect Senses*. New York: John Wiley & Sons, Inc., 1963.

Detwiler, S. R. *Neuroembryology*. New York: The Macmillan Company, 1936.

DeVore, I. *Primate Behavior*. New York: Holt, Rinehart and Winston, Inc., 1965.

Dice, L. R., and Bradley, R. M. Growth in the deer mouse, *Peromyscus maniculatus*. *J. Mammol.*, 23:416–427, 1942.

Dickinson, A. G. The effect of environment and heredity in cattle. Doctoral dissertation. University of Birmingham, 1954.

——— and Jinks, J. L. A generalized analysis of diallel crosses. *Genetics*, 41:65–78, 1956.

Dingman, W., and Sporn, M. B. The incorporation of 8-azaguanine into rat brain RNA and its effect on maze learning by the rats: an inquiry into the biochemical basis of memory. *J. Neurochem.*, 1:1–11, 1961.

Dixon, R. B. *The Racial History of Man*. New York: Charles Scribner's Sons, 1923.

Dobzhansky, T. *Genetics and the Origin of Species.* New York: Columbia University Press, 1937.

———. *Genetics and the Origin of Species,* 2d ed. New York: Columbia University Press, 1941.

———. Genetics of natural populations. XIV. A response of certain gene arrangements in the third chromosome of *Drosophila pseudoobscura* to natural selection. *Genetics,* 32:142–160, 1947.

———. The genetic nature of differences among men. In S. Persons (ed.), *Evolutionary Thought in America,* pp. 86–155. New Haven, Conn.: Yale University Press, 1950.

———. A review of some fundamental concepts and problems of population genetics. *Cold Spring Harbor Sympos. Quant. Biol.,* 20:1–15, 1955.

———. The biological concept of heredity as applied to man. In *The Nature and Transmission of the Genetic and Cultural Characteristics of Human Populations,* pp. 11–19. New York: Millbank Memorial Fund, 1957.

———. *Mankind Evolving: The Evolution of the Human Species.* New Haven, Conn.: Yale University Press, 1962.

——— and Montagu, M. F. A. Natural selection and the mental capacity of mankind. *Science,* 105:587–590, 1947.

——— and Pavlovsky, O. Indeterminate outcome of certain experiments on *Drosophila* populations. *Evolution,* 7:198–210, 1953.

——— and Spassky, B. Effects of selection and migration on geotactic and phototactic behavior of Drosophila. I. *Proc. Roy. Soc.,* (B), (in press), 1967.

Dodge, P. R., Mancall, E. L., Crawford, J. D., Knapp, J., and Paine, R. S. Hypoglycemia complicating treatment of phenylketonuria with a phenylalanine-deficient diet. *New Engl. J. Med.,* 260:1104–1110, 1959.

Donovan, B. T., and Harris, G. W. Pituitary and adrenal glands. *Ann. Rev. Physiol.,* 19:439–466, 1957.

Dorfman, R. I. Metabolism of adrenal cortical hormones. In H. D. Moon (ed.), *The Adrenal Cortex,* pp. 108–118. New York: Paul B. Hoeber, Inc., medical book department of Harper & Row, Publishers, Incorporated, 1961.

———. Biochemistry of the adrenocortical hormones. In O. Eichler and A. Farah (eds.), *Handbuch der Experimentellen Pharmakologie,* H. W. Deane (subed.), Part 1, The Adrenocortical Hormones, pp. 411–513. Berlin: Springer-Verlag OHG, 1962.

Dreger, R. M., and Miller, K. S. Comparative psychological studies of Negroes and whites in the United States. *Psychol. Bull.,* 57:361–402, 1960.

Drescher, W. The sex limited genetic load in natural populations of *Drosophila melanogaster. Amer. Nat.,* 98:167–171, 1964.

——— and Rothenbuhler, W. C. Sex determination in the honey bee. *J. Hered.,* 55:90–96, 1964.

Driscoll, S. G., and Hsia, D. Y.-Y. The development of enzyme system during early infancy. *Pediatrics,* 22:785–845, 1958.

Dunn, L. C. Cross currents in the history of human genetics. *Amer. J. Hum. Genet.,* 14:1–13, 1962.

Dunn, T. B. The importance of differences in morphology in inbred strains. *J. Natl. Cancer Inst.,* 15:573–589, 1954.

Dyrenfurth, I., Sybulski, S., Notchev, V., Beck, J. C., and Venning, E. H. Urinary corticosteroid excretion patterns in patients with adrenocortical dysfunction. *J. Clin. Endocrinol. Metab.*, 18:391–408, 1958.

Earl, R. W. Motivation, performance, and extinction. *J. Comp. Physiol. Psychol.*, 50:248–251, 1957.

Eayrs, J. T., and Goodhead, B. Postnatal development of the cerebral cortex in the rat. *J. Anat.*, 93:385–402, 1959.

Eberlein, W. R. The salt-losing form of congenital adrenal hyperplasia. *Pediatrics*, 21:667, 1958.

—— and Bongiovanni, A. M. Congenital adrenal hyperplasia with hypertension; unusual steroid pattern in blood and urine. *J. Clin. Endocrinol.*, 15:1531, 1955.

—— and ——. Plasma and urinary corticosteroids in the hypertensive form of congenital adrenal hyperplasia. *J. Biol. Chem.*, 223:85, 1956.

—— and ——. Pathophysiology of congenital adrenal hyperplasia. *Metabolism*, 9:326–340, 1960.

Eccles, J. C. *The Physiology of Synapses.* New York: Academic Press Inc., 1964.

Eckert, J. E., and Shaw, F. R. *Beekeeping.* New York: The Macmillan Company, 1960.

Edgar, R. S., and Epstein, R. H. Conditional lethal mutations in bacteriophage T4. *Genetics Today, Proc. XI Int. Congr. Genet.*, 2:1–16, 1965.

Egelhaaf, A., and Caspari, E. Ueber die Wirkungsweise und genetische Kontrolle des Tryptophanoxydasesystems bei *Ephestia kühniella. Z. Vererbungsl.*, 91:373–379, 1960.

Egyházi, E., and Hydén, H. Experimentally induced changes in the base composition of the ribonucleic acids of isolated nerve cells and their oligodendriglial cells. *J. Biophys. Biochem. Cytol.*, 10:403–410, 1961.

Ehrman, L. A genetic constitution frustrating the sexual drive of *Drosophila paulistorum. Science*, 131:1381–1382, 1960.

——. Courtship and mating behavior as a reproductive isolating mechanism in Drosophila. *Amer. Zool.*, 4:147–153, 1964.

Eisenhart, C. The assumptions underlying the analysis of variance. *Biometrics*, 3:1–21, 1947.

Eleftheriou, B. E. A comparative study on the cholinesterase activity in the brains of two subspecies of deermice, *Peromyscus maniculatus.* Master's thesis. University of Massachusetts, 1959.

Elmadjian, F. Aldosterone excretion in behavioral disorders. In S. R. Korey, A. Pope, and E. Robins (eds.), *Ultrastructure and Metabolism of the Nervous System*, vol. 40, pp. 414–420. Baltimore: The Williams & Wilkins Company, 1962.

Elton, C. *The Ecology of Animals.* London: Methuen & Co., Ltd., 1946.

——. *The Ecology of Invasions by Animals and Plants.* London: Methuen & Co., Ltd., 1958.

Emmelot, P., Bos, C. J., Benedetti, E. L., and Rumke, P. Studies on plasma membranes. I. Chemical composition and enzyme content of plasma membranes isolated from rat liver. *Biochem. Biophys. Acta*, 90:126–145, 1964.

Erlenmeyer-Kimling, L., and Hirsch, J. Measurement of the relations between chromosomes and behavior. *Science,* 134:1068–1069, 1961.

————, ————, and Weiss, J. M. Studies in experimental behavior genetics. III. Selection and hybridization analyses of individual differences in the sign of geotaxis. *J. Comp. Physiol. Psychol.,* 55:722–731, 1962.

———— and Jarvik, L. F. Genetics and intelligence: a review. *Science,* 142: 1477–1478, 1963.

Esch, H. Über die Schallerzeugung beim Werbetanz der Honigbiene. *Z. Vergl. Physiol.* 45:1–11, 1961.

————, Esch, I., and Kerr, W. E. Sound: an element common to communication of stingless bees and to dances of the honeybee. *Science,* 149:320–321, 1965.

Essig, C. F. Anticonvulsant effect of aminooxyacetic acid during barbiturate withdrawal in the dog. *Int. J. Neuropharmacol.,* 2:199–204, 1963.

Euler, U. S., von, Gemzell, C. A., Levi, L., and Strom, G. Cortical and medullary adrenal activity in emotional stress. *Acta Endocrinol.,* 30:567–573, 1959.

Ewing, A. W. Body size and courtship behaviour in *Drosophila melanogaster. Anim. Behav.,* 9:93–99, 1961.

————. Attempts to select for spontaneous activity in *Drosophila melanogaster. Anim. Behav.,* 11:369–378, 1963.

Eysenck, H. J. Criterion analysis—an application of the hypotheticodeductive method to factor analysis. *Psychol. Rev.,* 43:79–82, 1951.

———— and Press, D. B. The inheritance of neuroticism: an experimental study. *J. Ment. Sci.,* 97:441–465, 1951.

Falconer, D. S. Selection for large and small size in mice. *J. Genet.,* 51:470–501, 1953.

————. Patterns of response in selection experiments with mice. *Cold Spring Harbor Sympos. Quant. Biol.,* 20:178–196, 1955.

————. Breeding methods I. Genetic considerations. In A. N. Worden and W. Lane-Petter (eds.), *The UFAW Handbook on the Care and Management of Laboratory Animals,* 2d ed. London: Universities Federation for Animal Welfare, 1957.

————. *Introduction to Quantitative Genetics.* Edinburgh and London: Oliver & Boyd Ltd., 1960.

————. Maternal effects and selection response. *Genetics Today, Proc.* XI *Int. Congr. Genet.,* 3:763–774, 1965.

———— and Bloom, J. L. Inheritance of susceptibility to induced pulmonary tumours in mice. *Nature,* 191:1070–1071, 1961.

Fankhauser, G. The effects of changes in chromosome number on amphibian development. *Quart. Rev. Biol.,* 20:20–78, 1945.

————. The role of nucleus and cytoplasm. In B. H. Willier, P. A. Weiss, and V. Hamburger (eds.), *Analysis of Development,* pp. 126–150. Philadelphia: W. B. Saunders Company, 1955.

————, Vernon, J. A., Frank, W. H., and Slack, W. V. Effect of size and number of brain cells on learning in larvae of the salamander, *Triturus viridescens. Science,* 122:692–693, 1955.

Farrar, C. L. Large-cage design for insect and plant research. *U.S. Dept. Agr.,* ARS 33–77, 1963.

Felton, J. S. The Cornell Index used as an appraisal of personality by an industrial health service. I. The total score. *Industr. Med.*, 18:133–144, 1949.

Ferguson, G. A. On transfer and the abilities of man. *Canad. J. Psychol.*, 10:121–131, 1956.

Filmer, R. S. Bee investigations. *N.J. State Agr. Exp. Sta. 52nd Ann. Rep.*, pp. 228–238, 1931.

Fisher, R. A. The correlation between relatives on the supposition of Mendelian inheritance. *Trans. Roy. Soc. Edinburgh*, 52:399–433, 1918.

————. Limits to intensive production in animals. *Brit. Agr. Bull.*, 4:217–218, 1951.

————, Immer, F. R., and Tedin, O. The genetical interpretation of statistics of the third degree in the study of quantitative inheritance. *Genetics*, 17:107–124, 1932.

Fishman, J. R., Hamburg, D. A., Handlon, J. H., Mason, J. W., and Sachar, E. Emotional and adrenal cortical responses to a new experience. *Arch. Gen. Psychiat.*, 6:271–278, 1962.

Fiske, D. W., and Rice, L. Intra-individual response variability. *Psychol. Bull.*, 52:217–250, 1955.

Fitzpatrick, T. B. Albinism. In J. B. Stanburg, J. B. Wyngaarden, and D. S. Fredrickson (eds.), *The Metabolic Basis of Inherited Disease*, pp. 428–448. New York: McGraw-Hill Book Company, 1960.

Flexner, J. B., Flexner, L. B., and Stellar, E. Memory in mice as affected by intracerebral puromycin. *Science*, 141:51–59, 1963.

Flexner, L. B., Flexner, J. B., Roberts, R. B., and Haba, G. de la. Loss of recent memory in mice as related to regional inhibition of cerebral protein synthesis. *Proc. Natl. Acad. Sci.*, 52:1165–1169, 1964.

Folch-Pi, J. Composition of the brain in relation to maturation. In H. Waelsch (ed.), *Biochemistry of the Developing Nervous System*. New York: Academic Press Inc., 1955.

Ford, E. B. *Ecological Genetics*. London: Methuen & Co., Ltd., 1964.

Forsham, P. H. The adrenals. In R. H. Williams (ed.), *Textbook of Endocrinology*, 3d ed., pp. 282–394. Philadelphia: W. B. Saunders Company, 1962.

Forssman, H. On hereditary diabetes insipidus, with special regard to a sex-linked form. *Acta Med. Scandinav. (Suppl.)*, 159:1–96, 1945.

————. Two different mutations of the X-chromosome causing diabetes insipidus. *Amer. J. Hum. Genet.*, 7:21–27, 1955.

Fortier, C. Adenohypophysis and adrenal cortex. *Ann. Rev. Physiol.*, 24:223–258, 1962.

———— and de Groot, J. Neuroendocrine relationships. *Progr. Neurol. Psychiat.*, 14:256–269, 1959.

Foster, D. D. Differences in behavior and temperament between two races of the deer mouse. *J. Mammol.*, 40:495–513, 1959.

Fox, A. L. On taste blindness. *Science (Suppl.)*, 74:14, 1931.

————. The relationship between chemical constitution and taste. *Proc. Natl. Acad. Sci.*, 18:115–120, 1932.

Frank, F. The causality of Microtine cycles in Germany. *J. Wildlife Mgmt.*, 21:113–121, 1957.

Fredericson, E. Reciprocal fostering of two inbred mouse strains and its effect on the modification of inherited behavior. *Amer. Psychol.*, 7:241–242 (abstr.), 1952.

Free, J. B. The ability of worker honeybees (*Apis mellifera*) to learn a change in the location of their hives. *Anim. Behav.*, 6:219–223, 1958.

————. The attractiveness of geraniol to foraging honeybees. *J. Apicult. Res.*, 1:52–54, 1962.

———— and Butler, C. G. *Bumblebees*. London: Collins Press, 1959.

Friedman, S. B., Mason, J. W., and Hamburg, D. A. Urinary 17-hydroxy-corticosteroid levels in parents of children with neoplastic disease. *Psychosom. Med.*, 25:364–376, 1963.

Frings, H. The loci of olfactory end-organs in the blowfly, *Cynomyia cadaverina* Desvoidy. *J. Exp. Zool.*, 88:65–93, 1941.

————. The loci of olfactory end-organs in the honey-bee, *Apis mellifera* Linn. *J. Exp. Zool.*, 97:123–134, 1944.

———— and Frings, M. The production of stocks of albino mice with predictable susceptibilities to audiogenic seizures. *Behaviour*, 5:305–319, 1953.

Frisch, K. von. Der Farbensinn und Formensinn der Biene. *Zool. Jbr. Abt.*, 35:1–182, 1914.

————. *Bees—Their Vision, Chemical Senses, and Language.* Ithaca, N.Y.: Cornell University Press, 1950.

————. *The Dancing Bees.* New York: Harcourt, Brace & World, Inc., 1955.

————. Dialects in the language of the bees. *Sci. Amer.*, 207:78–80, 83–84, 86–87, 1962.

———— and Jander, R. Über den Schwänzeltanz der Bienen. *Z. Vergl. Physiol.*, 40:239–263, 1957.

———— and Rösch, G. A. Neue Versuche über die Bedeutung von Duftorgan und Pollenduft für die Verständigung im Bienenvolk. *Z. Vergl. Physiol.*, 4:1–21, 1926.

Fukushima, D. K., Bradlow, H. L., Hellman, L., and Gallagher, T. F. Peripheral metabolism of hormones in congenital adrenal hyperplasia. In A. R. Currie, T. Symington, and J. K. Grant (eds.), *The Human Adrenal Cortex*, pp. 371–382. Baltimore: The Williams & Wilkins Company, 1962.

Fulker, D. W. Mating speed in male *Drosophila melanogaster:* a psychogenetic analysis. *Science*, 153:203–205, 1966.

Fuller, J. L. Gene mechanisms and behavior. *Amer. Nat.*, 85:145–157, 1951.

————. Measurement of alcohol preference in genetic experiments. *J. Comp. Physiol. Psychol.*, 57:85–88, 1964a.

————. Physiological and population aspects of behavior genetics. *Amer. Zool.*, 4:101–109, 1964b.

————, Easler, C., and Smith, M. E. Inheritance of audiogenic seizure susceptibility in the mouse. *Genetics*, 35:622–632, 1950.

———— and Thompson, W. R. *Behavior Genetics.* New York: John Wiley & Sons, Inc., 1960.

Ganong, W. F. The central nervous system and the synthesis and release of adrenocorticotropic hormone. In A. V. Nalbandov (ed.), *Advances in Neuroendocrinology*, pp. 92–157. Urbana, Ill.: University of Illinois Press, 1963.

Gardner, L., and Migeon, C. Unusual plasma 17-ketosteroid pattern in a boy with congenital adrenal hyperplasia and periodic fever. *J. Clin. Endocrinol.*, 19:266, 1959.

Garmezy, N., and Rodnick, E. H. Premorbid adjustment and performance in schizophrenia. Implications for interpreting heterogeneity in schizophrenia. *J. Nerv. Ment. Dis.*, 129:450–466, 1959.

Garn, S. M. (ed.). *Readings on Race*. Springfield, Ill.: Charles C Thomas, Publisher, 1960.

—————. *Human Races*. Springfield, Ill.: Charles C Thomas, Publisher, 1961.

————— and Coon, C. S. On the number of races of mankind. *Amer. Anthrop.*, 57:996–1001, 1955.

Garrod, A. E. *Inborn Errors of Metabolism*. London: Henry Frowde, 1909.

Garth, T. R. *Race Psychology*. New York: McGraw-Hill Book Company, 1931.

Gary, N. E. Queen honey bee attractiveness as related to mandibular gland secretion. *Science*, 133:1479–1480, 1961.

—————. Chemical mating attractants in the queen honey bee. *Science*, 136:773–774, 1962.

—————. Observations of mating behaviour in the honeybee. *J. Apicult. Res.*, 2:3–13, 1963.

Gates, R. R. *Human Genetics*. New York: The Macmillan Company, 1946. 2 vols.

Gibbs, N. K., and Woolf, L. I. Tests for phenylketonuria: results of a one year programme for its detection in infancy and among mental defectives. *Brit. Med. J.*, 2:532–535, 1959.

Gibson, I. M., and McIlwain, H. Continuous recording of changes in membrane potential in mammalian cerebral tissues *in vitro;* recovery after depolarization by added substances. *J. Physiol.*, 176:261–283, 1965.

Gibson, J. B., and Thoday, J. M. An apparent 20 map-unit position effect. *Nature*, 196:661–662, 1962.

————— and —————. Effects of disruptive selection. VIII. Imposed quasi-random mating. *Heredity*, 18:513–524, 1963.

Ginsburg, B. E. Genetics and the physiology of the nervous system. Genetics and the inheritance of integrated neurological and psychiatric patterns. *Proc. Ass. Res. Nerv. Ment. Dis.*, 33:39–56, 1954.

—————. Genetics as a tool in the study of behavior. *Perspect. Biol. Med.*, 1:397–424, 1958.

—————. Genetics and personality. In J. M. Wepman and R. W. Heine (eds.), *Concepts of Personality*. Chicago: Aldine Publishing Company, 1963a.

—————. Causal mechanisms in audiogenic seizures. In *Psychophysiologie Neuropharmacologie et Biochimie de la Crise Audiogene*, Colloques Internationaux du Centre National de la Recherche Scientifique, No. 112, pp. 227–240, 1963b.

—————. Coaction of genetical and nongenetical factors influencing sexual behavior. In F. A. Beach (ed.), *Sex and Behavior*, pp. 53–75. New York: John Wiley & Sons, Inc., 1965a.

—————. Gene action in the central nervous system. *A.A.A.S. Symposium: Behavior, Brain, and Biochemistry*, 1965b.

————— and Allee, W. C. Some effects of conditioning on social dominance and subordination in inbred strains of mice. *Physiol. Zool.*, 15:485–506, 1942.

———— and Laughlin, W. S. The multiple bases of human adaptability and achievement: a species point of view. *Eugen. Quart.*, 13:240–257, 1966.

———— and Miller, D. S. Genetic factors in audiogenic seizures. In *Psychophysiologie Neuropharmacologie et Biochimie de la Crise Audiogene*, Colloques Internationaux du Centre National de la Recherche Scientifique No. 112, pp. 217–225, 1963.

Glickman, S. E. Perseverative neural processes and consolidation of the memory trace. *Psychol. Bull.*, 58:218–223, 1961.

Gluecksohn-Waelsch, S. Genetic factors and the development of the nervous system. In H. Waelsch (ed.), *Biochemistry of the Developing Nervous System.* New York: Academic Press Inc., 1955.

Gobineau, A. De. *Essai sur l'inégalité des races humaines.* Paris: Firmin-Didot et Cie, 1884.

Gold, J. J., and Frank, R. The borderline adrenogenital syndrome: an intermediate entity. *Amer. J. Obstet. Gynecol.*, 75:1034, 1958.

Goodall, J. Feeding behaviour of wild chimpanzees. In J. R. Napier and N. A. Barnicot (eds.), *Symposia of the Zoological Society of London*, No. 10, pp. 39–48. London: The Zoological Society of London, 1963.

————. Tool-using and aimed throwing in a community of free-living chimpanzees. *Nature*, 201:1264–1266, 1964.

Goodenough, F. L. The measurement of mental functions in primitive groups. *Amer. Anthrop.*, 38:1–11, 1936.

———— and Harris, D. B. Studies in the psychology of children's drawings. II. 1928–1949. *Psychol. Bull.*, 47:369–433, 1950.

Gottesman, I. I. Heritability of personality: a demonstration. *Psychol. Monogr.*, 27(572), 1963.

Green, J. D. Basic neuroendocrinology. In R. H. Williams (ed.), *Textbook of Endocrinology*, 3d ed., pp. 883–898. Philadelphia: W. B. Saunders Company, 1962.

Greenblatt, R. B., Barfield, W. E., and Lampros, C. P. Cortisone in the treatment of infertility. *Fertil. & Steril.*, 7:203, 1956.

Greer, M. A. Clinical neuroendocrinology. In R. H. Williams (ed.), *Textbook of Endocrinology*, 3d ed., pp. 899–907. Philadelphia: W. B. Saunders Company, 1962.

Gregory, I. Genetic factors in schizophrenia. *Amer. J. Psychiat.*, 116:961–972, 1960.

Griesel, R. D. The activity of rats reared by active and inactive foster mothers. *Psychol. African*, 10:189–196, 1964.

Grinfel'd, E. K. Origin and development of the apparatus for pollen collection in bees (Hymenoptera, Apoidea). *Entomol. Rev.*, 41:37–42, 1962.

Grout, R. A. (ed.). *The Hive and the Honeybee*, rev. ed. Hamilton, Ill.: Dadant and Sons, Inc., 1963.

Grüneberg, H. *Animal Genetics and Medicine.* New York: Paul B. Hoeber, Inc., medical book department of Harper & Row, Publishers, 1947.

————. *The Genetics of the Mouse*, 2d ed. The Hague: Martinus Nijhoff, 1952.

————. Variation within inbred strains of mice. *Nature*, 173:674, 1954.

Guhl, A. M., Craig, J. V., and Mueller, C. D. Selective breeding for aggressiveness in chickens. *Poult. Sci.*, 39:970–980, 1960.

Guillemin, R., and Schally, A. V. Recent advances in the chemistry of neuroendocrine mediators originating in the central nervous system. In A. V. Nalbandov (ed.), *Advances in Neuroendocrinology*, pp. 314–344. Urbana, Ill.: University of Illinois Press, 1963.

Guthrie, R. Blood screening for phenylketonuria. *J. Amer. Med. Ass.*, 178: 863, 1961.

Hadler, N. M. Heritability and phototaxis in *Drosophila melanogaster*. *Genetics*, 50:1269–1277, 1964.

Haggard, E. A. *Interclass Correlation and the Analysis of Variance*. New York: Holt, Rinehart and Winston, Inc., 1958.

Haldane, J. B. S. The interaction of nature and nurture. *Ann. Eugen.*, 13: 197–205, 1946.

————. In E. Goldschmidt (ed.), *The Genetics of Migrant and Isolate Populations*, p. 43. Baltimore: The Williams & Wilkins Company, 1963.

Hall, C. S. Emotional behavior in the rat: I. Defecation and urination as measures of individual differences in emotionality. *J. Comp. Psychol.*, 18:385–403, 1934.

————. The inheritance of emotionality. *Sigma Xi Quart.*, 26:17–27, 1938.

————. Genetic differences in fatal audiogenic seizures between inbred strains of house mice. *J. Hered.*, 38:3–6, 1947.

————. The genetics of behavior. In S. S. Stevens (ed.), *Handbook of Experimental Psychology*, pp. 304–329. New York: John Wiley & Sons, Inc., 1951.

Hall, K. R. L. Observational learning in monkeys and apes. *Brit. J. Psychol.*, 54:201–226, 1963a.

————. Tool-using performances as indicators of behavioral adaptability. *Current Anthrop.*, 4:479–494, 1963b.

Hamburg, D. A. Some issues in research on human behavior and adrenocortical function. *Psychosom. Med.*, 21:387–388, 1959.

————. Plasma and urinary corticosteroid levels in naturally occurring psychologic stresses. In *Ultrastructure and Metabolism of the Nervous System*, vol. 15, chap. 23. Research Publications, A.R.N.M.D., 1962.

————, Sabshin, M. A., Board, F. A., Grinker, R. R., Korchin, S. J., Basowitz, H., Heath, H., and Persky, H. Classification and rating of emotional experiences: special reference to reliability of observation. *Arch. Neurol. Psychiat.*, 79:415–426, 1958.

Hamburgh, M., and Flexner, L. B. Biochemical and physiological differentiation during morphogenesis. XXI. Effect of hypothyroidism and hormone therapy on enzyme activities of the developing cerebral cortex of the rat. *J. Neurochem.*, 1:279–288, 1957.

Hamilton, W. D. The genetical evolution of social behavior. II. *J. Theoret. Biol.*, 7:17–52, 1964.

Handlon, J. H., Wadeson, R. W., Fishman, J. R., Sachar, E. J., Hamburg, D. A., and Mason, J. W. Psychological factors lowering plasma 17-hydroxycorticosteroid concentration. *Psychosom. Med.*, 24:535–542, 1962.

Hara, S., and Yahiro, E. I. Phenylketonuria, first report of a Japanese-American child. *Acta Paediat. Jap.*, 62:1564–1569, 1958.

Harde, K. W. Das postnatale Wachstum cytoarchitektonischer Einheiten im Grosshirn der Weissen Maus. *Zool. Jbr., Abt. Anat.*, 70:225–268, 1949.

Harlow, H. F. The formation of learning sets. *Psychol. Rev.*, 56:51–65, 1949.

———. The evolution of learning. In A. Roe and G. G. Simpson (eds.), *Behavior and Evolution*, pp. 269–290. New Haven, Conn.: Yale University Press, 1958.

Harris, G. W. Central control of pituitary secretion. *Handb. Physiol.—Neurophysiol.*, 11:1007–1038, 1957.

———. Neuroendocrine relations. In S. R. Korey, A. Pope, and E. Robins (eds.), *Ultrastructure and Metabolism of the Nervous System*, vol. 40, pp. 380–405. Baltimore: The Williams & Wilkins Company, 1962.

Harris, H. *Human Biochemical Genetics*. London: Cambridge University Press, 1959.

———. Biochemical errors and mental defect. In D. Richter, J. M. Tanner, L. Taylor, and O. L. Zangwill (eds.), *Aspects of Psychiatric Research*, pp. 194–217. London: Oxford University Press, 1962.

——— and Kalmus, H. The measurement of taste sensitivity to phenylthiourea. *Ann. Eugen.*, 15:24–31, 1949.

Harris, V. T. An experimental study of habitat selection by prairie and forest races of the deer mouse, *Peromyscus maniculatus*. *Univ. Mich. Cont. Lab. Vert. Biol.*, no. 56, 1952.

Harrison, G. A., Weiner, J. S., Tanner, J. M., and Barnicot, N. A. *Human Biology*. Fair Lawn, N.J.: Oxford University Press, 1964.

Haskell, P. T. (ed.). *Insect Behaviour*. London: Royal Entomological Society, 1966.

Hassanein, M. H., and El-Banby, M. A. Studies on the honey production of certain races of the honeybee, *Apis mellifica* L. *Bull. Soc. Entomol. Egypte*, 44:1–11, 1960a.

——— and ———. Studies on the brood rearing activity of certain races of the honeybee, *Apis mellifica* L. *Bull. Soc. Entomol. Egypte*, 44:13–22, 1960b.

——— and ———. A contribution to the study on the flight activity of the honeybee. *Bull. Soc. Entomol. Egypte*, 44:367–375, 1960c.

Hayman, B. I. The theory and analysis of diallel crosses. *Genetics*, 39:789–809, 1954a.

———. The analysis of variance of diallel tables. *Biometrics*, 10:235–244, 1954b.

———. The theory and analysis of diallel crosses. II. *Genetics*, 43:63–85, 1958.

——— and Mather, K. The progress of inbreeding when homozygotes are at a disadvantage. *Heredity*, 7:165–183, 1953.

Hebb, D. O. Heredity and environment. *Brit. J. Anim. Behav.*, 1:43–47, 1953.

Heftmann, E., and Mosettig, E. *Biochemistry of Steroids*. New York: Reinhold Publishing Corp., 1960.

Hegmann, J. P., and DeFries, J. C. A quantitative genetic analysis of open-field behavior in a random mating population of mice. *J. Comp. Physiol. Psychol.* (in preparation), 1967.

Hein, G. Über richtungsweisende Bienentänze bei Futterplätzen in Stocknähe. *Experientia*, 6:142–144, 1950.

Heron, W. T. The inheritance of maze learning ability in rats. *J. Comp. Psychol.*, 19:77–89, 1935.

————. The inheritance of brightness and dullness in maze learning ability in the rat. *J. Genet. Psychol.*, 59:41–49, 1941.

Hess, E. H. Ethology, an approach towards the complete analysis of behavior. In R. Brown, E. Galanter, E. H. Hess, and G. Mandler, *New Directions in Psychology*, pp. 159–266. New York: Holt, Rinehart and Winston, Inc., 1962.

Hewes, G. W. Food transport and the origin of hominid bipedalism. *Amer. Anthrop.*, 63:687–710, 1961.

Hiernaux, J. (ed.). Proceedings of the expert meeting on the biological aspects of race, Moscow, August, 1964. *Int. Soc. Sci. J.*, 17:71–161, 1965.

Hinde, R. A. Behaviour and speciation in birds and lower vertebrates. *Biol. Rev.*, 34:85–128, 1959.

———— and Tinbergen, N. The comparative study of species-specific behavior. In A. Roe and G. G. Simpson (eds.), *Behavior and Evolution*, pp. 251–268. New Haven, Conn.: Yale University Press, 1958.

Hirsch, J. Studies in experimental behavior genetics. II. Individual differences in geotaxis as a function of chromosome variations in synthesized *Drosophila* populations. *J. Comp. Physiol. Psychol.*, 52:304–308, 1959.

————. Review of J. L. Fuller and W. R. Thompson, *Behavior Genetics*. New York: John Wiley & Sons, Inc.; *Amer. J. Psychol.*, 74:147–149, 1961a.

————. The role of assumptions in the analysis and interpretation of data. *Amer. J. Orthopsychiat.*, 31:474–480, 1961b.

————. Individual differences in behavior and their genetic basis. In E. L. Bliss (ed.), *Roots of Behavior: Genetics, Instinct, and Socialization in Animal Behavior*, pp. 3–23. New York: Paul B. Hoeber, Inc., medical book department of Harper and Row, Publishers, Incorporated, 1962.

————. Behavior genetics and individuality understood: behaviorism's counterfactual dogma blinded the behavioral sciences to the significance of meiosis. *Science*, 142:1436–1442, 1963.

————. Breeding analysis of natural units in behavior genetics. *Amer. Zool.*, 4:139–145, 1964.

————. Biopsychology comes of age. Review of J. P. Scott and J. L. Fuller, *Genetics and the Social Behavior of the Dog*. Chicago: University of Chicago Press; *Science*, 148:818–819, 1965.

————. Elegant, authoritative, important. Review of K. Mather, *Human Diversity: The Nature and Significance of Differences among Men*. New York: Free Press of Glencoe, 1965. *Contemp. Psychol.*, 11:261–262, 1966.

————. Behavior-genetic, or "experimental," analysis: the challenge of science versus the lure of technology. *Amer. Psychologist*, 22:118–130, 1967.

———— and Boudreau, J. C. Studies in experimental behavior genetics. I. The heritability of phototaxis in a population of *Drosophila melanogaster*. *J. Comp. Physiol. Psychol.*, 51:647–651, 1958.

———— and Erlenmeyer-Kimling, L. Sign of taxis as a property of the genotype. *Science*, 134:835–836, 1961.

———— and ————. Studies in experimental behavior genetics. IV. Chromosome analyses for geotaxis. *J. Comp. Physiol. Psychol.*, 55:732–739, 1962.

———— and Tryon, R. C. Mass screening and reliable individual measurement in the experimental behavior genetics of lower organisms. *Psychol. Bull.*, 53:402–410, 1956.

Hockett, C. F., and Ascher, R. The human revolution. *Current Anthrop.*, 5:135–168, 1964.

Hockman, C. H. Prenatal maternal stress in the rat: Its effect on emotional behavior in the offspring. *J. Comp. Physiol. Psychol.*, 54:679–684, 1961.

Hoffmann, R. S. The role of reproduction and mortality in population fluctuations of voles (*Microtus*). *Ecol. Monogr.*, 28:79–109, 1958.

Hogben, L. *Nature and Nurture.* New York: W. W. Norton & Company, Inc., 1939.

Holley, R. W., Apgar, J., Everett, G. A., Madison, J. T., Marquisee, M., Merrill, S. H., Penswick, J. R., and Zamir, A. Structure of a ribonucleic acid. *Science*, 147:1462–1465, 1965.

Hollingsworth, M. J. A gynandromorph segregating for autosomal mutants in *Drosophila subobscura*. *J. Genet.*, 53:131–135, 1955.

———. Observations on the sexual behaviour of intersexes in *Drosophila subobscura*. *Anim. Behav.*, 7:57–59, 1959.

Horigan, F. D. (ed.). *Psychiatric Studies on Adrenocortical Steroids: A Survey of the Literature.* Bethesda, Md.: National Institute of Mental Health, Clinical Investigations, Adult Psychiatry Branch, 1960.

Hörmann-Heck, S. von. Untersuchungen über den Erbgang einiger Verhaltensweisen bei Grillenbastarden (*Gryllus campestris* L. × *Gryllus bimaculatus* De Geer). *Z. Tierpsychol.*, 14:137–183, 1957.

Horner, B. E. Arboreal adaptations of *Peromyscus*, with special reference to use of the tail. *Univ. Mich. Cont. Lab. Vert. Biol.*, no. 61, 1954.

Horner, F. A., Streamer, C. W., Alejandrino, L. L., Reed, L. H., and Ibbott, F. Termination of dietary treatment of phenylketonuria. *New Engl. J. Med.*, 266:79–81, 1962.

Hovanitz, W. Polymorphism and evolution. *Sympos. Soc. Exp. Biol.*, 7:238–253, 1953.

Howell, D. E., and Usinger, R. L. Observations on the flight and length of life of drone bees. *Ann. Entomol. Soc. Amer.*, 26:239–246, 1933.

Hsia, D. Y.-Y. Recent developments in the study of hereditary diseases in children. *Postgrad. Med.*, 22:203–210, 1957.

———. Medical genetics. *New Engl. J. Med.*, 262:1172–1178, 1222–1227, 1273–1278, 1318–1323, 1960.

———. Phenylketonuria: a study in human biochemical genetics. *Pediatrics*, 38:173–184, 1966a.

———. Detection and treatment of inborn errors of metabolism associated with mental deficiency. In G. J. Martin and B. Kisch (eds.), *Enzymes in Mental Health*, pp. 121–151. Philadelphia: J. B. Lippincott Company, 1966b.

——— (ed.). Symposium on the treatment of amino acid disorders. *Amer. J. Dis. Child.*, 113:1–174, 1967.

———, Berman, J. L., and Slatis, H. M. A program of screening newborn infants for phenylketonuria. *J. Amer. Med. Ass.*, 188:203–206, 1964.

———, Driscoll, K. W., Troll, W., and Knox, W. E. Detection by Phenylalanine tolerance tests of heterozygous carriers of phenylketonuria. *Nature*, 178:1239–1240, 1956.

——— and Knox, W. E. A case of phenylketonuria with borderline intelligence. *Amer. J. Dis. Child.*, 94:33, 1957.

———, ———, Quinn, K. V., and Paine, R. S. A one-year controlled study of the effect of low-phenylalanine diet on phenylketonuria. *Pediatrics*, 21:178–202, 1958.

————, Nishimura, K., and Brenchley, Y. Mechanisms for the decrease of brain serotonin. *Nature*, 200:578, 1963.

————, Rowley, W. F., and Raskin, N. J. Clinical management of phenylketonuria. *Quart. Bull. Northwestern Univ. Med. Sch.*, 36:25–33, 1962.

Huang, I., Tannenbaum, S., Blaume, L., and Hsia, D. Y.-Y. Metabolism of 5-hydroxyindole compounds in experimentally produced phenylketonuric rats. *Proc. Soc. Exp. Biol. Med.*, 106:533–536, 1961.

Hudspeth, J. Strychnine: its facilitating effect on the solution of a simple oddity problem by the rat. *Science*, 145:1331–1333, 1964.

Hughes, K. R., Cooper, R. M., and Zubek, J. P. Effect of glutamic acid on the learning ability of bright and dull rats. III. Effect of varying dosages. *Canad. J. Psychol.*, 11:182–184, 1957.

———— and Zubek, J. P. Effect of glutamic acid on the learning ability of bright and dull rats. I. Administration during infancy. *Canad. J. Psychol.*, 10:132–138, 1956.

———— and ————. Effect of glutamic acid on the learning ability of bright and dull rats. II. Duration of the effect. *Canad. J. Psychol.*, 11:253–255, 1957.

Hutson, Ray. Bee investigations. *N.J. State Agr. Exp. Sta.*, 51st *Ann. Rep.*, pp. 176–182, 1930.

Huxley, J. S. *Problems of Relative Growth.* London: Methuen & Co., Ltd., 1932.

Huxley, T. H. *Man's Place in Nature* (originally published in 1863). Ann Arbor, Mich.: University of Michigan Press, 1959.

Hydén, H., and Egyházi, E. Nuclear RNA changes of nerve cells during a learning experiment in rats. *Proc. Natl. Acad. Sci.*, 48:1366–1373, 1962.

———— and ————. Changes in RNA content and base composition in cortical neurons of rats in a learning experiment involving transfer of handedness. *Proc. Natl. Acad. Sci.*, 52:1030–1035, 1964.

Imms, A. D. *A General Textbook of Entomology*, 9th ed., entirely revised by O. W. Richards and R. G. Davies. London: Methuen & Co., Ltd., 1957.

Ingram, V. M. A specific chemical difference between the globins of normal human and sickle cell anemia haemoglobin. *Nature*, 178:792–794, 1956.

————. Abnormal human haemoglobins. I. The comparison of normal human and sickle-cell haemoglobins by fingerprinting. *Biochim. Biophys. Acta*, 28:547, 1958.

Jacob, F., and Monod, J. On the regulation of gene activity. *Cold Spring Harbor Sympos. Quant. Biol.*, 26:193–211, 1961.

———— and ————. Genetic repression, allosteric inhibition and cellular differentiation. In M. Locke (ed.), *Cytodifferential and Macromolecular Synthesis: A Symposium*, pp. 30–64. New York: Academic Press Inc., 1963.

Jacobsohn, G. M. Isolation of 5-pregnanediol from the urine of a patient with congenital adrenocortical hyperplasia. *J. Clin. Endocrinol. Metab.*, 22:859–860, 1962.

Jander, R. Insect orientation. *Ann. Rev. Entomol.*, 8:95–114, 1963.

Jansen, G., Rosen, S., Schulze, J., Plester, D., and El-Mafty, A. Vegetative reactions to auditory stimuli. *Trans. Amer. Acad. Ophthal. Otolaryn.*, 68:445–455, 1964.

Jay, G. E., Jr. Genetic strains and stocks. In W. J. Burdette (ed.), *Methodology in Mammalian Genetics.* San Francisco: Holden-Day, Inc., 1963.

Jay, S. C. A bee flight and rearing room. *J. Apicult. Res.,* 3:41–44, 1964.

Jensen, A. R. Social class, race, genes and educational potential. Invited address to the Annual Meeting of the American Educational Research Association, New York City, Feb. 17, 1967.

Jervis, G. A. Phenylpyruvic oligophrenia: introductory study of 50 cases of mental deficiency association with the excretion of phenylpyruvic acid. *Arch. Neurol. Psychiat.,* 38:944–963, 1937.

――――. Phenylpyruvic oligophrenia deficiency of phenylalanine oxidizing system. *Proc. Soc. Exp. Biol. Med.,* 82:514–515, 1953.

――――. Phenylpyruvic oligophrenia. *Proc. Ass. Res. Nerv. Ment. Dis.,* 33:259–282, 1954.

Jinks, J. L. The F_{22} and backcross generations from a set of diallel crosses. *Heredity,* 10:1–30, 1956.

――――. *Extrachromosomal Inheritance.* Englewood Cliffs, N.J.: Prentice-Hall, Inc., 1964.

―――― and Broadhurst, P. L. Diallel analysis of litter size and weight in rats. *Heredity,* 18:319–336, 1963.

―――― and ――――. The detection and estimation of heritable differences in behaviour among individuals. *Heredity,* 20:97–116, 1965.

―――― and Jones, M. R. Estimation of the components of heterosis. *Genetics,* 43:223–234, 1958.

―――― and Mather, K. Stability in development of heterozygotes and homozygotes. *Proc. Roy. Soc. (London),* (B)143:561–578, 1955.

Joffe, J. M. Effect of foster-mothers' strain and prenatal experience on adult behaviour in rats. *Nature,* 208:815–816, 1965a.

――――. Genotype and prenatal and premating stress interact to affect adult behavior in rats. *Science,* 150:1844–1845, 1965b.

Johannsen, W. *Ueber Eblichkeit in Populationen und in reinen Linien.* Jena: Gustav Fischer, 1903. Summary and conclusions translated in J. A. Peters, *Classic Papers in Genetics.* Englewood Cliffs, N.J.: Prentice-Hall, Inc., 1959.

Johansson, I. *Genetic Aspects of Dairy Cattle Breeding.* Urbana, Ill.: University of Illinois Press, 1961.

Johnson, D. L., and Wenner, A. M. A relationship between conditioning and communication in honey bees. *Anim. Behav.,* 14:261–265, 1966.

Jones, G. E. S., Howard, J. E., and Langford, H. The use of cortisone in follicular phase disturbances. *Fertil. & Steril.,* 4:49, 1953.

Jones, H. E. Environmental influences on mental development. In L. Carmichael (ed.), *Handbook of Child Psychology.* New York: John Wiley & Sons, Inc., 1946.

Jones, R. L., and Rothenbuhler, W. C. Behaviour genetics of nest cleaning in honey bees. II. Responses of two inbred lines to various amounts of cyanide-killed brood. *Anim. Behav.,* 12:584–588, 1964.

Jost, H., and Sontag, L. W. The genetic factor in autonomic nervous-system function. *Psychosom. Med.,* 6:308–310, 1944. Reprinted in C. Kluckhohn and H. A. Murray (eds.), *Personality in Nature, Society, and Culture,* 2d ed., pp. 73–79. New York: Alfred A. Knopf, Inc., 1953.

Jumonville, J. E. Influence of genotype-treatment interactions in studies of emotionality in mice. Ph.D. thesis. University of Chicago, 1967.

Kagan, J., and Berkun, M. The reward value of running activity. *J. Comp. Physiol. Psychol.*, 47:108, 1954.

Kakihana, R. Developmental study of preference for and tolerance to ethanol in inbred strains of mice. Ph.D. thesis. University of California, Berkeley, 1965.

Kalmus, H. Vorversuche über die Orientierung der Biene im Stock. *Z. Vergl. Physiol.*, 24:166–187, 1937.

————. The resistance to desiccation of *Drosophila* mutants affecting body colour. *Proc. Roy. Soc. (London)*, (B)130:185–201, 1941.

————. Sun navigation of *Apis mellifica* L. in the Southern Hemisphere. *J. Exp. Biol.*, 33:554–565, 1956.

————. Improvements in the classification of the taster genotypes. *Ann. Hum. Genet.*, 22:222–230, 1958.

————. Genetical variation and sense perception. In G. E. W. Wolstenholme and C. M. O'Connor (eds.), *Biochemistry of Human Genetics*, pp. 60–72. London: J. & A. Churchill, Ltd., 1959.

———— and Ribbands, C. R. The origin of the odours by which honeybees distinguish their companions. *Proc. Roy. Soc. (London)*, (B)140:50–59, 1952.

Kaplan, A. R. Phenylketonuria: a review. *Eugen. Quart.*, 9:151–160, 1962.

Karon, B. D. *The Negro Personality: A Rigorous Investigation of the Effects of Culture*. New York: Springer Publishing Company, Inc., 1958.

Kathariner, L. Versuche über die Art der Orientierung bei der Honigbiene. *Biol. Zentr.*, 23:646–660, 1903.

Keeler, C. E. Rodless retina, an ophthalmic mutation in the house mouse. *J. Exp. Zool.*, 46:355–407, 1927.

————. The Caribe Cuna Moon-child and its heredity. *J. Hered.*, 44:163–171, 1953.

Keeley, K. Prenatal influence on behavior of offspring of crowded mice. *Science*, 135:44–45, 1962.

Kempthorne, O., and Tandon, O. B. The estimation of heritability by regression of offspring on parent. *Biometrics*, 2:90–100, 1953.

Kendall, J. W., Jr., Matsuda, K., Duyck, C., and Greer, M. Studies of the location of the receptor site for negative feedback control of ACTH release. *Endocrinology*, 74:279–283, 1964.

Kerr, W. E. Genetic determination of castes in the genus *Melipona*. *Genetics*, 35:143–152, 1950a.

————. Evolution of the mechanism of caste determination in the genus *Melipona*. *Evolution*, 4:7–13, 1950b.

————. Bases par o estudo da genetica de populacoes dos Hymenoptera em geral e dos Apinae sociaes em particular. *Anais Escola Super. Agr. "Luiz de Queiroz,"* 8:219–354, 1951.

————. Evolution of communication in bees and its role in speciation. *Evolution*, 14:386–387, 1960.

———— and Araujo, V. de P. Racas de abelhas de Africa. *Garcia de Orta*, 6:53–59, 1958.

———— and Kerr, L. S. Concealed variability in the X-chromosome of Drosophila melanogaster. Amer. Nat., 86:405–408, 1952.

———— and Laidlaw, H. H., Jr. General genetics of bees. Adv. Genet., 8:109–153, 1956.

Kessler, S. Courtship rituals and reproductive isolation between the races or incipient species of Drosophila paulistorum. Amer. Nat., 96:117–121, 1962.

Kety, S. S. Biochemical theories of schizophrenia. Science, 129:1528–1532, 1590–1596, 1959.

King, J. A. Social relations of the domestic guinea pig living under semi-natural conditions. Ecology, 37:221–228, 1956.

————. Maternal behavior and behavioral development in two subspecies of Peromyscus maniculatus. J. Mamm., 39:177–190, 1958.

————. Swimming and reaction to electric shock in two subspecies of deermice (Peromyscus maniculatus) during development. Anim. Behav., 9:142–150, 1961.

————. Social behavior and population homeostasis. Ecology, 46:210–211, 1965.

———— and Eleftheriou, B. E. Effects of social experience in the laboratory upon the adaptation of Peromyscus to a natural environment. Anat. Rec., 128:576 (abstr.), 1957.

———— and ————. Differential growth in the skulls of two subspecies of deermice. Growth, 24:179–192, 1960.

————, Maas, D., and Weisman, R. G. Geographic variation in nest size among species of Peromyscus. Evolution, 18:230–234, 1964.

———— and Shea, N. J. Subspecific differences in the responses of young deermice on an elevated maze. J. Hered., 50:14–18, 1959.

———— and Weisman, R. G. Sand digging contingent upon bar pressing in deermice (Peromyscus). Anim. Behav., 12:446–450, 1954.

King, J. C. The fourth moment of a character distribution as an index of the regulative efficiency of the genetic code. In R. J. Harris (ed.), Biological Organization at the Cellular and Supercellular Level, pp. 129–146. New York: Academic Press Inc., 1963.

Kish, G. B. Learning when the onset of illumination is used as reinforcing stimulus. J. Comp. Physiol. Psychol., 48:261–264, 1955.

Klein, D., and Franceschetti, A. Missbildungen und Krankheiten des Auges. In P. E. Becker (ed.), Humangenetik, pp. 1–247. Stuttgart: Georg Thieme Verlag KG, 1964.

Klineberg, O. Race Differences. New York: Harper & Row, Publishers, Incorporated, 1935.

————. Social Psychology, 2d ed., New York: Holt, Rinehart and Winston, Inc., 1954.

————. Race and psychology. In Race and Science, pp. 423–452. New York: Columbia University Press, 1961.

————. Negro-white differences in intelligence test performance: a new look at an old problem. Amer. Psychol., 18:198–203, 1963.

Klopfer, P. H. Behavioral Aspects of Ecology. Englewood Cliffs, N.J.: Prentice-Hall, Inc., 1962.

Knight, G. R., Robertson, A., and Waddington, C. H. Selection for sexual isolation within a species. *Evolution*, 10:14–22, 1956.

Knox, G. An estimation from pedigree data of gene frequency of color blindness. *Brit. J. Prevent. Soc. Med.*, 12:193–196, 1958.

Knox, W. E. The metabolism of phenyl-alanine and tyrosine. In W. D. McElroy and H. B. Glass (eds.), *A Symposium on Amino Acid Metabolism.* Baltimore: The Johns Hopkins Press, 1955.

————. An evaluation of the treatment of phenylketonuria with diets low in phenylalanine. *Pediatrics*, 26:1–11, 1960.

Komaromi, I., and Donhoffer, S. Effect of habituation and reward on adrenal ascorbic acid depletion in response to intravenous injection of physiological saline in the rat. *Acta Physiol. Acad. Sci. Hung.*, 23:293, 1963.

Koopman, K. F. Natural selection for reproductive isolation between *Drosophila pseudoobscura* and *Drosophila persimilis. Evolution*, 4:135–148, 1950.

Krnjevic, K. Micro-iontophoretic studies on cortical neurons. *Int. Rev. Neurobiol.*, 7:41–98, 1964.

Ksander, G. A. A chromosome assay for negative geotaxis in the fruit fly *Drosophila melanogaster.* Unpublished master's thesis. University of Illinois, Urbana, 1966.

Kuffler, S. W., and Potter, D. D. Glia in the leech central nervous system: physiological properties and neuron-glia relationship. *J. Neurophysiol.*, 27:290–320, 1964.

Kühn, A. *Vorlesungen über Entwicklungsphysiologie*, 2d ed. Berlin: Springer-Verlag OHG, 1965.

Kuriyama, K., Roberts, E., and Rubinstein, M. K. Elevation of γ-aminobutyric acid in brain with amino-oxyacetic acid and susceptibility to convulsive seizures in mice: a quantitative reevaluation. *Biochem. Pharmacol.*, 15:221–236, 1966.

Kurokawa, M., Kato, M., and Machiyama, Y. Choline acetylase activity in a convulsive strain of mouse. *Biochem. Biophys. Acta*, 50:385–386, 1961.

————, Machiyama, Y., and Kato, M. Distribution of acetylcholine in the brain during various states of activity. *J. Neurochem.*, 10:341–348, 1963.

Lack, D., and Southern, H. N. Birds on Tenerife. *Ibis*, 91:607–626, 1949.

Lade, B. I., and Thorpe, W. H. Dove songs as innately coded patterns of specific behaviour. *Nature*, 202:366–368, 1964.

Lagerspetz, K. Genetics and social causes of aggressive behavior in mice. *Scand. J. Psychol.*, 2:167–173, 1961.

———— and Wuorinen, K. A cross fostering experiment with mice selective bred for aggressiveness and non-aggressiveness. *Inst. Psychol., Univ. Turku*, no. 17, 1965.

Laidlaw, H. H., Jr. Development of precision instruments for artificial insemination of queen bees. *J. Econ. Entomol.*, 42:254–261, 1949.

Lanni, F. The biological coding problem. *Adv. Genet.*, 12:1–141, 1964.

Leahy, A. M. Nature-nurture and intelligence. *Genet. Psychol. Monogr.*, 17:235–308, 1935.

Leakey, L. S. B., Tobias, P. V., and Napier, J. R. A new species of the genus *Homo* from Olduvai Gorge. *Nature*, 202:7–9, 1964.

Lecomte, M. J. Recherches sur le comportement agressif des ouvrières d' *Apis mellifica. Behaviour*, 4:60–66, 1951.

Lederberg, J. Biological future of man. In G. Wolstenholme (ed.), *Man and His Future*, pp. 263–273. Boston: Little, Brown and Company, 1963.

Lee, B. F. A genetic analysis of taste deficiency in the American Negro. *Ohio J. Sci.*, 34:337–342, 1934.

Leeman, S. E., Glenister, D. W., and Yates, F. E. Characterization of a calf hypothalamic extract with adrenocorticotropin-releasing properties: evidence for a central nervous system site for corticosteroid inhibition of adrenocorticotropin release. *Endocrinology*, 70:249–262, 1962.

Lehrman, D. S. Hormonal regulation of parental behavior in birds and infrahuman mammals. In W. C. Young (ed.), *Sex and Internal Secretion*. Baltimore: The Williams & Wilkins Company, 1961.

———. A critique of Konrad Lorenz's theory of instinctive behavior. *Quart. Rev. Biol.*, 28:337–363, 1953.

Lennox, W. G., Gibbs, E. L., and Gibbs, F. A. Inheritance of cerebral dysrhythmia and epilepsy. *Arch. Neurol. Psychiat.*, 44:1155, 1940.

Leppik, E. E. The ability of insects to distinguish number. *Amer. Nat.*, 87:229–236, 1953.

———. Floral evolution in the Ranunculaceae. *Iowa State J. Sci.*, 39:1–101, 1964.

Lerner, I. M. *Genetic Homeostasis*. New York: John Wiley & Sons, Inc., 1954.

———. *The Genetic Basis of Selection*. New York: John Wiley & Sons, Inc., 1958.

Levine, S., and Broadhurst, P. L. Genetic and ontogenetic determinants of adult behavior in the rat. *J. Comp. Physiol. Psychol.*, 50:423–428, 1963.

——— and Treiman, D. M. Differential plasma corticosterone response to stress in four inbred strains of mice. *Endocrinology*, 75:142–144, 1964.

Levinger, E. L., and Escamilla, R. F. Hereditary diabetes insipidus report of 20 cases in seven generations. *J. Clin. Endocrinol.*, 15:547–552, 1955.

Lewis, P. R., Lobban, M. C., and Shaw, T. I. Patterns of urine flow in human subjects during a prolonged period of life on a 22-hour day. *J. Physiol.*, 133:659–669, 1956.

Li, C. C. *Population Genetics*. Chicago: University of Chicago Press, 1955.

Lidz, T., Cornelison, A., Terry, D., and Fleck, S. Intrafamilial environment of the schizophrenic patient. VI. The transmission of irrationality. *Arch. Neurol. Psychiat.*, 79:305, 1958.

Lieberman, S. Chemistry of adrenal corticosteroids. In H. D. Moon (ed.), *The Adrenal Cortex*, pp. 119–132. New York: Paul B. Hoeber, Inc., medical book department of Harper & Row, Publishers, Incorporated, 1961.

Lindauer, M. Ein Beitrag zur Frage der Arbeitsteilung im Bienenstaat. *Z. Vergl. Physiol.*, 34:299–345, 1952.

———. Division of labour in the honeybee colony. *Bee World*, 34:63–73, 85–90, 1953.

———. Communication among the honeybees and stingless bees of India. *Bee World*, 38:3–13, 34–39, 1957a.

———. Communication in swarm-bees searching for a new home. *Nature*, 179:63–66, 1957b.

———. Time-compensated sun orientation in bees. *Cold Spring Harbor Sympos. Quant. Biol.*, 25:371–377, 1960.

————. *Communication Among Social Bees.* Cambridge, Mass: Harvard University Press, 1961.

————. Ethology. *Ann. Rev. Psychol.*, 13:35–70, 1962.

————. Social behavior and mutual communication. In M. Rockstein (ed.), *The Physiology of Insecta*, vol. 2, pp. 123–186. New York: Academic Press Inc., 1965.

———— and Kerr, W. E. Communication between the workers of stingless bees. *Bee World*, 41:29–41, 65–71, 1960.

Lindquist, E. F. *Design and Analysis of Experiments in Psychology and Education.* Boston: Houghton Mifflin Company, 1953.

Lindzey, G. Emotionality and audiogenic seizure susceptibility in five inbred strains of mice. *J. Comp. Physiol. Psychol.*, 44:389–394, 1951.

————. *Projective Techniques and Cross Cultural Research.* New York: Appleton-Century-Crofts, Inc., 1961.

———— and Winston, H. Maze learning and effect of pretraining in inbred strains of mice. *J. Comp. Physiol. Psychol.*, 55:748–752, 1962.

Livingstone, F. B. On the non-existence of human races. *Current Anthropol.*, 3:279, 1962.

Lloyd, C. W. Central nervous system regulation of endocrine function in the human. In A. V. Nalbandov (ed.), *Advances in Neuroendocrinology*, pp. 460–500. Urbana, Ill.: University of Illinois Press, 1963.

Loeb, J. The significance of tropisms for psychology. Sixth International Congress of Psychology, Geneva, 1909. Reprinted in J. Loeb, *The Mechanistic Conception of Life*, edited by D. Fleming, Cambridge, Mass.: Harvard University Press, 1964.

Loeb, L., King, H. D., and Blumenthal, H. T. Transplantation and individuality differentials in inbred strains of rats. *Biol. Bull.*, 84:1–112, 1943.

Lopatina, N. G., and Chesnokova, E. G. Uslovnie reflexy u pchel na slozhnie razdrazhiteli (Conditioned reflexes of bees on complex stimuli). *Pchelovodstvo*, 36:35–38, 1959.

Lush, J. L. *Animal Breeding Plans*, 1st ed. Ames, Iowa: Collegiate Press, 1937; 2d ed. Ames, Iowa: Iowa State College Press, 1943.

————. Intra-sire correlations or regressions of offspring on dam as a method of estimating heritability of characteristics. *Thirty-third Ann. Proc. Amer. Soc. Anim. Prod.*, pp. 293–301, 1940.

————. Heritability of quantitative characters in farm animals. *Hereditas, Suppl. vol.:* 356–375, 1949.

Maas, J. W. Neurochemical differences between two strains of mice. *Science*, 137:621–622, 1962.

MacArthur, J. W. Selection for small and large body size in the house mouse. *Genetics*, 34:194–209, 1949.

Mackensen, O. Viability and sex determination in the honey bee (*Apis mellifera* L.). *Genetics*, 36:500–509, 1951.

————. Further studies on a lethal series in the honey bee. *J. Hered.*, 46:72–74, 1955.

———— and Roberts, W. C. A manual for the artificial insemination of queen bees. *U.S. Dept. Agri.*, ET-250, 1948.

Magnus, D. B. E. Experimentelle Untersuchungen zur Bionomie und Ethologie des Kaisermantels *Argynis paphia* L. (Lep. Nymph.). I. Über optische Auslöser von Anfliegereaktion und ihre Bedeutung für das Sich-finden der Geschlechter. Z. *Tierpsychol.*, 15:397–426, 1958.

Manning, A. The sexual isolation between *Drosophila melanogaster* and *Drosophila simulans. Anim. Behav.*, 7:60–65, 1959a.

————. The sexual behaviour of two sibling *Drosophila* species. *Behaviour*, 15:123–145, 1959b.

————. The effects of artificial selection for mating speed in *Drosophila melanogaster. Anim. Behav.*, 9:82–92, 1961.

————. Selection for mating speed in *Drosophila melanogaster* based on the behaviour of one sex. *Anim. Behav.*, 11:116–120, 1963.

————. *Drosophila*, and the evolution of behaviour. *Viewpoints in Biol.*, 4:123–167, 1965.

Manosevitz, M. Genotype, fear, and hoarding. *J. Comp. Physiol. Psychol.*, 60:412–416, 1965.

Marcuse, F. L., and Bitterman, M. E. Notes on the results of Army intelligence testing in World War I. *Science*, 104:231–232, 1946.

Marien, D. Selection for developmental rate in *Drosophila pseudoobscura. Genetics*, 43:3–15, 1958.

Markert, C. *Developmental Genetics.* Englewood Cliffs, N.J.: Prentice-Hall, Inc., 1965.

Markl, H., and Lindauer, M. Physiology of insect behavior. In M. Rochstein (ed.), *The Physiology of Insecta*, vol. 2, pp. 3–122. New York: Academic Press Inc., 1965.

Marler, P. R. Bird songs and mate selection. In W. E. Lanyon and W. N. Tavolga (eds.), *Animal Sounds and Communication.* Washington, D.C.: American Institute of Biological Sciences, 1960.

———— and Hamilton, W. J. III. *Mechanisms of Animal Behavior.* New York: John Wiley & Sons, Inc., 1966.

Martin, P. R., and Fernberger, S. W. Improvement in memory span. *Amer. J. Psychol.*, 41:91–94, 1929.

Marzke, M. R. W. Evolution of the human hand. Ph.D. thesis. University of California, Berkeley, 1964.

Mason, J. W. The central nervous system regulation of ACTH secretion. In H. W. Magoun (chairman), J. D. French, H. H. Jasper, R. N. DeLong, R. W. Gerald, A. A. Ward, Jr., R. Knighton, L. Proctor, W. Noshay, and R. T. Costello (eds.), *Henry Ford Hospital—International Symposium: Reticular Formation of the Brain*, pp. 645–662. Boston: Little, Brown and Company, 1958a.

————. Plasma 17-hydroxycorticosteroid response to hypothalamic stimulation in the conscious rhesus monkey. *Endocrinology*, 63:403–411, 1958b.

————. Visceral functions of the nervous system. *Ann. Rev. Physiol.*, 21: 353–380, 1959a.

————. Psychological influences on the pituitary–adrenal cortical system. *Recent Progr. Hormone Res.*, 15:345–378, 1959b.

————, Brady, J. V., and Sidman, M. Plasma 17-hydroxycorticosteroid levels and conditioned behavior in the rhesus monkey. *Endocrinology*, 60:741–752, 1957.

———— and Hamburg, D. Unpublished observations.

Mather, K. Variation and selection of polygenic characters. *J. Genet.*, 41: 159–193, 1941.

————. The genetical requirements of bio-assays in higher organisms. *Analyst*, 71:407–411, 1946.

————. *Biometrical Genetics: The Study of Continuous Variation*. London: Methuen & Co., Ltd., 1949.

————. Genetic apocalypse? Review of J. D. Roslansky (ed.), *Genetics and the Future of Man*. Amsterdam: North Holland Publishing Company, 1966. *Nature*, 213:126, 1967.

———— and Harrison, B. J. The manifold effects of selection. *Heredity*, 3:1–52, 131–162, 1949.

Matsuda, K., Duyck, C., and Greer, M. A. Restoration of the ability of rat pituitary homotransplants to secrete ACTH if placed under the hypothalamic median eminence. *Endocrinology*, 74:939–943, 1964a.

————, ————, Kendall, J. W., Jr., and Greer, M. A. Pathways by which traumatic stress and ether induce increased ACTH release in the rat. *Endocrinology*, 74:981–985, 1964b.

Maurizio, A. Factors influencing the lifespan of bees. In G. E. W. Wolstenholme and M. O'Connor (eds.), *The Lifespan of Animals*, vol. 5, pp. 231–246. Boston: Little, Brown and Company, 1959.

Maxson, S. C. Effect of genotype on brain mechanisms involved in audiogenic seizures. Ph.D. thesis. University of Chicago, 1966.

Mayr, E. *Systematics and the Origin of Species*. New York: Columbia University Press, 1942.

————. Behavior and systematics. In A. Roe and G. G. Simpson (eds.), *Behavior and Evolution*, pp. 341–362. New Haven, Conn.: Yale University Press, 1958.

————. Darwin and the evolutionary theory in biology. In B. J. Meggers (ed.), *Evolution and Anthropology: A Centennial Appraisal*, pp. 1–10. Washington, D.C.: The Anthropology Society of Washington, 1959.

————. *Animal Species and Evolution*. Cambridge, Mass.: Harvard University Press, 1963.

————. Races in animal evolution. *Int. Soc. Sci. J.*, 17:121–122, 1965.

McCabe, T. T., and Blanchard, B. D. *Three Species of Peromyscus*. Santa Barbara, Calif.: Rood Associates, 1950.

McCann, S. M., Fruit, A., and Fulford, B. D. Studies on the loci of action of cortical hormones in inhibiting the release of adrenocorticotrophin. *Endocrinology*, 63:29–42, 1958.

McClearn, G. E. Performance differences among mouse strains in a learning situation. *Amer. Psychol.*, 13:405 (abst.), 1958.

————. Genotype and mouse activity. *J. Comp. Physiol. Psychol.*, 54:674–676, 1961.

————. Genetic differences in the effect of alcohol upon behaviour of mice. In J. D. J. Havard (ed.), *Proceedings of the Third International Conference on Alcohol and Road Traffic*. London: British Medical Association, 1962.

————. The inheritance of behavior. In L. Postman (ed.), *Psychology in the Making*, pp. 144–252. New York: Alfred A. Knopf, Inc., 1963.

————. Genotype and mouse behaviour. In S. J. Geerts (ed.), *Genetics Today, Proc. XI Int. Congr. Genet.*, 3:795–805, 1965.

———— and Meredith, W. Dimensional analyses of activity and elimination in a genetically heterogeneous group of mice (*Mus musculus*). *Anim. Behav.*, 12:1–10, 1964.

———— and Rodgers, D. A. Differences in alcohol preference among inbred strains of mice. *Quart. J. Stud. Alcohol*, 20:691–695, 1959.

———— and ————. Genetic factors in alcohol preference of laboratory mice. *J. Comp. Physiol. Psychol.*, 54:116–119, 1961.

McClintock, B. Some parallels between control system in maize and in bacteria. *Amer. Nat.*, 95:265–277, 1961.

McConaghy, N. The use of an object sorting test in elucidating the hereditary factor in schizophrenia. *J. Neurol. Neurosurg. Psychiat.*, 22:243, 1959.

McDougall, W. Cutaneous sensation. In *Report of the Cambridge Anthropological Expedition to Torres Straits*, vol. 2, pt. II, pp. 189–195. London: Cambridge University Press, 1903a.

————. Muscular sense. In *Report of the Cambridge Anthropological Expedition to Torres Straits*, vol. 2, pt. II, pp. 196–200. London: Cambridge University Press, 1903b.

McGaugh, J. L. Some neurochemical factors in learning. Unpublished doctoral dissertation. University of California, Berkeley, 1959.

————, Jennings, R. D., and Thompson, C. W. Effect of distribution of practice on the maze learning of descendants of the Tryon maze bright and maze dull strains. *Psychol. Rep.*, 9:147–150, 1962.

————, Westbrook, W., and Burt, G. Strain differences in the facilitative effects of 5-7-diphenyl-1-3-diazadamantan-6-OL (1757 I.S.) on maze learning. *J. Comp. Physiol. Psychol.*, 54:502–505, 1961.

————, ————, and Thomson, C. W. Facilitation of maze learning with post-trial injections of 5-7-diphenyl-1-3-diazadamantan-6-OL (1757 I.S.). *J. Comp. Physiol. Psychol.*, 55:710–713, 1962.

McGill, T. E. Sexual behavior in three inbred strains of mice. *Behaviour*, 19:341–350, 1962.

McIntosh, W. B. The applicability of covariance analysis for comparison of body and skeletal measurements between two races of the deer mouse, *Peromyscus maniculatus. Cont. Lab. Vert. Biol.*, no. 72, 1955.

McKean, C. M., Shanberg, S. M., and Giarman, N. J. A mechanism of the indole defect in experimental phenylketonuria. *Science*, 137:604–605, 1962.

McKusick, V. A. *Mendelian Inheritance in Man: Catalogs of Autosomal Recessive, and X-Linked Phenotypes.* Baltimore: The Johns Hopkins Press, 1966.

McLaren, A., and Michie, D. Are inbred strains suitable for bioassay? *Nature*, 173:686–687, 1954.

———— and ————. Studies on the transfer of fertilized mouse eggs to uterine foster-mothers. I. Factors affecting the implantation and survival of native and transferred eggs. *J. Exp. Biol.*, 33:394–416, 1956.

———— and ————. Studies on the transfer of fertilized mouse eggs to uterine foster-mothers. II. The effect of transferring large number of eggs. *J. Exp. Biol.*, 36:40–50, 1959.

McLennan, H., and Elliott, K. A. C. Effects of convulsant and narcotic drugs on acetylcholine synthesis. *J. Pharmacol.*, 103:35–43, 1951.

McQuiston, M. D. Effects of maternal restraint during pregnancy upon offspring behavior in rats. *Dissert. Abstr.*, 24:853–854, 1963.

Meakin, J. W., Nelson, D. H., and Thorn, G. W. Addison's disease in two brothers. *J. Clin. Endocrinol.*, 19:726–731, 1959.

Meier, H. *Experimental Pharmacogenetics*. New York: Academic Press Inc., 1963.

Mendel, G. 1866. See Bennett, 1965.

Merrell, D. J. Selective mating in *D. melanogaster*. *Genetics*, 34:370–389, 1949.

————. Selective mating as a cause of gene frequency changes in laboratory populations of *Drosophila melanogaster*. *Evolution*, 7:287–296, 1953.

Merriam, R. W., and Ris, H. Size and DNA content of nuclei in various tissues of male, female and worker honeybees. *Chromosoma*, 6:522–538, 1954.

Merton, B. B. Taste sensitivity to PTC in 60 Norwegian families with 176 children. Confirmation of the hypothesis of single gene inheritance. *Acta Genet. Stat. Med.*, 8:114–138, 1958.

Mettler, F. A. Culture and the structural evolution of the neural system. (James Arthur Lecture on the Evolution of the Human Brain.) New York: The American Museum of Natural History, 1955.

Michael, R. P., and Gibbons, J. L. Interrelationships between the endocrine system and neuropsychiatry. *Int. Rev. Neurobiol.*, 5:243–302, 1963.

Michener, C. D. The evolution of social behavior in bees. *Proc. 10th Int. Congr. Entomol.*, 2:441–447, 1958.

————. Social polymorphism in Hymenoptera. In J. S. Kennedy (ed.), *Insect Polymorphism*, pp. 45–56. London: Royal Entomological Society, 1961.

————. Division of labor among primitively social bees. *Science*, 141:434–435, 1963.

————. Evolution of the nests of bees. *Amer. Zool.*, 4:227–239, 1964.

———— and Michener, M. H. *American Social Insects*. Princeton, N.J.: D. Van Nostrand Company, Inc., 1951.

Milam, A. T., and Sumner, F. C. Spread and intensity of vocational interests and evaluative attitudes in first year Negro medical students. *J. Psychol.*, 37:31–38, 1954.

Milani, R., and Rivosecchi, L. Gynandromorphism and intersexuality in *M. domestica*. *Drosophila Information Service*, 28:135–136, 1954.

Miller, D. D. Mating behavior in *Drosophila affinis* and *Drosophila algonguin*. *Evolution*, 4:123–134, 1950.

Miller, D. S. Effects of low-level radiation on audiogenic convulsive seizures in mice. In *Psychophysiologie Neuropharmacologie et Biochimie de la Crise Audiogene*, Colloques Internationaux du Centre National de la Recherche Scientifique, no. 112, pp. 289–305, 1963.

Miller, R. S. Activity rhythms in the wood mouse *Apodemus sylvaticus* and the bank vole *Clethrionomys glarcolus*. *Proc. Zool. Soc. London*, 125:505–519, 1955.

Misra, R. K., and Reeve, E. C. R. Genetic variation of relative growth rates in *Notonecta undulata*. I. The relation of femur length to body length. *Genet. Res.*, 5:384–396, 1964.

Mitchell, R. G., and Rhaney, K. Congenital adrenal hypoplasia in siblings. *Lancet*, 1:488–492, 1959.

Mitoma, C., Auld, R. M., and Udenfriend, S. On the nature of enzymatic defect in phenylpyruvic oligophrenia. *Proc. Soc. Exp. Biol. Med.*, 94:634–635, 1957.

Montagu, M. F. A. *Man's Most Dangerous Myth: The Fallacy of Race*, 3d ed. New York: Harper & Row, Publishers, Incorporated, 1952.

Montgomery, K. C., and Segall, M. Discrimination learning based upon the exploratory drive. *J. Comp. Physiol. Psychol.*, 48:225–228, 1955.

Moray, N., and Arnold, P. Confirmation of apparent genetical assimilation of behaviour in *Drosophila melanogaster*. *Nature*, 204:504, 1964.

———— and Connolly, K. A possible case of genetic assimilation of behaviour. *Nature*, 199:358–360, 1963.

Mordkoff, A. M., and Fuller, J. L. Variability in activity within inbred and crossbred mice. *J. Hered.*, 50:6–8, 1959.

Morgan, C. L. *Animal Behavior*. London: E. Arnold (Publishers) Ltd., 1900.

Morgan, P. A study in perceptual differences among cultural groups in Southern Africa, using tests of geometric illusion. *J. Natl. Personnel Res.*, 8:39–43, 1959.

Morgan, T. H., and Bridges, C. B. The origin of gynandromorphs. *Carnegie Inst. Wash. Publ.*, no. 278, pp. 22–23, 1919.

Morrell, F. Lasting changes in synaptic organization produced by continuous neural bombardment. In A. Fessard, R. W. Gerard, J. Konovski, and J. F. Delafresnaye (eds.), *Brain Mechanisms and Learning*, pp. 375–389. Oxford: Blackwell Scientific Publications, Ltd., 1961.

Morris, D. "Typical intensity" and its relationship to the problem of ritualization. *Behaviour*, 11:1–13, 1957.

Morton, N. E. The components of genetic variability. In E. Goldschmidt (ed.), *The Genetics of Migrant and Isolate Populations*, pp. 225–236. Baltimore: The Williams & Wilkins Company, 1963.

———— and Chung, C. S. Formal genetics of muscular dystrophy. *Amer. J. Hum. Genet.*, 11:360–379, 1959.

Mosier, H. D. Hypoplasia of the pituitary and adrenal cortex: report of occurrence in twin siblings and autopsy findings. *J. Pediat.*, 48:633–639, 1956.

Motulsky, A. G. Genetics and endocrinology. In R. H. Williams (ed.), *Textbook of Endocrinology*, 3d ed., pp. 1087–1120. Philadelphia: W. B. Saunders Company, 1962a.

————. Controller genes in synthesis of human haemoglobin. *Nature*, 194:607–609, 1962b.

————. Current concepts of the genetics of the thalassemias. *Cold Spring Harbor Sympos. Quant. Biol.*, 29:399–412, 1964.

Mourant, A. E. *The Distribution of Human Blood Groups*. Oxford: Blackwell Scientific Publications, Ltd., 1954.

Mourer, S. L. Effects of incoming liquid food on the expression of hygienic behavior of honey bees *Apis mellifera* L. Master's thesis. Ohio State University, 1964.

Müller, H. J. Mental traits and heredity. *J. Hered.*, 16:433–448, 1925.

————. Isolating mechanisms, evolution and temperature. *Biol. Sympos.*, 6:71–122, 1942.

Myers, C. S. A study of Papuan hearing. *Arch. Otology*, 31:283–288, 1902.

————. Hearing. In *Report of the Cambridge Anthropological Expedition to Torres Straits*, vol. 2, pt. II, pp. 141–168. London: Cambridge University Press, 1903a.

————. Smell. In *Report of the Cambridge Anthropological Expedition to Torres Straits*, vol. 2, pt. II, pp. 169–185. London: Cambridge University Press, 1903b.

————. Taste. In *Report of the Cambridge Anthropological Expedition to Torres Straits*, vol. 2, pt. II, pp. 186–188. London: Cambridge University Press, 1903c.

————. Reaction times. In *Report of the Cambridge Anthropological Expedition to Torres Straits*, vol. 2, pt. II, pp. 205–223. London: Cambridge University Press, 1903d.

Nachman, M. The inheritance of saccharin preference. *J. Comp. Physiol. Psychol.*, 52:451–457, 1959.

Nadler, H. L., Justice, P., and Hsia, D. Y.-Y. Studies on the tryptophan metabolism in phenylketonuria. To be published.

Nalbandov, A. V. (ed.). *Advances in Neuroendocrinology*. Urbana, Ill.: University of Illinois Press, 1963.

Napier, J. R. The evolution of the hand. *Sci. Amer.*, 207:56–62, 1962.

———— and Barnicot, N. A. (eds.). *The Primates*. Symposia of the Zoological Society of London, no. 10. London: The Zoological Society of London, 1963.

Naruse, H., Kato, M., Kurokawa, M., Haba, R., and Yabe, T. Metabolic defects in a convulsive strain of mouse. *J. Neurochem.*, 5:359–369, 1960.

Nauta, W. J. H. Central nervous organization and endocrine motor system. In A. V. Nalbandov (ed.), *Advances in Neuroendocrinology*, pp. 5–28. Urbana, Ill.: University of Illinois Press, 1963.

———— and Kuypers, H. G. J. M. Some ascending pathways in the brain stem reticular formation. In H. W. Magoun (chairman), J. D. French, H. H. Jasper, R. N. Delong, R. W. Gerard, A. A. Ward, Jr., R. Knighton, L. Proctor, W. Noshay, and R. T. Costello (eds.), Henry Ford Hospital—International Symposium: *Reticular Formation of the Brain*, pp. 3–30. Boston: Little, Brown and Company, 1958.

Naylor, A. F. Mating systems which could increase heterozygosity for a pair of alleles. *Amer. Nat.*, 96:51–60, 1962.

Needham, J. Chemical heterogony and the ground plan of animal growth. *Biol. Rev.*, 9:79–109, 1934.

Neel, J. V., and Post, R. H. Transitory "positive" selection for color-blindness. *Eugen. Quart.*, 10:33–35, 1963.

Nelson, E. B., and Jay, S. C. Installation and maintenance of miniature colonies. *Bull. Entomol. Soc. Amer.*, 10:170, 1964.

Newman, A. E., Redgate, E. S., and Farrell, G. The effects of diencephalic-mesencephalic lesions on aldosterone and hydrocortisone secretion. *Endocrinology*, 63:723–737, 1958.

Newman, H. H., Freeman, F. N., and Holzinger, K. J. *Twins: A Study of Heredity and Environment*. Chicago: University of Chicago Press, 1937.

Nice, M. M. The role of territory in bird life. *Amer. Mid. Nat.*, 26:441–487, 1941.

Nicholls, J. G., and Kuffler, S. W. Extracellular space as a pathway for exchange between blood and neurons in the central nervous system of the leech: ionic composition of glial cells and neurons. *J. Neurophysiol.*, 27:645–671, 1964.

———— and ————. Na and K content of glial cells and neurons determined by flame photometry in the central nervous system of the leech. *J. Neurophysiol.*, 28:519–525, 1965.

Nichols, J. R. The effect of cross-fostering on addiction-prone and addiction-resistant strains of rats. *Amer. Psychol.* (abstr.), 19:529, 1964.

Nichols, R. C. The inheritance of general and specific ability. *Natl. Merit Scholarship Corp. Res. Rep.*, 1:1–9, 1965.

Nye, W. P. Observations on the behaviour of bees in a temperature-controlled-environment room. *J. Apicult. Res.*, 1:28–32, 1962.

————. Personal communication. 1965.

———— and Mackensen, O. Preliminary report on selection and breeding of honeybees for alfalfa pollen collection. *J. Apicult. Res.*, 4:43–48, 1965.

Oberly, H. S. A comparison of the spans of "attention" and memory. *Amer. J. Psychol.*, 40:295–302, 1928.

O'Connor, N. Imbecility and color blindness. *Amer. J. Ment. Def.*, 62:83–87, 1957.

———— and Hermelin, B. *Speech and Thought in Severe Subnormality* (an experimental study). New York: Pergamon Press, 1963.

Oehninger, M. Ueber Kerngrössen bei Bienen. *Verh. Phys.-Med. Ges. Wurzburg*, 42:135–140, 1913.

Opfinger, E. Über die Orientierung der Bienen an der Futterquelle. *Z. Vergl. Physiol.*, 15:431–487, 1931.

————. Zur Psychologie der Duftdressuren bei Bienen. *Z. Vergl. Physiol.*, 31:441–453, 1949.

Ottinger, D. R., Denenberg, V. H., and Stephens, M. W. Maternal emotionality, multiple mothering, and emotionality in maturity. *J. Comp. Physiol. Psychol.*, 56:313–317, 1963.

Pace, N., Schaffer, F. L., Elmadjian, F., Minard, D., Davis, S. W., Kilbuck, J. H., Walker, E. L., Johnston, M. E., Zilinsky, A., Gerard, R. W., Forsham, P. H., and Taylor, J. G. *Physiological Studies on Infantrymen in Combat.* Berkeley, Calif.: University of California Press, 1956.

Paigen, K., and Ganschow, R. Genetic factors in enzyme realization. *Brookhaven Sympos. Biol.*, 18:99–115, 1965.

Paine, R. S. The variability in manifestations of untreated patients with phenylketonuria (phenylpyruvic oligophrenia). *Pediatrics*, 20:290–301, 1957.

———— and Hsia, D. Y.-Y. The dietary phenylalanine requirements and tolerances of phenylketonuric patients. *Amer. Med. Ass. J. Dis. Child.*, 94:224–230, 1957.

Pare, C. M. B., Sandler, M., and Stacey, R. S. 5-hydroxytryptamine deficiency in phenylketonuria. *Lancet*, 1:551–553, 1957.

Park, O. W. Disease resistance and American foulbrood. *Amer. Bee J.*, 76:12–14, 1936.

————. Testing for resistance to American foulbrood in honeybees. *J. Econ. Entomol.*, 30:504–512, 1937.

————. Is there a best race of bees? *Amer. Bee J.*, 78:366–368, 377, 414–417, 1938.

————. The honey-bee colony-life history. In R. A. Grout (ed.), *The Hive and the Honey Bee*, pp. 21–78. Hamilton, Ill.: Dadant and Sons, Inc., 1949.

Parkes, A. S., and Bruce, H. M. Olfactory stimuli in mammalian reproduction. *Science*, 134:1049–1054, 1961.

Parsons, P. A. A diallel cross for mating speeds in *Drosophila melanogaster*. *Genetica*, 35:141–151, 1964.

Partington, M. W. The early symptoms of phenylketonuria. *Pediatrics*, 27:465–473, 1961.

Patterson, J. T., and Stone, W. S. *Evolution in the Genus Drosophila*. New York: The Macmillan Company, 1952.

Patton, R. L. *Introductory Insect Physiology*. Philadelphia: W. B. Saunders Company, 1963.

Pauling, L. Abnormality of hemoglobin molecules in hereditary hemolytic anemias. *Harvey Lectures*, 49:218, 1955.

————, Itano, H. A., Singer, S. J., and Wells, I. C. Sickle cell anemia, a molecular disease. *Science*, 110:543–548, 1949.

Pearson, E. S., and Hartley, H. O. *Biometrika Tables for Statisticians*, vol. I. London: Cambridge University Press, 1958.

Peer, D. F. Multiple mating of queen honey bees. *J. Econ. Entomol.*, 49:741–743, 1956.

————. Further studies on the mating range of the honey bee, *Apis mellifera* L. *Canad. Entomol.*, 89:108–110, 1957.

Penfield, W., and Rasmussen, T. *The Cerebral Cortex of Man*. New York: The Macmillan Company, 1950.

———— and Roberts, L. *Speech and Brain-Mechanisms*. Princeton, N.J.: Princeton University Press, 1959.

Penrose, L. S. Discussion. In *The Race Concept*. Paris: UNESCO, 1951.

Perdeck, A. C. The isolating value of specific song patterns in two sibling species of grasshoppers (*Chorthippus brunneus* Thunb. and *C. biguttulus* L.). *Behaviour*, 12:1–75, 1958.

Persky, H. Adrenocortical function in anxious human subjects: the disappearance of hydrocortisone from plasma and its metabolic fate. *J. Clin. Endocrinol. Metab.*, 17:760–765, 1957.

————, Hamburg, D., Basowitz, H., Grinker, R. R., Sabshin, M., Korchin, S. J., Herz, M., Board, F. A., and Heath, H. A. Relation of emotional responses and changes in plasma hydrocortizone levels after stressful interview. *Arch. Neurol. Psychiat.*, 79:434, 1958.

————, Korchin, S. J., Basowitz, H., Board, F. A., Sabshin, M., Hamburg, D. A., and Grinker, R. R. Effect of two psychological stresses on adrenocortical function. *Arch. Neurol. Psychiat.*, 81:219–226, 1959.

Pessotti, I. Alcune misure di relazioni temporali in una discriminazione in *Mellipona seminigra merrillae*. *Rassegna di Psicologia Generale e Clinica*, 6:3–18, 1963.

————. Estudo sôbre a aprendizagem e extrinção de uma discriminação em *Apis mellifera*. *Jornal Brasileiro Psicologia*, 1:77–93, 1964.

Petit, C. La déterminisme génétique et psycho-physiologique de la compétition sexuelle chez *Drosophila melanogaster*. *Bull. Biol.*, 92:248–329, 1958.

Pettigrew, T. F. *The Psychology of the Negro American*. Princeton, N.J.: D. Van Nostrand Company, Inc., 1964.

Phenylketonuria: Low phenylalanine dietary management with Lofenalac. Evansville, Ind.: Mead Johnson and Company, 1958.

Phillips, E. F. Variation and correlation in the appendages of the honeybee. *Cornell University Agr. Exp. Sta.*, Memoir 121, Ithaca, N.Y., 1929.

Pittendrigh, C. S. On temperature independence in the clock system controlling emergence in *Drosophila*. *Proc. Natl. Acad. Sci.*, 40:1018–1029, 1954.

————. Adaptation, natural selection and behavior. In A. Roe and G. G. Simpson (eds.), *Behavior and Evolution*, pp. 390–416. New Haven, Conn.: Yale University Press, 1958.

Plateaux-Quénu, C. Biology of *Halictus marginatus* Brullé. *J. Apicult. Res.*, 1:41–51, 1962.

Plath, O. E. *Bumblebees and Their Ways*. New York: The Macmillan Company, 1934.

Polani, P. E. Chromosome abnormalities as a cause of defective development. In D. Richter, J. M. Tanner, L. Taylor, and O. L. Zangwill (eds.), *Aspects of Psychiatric Research*, pp. 154–193. London: Oxford University Press, 1962.

Polidora, V. J., Boggs, D. E., and Waisman, H. A. A behavioral defect associated with phenylketonuria in rats. *Proc. Soc. Exp. Biol. Med.*, 113:817–820, 1963.

Pons, J. A contribution to the heredity of the PTC taste character. *Ann. Hum. Genet.*, 24:153–160, 1960.

Pontecorvo, G. *Trends in Genetic Analysis*. New York: Columbia University Press, 1958.

Post, R. H. Population differences in red and green color vision deficiency: a review, and a query on selection relaxation. *Eugen. Quart.*, 9:131–146, 1962.

Power, M. E. The effect of reduction in numbers of ommatidia upon the brain of *Drosophila melanogaster*. *J. Exp. Zool.*, 94:33–71, 1943.

Prader, A. Die Haufigkeit des kongenitalen adrenogenitalen Syndroms. *Helvet. Paediat. Acta*, 13:426–431, 1958.

———— and Siebenmann, R. E. Nebennereninsuffizienz bei kongenitaler Lipoidhyperplasie der Nebennieren. *Helvet. Paediat. Acta*, 11:569–595, 1957.

Pratt, R. T. C., Gardiner, D., Curzon, G., Piercy, M. F., and Cumings, J. N. Phenylalanine tolerance in endogenous depression. *Brit. J. Med. Psychol.*, 109:624, 1963.

Premack, D. Toward empirical behavior laws. I. Positive reinforcement. *Psychol. Rev.*, 66:219–233, 1959.

————. Reversibility of the reinforcement relation. *Science*, 136:255–257, 1962.

————. Prediction of the comparative reinforcement values of running and drinking. *Science*, 139:1062–1063, 1963.

Pressey, S. L., and Pressey, L. C. Development of the interest-attitude test. *J. Appl. Psychol.*, 17:1–16, 1933.

Price, B. Primary biases in twin studies. *Amer. J. Hum. Genet.*, 2:293–352, 1950.

Race and Science. New York: Columbia University Press, 1961.

The Race Concept. Paris: UNESCO, 1952.

Reed, S. C. Speculations about human albinism. *J. Hered.*, 56:64–66, 1965.

———— and Reed, E. W. Natural selection in laboratory populations of *Drosophila*. II. Competition between white-eye gene and its wild type allele. *Evolution*, 4:34–42, 1950.

Reeve, E. C. R. The variance of the genetic correlation coefficient. *Biometrics*, 11:357–374, 1955.

Reichlin, S. Neuroendocrinology. *New Engl. J. Med.*, 269:1128–1182, 1246–1250, 1296–1303, 1963.

Reinhardt, J. F. Some responses of honey bees to alfalfa flowers. *Amer. Nat.*, 86:257–275, 1952.

Rendel, J. M. Mating of ebony, vestigial and wild type *Drosophila melanogaster* in light and dark. *Evolution*, 5:226–230, 1951.

Renner, M. Über die Haltung von Bienen in geschlossenen künstlich beleuchteten Räumen. *Naturwissenschaften*, 42:539, 1955.

————. The contribution of the honey bee to the study of time-sense and astronomical orientation. *Cold Spring Harbor Sympos. Quant. Biol.*, 25:361–367, 1960.

Rensch, B. Die Abhangigkeit der Struktur und der Leistungen tierischer Gehirne von ihrer Grosse. *Naturwissenschaften*, 45:145–154, 1958.

————. Trends towards progress of brains and sense organs. *Cold Spring Harbor Sympos. Quant. Biol.*, 24:291–303, 1959.

Ressler, R. H. Parental handling in two strains of mice reared by foster parents. *Science*, 137:129–130, 1962.

————. Genotype-correlated parental influence in two strains of mice. *J. Comp. Physiol. Psychol.*, 56:882–886, 1963.

————. Environmental inheritance of exploratory behavior in mice. *Amer. Psychol.* (abstr.), 19:505, 1964.

Reynolds, J. W., and Ulstrom, R. A. Studies of cortisol metabolism in a case of the hypertensive form of congenital adrenal hyperplasia: demonstration of the absence of 11β-dehydroxylation. *J. Clin. Endocrinol. Metab.*, 23:191–196, 1963.

Reynolds, V., and Reynolds, F. Chimpanzees of the Budongo Forest. In I. DeVore (ed.), *Primate Behavior: Field Studies of Monkeys and Apes*. New York: Holt, Rinehart and Winston, Inc., 1965.

Ribbands, C. R. *The Behaviour and Social Life of Honeybees*. London: Bee Research Association, Ltd., 1953.

———— and Speirs, N. The adaptability of the homecoming honeybee. *Brit. J. Anim. Behav.*, 1:59–66, 1953.

Richter, C. The effects of domestication and selection on the behavior of the Norway rat. *J. Natl. Cancer Inst.*, 15:727, 1954.

Rife, D. C. Genetic studies of monozygotic twins. V. *J. Hered.*, 29:83–90, 1938.

Rikimaru, J. Taste deficiency for PTC with special reference to its hereditary nature. *Shinri-gaku Kenkyu*, 12:33–54, 1937.

Rivers, W. H. R. Physiology and psychology. In *Report of the Cambridge Anthropological Expedition to Torres Straits*, vol. 2, pt. I, London: Cambridge University Press, 1901.

————. Observations on the senses of the Todas. *Brit. J. Psychol.*, 1:321–396, 1905.

Roberts, E. The synapse as a biochemical self-organizing cybernetic unit. In K. Rodahl (ed.), *Nerve as a Tissue*. New York: Paul B. Hoeber, Inc., medical book department of Harper & Row, Publishers, Incorporated, 1966.

———— and Baxter, C. F. Neurochemistry. *Ann. Rev. Biochem.*, 32:513–552, 1963.

————, Harman, P. J., and Frankel, S. γ-Aminobutyric acid content and glutamic decarboxylase activity in developing mouse brain. *Proc. Soc. Exp. Biol. Med.*, 78:799, 1951.

————, Wein, J., and Simonsen, D. G. γ-Aminobutyric acid (γABA), vitamin B₆, and neuronal function. *Vitamins and Hormones*, 22:503–559, 1964.

Roberts, W. C. Multiple mating of queen bees proved by progeny and flight tests. *Gleanings Bee Culture*, 72:255–259, 303, 1944.

Robertson, A. The effect of inbreeding on the variation due to recessive genes. *Genetics*, 37:189–207, 1952.

————. Experimental design in the evaluation of genetic parameters. *Biometrics*, 15:219–226, 1959.

————. A theory of limits in artificial selection. *Proc. Roy. Soc. (London)*, (B)153:234–249, 1960.

Robertson, F. W. Studies in quantitative inheritance. V. Chromosome analyses of crosses between selected and unselected lines of different body size in *Drosophila melanogaster*. *J. Genet.*, 52:494–520, 1954.

————. Changing the relative size of the body parts of *Drosophila* by selection. *Genet. Res.*, 3:169–180, 1962.

Robinson, J. T. The genera and species of the Australopithecinae. *Amer. J. Phys. Anthrop.*, 12:181–200, 1954.

————. The origin and adaptive radiation of the Australopithecines. In G. Kurth (ed.), *Evolution and Hominisation*, pp. 120–140. Stuttgart: Gustav Fischer Verlag KG, 1962.

————. Adaptive radiation in the Australopithecines and the origin of man. In F. C. Howel and F. Bourlière (eds.), *African Ecology and Human Evolution*, pp. 385–416. Chicago: Aldine Publishing Company (Wenner-Gren Foundation for Anthropological Research, Viking Fund Publications in Anthropology, no. 36), 1963.

Rodgers, D. A., and McClearn, G. E. Alcohol preference of mice. In E. L. Bliss (ed.), *Roots of Behavior*, pp. 68–95. New York: Paul B. Hoeber, Inc., medical book department of Harper & Row, Publishers, Incorporated, 1962.

Roeder, K. D. An experimental analysis of the sexual behavior of the praying mantis. *Biol. Bull.*, 69:203–220, 1935.

———— (ed.). *Insect Physiology*. New York: John Wiley & Sons, Inc., 1953.

————. The nervous system. *Ann. Rev. Entomol.*, 3:1–18, 1958.

————. *Nerve Cells and Insect Behavior*. Cambridge, Mass.: Harvard University Press, 1963.

————, Tozian, L., and Weiant, E. A. Endogenous nerve activity and behavior in the mantis and cockroach. *J. Insect Physiol.*, 4:45–62, 1960.

Root, A. I. *The ABC and XYZ of Bee Culture*, rev. 32d ed. Medina, Ohio: A. I. Root Co., 1962.

Roots of Behavior: Animal Behavior. American Psychiatric Association Symposium. *Science*, 130:1344, 1959.

Rosenzweig, M. R., Krech, D., and Bennett, E. L. Brain chemistry and adaptive behavior. In H. F. Harlow and C. N. Woolsey (eds.), *Biological and*

Biochemical Bases of Behavior. Madison, Wis.: The University of Wisconsin Press, 1958.

————, ————, and ————. A search for relations between brain chemistry and behavior. *Psychol. Bull.,* 57:476–492, 1960.

Ross, H. H. *A Textbook of Entomology,* 2d ed. New York: John Wiley & Sons, Inc., 1956.

Rothenbuhler, W. C. Diploid male tissue as new evidence on sex determination in honey bees. *J. Hered.,* 48:160–168, 1957.

————. Genetics of a behavior difference in honeybees. *Proc. 10th Int. Congr. Genet.,* 2:252, 1958.

————. Unpublished data, 1959.

————. A technique for studying genetics of colony behavior in honey bees. *Amer. Bee J.,* 100:176, 198, 1960.

————. Behaviour genetics of nest cleaning in honey bees. I. Response of four inbred lines to disease-killed brood. *Anim. Behav.,* 12:578–583, 1964a.

————. Behavior genetics of nest cleaning in honey bees. IV. Response of F_1 and backcross generations to disease-killed brood. *Amer. Zool.,* 4:111–123, 1964b.

————, Gowen, J. W., and Park, O. W. Recent work in honey bee genetics with preliminary consideration on the breeding of bees for pollination. *North Central States Entomol. Proc.,* 32:45–46, 1953.

Royce, J. B. Factor theory and genetics. *Educ. Psychol. Measmt.,* 17:361–376, 1957.

Royce, P. C., and Sayers, G. Corticotropin releasing activity of a pepsin labile factor in the hypothalamus. *Proc. Soc. Exp. Biol. Med.,* 98:677–680, 1959.

Rulon, P. A semantic test of intelligence. *Proc. 1952 Cont. Test. Probl.,* pp. 84–92. Educational Testing Service, 1953.

Rümler, P. *Die Leislungen des Geschmacksinnes bei.* Zwillingen, Inaug. Diss., Iena, 1943. Cited in Piéron, H., *La Psychologie Différentielle,* 2d ed., p. 97. Paris: Presses Universitaires de France, 1962.

Rundquist, E. A. Inheritance of spontaneous activity in rats. *J. Comp. Psychol.,* 16:415–438, 1933.

Rushton, W. A. H. *Visual Pigments in Man.* Liverpool: Liverpool University Press, 1962.

Russell, W. L. Inbred and hybrid animals and their value in research. In G. D. Sneel (ed.), *Biology of the Laboratory Mouse,* pp. 325–348. New York: McGraw-Hill Book Company, 1941.

Ruttner, F. The mating of the honey bee. *Bee World,* 37:2–15, 23–24, 1956.

————. Races of bees. In R. A. Grout (ed.), *The Hive and the Honey Bee,* pp. 19–34. Hamilton, Ill.: Dadant and Sons, Inc., 1963.

Saffran, M. Mechanisms of adrenocortical control. *Brit. Med. Bull.,* 18:122–126, 1962.

———— and Saffran, J. Adenohypophysis and adrenal cortex. *Ann. Rev. Physiol.,* 21:403–444, 1959.

———— and Schally, A. V. The release of corticotrophin by anterior pituitary tissue in vitro. *Canad. J. Biochem. Physiol.,* 33:408–415, 1955.

Sakagami, S. F. Untersuchungen über die Arbeitsteilung in einem Zwergvolk der Honigbiene. Beiträge zur Biologie des Bienenvolkes, *Apis mellifera* L. I. *Japan. J. Zool.*, 11:117–185, 1953.

———— and Michener, C. D. *The Nest Architecture of the Sweat Bees (Halictinae).* Lawrence, Kans.: The University of Kansas Press, 1962.

———— and Takahashi, H. Beobachtungen über die gynandromorphen Honigbienen, mit besonderer Berücksichtigung ihrer Handlungen innerhalb des Volkes. *Ins. Soc.*, 3:513–529, 1956.

Saldanha, P. H., and Nacrur, J. Taste thresholds for phenylthiourea among Chileans. *Amer. J. Phys. Anthrop.*, 21:113–119, 1963.

Sato, M. Statistische Beobachtung der angebornen Farbensinnstörung. *Acta Soc. Ophthal. Jap.*, 38:2227–2230, 1935.

Sawin, P. B., and Crary, D. D. Genetic and physiological background of reproduction in the rabbit. II. Some racial differences in the pattern of maternal behavior. *Behaviour*, 6:128–146, 1953.

Sawrey, W. L., and Long, D. Strain and sex differences in ulceration in the rat. *J. Comp. Physiol. Psychol.*, 55:603–605, 1962.

Sax, K. The association of size differences with seed-coat pattern and pigmentation in *Phaseolus vulgaris. Genetics*, 8:552–560, 1923.

Sayers, G. Control and inhibition of adrenocortical secretion: introductory review. In A. R. Currie, T. Symington, and J. K. Grant (eds.), *The Human Adrenal Cortex*, pp. 181–184. Baltimore: The Williams & Wilkins Company, 1962.

————, Redgate, E. S., and Royce, P. C. Hypothalamus, adenohypophysis and adrenal cortex. *Ann. Rev. Physiol.*, 20:243–274, 1958.

Schaller, G. B. The orang-utan in Sarawak. *Zoologica*, 46:73–82, 1961.

————. *The Mountain Gorilla.* Chicago: The University of Chicago Press, 1963.

Schally, A. V., Saffran, M., and Simmerman, B. A corticotrophin-releasing factor (CRF). Partical purification and amino acid composition. *Biochem. J.*, 70:97–103, 1958.

Scharrer, E., and Scharrer, B. *Neuroendocrinology.* New York: Columbia University Press, 1963.

Schlesinger, K., and Mordkoff, A. M. Locomotor activity and oxygen consumption: variability in two inbred strains of mice and their F_1 hybrids. *J. Hered.*, 54:177–182, 1963.

Schmalhausen, I. I. *Factors of Evolution.* New York: McGraw-Hill Book Company, 1949.

Schmidt, J. La valeur de l'individu à titre de générateur appreciées suivant le methode du croisement diallèle. *C. R. Lab. Calsberg*, 14:1–33, 1919.

Schmidt, R. S. The evolution of nest building in *Apicotermes* (Isoptera). *Evolution*, 9:157–181, 1955a.

————. Termite (*Apicotermes*) nests—important ethological material. *Behaviour*, 8:344–356, 1955b.

————. *Apicotermes* nests. *Amer. Zool.*, 4:221–225, 1964.

Schmieder, R. G., and Whiting, P. W. Reproductive economy in the chalcidoid wasp *Melittobia. Genetics*, 32:29–37, 1947.

Schmitt, J. B. The comparative anatomy of the insect nervous system. *Ann. Rev. Entomol.*, 7:137–156, 1962.

Schneirla, T. C. Modifiability in insect behavior. In K. D. Roeder (ed.), *Insect Physiology*, pp. 723–747. New York: John Wiley & Sons, Inc., 1953.

————. Psychological comparison of insect and mammal. *Psychol. Beiträge*, 6:509–520, 1962.

Schöler, A. W. Z. *Ethnol.* 1880, bd XII, s. 67 (summarized in Rivers, 1901).

Schrödinger, E. *What Is Life? The Physical Aspect of the Living Cell.* New York: The Macmillan Company, 1946.

Schultz, A. H. Age changes, sex differences, and variability as factors in the classification of primates. In S. L. Washburn (ed.), *Classification and Human Evolution*, pp. 146–177. Chicago: Aldine Publishing Company (Wenner-Gren Foundation for Anthropological Research. Viking Fund Publications in Anthropology, no. 37), 1963.

Schwidetzky, I. (ed.). *Die neue Rassenkunde.* Stuttgart: Gustav Fischer Verlag KG, 1962.

Scott, J. P. Genetic differences in the social behavior of inbred strains of mice. *J. Hered.*, 33:11–15, 1942.

————. Social behavior, organization and leadership in a small flock of domestic sheep. *Comp. Psychol. Monogr.*, 18, 1945.

————. *Aggression.* Chicago: The University of Chicago Press, 1958.

————. Critical periods in behavioral development. *Science*, 138:949–958, 1962.

————. Genetics and the development of social behavior in dogs. *Amer. Zoologist*, 4:161–168, 1964.

———— and Fredericson, E. The cause of fighting in rats and mice. *Physiol. Zool.*, 24:273–309, 1951.

———— and Fuller, J. L. *Genetics and the Social Behavior of the Dog.* Chicago: The University of Chicago Press, 1965.

————, ————, and King, J. A. The inheritance of animal breeding cycles in hybrid Basenji and Cocker Spaniel dogs. *J. Hered.*, 50:254–261, 1959.

Scott-Moncrieff, R. The genetics and biochemistry of flower colour variation. *Ergebn. Enzymforsch.*, 8:277–306, 1939.

Segall, M. H., Campbell, D. T., and Herskovits, M. J. Cultural differences in the perception of geometric illusions. *Science*, 139:769–771, 1963.

————, ————, and ————. *The Influence of Culture on Visual Perception.* Indianapolis: The Bobbs-Merrill Company, Inc., 1966.

Seiger, M., and Kemperman, J. H. B. Unpublished results.

Selander, R. K. On mating systems and sexual isolation. *Amer. Nat.*, 99:129–141, 1965.

Semon, R. Die Mneme als erhaltendes Prinzip im Wochsel des organischen Geschehens. (Four editions between 1904 and 1914.)

Setekleiv, J., Skaug, O. E., and Kaada, B. R. Increase of plasma 17-hydroxy-cortico-steroids by cerebral cortical and amygdaloid stimulation in the cat. *J. Endocrinol.*, 22:119–127, 1961.

Setterfield, W., Schott, R. G., and Snyder, L. H. Studies in human inheritance. XV. The bimodality of the threshold curve for taste of phenyl-thio-carbamide. *Ohio J. Sci.*, 36:231–234, 1936.

Shaw, E. The development of schooling behavior in fishes. *Physiol. Zool.*, 33:79–86, 1960.

————. The development of schooling behavior in fishes. *Physiol. Zool.*, 34:263–273, 1961.

Shepard, T. H., Landing, B. H., and Mason, D. G. Familial Addison's disease: case reports of two sisters with corticoid deficiency unassociated with hypoaldosternoism. *Amer. J. Dis. Child.*, 97:154–162, 1959.

Sheppard, P. M. *Natural Selection and Heredity.* London: Hutchinson & Co. (Publishers), Ltd., 1958.

————. Some contributions to population genetics resulting from the study of the hepidoptera. *Adv. Genet.*, 10:165–216, 1961.

Sheppe, W. Systematic and ecological relations of *Peromyscus oreas* and *P. maniculatus. Proc. Amer. Phil. Soc.*, 105:421–446, 1961.

Shuey, A. M. *The Testing of Negro Intelligence,* 2d ed. Lynchburg, Va.: J. B. Bell Company, 1966.

Siegel, F. L., Roach, M. K., and Pomeroy, L. R. Plasma amino acid patterns in alcoholism: the effects of ethanol loading. *Proc. Natl. Acad. Sci.*, 51:605–611, 1964.

Siegel, P. B. Genetics of behavior: selection for mating ability in chickens. *Genetics,* 52:1269–1277, 1965.

Simons, E. L. A critical reappraisal of tertiary primates. In J. Buettner-Janusch (ed.), *Evolutionary and Genetic Biology of Primates,* vol. I, pp. 65–129. New York: Academic Press Inc., 1963a.

————. Some fallacies in the study of hominid phylogeny. *Science,* 141:879–889, 1963b.

————. The early relatives of man. *Sci. Amer.*, 211:50–62, 1964.

Simpson, G. G. *The Meaning of Evolution.* New Haven, Conn.: Yale University Press, 1949.

————. The Baldwin effect. *Evolution,* 7:110–117, 1953a.

————. *The Major Features of Evolution.* New York: Columbia University Press, 1953b.

————. *Principles of Animal Taxonomy.* New York: Columbia University Press, 1961.

————. The meaning of taxonomic statements. In S. L. Washburn (ed.), *Classification and Human Evolution,* pp. 1–31. Chicago: Aldine Publishing Company (Wenner-Gren Foundation for Anthropological Research, Viking Fund Publications in Anthropology, no. 37), 1963.

Sinnott, E. W., Dunn, L. C., and Dobzhansky, T. *Principles of Genetics,* 5th ed. New York: McGraw-Hill Book Company, 1958.

Skeller, E. Anthropological and ophthalmological studies on the Angmassalik Eskimos. Copenhagen, 1954.

Slobodkin, L. B. *Growth and Regulation of Animal Populations.* New York: Holt, Rinehart and Winston, Inc., 1961.

Sloper, J. C. Morphological aspects of the hypothalamic control of anterior pituitary function. In A. R. Currie, T. Symington, and J. K. Grant (eds.), *The Human Adrenal Cortex,* pp. 230–247. Baltimore: The Williams & Wilkins Company, 1962.

Slusher, M. A. Dissociation of adrenal ascorbic acid and corticosterone responses to stress in rats with hypothalamic lesions. *Endocrinology,* 63:412–419, 1958.

———— and Hyde, J. E. Inhibition of adrenal corticosteroid release by brain stem stimulation in cats. *Endocrinology,* 68:773–782, 1961a.

————— and —————. Effect of limbic stimulation on release of corticosteroids into the adrenal venous effluent of the cat. *Endocrinology*, 69:1080–1084, 1961b.

Smelik, P. G. Modifying effect of hypothalamic lesions on the adrenal stress response. *Acta Endocrinol. (Suppl.* 38), 28:84, 1958.

————— and Sawyer, C. H. Effects of implantation of cortisol into the brain stem or pituitary gland on the adrenal response to stress in the rabbit. *Acta Endocrinol.*, 41:561–570, 1962.

Smith, C., King, J. W. B., and Gilbert, N. Genetic parameters of British large white bacon pigs. *Anim. Prod.*, 4:128–143, 1962.

Smith, J. M. Sexual selection. In S. A. Barnett (ed.), *A Century of Darwin*, pp. 231–244. London: William Heinemann, Ltd., 1958.

Snell, G. D. Dwarf, a new Mendelian recessive character in the house mouse. *Proc. Natl. Acad. Sci.*, 15:733–734, 1929.

Snodgrass, R. E. *Principles of Insect Morphology*. New York: McGraw-Hill Book Company, 1935.

—————. *Anatomy of the Honey Bee*. Ithaca, N.Y.: Comstock Publishing Associates, 1956.

Snyder, D. P. Survival rates, longevity, and population fluctuations in the white-footed mouse, *Peromyscus leucopus*, in southeastern Michigan. *Univ. Mich., Mus. Zool., Misc. Publ.*, 95, 1956.

Snyder, L. H. Inherited taste deficiency. *Science*, 74:151–152, 1931.

—————. Studies in human inheritance. IX. The inheritance of taste deficiency in man. *Ohio J. Sci.*, 32:436–440, 1932.

Soffer, L. J., Dorfman, R. I., and Gabrilove, J. L. *The Human Adrenal Gland*. Philadelphia: Lea & Febiger, 1961.

Sonneborn, T. M. Nucleotide sequence of a gene: first complete specification. *Science*, 148:1410, 1965.

Southwick, C. H. Regulatory mechanisms of house mouse populations: social behavior affecting litter survival. *Ecology*, 36:627–634, 1955.

Speicher, B. R. Cell size and chromosomal types in *Habrobracon*. *Amer. Nat.*, 69:79–80, 1935.

————— and Speicher, G. K. The occurrence of diploid males in *Habrobracon brevicornis*. *Amer. Nat.*, 74:379–382, 1940.

Sperry, R. W. Development basis of behavior. In A. Roe and G. G. Simpson (eds.), *Behavior and Evolution*, pp. 128–139. New Haven, Conn.: Yale University Press, 1958.

—————. Lipides of the brain during early development in the rat. In H. Waelsch (ed.), *Biochemical Development of the Nervous System*. New York: Academic Press Inc., 1955.

Spickett, S. G. Genetic and developmental studies of a quantitative character. *Nature*, 199:870–873, 1963.

Spiess, E. B., and Langer, B. Chromosomal adaptive polymorphism in *Drosophila persimilis*. III. Mating propensity of homokaryotypes. *Evolution*, 15:535–544, 1961.

————— and —————. Mating speed control by gene arrangement carriers in *Drosophila persimilis*. *Evolution*, 18:430–444, 1964.

Spieth, H. T. Mating behavior within the genus *Drosophila* (Diptera). *Bull. Amer. Mus. Nat. Hist.*, 99:395–474, 1952.

————— and Hsu, T. C. The influence of light on seven species of the *Drosophila melanogaster* group. *Evolution,* 4:316–325, 1950.

Spuhler, J. N. Some genetic variations in American Indians. In W. S. Laughlin (ed.), *The Physical Anthropology of the American Indian,* pp. 177–202. New York: The Viking Fund, Inc., 1951.

—————. Some problems in the physical anthropology of the American Southwest. *Amer. Anthrop.,* 56:604–625, 1954.

—————. Somatic paths to culture. In J. N. Spuhler (ed.), *The Evolution of Man's Capacity for Culture,* pp. 1–13. Detroit: Wayne State University Press, 1959.

Stebbins, G. L. From gene to character in higher plants. *Amer. Scient.,* 53:104–126, 1965.

Stefferud, A. (ed.). *Insects—The Yearbook of Agriculture.* Washington, D.C.: U.S. Department of Agriculture, 1952.

Stempfel, R. S., Jr., and Engel, F. L. A congenital, familial syndrome of adrenocortical insufficiency without hypoaldosteronism. *J. Pediat.,* 57:443–451, 1960.

Stern, C. Le daltonisme lié au chromosome X a-t-il une localisation unique ou double? *J. Genet. Hum.,* 7:302–307, 1958.

—————. *Principles of Human Genetics,* 2d ed. San Francisco: W. H. Freeman and Company, 1960.

Stewart, H. F., and Keeler, C. E. A comparison of the intelligence and personality of moon-child albino and control Cuna Indians. *J. Genet. Psychol.,* 106:319–324, 1965.

Stockard, C. R. The artificial production of a single median cyclopian eye in the fish embryo by means of sea water solutions of magnesium chlorid. *Roux' Arch. Entw.-Mech.,* 23:249–258, 1907. Cited in Rudnick, D., Teleosts and birds. In B. H. Willier, P. A. Weiss, and V. Hamburger (eds.), *Analysis of Development,* pp. 297–314. Philadelphia: W. B. Saunders Company, 1955.

—————, Anderson, O. D., and James, W. T. *The Genetic and Endocrinic Basis for Differences in Form and Behavior.* Philadelphia: Wistar Institute, 1941.

Straus, W. L., Jr. The riddle of man's ancestry. *Quart. Rev. Biol.,* 24:200–223, 1949.

Stride, G. O. Investigations into the courtship behaviour of the male of *Hypolimnas misippus* (Lepidoptera, Nymphalidae) with special reference to the role of visual stimuli. *Brit. J. Anim. Behav.,* 5:153–167, 1957.

Strom, G. Visceral functions of the nervous system. *Ann. Rev. Physiol.,* 23:419–450, 1961.

Sturtevant, A. H. Intersexes in *Drosophila simulans. Science,* 51:352–353, 1920.

—————. A gene in *Drosophila melanogaster* that transforms females into males. *Genetics,* 30:297–299, 1945.

Suomalainen, E. Parthenogenesis in animals. *Adv. Genet.,* 3:194–253, 1950.

Sutton, H. E. *Genes, Enzymes, and Inherited Diseases.* New York: Holt, Rinehart and Winston, Inc., 1961.

Suwa, N., Yamashita, I., Owada, H., Shinohara, S., and Nakazawa, A. Psychic state and adrenocortical function: a psychophysiologic study of emotion. *J. Nerv. Ment. Dis.,* 134:268–276, 1962.

Synge, A. D. Pollen collection by honeybees (*Apis mellifera*). *J. Anim. Ecol.,* 16:122–138, 1947.

Taber, S. III. The frequency of multiple mating queen honey bees. *J. Econ. Entomol.*, 47:995–998, 1954.

———. Sperm distribution in the spermathecae of multiple-mated queen honey bees. *J. Econ. Entomol.*, 48:522–525, 1955.

———. Factors influencing the circadian flight rhythm of drone honey bees. *Ann. Entomol. Soc. Amer.*, 57:769–775, 1964.

——— and Wendel, J. Concerning the number of times queen bees mate. *J. Econ. Entomol.*, 51:786–789, 1958.

Takeda, K. Classical conditioned response in the honey bee. *J. Insect Physiol.*, 6:168–179, 1961.

Tanaka, K., Matsunaga, E., Murata, T. Y., and Takehara, K. Phenylketonuria in Japan. *Japan. J. Hum. Genet.*, 6:65–77, 1961.

Taylor, A. N., and Farrell, G. Effects of brain stem lesions on aldosterone and cortisol secretion. *Endocrinology*, 7:556–566, 1962.

Tepperman, J. *Metabolic and Endocrine Physiology.* Chicago: The Year Book Medical Publishers, Inc., 1962.

Terman, C. R. The influence of differential early social experience upon spatial distribution within population of prairie deermice. *Anim. Behav.*, 11:246–262, 1963.

Tessman, I. Genetic ultrafine structure in the T4 rII region. *Genetics*, 51:63–75, 1965.

Thiessen, D., and Nealey, V. Adrenocortical activity, stress, response and behavioral reactivity of five inbred mouse strains. *Endocrinology*, 71:227–232, 1962.

Thoday, J. M. Location of polygenes. *Nature*, 191:368–370, 1961.

Thompson, D'A. W. *On Growth and Form.* London: Cambridge University Press, 1917.

Thompson, V. C. Behaviour genetics of nest cleaning in honeybees. III. Effect of age of bees of a resistant line on their response to disease-killed brood. *J. Apicult. Res.*, 3:25–30, 1964.

Thompson, W. R. The inheritance of behavior: behavioral differences in fifteen mouse strains. *Canad. J. Psychol.*, 7:145–155, 1953.

———. The inheritance and development of intelligence. *Proc. Ass. Res. Nerv. Ment. Dis.*, 33:209–231, 1954.

———. Traits, factors and genes. *Eugen. Quart.*, 4:8–16, 1957a.

———. The significance of personality and intelligence tests in evaluation of population characteristics. In *The Nature and Transmission of the Genetic and Cultural Characteristics of Human Populations*, pp. 37–50. New York: Milbank Memorial Fund, 1957b.

———. Multivariate analysis in behavior genetics. In R. B. Cattell (ed.), *Handbook of Multivariate Experimental Psychology.* Chicago: Rand McNally & Company, 1966.

——— and Goldenberg, L. Some physiological effects of maternal adrenalin injection during pregnancy in rat offspring. *Psychol. Rep.*, 10:759–774, 1962.

———, ———, Watson, J., and Watson, M. Behavioral effects of maternal adrenalin injection during pregnancy in rat offspring. *Psychol. Rep.*, 12:279–284, 1963.

——— and Olian, S. Some effects on offspring behavior of maternal adrenalin injections during pregnancy in three inbred mouse strains. *Psychol. Rep.*, 8:87–90, 1961.

————, Watson, J., and Charlesworth, W. R. The effects of prenatal maternal stress on offspring behavior in rats. *Psychol. Monogr.*, 76(38), 1962.

Thorpe, W. H. Further experiments on olfactory conditioning in a parasitic insect. The nature of the conditioning process. *Proc. Roy. Soc. (London)*, (B)126:370–397, 1938.

————. Further experiments on pre-imaginal conditioning in insects. *Proc. Roy. Soc. (London)*, (B)127:424–433, 1939.

————. *Bird-song*. London: Cambridge University Press, 1961.

———— and Jones, F. G. W. Olfactory conditioning and its relation to the problem of host selection. *Proc. Roy. Soc. (London)*, (B)124:56–81, 1937.

———— and Zangwill, O. L. *Current Problems in Animal Behaviour*. London: Cambridge University Press, 1961.

Thouless, R. H. A racial difference in perception. *J. Soc. Psychol.*, 4:330–339, 1933.

————. Test unreliability and function fluctuation. *Brit. J. Psychol.*, 26:325–343, 1936.

Thurstone, L. L. *Multiple Factor Analysis*. Chicago: The University of Chicago Press, 1947.

Tinbergen, N. *The Study of Instinct*. London: Oxford University Press, 1951.

————. "Derived" activities; their causation, biological significance, origin and emancipation during evolution. *Quart. Rev. Biol.*, 27:1–32, 1952.

————. Behaviour, systematics and natural selection. *Ibis*, 101:318–330, 1959a.

————. Comparative studies of the behaviour of gulls (Laridae): a progress report. *Behaviour*, 15:1–70, 1959b.

————, Meeuse, B. J. D., Boerema, L. K., and Varossieau, W. W. Die Balz des Samtfalters, *Eumenis* (=*Satyrus*) *semele* (L.). *Z. Tierpsychol.*, 5:182–226, 1942.

Tobias, P. V. The Olduvai Bed 1 hominine with special reference to its cranial capacity. *Nature*, 202:3–4, 1964.

Todd, F. E., and Bishop, R. K. Trapping honeybee-gathered pollen and factors affecting yields. *J. Econ. Entomol.*, 33:866–870, 1940.

Tolman, E. C. The inheritance of maze-learning ability in rats. *J. Comp. Psychol.*, 4:1–18, 1924.

————, Tryon, R. C., and Jeffress, L. A. A self-recording maze with an automatic delivery table. *Univ. Calif. Publ. Psychol.*, 4:99–112, 1929.

Tomkins, G., and McGuire, J. The adrenogenital syndrome. In J. Stanbury, J. Wyngaarden, and D. Fredrickson (eds.), *The Metabolic Basis of Inherited Disease*, pp. 637–676. New York: McGraw-Hill Book Company, 1960.

Tomkins, S. S., and Miner, J. B. *The Picture Arrangement Test*. New York: Springer Publishing Company, Inc., 1957.

Townsend, G. F., and Shuel, R. W. Some recent advances in apicultural research. *Ann. Rev. Entomol.*, 7:481–500, 1962.

Triasko, V. V. Signs indicating the mating of queens. *Pchelovodstvo*, 11:25–31 (in Russian); English abstract by M. Simpson in *Bee World*, 34:14, 1951.

Trump, R. Unpublished data, 1961.

Tryon, R. C. The genetics of learning ability in rats. *Univ. Calif. Publ. Psychol.*, 4:71–89, 1929.

————. Genetic differences in maze-learning ability in rats. *Yearbook Natl. Soc. Stud. Educ.*, 39(I):111–119, 1940.

————. Behavior genetics in social psychology. *Amer. Psychol.*, 12:453 (title only), 1957.

Tschumi, P. Über den Werbetanz der Bienen bei nahen Trachtquellen. *Schweiz. Bienen-Ztg.*, 73:129–134, 1950.

Tucker, K. W., and Laidlaw, H. H. Compound inseminations to abbreviate tests for allelism. *J. Hered.*, 56:127–130, 1965.

Tuttle, R. H. A study of the chimpanzee hand with comments on hominoid evolution. Ph.D. dissertation. University of California, Berkeley, 1965.

Tyler, L. *The Psychology of Human Differences.* New York: Appleton-Century-Crofts, Inc., 1956.

Uyeno, E. T. Hereditary and environmental aspects of dominant behavior in the albino rat. *J. Comp. Physiol. Psychol.*, 53:138–141, 1960.

Valls, A. M. Estudio antropogenetica de la capacidad gustatira para la fenitiocarbamida. *Fac. Cienc. Univ. Madrid*, 1958.

Vandenberg, S. G. The hereditary abilities study. *Eugen. Quart.*, 3:94–99, 1956.

————. The hereditary abilities study: hereditary components in a psychological test battery. *Amer. J. Hum. Genet.*, 14:220–237, 1962.

———— (ed.). *Methods and Goals in Human Behavior Genetics.* New York: Academic Press Inc., 1965.

Vanderdonck, R., and Verriest, G. Femme protanomale et hétérozygote mixte (gènes de la protanomalie et de la deutéranopie en position de repulsion) ayant deux fils deutéranopes, un fils protanomale et deux fils normaux. *Biotypologie*, 21:110–120, 1960.

Velthuis, H. H. W., and van Es, J. Some functional aspects of the mandibular glands of the queen honeybee. *J. Apicult. Res.*, 3:11–16, 1964.

Vernon, J. A., and Butsch, J. Effect of tetraploidy on learning and retention in the salamander. *Science*, 125:1033–1034, 1957.

Verplanck, W. S. Since learned behaviour is innate, and vice versa, what now? *Psychol. Rev.*, 62:139–144, 1955.

Vicari, E. M. Mode of inheritance of reaction time and degree of learning in mice. *J. Exp. Zool.*, 54:139–144, 1929.

————. Fatal convulsive seizures in the DBA mouse strain. *J. Psychol.*, 32:79–97, 1951.

Vogt, M. The control of the secretion of corticosteroids. In F. Clark and J. K. Grant (eds.), *The Biosynthesis and Secretion of Adrenocortical Steroids*, pp. 85–95. London: Cambridge University Press, 1960.

Von Békésy, G., and Rosenblith, W. A. The mechanical properties of the ear. In S. S. Stevens (ed.), *Handbook of Experimental Psychology*, pp. 1075–1115. New York: John Wiley & Sons, Inc., 1951.

Von Skramlik, E. Verebungsforschungen auf dem Gebiete des Geschmacksinnes. *Ienaische Z. Med. Naturwiss.*, 50:1943. Cited in Piéron, H., *La Psychologie Différentielle*, 2d ed., p. 97. Paris: Presses Universitaires de France, 1962.

Vowles, D. M. Neural mechanisms in insect behaviour. In W. H. Thorpe and O. L. Zangwill (eds.), *Current Problems in Animal Behaviour*, pp. 5–29. London: Cambridge University Press, 1961a.

————. The physiology of the insect nervous system. *Int. Rev. Neurobiol.*, 3:349–373, 1961b.

Waardenburg, P. J., Franceschetti, A., and Klein, D. *Genetics and Ophthalmology*, vol. 1. Springfield, Ill.: Charles C Thomas, Publisher, 1961.

————, ————, and ————. *Genetics and Ophthalmology*, vol. 2. Springfield, Ill.: Charles C Thomas, Publisher, 1963.

Waddington, C. H. Genetic assimilation of an acquired character. *Evolution*, 7:118–126, 1953.

————. *The Strategy of the Genes*. New York: The Macmillan Company, 1957.

————. Canalization of development and genetic assimilation of acquired characters. *Nature*, 183:1654–1655, 1959.

————. Genetic assimilation. *Adv. Genet.*, 10:257–293, 1961.

————, Woolf, B., and Perry, M. Environment selection by *Drosophila* mutants. *Evolution*, 8:89–96, 1954.

Wadeson, R. W., Mason, J. W., Hamburg, D. A., and Handlon, J. H. Plasma and urinary 17-OHCS responses to motion pictures. *Arch. Gen. Psychiat.*, 9:146–156, 1963.

Walker, M. C. Cytoplasmic bud formation in Hymenopteran spermatogenesis. *Nature*, 163:645–646, 1949.

Wallace, B., and Dobzhansky, T. Experiments on sexual isolation in *Drosophila*. VIII. Influence of light on the mating behavior of *Drosophila subobscura*, *Drosophila persimilis* and *Drosophila pseudoobscura*. *Proc. Natl. Acad. Sci.*, 32:226–234, 1946.

Wallace, H. W., Moldave, K., and Meister, A. Studies on conversion of phenylalanine to tyrosine in phenylpyruvic oligophrenia. *Proc. Soc. Exp. Biol. Med.*, 94:632–633, 1957.

Wallace, M. E. The relative homozygosity of inbred lines and closed colonies. *J. Theoret. Biol.*, 9:93–116, 1965.

Washburn, S. L. (ed.). *Classification and Human Evolution*. Chicago: Aldine Publishing Company (Wenner-Gren Foundation for Anthropological Research, Viking Fund Publications in Anthropology, no. 37), 1963.

———— and Avis, V. Evolution and human behavior. In A. Roe and G. G. Simpson (eds.), *Behavior and Evolution*, pp. 421–436. New Haven, Conn.: Yale University Press, 1958.

———— and Howell, F. C. Human evolution and culture. In S. Tax (ed.), *Evolution After Darwin*, vol. 2, *The Evolution of Man*, pp. 33–56. Chicago: The University of Chicago Press, 1960.

Watson, J. B. *Behavior: An Introduction to Comparative Psychology*. New York: Holt, Rinehart and Winston, Inc., 1914.

Watson, J. D., and Crick, F. H. C. The structure of DNA. *Cold Spring Harbor Sympos. Quant. Biol.*, 17:123–131, 1953.

Watson, L. R. Controlled mating in honeybees. *Quart. Rev. Biol.*, 3:377–390, 1928.

Watson, M. L. The inheritance of epilepsy and waltzing in *Peromyscus*. *Univ. Mich. Cont. Lab. Vert. Genet.*, no. 11, pp. 1–24, 1939.

Wearden, S. Alternative analyses of the diallel cross. *Heredity*, 19:669–680, 1964.

Weaver, N. The foraging behavior of honeybees on hairy vetch. Foraging methods and learning to forage. *Ins. Soc.* 3:537–549, 1956.

———. The foraging behavior of honeybees on hairy vetch. II. Foraging methods and learning to forage. *Ins. Soc.*, 4:43–57, 1957a.

———. Effects of larval age on dimorphic differentiation of the female honeybee. *Ann. Entomol. Soc. Amer.*, 50:283–294, 1957b.

———. The physiology of caste determination. *Ann. Rev. Entomol.*, 11:79–102, 1966.

Wechsler, D. *The Range of Human Capacities*, rev. ed. Baltimore: The Williams & Wilkins Company, 1952.

Wecker, S. C. The role of early experience in habitat selection by the prairie deer mouse, *Peromyscus maniculatus bairdii*. *Ecol. Monogr.*, 33:307–325, 1963.

Weesner, F. M. Evolution and biology of the termites. *Ann. Rev. Entomol.*, 5:153–170, 1960.

Weidenreich, F. The Skull of Sinanthropus Pekinensis Palaeontologia Sinica. New Series D, no. 10, Whole Series no. 127. Pehpei, Chungking: Geological Survey of China, 1943.

Weir, M. W., and DeFries, J. C. Prenatal maternal influence on behavior in mice: evidence of a genetic basis. *J. Comp. Physiol. Psychol.*, 58:412–417, 1964.

Weiss, K. Versuche mit Bienen und Wespen in farbigen Labyrinthen. *Z. Tierpsychol.*, 10:29–44, 1953.

Weitzman, M. (ed.). *Bibliographia Neuroendocrinologica*, vol. 1, no. 1. New York: Albert Einstein College of Medicine, Department of Anatomy, 1964.

Wenner, A. M. Division of labor in a honey bee colony—a Markov process? *J. Theoret. Biol.*, 1:324–327, 1961.

———. Sound production during the waggle dance of the honey bee. *Anim. Behav.*, 10:79–95, 1962a.

———. Communication with queen honey bees by substrate sound. *Science*, 138:446–447, 1962b.

———. Sound communication in honeybees. *Sci. Amer.*, 210:116–122, 124, 1964.

——— and Johnson, D. L. Simple conditioning in honey bees. *Anim. Behav.*, 14:149–155, 1966.

Wheeler, W. M. *Social Life Among the Insects*. New York: Harcourt, Brace & World, Inc., 1923.

———. *The Social Insects*. New York: Harcourt, Brace & World, Inc., 1928.

Wherry, R. J. Determination of the specific components of maze-ability for Tryon's bright and dull rats by means of factorial analysis. *J. Comp. Psychol.*, 32:237–252, 1941.

White, M. J. D. *Animal Cytology and Evolution*. London: Cambridge University Press, 1954.

Whiting, P. W. Reproductive reactions of sex mosaics of a parasitic wasp, *Habrobracon juglandis*. *J. Comp. Psychol.*, 14:345–363, 1932.

———. Multiple alleles in complementary sex determination of *Habrobracon*. *Genetics*, 28:365–382, 1943.

Whitten, W. K. Modification of the oestrous cycle of the mouse by external stimuli associated with the male. *J. Endocrinol.*, 13:399–404, 1956.

————. Occurrence of an oestrus in mice caged in groups. *J. Endocrinol.,* 18:102–107, 1959.

Wigglesworth, V. B. *The Principles of Insect Physiology,* 4th ed. New York: E. P. Dutton & Co., Inc., 1950.

Wilkins, L. Congenital virilizing adrenal hyperplasia. In A. R. Currie, T. Symington, and J. K. Grant (eds.), *The Human Adrenal Cortex,* pp. 360–370. Baltimore: The Williams & Wilkins Company, 1962.

Willham, R. L., Cox, D. F., and Karas, G. G. Genetic variation in a measure of avoidance learning in swine. *J. Comp. Physiol. Psychol.,* 56:294–297, 1963.

Williams, C. M., and Reed, S. C. Physiological effects of genes: the flight of Drosophila considered in relation to gene mutations. *Amer. Nat.,* 78:214–223, 1944.

Williams, R. H., and Bakke, J. L. The thyroid. In R. H. Williams (ed.), *Textbook of Endocrinology,* 3d ed., pp. 96–281. Philadelphia: W. B. Saunders Company, 1962.

Wilson, E. O. The social biology of ants. *Ann. Rev. Entomol.,* 8:345–368, 1963.

———— and Bossert, W. H. Chemical communication among animals. *Recent Progr. Hormone Res.,* 19:673–718, 1963.

Winston, H. D. Heterosis and learning in the mouse. *J. Comp. Physiol. Psychol.,* 57:279–283, 1964.

Witt, G., and Hall, C. S. The genetics of audiogenic seizure in the house mouse. *J. Comp. Physiol. Psychol.,* 42:58–63, 1949.

Wolf, E. Über das Heimkehrvermögen der Bienen. *Z. Vergl. Physiol.,* 3:615–691, 1926.

————. Über das Heimkehrvermögen der Bienen. *Z. Vergl. Physiol.,* 6:221–254, 1927.

Wolstenholme, D. R., and Thoday, J. M. Effects of disruptive selection. VII. A third polymorphism. *Hered.,* 18:413–431, 1963.

Woodger, J. H. What do we mean by "inborn"? *Brit. J. Phil. Sci.,* 3:319–326, 1953.

Wood-Gush, D. G. M. The courtship of the brown leghorn cock. *Brit. J. Anim. Behav.,* 2:95–102, 1954.

Woodrow, A. W., and Holst, E. C. The mechanism of colony resistance to American foulbrood. *J. Econ. Entomol.,* 35:892–895, 1942.

Woodworth, R. S. Racial differences in mental traits. *Science,* 31:171–186, 1910.

————. Heredity and environment: a critical survey of recently published materials on twins and foster children. *Soc. Sci. Res. Council., Bull.* 47, 1941.

Wooley, D. W., and van der Hoeven, T. Alteration in learning ability caused by changes in cerebral serotonin and catechol amines. *Science,* 139:610–611, 1963.

———— and ————. Serotonin deficiency in infancy as one cause of a mental defect in phenylketonuria. *Science,* 144:883–884, 1964.

Woolf, C. M. Albinism among Indians in Arizona and New Mexico. *Amer. J. Hum. Genet.,* 17:23–35, 1965.

Woolf, L. I. Excretion of conjugated phenylacetic acid in phenylketonuria. *Biochem. J.,* 49:IX, 1951.

————, Ounsted, C., Lee, D., Humphrey, M., Cheshire, N. M., and Steed, G. R. Atypical phenylketonuria in sisters with normal offspring. *Lancet,* 2:464–465, 1961.

Woolpy, J. H., and Schaefer, T., Jr. Exploration in the cockroach. Paper presented at Midwest Psychological Assoc. meeting, Chicago, 1962.

Woyke, J. Multiple mating of the honeybee queen (*Apis mellifica* L.) in one nuptial flight. *Bull. L'Acad. Polonaise Sci.,* II, 3:175–180, 1955.

————. The hatchability of "lethal" eggs in a two sex-allele fraternity of honeybees. *J. Apicult. Res.,* 1:6–13, 1962.

————. Drone larvae from fertilized eggs of the honeybee. *J. Apicult. Res.,* 2:19–24, 1963a.

————. What happens to diploid dronę larvae in a honeybee colony. *J. Apicult. Res.,* 2:73–75, 1963b.

————. Rearing and viability of diploid drone larvae. *J. Apicult. Res.,* 2:77–84, 1963c.

Wright, S. The relative importance of heredity and environment in determining the piebald pattern of guinea-pigs. *Proc. Natl. Acad. Sci.,* 6:320–332, 1920.

————. Systems of mating. I–V. *Genetics,* 6:111–123, 1921. Reprinted in J. L. Lush (ed.), *Systems of Mating and Other Reports.* Ames, Iowa: Iowa State University Press, 1958.

————. Evolution in Mendelian populations. *Genetics,* 16:97–159, 1931.

————. Adaptation and selection. In G. L. Jepsen, E. Mayr, and G. G. Simpson (eds.), *Genetics, Paleontology and Evolution,* pp. 365–389. Princeton, N.J.: Princeton University Press, 1949.

Wynne-Edwards, V. C. *Animal Dispersion in Relation to Social Behaviour.* Edinburgh and London: Oliver & Boyd Ltd., 1962.

Yamaguchi, H. G., Hull, C. L., Felsinger, J. M., and Gladstone, A. I. Characteristics of dispersions based on the pooled momentary reaction potentials ($_sE_R$) of a group. *Psychol. Rev.,* 55:216–238, 1948.

Yerkes, R. M. (ed.). Psychological examining in the U.S. Army. *Mem. Natl. Acad. Sci.,* 15, 1921.

Yuwiler, A., and Louttit, R. T. Effects of phenylalanine diet on brain serotonin in the rat. *Science,* 134:831–832, 1961.

Zmarlicki, C., and Morse, R. A. Drone congregation areas. *J. Apicult. Res.,* 2:64–66, 1963.

Zuckerkandl, E. Perspectives in molecular anthropology. *Viking Fund Publ. Anthrop.,* 37:243–272, 1963.

NAME INDEX

Waddington, C. H., 40–42, 55, 59, 129, 342, 486
Wadeson, R. W., 158, 159, 455, 486
Wahl, O., 87
Waisman, H. A., 438, 474
Walker, E. L., 472
Walker, M. C., 68, 486
Wallace, B., 53, 486
Wallace, H. W., 182, 486
Wallace, M. E., 312, 486
Walshe, J. M., 193
Washburn, S. L., xiii, xiv, 9, 10, 11n., 12, 486
Watson, J. B., 486
Watson, J. D., 112, 293, 419, 483, 484, 486
Watson, L. R., 96, 486
Watson, M. L., 135, 293, 483, 486
Wearden, S., 297, 486
Weaver, N., 70, 87, 487
Wechsler, D., 430, 487
Wecker, S. C., 29, 30, 41, 487
Weesner, F. M., 62, 487
Weiant, E. A., 50, 476
Weidenreich, F., 12, 487
Wein, J., 476
Weiner, J. S., 11n.
Weir, M. W., 309, 323, 339, 447, 487
Weisman, R. G., 39, 40, 284, 462
Weiss, J. M., 450
Weiss, K., 88, 320, 341, 487
Weitzman, M., 154, 487
Wells, I. C., 473
Wendel, J., 67, 483
Wenner, A. M., 79, 85, 88, 460, 487
Westall, R. G., 192, 193
Westbrook, W., 288, 468
Wheeler, W. M., 64, 74, 75, 487
Wherry, R. J., 363, 364, 487
Whimbey, A. E., 293, 447
White, M. J. D., 68, 71, 487
Whiting, P. W., 50, 51, 69, 487
Whitten, W. K., 24, 487
Wigglesworth, V. B., 62, 488
Wilkins, L., 163–165, 488
Willham, R. L., 326, 488
Williams, C. M., 48, 488
Williams, R. H., 163, 172, 488
Wilson, E. O., 62, 78, 488

Wilson, J. R., 293, 439
Wilson, S. A. K., 191
Wilson, V. K., 192
Winston, H., 308, 362, 406, 465, 488
Witt, G., 135, 488
Wolf, E., 83, 84, 488
Wolstenholme, D. R., 269, 488
Wood-Gush, D. G. M., 343, 488
Woodger, J. H., 303, 488
Woodrow, A. W., 94, 488
Woodworth, R. S., 378, 405, 488
Wooley, D. W., 185, 488
Woolf, B., 59, 486
Woolf, C. M., 374, 488
Woolf, L. I., 181, 182, 188, 191, 407, 443, 446, 453, 488
Woolf, O. H., 193
Woolpy, J. H., 27, 489
Woyke, J., 67, 69, 489
Wright, S., 6, 28, 31–33, 257, 324, 325, 423, 489
Wuorinen, K. A., 293, 463
Wyeokoff, C. B., Jr., 438
Wynne-Edwards, V. C., 23, 26, 30, 489

Yabe, T., 471
Yahiro, E. I., 408, 455
Yamaguchi, H. G., 432, 489
Yamashita, I., 482
Yates, F. E., 464
Yerkes, R. M., 395, 396, 489
Yeudall, L. T., 443
Young, D., 193
Young, W. J., 170, 444
Yuwiler, A., 185, 489

Zamir, A., 458
Zangwill, O. L., 358, 484
Zannoni, V. G., 192
Zarrow, M. X., 293, 447
Zeigler, L. K., 192
Zilinsky, A., 472
Zmarlicki, C., 67, 489
Zubek, J. P., 146, 309, 360, 445, 459
Zuckerkandl, E., 373, 489

SUBJECT INDEX